Proceedings of the International Conference on Quantitative Genetics

PROCEEDINGS OF THE
International Conference on Quantitative Genetics

AUGUST 16-21, 1976

EDITED BY

Edward Pollak, Oscar Kempthorne
and *Theodore B. Bailey, Jr.*

THE IOWA STATE UNIVERSITY PRESS / *Ames*
1 9 7 7

EDWARD POLLAK is Professor of Statistics, OSCAR KEMPTHORNE is Distinguished Professor of Statistics and THEODORE B. BAILEY, JR., is Assistant Professor of Statistics, Iowa State University, Ames, Iowa.

© 1977 The Iowa State University Press
Ames, Iowa 50010. All rights reserved

Printed by
Cushing-Malloy, Inc.

First edition, 1977
Second printing, 1978
Third printing, 1979

International Standard Book Number 0-8138-1895-8

Library of Congress Cataloging in Publication Data

International Conference on Quantitative Genetics,
    Ames, Iowa, 1976.
    Proceedings of the International Conference on
Quantitative Genetics, August 16-21, 1976.

    1. Quantitative genetics--Congresses. 2. Breeding--Congresses. I. Pollak, Edward. II. Kempthorne, Oscar. III. Bailey, Theodore B., 1940-

QH452.7.I58   1976   575.1   77-8513

ISBN 0-8138-1895-8

Preface

This volume contains the full proceedings of the International Conference on Quantitative Genetics, held at Ames, Iowa, August 16 through August 21, 1976.

The principal motivation for organizing the conference came from the following considerations:

(1) Almost every attribute of an individual is in the last resort quantitative, as opposed to qualitative, and is influenced by the environment, so that a complete theory of genetic populations must take account of these basic facts.

(2) Quantitative genetics provides the theoretical foundation for the genetic improvement of domestic plants and animals and has implications for the better understanding of the dynamics of natural and human populations.

(3) In recent years there have been considerable feelings of unease among some quantitative geneticists and expressions of deep uncertainty about the quality and utility of the available theory of quantitative genetics and whether the available theory agrees with results of applied selection.

The organizers felt, then, that it was high time to have a wide-ranging conference in which leading researchers attempt to spell out in some detail the strengths and shortcomings of the existing theory and give their views about what is to be done to remedy the latter. Since we found that several people in the field shared this sentiment, we went ahead with our plans. We asked leading researchers in quantitative genetics to be on a program committee which would aid us in deciding upon suitable subjects and speakers. We are most grateful to members of the program committee, without whose help we could not have obtained as clear an appreciation as we did for the basic relevant lines of research. The committee consisted of the following individuals: R. E. Comstock (USA), J. F. Crow (USA), G. E. Dickerson

(USA), D. S. Falconer (UK), J. Felsenstein (USA), C. O. Gardner (USA), M. Gillois (France), B. Griffing (USA), A. Gustafsson (Sweden), C. R. Henderson (USA), W. G. Hill (UK), A. Jacquard (France), S. Karlin (USA), R. C. Lewontin (USA), J. L. Lush (USA), N. E. Morton (USA), A. W. Nordskog (USA), A. Robertson (UK), W. J. Schull (USA), W. Seyffert (Germany).

We also owe thanks to colleagues at Iowa State University, including K. J. Frey, A. R. Hallauer, W. A. Russell, A. E. Freeman and R. L. Willham, who gave us valuable suggestions.

The consensus reached was that the conference should be directed toward quantitative genetics, with papers on population genetics being solicited only if they had a bearing on quantitative genetics. There is perhaps no sharp line of demarcation between these two areas of research. However, one can say in general that population genetics is concerned with the study of the effects on the composition of natural populations of natural selection, systems of mating, mutation, and so on. As such, population genetics is concerned with the dynamics over time of genetic populations under natural processes with respect to frequencies of genes and of genotypes. In contrast, quantitative genetics is primarily a study of changes in arithmetic attributes in populations that are closely observable, and even manipulable, by man. Natural populations may be sampled or surveyed, whereas many of the populations of concern to quantitative geneticists may have experiments done upon them, with controlled matings, under carefully controlled conditions, with use being made of randomization and other principles of experimental design. It may be noted that one is in a better position to validate a theory if experimentation can be done than if it cannot be done.

The upshot was that the organizers made the decision that the conference should be directed towards the general theory of quantitative genetics and to the application of that theory to species that are of basic importance to man and can be manipulated. The use of quantitative genetics ideas in relation to humans has increased markedly in the past decade or so. The organizers felt that a mere

PREFACE

session or two could only give a very narrow, and even biased, picture, and that the area merits a conference totally directed to it. The one exception we made was to have the presentation of Feldman and Cavalli-Sforza, which discusses to some extent the role of cultural inheritance, which may be no less important for human populations than biological inheritance. The general idea will surely have impact on behavioral genetics in experimental species.

In spite of the great difficulty of developing a program that gives proper weight to the more important aspects of the theory and application of quantitative genetics, we feel that we achieved some reasonable degree of success. In general, with the exception of human genetics, we obtained some coverage of the most important avenues in quantitative genetics from a variety of viewpoints.

An attempt was made to have discussion of the implications of the huge developments in molecular biology over the past two decades and of some general developments of population genetics. The papers of Lewontin, Hartl and of Gillois and his collaborators have a bearing on this. Undoubtedly, more discussion would have been valuable, but we had severe time limitations.

The papers of Falconer, Eisen, Thoday, Jayakar and his colleagues, and Enfield and Berger report some results from laboratory species that, for various reasons, are more convenient to study than species that are of importance to human affairs, and can lead, perhaps, to the discovery of strengths and weaknesses of biological and mathematical theory.

In the past decade there have been extensive mathematical efforts to advance the theory of genotypic selection and it is appropriate to consider possible implications of this work to quantitative genetics. In this direction, Karlin has a paper in which multilocus genotypic selection theory for infinite populations is discussed in detail. J. Felsenstein considers a model with a finite number of loci, at each of which there are an infinite number of alleles with normally distributed effects on the phenotype. Weir and Cockerham give a development of the two-locus theory of identity by descent and its relationship to genetic means and variances and covariances between relatives. Chevalet and Gillois discuss the implications of the theory of small populations on the estimation of genetic variance components. The concluding paper of this section, by Robertson, is on non-linearity of offspring-parent regression as a possible cause of observed asymmetry of responses from selection when selection is done in opposite directions.

There is then a section on the theory of response and limits to selection, a topic of crucial importance both theoretically and in the development of actual selection processes to be applied to important species. Robertson discusses theory for the case of a large number of linked loci. Dickerson and Lindhé examine whether it is possible for response to selection to be increased if selection is alternated with inbreeding for the purpose of increasing the variance between lines among which there is selection. In one of his papers, Hill considers variation in response to selection, a very important topic because any actual selection experiment leads to variable responses over time, because of finite sampling of possible gametes. Hill, in his other paper, and Pollak explore the role of overlapping generations, in attempts to overcome a striking defect of most theory, which assumes that generations do not overlap. Bailey considers selection limits in self-fertilizing populations following the cross of homozygous lines. Griffing develops a theory for populations, such as those of plants in fields, in which the genotype of a plant influences the performance of its neighbors, as well as of itself, making it necessary to select among groups rather than among individuals or to select on the basis of individual and group performances. The concluding paper of the section, by Sved, explores some effects of opposition to artificial selection caused by natural selection at linked loci. This topic is important because standard quantitative genetic theory is based on selection only for measured attributes, with no variability in reproductive fitness of selected individuals.

PREFACE

In the next two sections there are several papers discussing theory and results in plants and animals. It is important to raise and answer the question of whether the results of artificial selection are in conformity with the theory we have. Dudley discusses continuing results from the celebrated Illinois long-term selection experiment on corn. Other results on corn and sorghum are presented by Gardner and on corn by Eberhart, while Matzinger, Cockerham and Wernsman report data on tobacco. Gallais has an experimental check on quantitative genetics theory of autotetraploids using data on alfalfa. Data on dairy cattle and poultry are discussed by Van Vleck and Nordskog, respectively. The evaluation of the success of theory that is given in these papers and the one by D. L. Harris varies from author to author and species to species.

Much of applied quantitative genetics consists of using performance records to make judgments about the relative values of potential contributors to succeeding generations. The main tool here is the use of linear models, design of experiments and analysis of variance. The reader will realize that this is an important subarea and that development of statistical methods is a critical component of selection techniques. It was important then to obtain the contributions of Henderson, Thompson and Hinkelmann.

In the special invited paper by Sewall Wright, who is surely one of the founding fathers of the whole of population genetics and selection theory, there is a broad and general discussion of the factors that influence the transformation of characters. Included in this is a historical survey of ideas on evolutionary change under the assumption of Mendelian heredity.

The penultimate section of the invited papers consists of general papers by Comstock, Kempthorne and Feldman and Cavalli-Sforza. Comstock considers the problem of synthesizing ideal genotypes for all of the environment populations in which an agricultural species has utility and doing it in the least possible time. Various factors influencing this are mentioned. Kempthorne discusses the history and status of the theoretical underpinning of quantitative genetic theory. Feldman and Cavalli-Sforza consider a theory of stabilizing selection on a phenotype. It incorporates a form of phenotypic assortative mating, and gives consideration to general transmission rules that possibly apply to cultural phenomena.

The last section of invited papers includes short discussion statements by Gillois, Jacquard, Harris and Gowe.

Our volume also contains some contributed papers which we believe to bear usefully on various aspects of the whole area of quantitative genetics.

In addition to the speakers there were other people without whose help the conference could not have gone so well. The organizers wish to most heartily thank those people connected with Iowa State University, without whose help the speakers and their families could not have been as well cared for as they were. In particular, Mrs. Kathy Shaver and Mrs. Carolyn Cornette organized daily activities for the families of conference participants and arranged tours to various places of interest throughout Iowa. Mrs. Shaver was also very helpful with the preconference publicity. We also greatly appreciate the efficient secretarial work of Mrs. Phyllis Carr, before and during the conference. It was she who kept our records in order and saw to it that the headquarters office ran smoothly during the conference.

We are grateful to Dr. George H. Ebert, Leader, Extension Courses and Conferences, and to Mr. Tom McCormick, Conference Coordinator in the Office of Extension Courses and Conferences. As well as placing fine convention facilities at our disposal, they were unfailingly efficient, courteous and helpful during the course of the conference in settling the many problems that arose.

After the conference was over there was a considerable amount of work to do. We are very grateful for the skillful typing of Mrs. Phyllis Carr, Mrs. Kathy Shaver and Mrs. Marlene Sposito. Thanks are also due to J. R. Sedcole and J. Lewis for help with the considerable amount of editoral work.

PREFACE

We are very grateful also to Professor H. A. David, the head of our department, for his encouragement and occasional gentle prodding when we lacked faith that the undertaking could be accomplished. Dr. Daniel Zaffarano, the dean of the Graduate School of Iowa State University, also gave us encouragement.

Finally we thank the National Science Foundation, the National Institutes of Health, Pioneer Hi-Bred International, Inc. and the Iowa Beef Industry Council for their necessary financial support.

> Edward Pollak
> Oscar Kempthorne
> Theodore B. Bailey, Jr.

Table of Contents

Preface v

Senior Authors of Invited Papers xix

Authors of Contributed Papers xxi

Full List of Registrants xxiii

I. INTRODUCTION

The International Conference on Quantitative Genetics:
Introduction 3
    O. Kempthorne

Why are mice the size they are? 19
    D. S. Falconer

Canalisation in quantitative genetics 23
    J. M. Rendel

Quantitative genetic research in plants: Past accomplishments and research needs 29
    C. O. Gardner

The general area: Setting the stage 39
    D. L. Harris

Introductory statement: Poultry 47
    A. W. Nordskog

II. IMPLICATIONS OF POPULATION GENETICS AND MOLECULAR BIOLOGY

The relevance of molecular biology to plant and animal breeding 55
    R. C. Lewontin

Applications of meiotic drive in animal breeding and
population control 63
    D. L. Hartl

On gene action 89
    M. Gillois, Cl. Chevalet, A. Micali, R. Baron
    and R. Costa

III. WHAT HAVE WE LEARNED FROM LABORATORY SPECIES?

Some results of the Edinburgh selection experiments with
mice 101
    D. S. Falconer

Antagonistic selection index results with mice 117
    E. J. Eisen

Effects of specific genes 141
    J. M. Thoday

A genetic linkage study of a quantitative trait in *Drosophila
melanogaster* 161
    S. D. Jayakar, L. Della Croce, M. Scacchi and
    G. Guazzotti

Selection experiments in Tribolium designed to look at
gene action issues 177
    F. D. Enfield

Multiple-trait selection experiments: Current status,
problem areas and experimental approaches 191
    P. J. Berger

IV. MATHEMATICAL AND STATISTICAL GENETICS

Selection with many loci and possible relations to
quantitative genetics 207
    S. Karlin

Multivariate normal genetic models with a finite number of
loci 227
    J. Felsenstein

Two-locus theory in quantitative genetics 247
    B. S. Weir and C. C. Cockerham

Estimation of genotypic variance components with
dominance in small consanguineous populations 271
    Cl. Chevalet and M. Gillois

The non-linearity of offspring-parent regression 297
    A. Robertson

V. THEORY OF RESPONSE AND LIMITS TO SELECTION

Artificial selection with a large number of linked loci 307
    A. Robertson

Potential uses of inbreeding to increase selection response . 323
    G. E. Dickerson and N. B. H. Lindhé

Variation in response to selection 343
    W. G. Hill

Selection with overlapping generations 367
    W. G. Hill

Selective advance in populations with overlapping
generations 379
    E. Pollak

Selection limits in self-fertilizing populations following
the cross of homozygous lines 399
    T. B. Bailey, Jr.

Selection for populations of interacting genotypes 413
    B. Griffing

Opposition to artificial selection caused by natural
selection at linked loci 435
    J. A. Sved

VI. RESULTS AND THEORY WITH PLANTS

76 generations of selection for oil and protein percentage
in maize 459
    J. W. Dudley

Quantitative genetic studies and population improvement
in maize and sorghum 475
    C. O. *Gardner*

Quantitative genetics and practical corn breeding 491
    S. A. *Eberhart*

Single character and index mass selection with random
mating in a naturally self-fertilizing species 503
    D. F. *Matzinger*, C. C. *Cockerham* and
    E. A. *Wernsman*

An experimental check of quantitative genetics on an
autotetraploid plant, *Medicago sativa*, L., with special
reference to the identity by descent relationship 519
    A. *Gallais*

VII. RESULTS AND THEORY WITH ANIMALS

Theoretical and actual genetic progress in dairy cattle ... 543
    L. D. *Van Vleck*

Success and failure of quantitative genetic theory in poultry. 569
    A. W. *Nordskog*

Past, present and potential contributions of quantitative
genetics to applied animal breeding 587
    D. L. *Harris*

VIII. MIXED MODEL THEORY IN QUANTITATIVE GENETICS

Prediction of future records 615
    C. R. *Henderson*

Estimation of quantitative genetic parameters 639
    R. *Thompson*

Diallel and multi-cross designs: What do they achieve? .. 659
    K. *Hinkelmann*

IX. SPECIAL INVITED PAPER

Modes of evolutionary change of characters 679
    S. Wright

X. GENERAL PAPERS

Quantitative genetics and the design of breeding programs . 705
    R. E. Comstock

Status of quantitative genetic theory 719
    O. Kempthorne

Quantitative inheritance, stabilizing selection and cultural evolution 761
    M. W. Feldman and L. L. Cavalli-Sforza

XI. INVITED DISCUSSIONS

Somatic cell genetics and genetic improvement 781
    M. Gillois

Transmission of genes and transmission of characters ... 787
    A. M. Jacquard

What's different about chickens? 793
    D. L. Harris

Comments on the conference 799
    R. S. Gowe

XII. CONTRIBUTED PAPERS

Genetic analysis of quantitative data from the families of identical twins 805
    W. E. Nance and L. A. Corey

Genetic analysis of human serum cholesterol 809
    C. F. Sing and J. D. Orr

Likelihoods on complex pedigrees for quantitative traits .. 815
    E. A. Thompson and M. H. Skolnick

Genetic integration of morphometric traits in randombred house mice 819
    L. Leamy

Coefficients of constraint as a generalization of coefficients of identity 823
    M. Gillois and Cl. Chevalet

A general approach to dependence relationships among genes with some applications 829
    A. Gallais

Diallel crosses in relation to breeding system 837
    S. O. Fejer

Model building in quantitative genetics 843
    W. Y. Tan and M. P. Mi

Response and variance of response to selection 847
    Y. Park and R. Nassar

Variance-covariance structure of group means with overlapping generations 851
    D. L. Johnson

Errors in predicting genetic gain from mass selection 859
    F. H. Kung

Analysis of binomial data in quantitative genetics experiments 865
    W. A. Becker

A new interpretation of recombination frequencies in genetics 869
    J. H. Ursell

Senior Authors of Invited Papers

T. B. Bailey, Jr., Iowa State University, Ames, Iowa 50011

P. J. Berger, Iowa State University, Ames, Iowa 50011

Cl. Chevalet, Centre de Recherches de Toulouse, I. N. R. A., B. P. 12, 31320 Castanet-Tolosan, France

R. E. Comstock, University of Minnesota, St. Paul, Minnesota 55108

G. E. Dickerson, University of Nebraska, Lincoln, Nebraska 68583

J. W. Dudley, University of Illinois, Urbana, Illinois 61801

S. A. Eberhart, Funk Seeds International, Bloomington, Illinois 61701

E. J. Eisen, North Carolina State University, Raleigh, North Carolina 27607

F. D. Enfield, University of Minnesota, St. Paul, Minnesota 55108

D. S. Falconer, University of Edinburgh, Edinburgh, Scotland

M. W. Feldman, Stanford University, Stanford, California 94305

J. Felsenstein, University of Washington, Seattle, Washington 98195

A. Gallais, Institut National de la Recherche Agronomique, 86600 Lusignan, France

C. O. Gardner, University of Nebraska, Lincoln, Nebraska 68583

M. Gillois, Centre de Recherches de Toulouse, I. N. R. A., B. P. 12, 31320 Castanet-Tolosan, France

R. S. Gowe, Agriculture Canada, Ottawa, Ontario, Canada

B. Griffing, Ohio State University, Columbus, Ohio 43210

D. L. Harris, Purdue University, West Lafayette, Indiana 47906

D. L. Hartl, Purdue University, West Lafayette, Indiana 47907

C. R. Henderson, Cornell University, Ithaca, New York 14853

W. G. Hill, Institute of Animal Genetics, Edinburgh EH9 3JN Scotland

K. Hinkelmann, Virginia Polytechnic Institute and State University, Blacksburg, Virginia 24061

A. M. Jacquard, Institut National d'Études Demographiques, Paris, France

SENIOR AUTHORS OF INVITED PAPERS

S. D. Jayakar, Laboratorio di Genetica Biochimica ed Evoluzionistica C.N.R., Pavia, Italy

S. Karlin, Stanford University, Stanford, California 94301

O. Kempthorne, Iowa State University, Ames, Iowa 50011

R. C. Lewontin, Harvard University, Cambridge, Massachusetts 02138

D. F. Matzinger, North Carolina State University, Raleigh, North Carolina 27607

A. W. Nordskog, Iowa State University, Ames, Iowa 50011

E. Pollak, Iowa State University, Ames, Iowa 50011

J. M. Rendel, CSIRO, Epping, N. S. W., Australia 2121

A. Robertson, Institute of Animal Genetics, Edinburgh EH9 3JN Scotland

J. A. Sved, University of Sydney, Australia

J. M. Thoday, University of Cambridge, Cambridge CB2 3EH England

R. Thompson, University of Edinburgh, Edinburgh EH9 3JZ Scotland

L. D. Van Vleck, Cornell University, Ithaca, New York 14853

B. S. Weir, North Carolina State University, Raleigh, North Carolina 27607

S. Wright, University of Wisconsin, Madison, Wisconsin 53711

Authors of Contributed Papers

W. A. Becker, Washington State University, Pullman, Washington 99163

Cl. Chevalet, Centre de Recherches de Toulouse, I. N. R. A., B. P. 12, Castanet-Tolosan, France

L. A. Corey, Medical College of Virginia, Richmond, Virginia 23219

S. O. Fejer, Agriculture Canada, Ottawa, Ontario K1A 0C6, Canada

A. Gallais, Institut National de la Recherche Agronomique, 86600 Lusignan, France

M. Gillois, Centre de Recherches de Toulouse, I. N. R. A., B. P. 12, Castanet-Tolosan, France

D. L. Johnson, Ruakura Agricultural Research Centre, Hamilton, New Zealand

F. H. Kung, Southern Illinois University, Carbondale, Illinois 62901

L. Leamy, California State University, Long Beach, California 90840

M. P. Mi, University of Hawaii, Honolulu, Hawaii 96825

W. E. Nance, Medical College of Virginia, Richmond, Virginia 23219

R. Nassar, Kansas State University, Manhattan, Kansas 66502

J. D. Orr, University of Michigan Medical School, Ann Arbor, Michigan 48106

Y. Park, Kansas State University, Manhattan, Kansas 66502

C. F. Sing, University of Michigan Medical School, Ann Arbor, Michigan 48106

M. H. Skolnick, University of Utah, Salt Lake City, Utah 84101

W. Y. Tan, Memphis State University, Memphis, Tennessee 38111

E. A. Thompson, University of Cambridge, England

J. H. Ursell, Queen's University, Kingston, Ontario, Canada

Full List of Registrants

A. O. Adegoke, C. E. Ago, J. Alika, F. Aloeffel, R. D. Anderson,
J. A. Arthur, O. J. Ayala, T. B. Bailey, Jr., R. J. Baker,
M. U. Ball, L. Barrales, J. Beck, W. A. Becker, E. Bell, L. Benyshek,
P. J. Berger, B. Binsika, A. Blakely, N. Blakely, H. R. Boerma,
T. P. Bogyo, B. B. Bohren, B. Bradley, J. Bradley, E. Brannang,
C. Brown, A. Bueno, D. Butcher, A. R. Campbell, M. S. Campos,
J. Cedeno, A. B. Chapman, Cl. Chevalet, T. Choo, J. C. Christian,
B. R. Christie, T. Colbert, R. E. Comstock, L. A. Corey,
P. L. Cornelius, J. L. Cornette, H. Cortez, D. F. Cox, L. Craymer,
C. W. Crum, P. J. Cunningham, M. Curie-Cohen, K. Daly, H. A. David,
J. Demopulos-Rodriguez, G. E. Dickerson, D. P. Doolittle,
J. W. Dudley, D. N. Duvick, S. A. Eberhart, F. Eftekhari,
E. J. Eisen, T. H. Emigh, F. Enfield, W. Engels, M. A. B. Fakorede,
D. S. Falconer, S. O. Fejer, M. W. Feldman, J. Felsenstein,
D. K. Flock, K. Foster, R. Frankham, I. Franklin, K. J. Frey,
G. Friars, C. E. Fuchs, A. Gallais, N. Galvin, E. E. Gama,
C. O. Gardner, H. H. Geiger, D. Gianola, P. T. Gibson, J. Gill,
M. Gillois, J. Gonella, V. Gonzalez, R. S. Gowe, B. Griffing,
M. Grossman, A. R. Grunst, J. C. Halinar, R. B. Hall, A. Hallauer,
J. Hammond, J. Hardiman, G. L. Hargrove, D. L. Harris, D. L. Hartl,
N. Hartung, D. A. Harville, P. Hedrick, C. R. Henderson,
R. R. Hill, Jr., W. G. Hill, K. Hinkelmann, W. Hohenboken,
A. M. Hopkins, D. Hotchkiss, L. L. Hulbert, D. Ivers, A. Jacquard,
S. Jana, S. Jayakar, L. Jensen, D. L. Johnson, G. R. Johnson,
R. K. Johnson, R. Jondle, P. J. Kaltsikes, K. Kang, L. W. Kannenberg,
S. Karlin, A. J. Katz, O. Kempthorne, K. K. Kidd, J. F. Kidwell,
M. G. Kidwell, J. F. Kiekebusch, C. Kimura, D. D. Kress, F. H. Kung,
R. C. Laben, L. Leamy, A. J. Lee, P. D. Legg, B. Levikson, J. Lewis,
R. Lewontin, G. H. Liang, P. J. Loesch, V. A. Logan, E. Lopez,
D. Luizzi, J. Lush, U. R. Maag, M. MacNeil, J. Martin, T. Maruyama,

D. F. Matzinger, M. McNeill, R. McNew, B. W. Menzel, W. R. Meredith,
A. Micali, K. Miezan, J. Miller, R. H. Miller, J. B. Miranda, Filho,
J. Mitchell, W. M. Muir, M. A. Mutschler, W. E. Nance, D. J. Nash,
R. Nassar, J. E. Ngam, M. Ngandu, S. Noble, A. W. Nordskog,
W. E. Nyquist, A. T. Obilana, Y. C. Park, B. Pasadente, I. Y. Pevzner,
J. H. Pfund, E. Pollak, C. O. Qualset, J. Rapp, J. M. Rendel,
M. Rezai, A. Robertson, D. S. Robertson, D. M. Rodgers,
R. W. Rosenbrook, K. E. Rowe, J. J. Rutledge, M. J. Samann,
E. Scheinberg, J. G. Schimmel, J. R. Sedcole, R. L. Segebart,
R. P. Seifert, H. I. Sellers, R. Shanks, R. Shorter, C. Sing,
W. D. Slanger, R. Smith, A. F. Soldati, M. Solh, M. Soller, Sridodo,
C. Strobeck, C. W. Stuber, R. E. Stucker, D. D. Stuthman, J. Sved,
O. Syrstad, W. Y. Tan, N. M. Teixeira, W. V. Thayne, J. M. Thoday,
A. E. Thompson, E. A. Thompson, R. Thompson, J. F. Tierce,
G. E. Tolla, G. Torres, C. Trowbridge, A. F. Troyer, M. Turelli,
W. Uitdewilligen, J. Ursell, M. Uyenoyama, L. D. Van Vleck,
R. Vencovsky, R. Walejko, J. L. Walters, A. B. S. Wang, B. S. Weir,
R. L. Willham, W. Williams, J. A. Williamson, S. Wright,
A. J. Wyatt, Y. Yamada, S. Yokoyama, C. Young.

I
Introduction

The International Conference on Quantitative Genetics: Introduction

Oscar Kempthorne
STATISTICAL LABORATORY
IOWA STATE UNIVERSITY, AMES, IA 50011

Iowa State University has had a tradition for more than fifty years of extensive effort in the areas of plant and animal breeding. It is worth recalling that Henry A. Wallace, former Vice-President of the United States and founder of Pioneer hybrid seed corn company, developed an association with Snedecor around 1920. Wallace was interested in what statistical methods could contribute to plant breeding and Snedecor was interested in applications of statistical methods generally and particularly to biology and agriculture. From that time, there has been a deep association at Iowa State between statistics on the one hand and plant and animal breeding--or applied quantitative genetics--on the other hand.

The viability of this association was based on outstanding strengths. On the statistical side was G. W. Snedecor, the most effective expositor of R. A. Fisher in the United States, with

---

Journal Paper No. J-8675 of the Iowa Agriculture and Home Economics Experiment Station, Ames, Iowa. Project 1669. This work was supported by grants GM13827-09 and GM23339-01 from The National Institutes of Health.

impact on biological and agricultural research that is almost immeasurable. On the animal side was J. L. Lush, who from the early 1930's developed the basis of current animal breeding practice. In the genetics department were E. W. Lindstrom and J. W. Gowen, with very strong interests in quantitative genetics in plants and animals, respectively. The agronomy department had a very strong sustained effort in plant breeding with several of the leaders of the field.

These early efforts have continued over the decades with a very large sustained effort in applied quantitative genetics. Students from the animal side became leaders of animal breeding to the point that animal breeding in the United States and over much of the whole world has been dominated by what may justifiably be termed the Lush-Iowa School. Students of Iowa State University in plant breeding and poultry breeding have spread over the United States, and indeed, over the whole world.

It was natural, then, for Iowa State University to undertake the effort of organizing the present conference. Ideas of mathematical modelling of the consequences of the extremely complicated Mendelian probability models, of statistical modelling and of statistical analysis of data permeate the area of quantitative genetics. On this campus the Statistical Laboratory has always had a deep commitment to attempt to provide knowledge and technical interaction and help. One might even say that without the efforts of the statistical group, the strengths of quantitative genetics here would not have been achieved, and without a strong effort by statistics in teaching and research, it would be very difficult to continue the strengths into the future. So the Statistical Laboratory felt a strong obligation to take the initiative.

It will be of interest, I think, to run over quickly the processes we followed. This will explain to some extent the structure of the final program. Our first step was to check around the world on whether a conference on quantitative genetics would be a good idea. Obviously the answer on this was in the affirmative.

INTRODUCTION

We then made up a program committee trying to get representation of the subareas and of the various parts of the world, and solicited suggestions for the program. We then obtained reactions of this program committee to the various suggestions, and it was immediately obvious that there were difficulties at the conceptual level.

The initial difficulty in the preliminary discussions arose from the not-uncommon tendency to regard a conference on quantitative genetics as a conference on population genetics, as though the former is merely a branch of the latter using almost exactly the same ideas and techniques. In the minds of many, however, this is a gross misunderstanding which has led and will lead to unfortunate consequences, for example in the evaluation of and support for research. Appreciation of the similarities and the differences of the two very highly related fields is critical, and the following remarks are aimed at a needed clarification.

The sequence of discoveries and of modelling in genetics needs little discussion. Mendelism is the basis, of course, and all that has ensued is really summed up as the application of scientific technology, hardware and software, to this basis. The words 'hardware' and 'software' are very overworked and perhaps of uncertain meaning. By 'hardware' I mean the variety of physical and chemical apparatuses and techniques that have been found informative. By 'software' I mean the probability and statistical modelling and the processes of data analysis, using almost the full breadth of statistical techniques. Both the hardware and software are inventions of the present century, and as throughout science, both have developed by bootstrapping and iteration. The whole arena of genetics has become partitioned in many ways, giving, for instance, molecular genetics, physiological genetics, developmental genetics, behavior genetics, and so on. One recognized branch consists of ideas of population genetics and ideas of quantitative genetics. The latter is sometimes given the name, statistical genetics, or the name biometrical genetics, both of which are misnomers

because the whole history of genetics of this century is replete
with statistical methods, and necessarily so because observations
must be given the status of random variables, and biometry is
merely a name, useful, of course, for statistics applied to biological systems.

The existence of the different subareas of population genetics
and quantitative genetics may be recognized, it seems, merely by
examining books, the one by Li (1976) or the one by Crow and Kimura
(1970) on population genetics, and the one by Falconer (1960) on
quantitative genetics. It is obvious that basic ideas of Mendelism
and their consequences are shared by both subareas, but it is
equally obvious that the student who has assimilated the knowledge
in one subarea only will have little understanding and competence
in the other. Some discussion of the similarities and differences
is necessary. The two subareas start from a common basis, but
after some distance diverge with respect to ideas, formulation,
mathematical and statistical techniques. It is difficult to be
expert in population genetics theory without a considerable background in applied stochastic processes. It is difficult to be
expert in quantitative genetics without a deep understanding of
general statistical processes, particularly the analysis of variance and linear models, as we can see, for instance, from my own
book on genetic statistics (Kempthorne, 1957), or from the work of
Henderson, some of which is to be presented in the conference.
Such divergence from a common basis is common to all the basic
sciences. The point of stating these facts is to make clear that
an expert in population genetics may not be at all an expert in
quantitative genetics and vice versa. As soon as experts go outside their areas of expertise, trouble occurs; improper transfer
of concepts is made, the differing roles of the subareas are not
properly appreciated, and each subarea partially loses the recognition it should justly receive. We who think we have some understanding of statistics are constantly appalled by the misuses, and,
contrariwise, the biologist who knows his biology is often appalled

INTRODUCTION

by the naiveté of the statistician in a biological problem.

How may one briefly characterize population genetics? It is dominated by the study of natural populations and, on the theoretical side, by the development of mathematical and particularly probabilistic models for genetic processes at all levels. It is to say that in recent years the theory has been directed almost totally to natural evolutionary processes, and the roles of mutation, selection and migration have been the main topics of work. The dominant current interest appears to be explanation of the huge amount of polymorphism in natural populations. Population genetics takes the long-time look, as we see in the work of Kimura, for instance. The main ingredient in the broth has been a very natural idea of fitness, which is then embedded in the mathematical theory, usually as an s-value. Crow and Kimura, in their well-known book on population genetics state:

> "We deal primarily with natural populations... we emphasize more the behavior of genes and population attributes under natural selection where the most important measure is Darwinian fitness."

Interestingly, perhaps, the notion of Darwinian fitness seems not to have been defined adequately except for a population whose reproductive processes do not involve mating, and is extremely difficult to measure even in the simplest circumstances, when it involves only viability of single individuals with a life cycle rather than merely an unrealistic two-stage life. On the nontheoretical side, population genetics consists of well-directed observation of natural populations and then model building and statistical data analysis to interpret the results. The field of population genetics is primarily one based on observation rather than "controlled" experimentation.

This is not to assert that there is no controlled observation in population genetics. One merely needs to examine Wallace's (1968) book to see the contrary. However, the sort of experimentation Wallace describes is of rather specialized types, involving in many cases, very specific genetic mechanisms, or simply, crosses of different natural genetic populations.

The field of quantitative genetics is a natural outgrowth of classical qualitative genetics, which can be characterized reasonably, perhaps, as the Mendelian modelling of phenotypic expressions that are categorical, with essentially disjoint classes, that are under the control of a few Mendelian loci, and that are essentially constant under controlled conditions. Very early, the question of whether "continuous" or quantitative traits could be explained by Mendelism came up and led to very deep controversies. I shall not give here a detailed history and merely mention some early workers, Bateson, Karl Pearson, G. Udny Yule and W. Weinberg. The leading effort was made by R. A. Fisher, who addressed the question of whether one could reconcile classical categorical or qualitative genetics with the obvious phenomenon of continuous variation. This phenomenon necessitated the introduction of two extremely complex aspects into the modelling. On the one hand, it became necessary to envisage a very large number of segregating Mendelian loci. Every locus one had discovered in humans could have effect on height, say, and every locus discovered in maize could have effect on yield obviously. To add to this basis, it was obvious that the number of known loci, in the classical genetics context, was to a considerable extent a monotone increasing function of the amount of investigatory effort. Finally, jumping ahead by some decades, we now have a model with a quite fantastic number of "loci," based on C. G.A, and T, a model which, judging from some of the recent literature, presents horrendously difficult, if not impossible, problems of interpretation of "ordinary" attributes. If, for instance, we have 1000 loci and 2 alleles per locus, we have 3^{1000} genotypes, a number so huge that the possibility of examining Mendelian combinations does not exist. On the other hand, uncontrollable variation in environment outside the laboratory (and even within it) produces considerable variation in phenotype. So the reconciliation of Mendelism and

INTRODUCTION

continuous variation requires, unlike classical qualitative genetics, some "decent respect" for the environment. Whether this is accomplished by treating the effect of environment as a $N(0, \sigma^2)$ random variable is another matter, but to do this is at least one primitive step in a necessary direction.

The aim of quantitative genetics is then to develop validated models for phenotypic expression in the face of partial nonidentifiability of genotypes and partial nonidentifiability of environments.

The use of the adjective "partial" should be noted and appreciated. Obviously, we may have knowledge of part of the genotype as in knowledge of marker genes or genes with very large effect. On the environmental side, we may recognize climatic regions, or different controlled laboratory environment factors, or different systems of management, as in the case of cage and floor management of commercial poultry. However, in all such cases, when subpopulations are identified whether genetically or environmentally, one reaches the point at which residual genotypes or residual environments are not identifiable. In this connection we may also note, as Thoday will discuss in this conference, that a situation in which genotypic variability is thought not to be identifiable can be partially resolved by identifying genetic segregations with large effects. Failure to appreciate this possibility can impede strongly the development of understanding of phenotypic variability.

Fisher used the data base of Karl Pearson, which dealt with human physical measurements. It is interesting that there has been in recent years considerable activity in the interpretation of human mental measurements. In terms of total effort, experimental, theoretical and data analytical, however, the big thrust has been towards the species of direct relevance to the food needs of the human species, and this explains why quantitative genetics is dominated by animal and plant breeding. The amount of experimental quantitative genetics investigation in food species over the past half-century has been huge. Obviously, it will continue for the foreseeable future.

A natural question is whether the past efforts led to improvement, and I think, in spite of some strongly voiced statements which we have heard, that the only rational interpretation of the historical record is that with some exceptions, which are, indeed, puzzling, the efforts have produced superior populations. But could more progress have been achieved and will the methodology of the past be successful in the future? It was natural, with this background, that a considerable portion of the program should be devoted to the dominant food species. And, it seems, one needs no longer to apologize to the people of the Western World, and particularly of the United States, for working to improve yields of food crops.

We wished to organize a conference, then, in quantitative genetics and not in population genetics as it is conventionally understood. The distinction is in some ways very much one of emphasis rather than intrinsic basis. The ideas of Mendelism and the associated probabilistic structure are, of course, common to both areas. Many of the ideas of conventional population genetics are important to quantitative genetics. And, contrariwise, it seems clear that many of the ideas of conventional quantitative genetics are relevant to population genetics; obviously even viability fitness is a multi-loci trait and is influenced by environment, though to take account in the theory of these facts is extremely difficult, as examination of the literature shows. Part of the distinction between the two areas is simply that quantitative genetics should be called experimental population genetics, with emphasis on the word 'experimental,' connoting that we make genetic populations by controlled operations, while conventional population genetics is primarily observational population genetics, trying to understand populations that have arisen by natural and not humanly directed processes. We may also note that quantitative genetics is directed towards the making of quite "unnatural"

INTRODUCTION 11

genetic populations, some of which, like hybrids from inbred lines, have a lifetime of just one season. To this we may add that the applied quantitative geneticist cannot consider as deeply relevant any argument that uses the natural evolutionary time scale. The upshot of this deliberation was that we removed from consideration any population genetics work that does not possess a strong quantitative genetic component. Also, we eliminated any population genetics and ecology, to which the Israel conference was specifically addressed.

One aspect of quantitative genetics on which we need to give an explanation is the nearly complete absence of human quantitative genetics. Surely no more appropriate area of application exists, and the methodology in it has very close relations to some conventional quantitative genetics. The subarea is almost totally non-experimental with a very few very notable exceptions. The development is necessarily observational.

Human quantitative genetics may be partitioned conveniently into physico-biologic and mental moieties. In the former moiety one may place the inheritance of physical morphic attributes, which was of interest, it will be recalled, to the early biometric school. Here one would also place aspects of physiological activity, though not a large effort has been made in this direction. Another physico-biologic interest is in the area of modelling ordered categorical attributes by means of an assumed underlying continuous or quantitative variable, which leads to the ordered categorical variable by cuts in this underlying continuous variable. We would have liked to devote some of the conference activity to a critical and not necessarily favorable or unfavorable examination of this methodology.

Related, of course, to this topic is the matter of genetic counselling. This may be rather easy in terms of logic for the case of readily identified major human problems which have been shown to be largely under the control of a single Mendelian locus. But many distressing human ailments are attributes for which such a simple Mendelian model is not tenable, and recourse is made to the idea of an underlying variable, which may be called vulnerability or tolerance, and then to the same sort of methodology of quantitative genetics that is used to predict a metric trait of, say a bovine male. But a mere one or two 2-hour session on this would have been quite inadequate and would have resulted, we thought, only in superficiality. I note that this area has been marked by mere polemics, which serve no useful purpose.

It is necessary to mention a topic of human quantitative genetics, which has generated a huge amount of controversy, perhaps more than any other area of the whole of biology. I refer to human mental attributes, and particularly, of course, to IQ in humans. This is not the time or place to attempt an objective evaluation. The general topic of IQ in humans has been of deep interest going back to antiquity and more deeply to the time of Galton. My view, which I believe to be relevant even to present-day discussions, is that even the earliest serious workers were quite unable to separate their objective scientific evaluation from their very natural subjective views about the nature of the observed variability in expression of mental attributes in human populations. It was and is all too easy to attribute the advantages or disadvantages of environment to genetic factors, as we can see, if we look carefully, in literature before, say, the 1930's. It is all too easy, contrariwise, to attribute all the variability in human expression, whether in a traditional academic discipline or in social behavior or misbehavior to variability in environment alone. This has been popular over the ages and perhaps particularly so in very recent times. In the case of psychiatric disorders, for instance, one may easily be led, by the human values and prejudices one inevitably accumulates, to the view that they are largely genetic, or to the view that they are largely environmental. Interestingly enough, insofar as intervention via drugs produces alleviation, one is still in a very uncertain situation vis-a-vis causation, because a drug may, when effective, overcome

INTRODUCTION   13

some inborn genetic biochemical deficiency, or it may be negating
an undetermined environmental insult. The point, without even
entering the IQ controversy, is that the separation of causation
into genetic and environmental components poses extremely difficult
questions of definition, and even if these are resolved, a resolution from observational studies is extremely difficult, if not
impossible. To mount a separation from observational data is
formally not possible logically. There is no point in "sugar-
coating" this harsh epistemological fact. One need only look at
the logic of assigning causality, as it was laid out in a serious
way--our philosophers over the centuries having contributed
essentially nothing to the resolution except some obvious
platitudes--by Sewall Wright in his fundamental philosophic (though
I do not know that he would so term it) approach to the analysis
of causation in his development of path analysis. Inferring of
causality from observational studies is extremely hazardous. One
may look at economics, a most difficult area, in which the amount
of observational data is huge compared with any biological or
psychological area. Inferrring causality, except in a highly
developed science with highly validated laws, empirical at base to
be sure, can be erroneous very easily. The only way to infer
causality, and then only perhaps a very limited sort of causality,
is by way of controlled experiments, which are essentially impossible in the human genetic-environment milieu. The only effective
advances occur, I believe, with experiments like the now famous
Milwaukee experiment, which should be so well-known as not to need
referencing. In general, the so-called ex post facto experiment
is not an experiment at all, though tentatively we may be forced
to regard it so.

The bearing of these considerations on the human IQ matter is
obvious. Study in this field has, I believe, consisted of extremely
naive application of ideas of quantitative genetics for metric
traits, which have been found useful in animal breeding. The
literature, which now numbers untold pages, is replete with simple

models, with a data base that is inadequate to enable anything more than weakly sensitive tests of validity of the simple models. Inferences with respect to causality can be made from these simple models, and the inferences have been advanced as strongly established by some of the protagonists in the topic. These inferences have been attacked on a scientific basis to some extent, and some of the attacks will, I believe, be found on purely objective analysis to be valid. But the discussion has degenerated very, very often into mere polemics, in which attackers confuse objective science and their rational scientific criticisms with their own perfectly valid human ethical values, leaving open those who are attacked an obvious, rational, justified, and scientific defense, which again occupies pages and pages, but which tends strongly not to discuss the real issues of scientific logic and method. It is rare, in my opinion, that the discussions achieve the high degree of scientific competence and objectivity that this topic, of obviously critical importance to the whole panorama of human individual and societal life, merits. One may rationally infer that brief attention to this topic would have degenerated into polemics, mud slinging, and poor scientific discussion, which would have achieved nothing. One may only hope that a conference on the matter that is really scientific will be mounted by some group, in which the past protagonists are allowed only a limited expository role. In view of all this "murky" background, our attention in the conference will be essentially negligible. We made this decision with considerable regret.

We aimed then at a conference that would be concerned with the theory of quantitative genetics and of its associated mathematical, probabilistic and statistical aspects, with the adequacy of present theory, with discussions of directions future theory should take, and critically with the relationships of theory of applied genetic processes, principally selection, of course, to the actual results of applications of these processes. Is our present theory working in the sense of predicting successfully the results of selection experiments? Can we predict response from long-term selection? The theory of quantitative inheritance must be "causal" theory in that it predicts successfully the outcome of humanly chosen operations, rather than a theory that gives ideas of what may have happened but that cannot be confronted by repetition of experiment.

I wish to mention specifically two topics that we felt should be addressed very seriously. Given that we accept the validity of a distinction between conventional population genetics and quantitative genetics, we must ask if there are developments in population genetics which are relevant to quantitative genetics and are not being utilized. We hope that this will be addressed. A more challenging question is the following. We have seen in the past two decades huge developments in understanding by way of molecular biology. This field has shown us that the Fisher idea of many influential loci is true, probably beyond Fisher's wildest dreams in 1918. I think it is fair to say that molecular biology in spite of its great advances, has had essentially no impact on theoretical quantitative genetics and surely nearly no impact on applied quantitative genetics. Why is this? Have we, as workers in quantitative genetics, "missed the boat" through ignorance (in my own case, I admit) or through parochiality? Or is there "a shoe on both?" We have all seen the claims of some molecular biologists that molecular biology is the one and only key to understanding phenotype. The nonmolecular biologist received short shrift for a considerable period. He was not getting to the real basis. As a statistician, I cannot but wonder how one would tackle the explanation of phenotype, e.g., IQ in humans, if one were given a complete listing of the DNA content of the humans we measure. If I may make an imperfect analogy, the problem is a bit like deducing or inferring a global property of a container of gases from knowledge of the position and momentum of each particle in the container. It would seem that the step from DNA to the organism is rather a long one, which will take centuries to unravel. In the meantime, we have to seek partial understanding and we have to tackle prosaic aims such as feeding the world population. We cannot sit on our thumbs until the "ultimate" or "final" answer is achieved. Also, on a human note, we have to do the best we can in genetic counselling of humans. We cannot merely say, as some biologists with molecular ideas appear to be saying, that there can be no real progress until the molecular biology is "stripped out." Some applied quantitative geneticists tend to denigrate the pure geneticist as a mere "gene chaser," who will reach the total answer only after millenia, if humanity survives. I do not support this view, but it is not a problem to find writings that tend to validate this view. Regardless of such partially polemic views, the question remains of whether we are missing ideas from molecular biology and we hope to be given a partial answer.

Another rather different aspect I would like the field of quantitative genetics to react to is the development of biological technique. What could be the role of ovum splitting, of ovum transplantation, of development of clonal propagation for organisms that do not clone, of cell fusion, of wandering genes, and so on? Is our field locked into a particular mode of statistical thinking? I think it fair to say that in quantitative genetics technology has dominated science, whereas science and technology should be the two arms of a complete attack.

Whether our conference addresses the questions raised above or more important and more relevant questions will be seen. Will our conference be like most conferences, a collection of platitudes in very scientific jargon, coupled with a mutual "back-slapping" for some of the real advances that have been made, but with little attention to the deep obscurities? Will our young workers come with fresh minds that have not accepted totally the indoctrination that our college professors naturally and inevitably give? Can they learn enough about the background of past thinking to pick the good from the bad?--not an easy task, to be sure. Or will they proceed, as have the long-term workers in the field, by

INTRODUCTION 17

discovering their own "hobby-horses," which they promulgate as the "real" truth? Will they take a small observational basis, and then proceed to the fascinating mathematical adventures that simplistic models open up? Mathematics has necessarily invaded biology, and this will advance the human condition if and only if we have first good biology and then good mathematics. The mere simulation of biology with the differential equations of mathematical physics is not likely to lead to significant advances in biology. It is becoming clear that biology will require a different sort of mathematics, as we see already with the prominence of stochastic process thinking, and as we may surmise because of the complexity resulting from ploidy and mating to produce offspring. Nor will the naive transplantation of basic and fundamental statistical data procedures necessarily lead us towards the hidden truth, as we see in the human behavioral and mental contexts. Will the 'establishment' of older minds give freedom to our young workers only when they pursue hackneyed paths? We see how Fisher's 1918 paper was evaluated by Karl Pearson and R. C. Punnett, a "cautionary tale" in the words of my colleague, H. A. David. The understanding of phenotypic variation, particularly in relation to variable exhibiting so-called continuous variation, is a very difficult problem. The development of populations that will serve human food needs is a very difficult problem, one worthy of the best efforts of science, and not one to be dismissed as pure technology beneath the dignity of the serious scientist. One may note, finally, that the applied quantitative geneticist is subject to hard judgment; either he produces superior stocks or he does not, and no amount of interesting theory will be accorded respect unless it leads towards the achievment of a basic human need. So one question that will be prominent, I hope, is a clear definitive substantiation of progress that has been made. In attempting to do this, it is critical, of course, to take cognizance of the fact that considerable environmental modifications have been made with many species, which alone without genetic change would give improvement.

These are the 'hard' questions, and the proceedings of the conference will be available to all to pronounce judgment on whether they have been addressed.

BIBLIOGRAPHY

Crow, J. F. and Kimura, M. (1970). *An Introduction to Population Genetics Theory*. New York: Harper and Row.

Falconer, D. S. (1960). *Introduction to Quantitative Genetics*. Edinburgh and London: Oliver and Boyce.

Kempthorne, O. (1957). *An Introduction to Genetic Statistics*. Ames: Iowa State University Press. (Reprint of New York: John Wiley and Son.)

Li, C. C. (1976). *First Course in Population Genetics*. 3rd edition. Pacific Grove, California: Boxwood Press.

Wallace, B. (1968). *Topics in Population Genetics*. New York: W. W. Norton.

Why are mice the size they are?

D. S. Falconer
AGRICULTURAL RESEARCH COUNCIL, UNIT OF ANIMAL GENETICS
DEPARTMENT OF GENETICS, UNIVERSITY OF EDINBURGH

I want to devote my allotted space here to pointing out one of the problems in quantitative genetics where we lack understanding, and where I believe theory and experiment would be profitable.

Population genetics is enlivened by the arguments about the origin and maintenance of genetic variation at polymorphic loci. The analogous problem in quantitative genetics, though it has not yet engendered much argument, is why any particular metric character has the mean value that it does. Why are mice the size they are? Or rats, or men? The ease with which artificial selection can change almost any character proves that natural selection could do so too, if it wanted to. The conclusion seems inescapable that the mean is for some reason the best value for that organism's particular way of life; not necessarily because the character itself matters, but because of pleiotropic, or correlated, effects of the genes on the character and on fitness. I think most biologists would expect most metric characters to be subject to stabilizing selection, and indeed many examples have been found that do at least seem to support this. Artificial selection in the laboratory, applied to almost any character, leads to a reduction of reproductive performance, again suggesting that natural selection would be stabilizing. But showing that extremes are selected against does not tell us how selection is acting on the genes, and I do not think we have got much further toward understanding this problem since Alan Robertson's paper of 1956. (J. Genet., 54, 236-248). For example, when mice selected for large size approach the selection limit they become so infertile that the strain can often be maintained only with great difficulty. But it is not because they are large that they are infertile, since crossing two such strains fully restores the fertility though the cross-bred mice are even larger.

The situation at the selection limit may not be relevant to the problem of the mean, except to illustrate the difficulty of interpreting stabilizing selection. Within the normal range of variation, the larger the mouse the more fertile it is - as a female at least. Why therefore does the fertility component of fitness not lead to an increase of size? There must be an antagonistic correlation of size with some other component of fitness. In the wild this might possibly be survival, since larger mice are less reactive and would be more easily caught by predators. Thus stabilizing selection may act through larger mice having reduced survival and smaller mice having reduced fertility. A different antagonistic correlation is of course needed for laboratory mice, but perhaps the absence of predators does explain why laboratory mice are larger than wild mice.

We need a theory to tell us whether antagonistic correlations with components of fitness can lead to stabilizing selection, and whether we can learn anything from the nature of the stabilizing selection about the action of natural selection on the individual loci. In particular, does it give rise to overdominance? Overdominance with respect to fitness, that is. There seems to be little overdominance in the effects of genes on metric characters or on the components of fitness, but we know little about fitness

WHY ARE MICE THE SIZE THEY ARE?

itself. It is easy to construct a very simple model of a locus with no dominance in its separate effects on two components of fitness, but overdominance in its effects on fitness. For example, a gene in mice might, hypothetically but realistically, have the following genotypic values:

	A_1A_1	A_1A_2	A_2A_2
No. of young per litter	10	9	8
No. of litters per life-time	8	9	10
Total offspring produced	80	81	80

The overdominance of the sickle-cell gene is of this sort, overdominance with respect to fitness but not with respect to either of the two pleiotropic effects, and so is that of the gene that makes rats resistant to warfarin. Can theory clarify the connections between antagonistic correlations with components of fitness, stabilizing selection and overdominance for fitness?

Canalisation in quantitative genetics

Dr. J. M. Rendel
CSIRO, DIVISION OF ANIMAL PRODUCTION, P.O. BOX 90
EPPING, N.S.W. 2121

Mass selection, progeny testing and family selection aim to increase the frequency of numerous genes with small indistinguishable additive effects; these can eventually be built up to bring about the sort of difference in phenotype which Lerner achieved when he increased the hen-housed average of his flock of chickens from 120 to 240 eggs by 12 generations of family selection. Selection techniques are very successful. There are sometimes limits to what can be done; it has not for example been possible to go much further than Lerner did with egg production in chickens. But any disappointment which may be felt ought not to be levelled at the techniques or their theoretical basis. I believe when these techniques cease to make progress it is usually because additive genetic variance has given up all it has to give or because the character selected is the wrong one.

Quantitative genetic theory on which animal breeding practices are based does consider non-additive genetic variance - it does consider interactions of various kinds. But the interactions are seen as interactions between overt effects. It is certainly the overt effects that animal breeding plans aim to use; indeed breeding plans and selection routines are usually based on genetic relationships which can be measured in the foundation populations. But embryological development is very well ordered; it has feedback controls built into it which prevent it from wandering too far from the common track. These controls supress and hide genetic variations which are not therefore overt. Overt genetic variation is variation which for one reason or another is not subject to feedback controls. Once overt genetic variation is exhausted it is the hidden sources of genetic change that have to be tapped and so far there has been little serious effort made to tap them.

I can best illustrate the kind of genetic variance I have in mind by reference to the main features of the canalised character scute in Drosophila Melanogaster.

Melanogaster has 4 bristles on the scutellum; in most unselected populations only a few flies in a thousand have fewer or more than 4. The character is strongly canalised at 4.

The sex linked mutant 'scute' reduces the mean bristle number to about 1, and there is variation about this mean which responds to selection.

Response is steady until flies with 4 bristles appear after which further selection continues to pile up the proportion of flies with 4 bristles until a mean of about 3.6 is reached. Very rarely in very large counts flies with more that 4 bristles appear. Thus despite the substitution of the wild type by the scute gene canalisation is still at 4 bristles; nor has selection of the overt genetic variation, present in scute flies, made any impact on the regulatory genotype.

The effects of selection in scute are manifest in wild type as well as in scute. By the time scute flies have

CANALISATION IN QUANTITATIVE GENETICS

> a mean of 3.5 bristles their wild type sibs in segregating populations have a proportion of flies with 5, 6 and occasionally 7 bristles with a mean of about 4.2.
>
> Finally, despite repeated attempts over hundreds of generations, it has not so far been possible to select a scute population with a mean of more than 3.6 by making use of variation manifest at levels below 4. I think it is clear that the low selection differential is not the reason for this. I think the necessary genotype is just not expressed below 4.

These findings gave rise to the idea that the development of this character is directed by the structural gene at the scute locus against a background of numerous genes with small indistinguishable effects. These background genes can be selected up and down the scale; canalisation is brought about by a regulator genotype that acts on the structural gene to regulate its activity to a level such that together with the background genotype it is just sufficient to produce 4 bristles.
Selection for increased bristle number in wild type flies follows a course quite different from selection in scute.

> At first there is very little overt genetic variance – just a few flies in a thousand with 5 bristles.
>
> Once selection begins to take effect bristle number increases by rather rapid steps interrupted by pauses at means of $5\frac{1}{2}$, $7\frac{1}{2}$ to 8, and 10 to 11.
>
> Canalisation remains at 4 in populations with a mean low enough to measure canalisation at 4, but is drastically reduced.

The effects of selection in wild type are little manifested in scute sibs. What effects there are on scute flies could be accounted for by supposing that some selection had taken place on the effects of genes of the background genotype - the ones selected in scute populations.

Finally there is as yet no limit in sight to increases in bristle number. Sheldon and Milton have reached 20 bristles so far. To judge from the second chromosome gene tufted in which the number of scutellar bristles is of the order of 50 - selection still has a long way to go. Thus selecting in scute we seem to reach an impass after increasing bristles by about $2\frac{1}{2}$, selecting in wild type the number has been increased by 16 so far and is still going up.

We have to account for the appearance of a lot of genetic variation not present when selection started, which does not act or manifest itself in scute flies, which reduces canalisation at 4 bristles and which is so much more powerful than the overt variation present in scute.

The obvious point to explore first is the possibility that we are confronted with failure or partial failure of the genotype responsible for canalisation. Regulator genes are failing to suppress overaction of the wild type gene at the scute locus. The great increase in bristle number is not due to the selection up of a background genotype containing many small genes but to the unfettered action of the structural gene at the scute locus. Increase in variance is due not only to failure of the regulators to compensate for differences in the background genotype but also to variability in the degree of failure of regulation, and hence the degree of activity of the structural gene. Canalisation, since it depends on regulation of the structural gene, will be reduced if the precision of regulation is reduced.

I have only one test of this interpretation. Suppose regulators of the gene at the scute locus which come into action to keep bristle number to 4 will not act in scute flies because the bristle number of these is already below 4. Then in high scute lines in which a large proportion of flies have 4 bristles, failure to regulate the scute gene properly should result in the appearance of flies with 5 or even 6 bristles. In a high wild type line selected by latter it has been possible to locate one or two presumed regulators in the right arm of chromosome 3. When one of these was crossed into a high scute line flies with 5 and 6 bristles did in fact appear in numbers very significantly greater than normally found in high scute stocks. Increase in 5s and 6s took place without any apparent effect on the numbers of flies with 4 bristles and less. Thus interaction is between the regulator and the phenotype, not the regulator and the scute gene; failure of regulation takes effect on all populations with a high enough bristle count.

The account I have given you is not conclusive proof that we are dealing with regulators, but it is exactly what you would expect to find if we were. Whether or not I have given a correct interpretation of the development of the scute phenotype there can be no doubt that regulators are a feature of embryological development and that failure or partial failure would result in the appearance of new variation which might well be more potent than overt variations for which natural selection has not elaborated a control. The hypothesis is then that the development of the scute phenotype takes place as the result of the action of a structural gene acting under the control of a regulator system against the background of additive genetic variance whose effect is quite small by comparison with the effects which can be produced when the regulator genes fail to function properly. Neither quantitative genetic theory which adopts a purely statistical approach and is not generally concerned with mechanisms nor animal breeding practice which is based on theory has paid much attention to the possibilities of this source of genetic variation.

I suggest that overt genetic variance, the raw material of animal breeding and of selection experiments in the laboratory, corresponds to the variance for bristle number found in scute populations and that a large and powerful source of variation corresponding to the breakdown of canalisation of the scute phenotype remains untapped. This untapped source is expected to have the following characteristics.

It will depend on a relatively small number of genes with predominantly recessive effects. These will be rare affecting one or two individuals in 500 in unselected populations. The result of selecting for them will be a large increase in the amount of expressed genetic variance and in the first instance, a deterioration of fitness. Finally there is a suggestion notably from Thomson's work with wing veins that there is a hierarchy of regulated stages in development that come into action in sequence so that as selection for unusual phenotypes proceeds what appear at first as isolated processes prove to be related at more remote levels. In scute, effects at the level of 7 or so bristles are shared with bristles typically associated with achaete.

I conclude that epistatic effects, rare events and the mechanisms that lie behind them are now ripe for theoretical and practical attention.

Quantitative genetic research in plants: Past accomplishments and research needs

C. O. Gardner
DEPARTMENT OF AGRONOMY
UNIVERSITY OF NEBRASKA, LINCOLN, NE 68583

The two decades of the 1950's and 1960's might be called the "golden years" for plant quantitative genetics and for the application of quantitative genetic theory in solving some of the problems of plant breeders. The work was built on the foundations laid by R. A. Fisher (1918), Sewell Wright (1935) and J. B. S. Haldane (1924), and was sparked by publication and research of K. Mather (1949) and his co-workers in England and by R. E. Comstock and H. F. Robinson and their co-workers in North Carolina, U.S.A. Important contributions incorporating the theory of epistasis into genetic models were made by C. C. Cockerham (1954) and Oscar Kempthorne (1954) to further stimulate quantitative genetic research. Schnell (1961, 1963) developed the theory of linkage as related to quantitative genetic models, but relatively little use of his work has been made in plant quantitative genetics. Contributions of Comstock and Robinson (1952) on the effects of linkage on degree of dominance have played an important role.

In the 1970's quantitative genetics still plays an important role in plant breeding, but emphasis in many departments has shifted to physiological and biochemical approaches to plant breeding. As activities in these fields has progressed, the need for quantitative geneticists to be working with physiologists and biochemists has become more and more obvious. Many physiological processes and the chemical composition of plants, particularly grain and forage quality traits, are under multigenic control and will ultimately need to be studied using quantitative genetic techniques. We look to the physiologists and biochemists to develop measurement techniques which will permit evaluation of large numbers of individuals or families for quantitative genetic study.

Theoretical developments in quantitative genetics have stayed well ahead of applications in plant breeding. The plant is a very complex organism which responds differently to the infinite numbers of environmental variations that are possible during the lifetime of the plant. In many instances we have been forced to use relatively simple genetic models. Yet a great deal has been accomplished.

PAST ACCOMPLISHMENTS

Let me mention some of the advances made using quantitative genetic theory in maize breeding and in plant breeding in general.
1. Mating designs and genetic models along with appropriate statistical methods for analysis and genetic interpretations have been developed by Comstock and Robinson (1948, 1952) and Griffing (1956) and were summarized and elaborated upon by Cockerham (1963) in a very excellent paper. Books by Kempthorne (1957), Mather (1949), Falconer (1960) and more recently by Mather and Jinks (1971) have been valuable aids to the plant breeder.

QUANTITATIVE GENETIC RESEARCH IN PLANTS

2. Genetic variances have been estimated in a host of plants and have served as a basis for major plant breeding decisions. In maize, we have clearly demonstrated the existence of substantial additive genetic variance for grain yield and other traits in open-pollinated varieties indicating that intra-population selection schemes should be effective, contrary to the beliefs of many early maize breeders. Success attained in improving populations through various recurrent selection schemes provides further evidence that these estimates of additive genetic variance are valid ones.

3. The overdominance hypothesis advanced by Hull as the primary cause of heterosis in corn has been refuted by work at North Carolina State University and at the University of Nebraska. The effect of linkage in causing apparent overdominant effects in genes controlling yield in maize has also been clearly demonstrated. (See Gardner (1963) for summary).

4. The theoretical basis for predicting rates of genetic improvement in plants by almost any recurrent selection systems has been developed. Major credit goes to R. E. Comstock and C. C. Cockerham, although their work and teaching is based on that of J. L. Lush, which in turn traces to the early work of Sewell Wright and R. A. Fisher. Many prediction formulas have been developed in detail and published for specific selection systems. Empig, Gardner and Compton (1972) summarized prediction equations for intra- and inter-population selection systems, which are widely used by plant breeders in choosing among alternative breeding systems.

5. Approximate theory for the probability of fixation of favorable alleles has been developed by M. Kimura (1957) which provides insight into optimum effective population size in recurrent selection programs.

6. Selection index theory has been developed which provides some insight into optimum weighting of traits to be used in the

selection criterion. However, relatively little use of
selection indices has been made by plant breeders.

7. New breeding systems have been developed and old ones
have been modified and improved as a result of quantitative genetic studies. Reciprocal recurrent selection of
Comstock, Robinson and Harvey (1949) utilizes all kinds
of genetic variability. Full-sib reciprocal recurrent
selection (Lonnquist and Williams (1967) and Hallauer (1967))
looks even more promising from a commercial corn-breeding
standpoint because of the natural spin-off of new lines and
hybrids each generation. Full-sib reciprocal recurrent
selection using an inbred tester (Eberhart, Debela and
Hallauer (1973); Eberhart and Russell (1975)) provides
for more precise evaluation of genotypes and likewise
looks promising. Modification of mass selection by using
the grid system and other environmental controls (Gardner
(1961)) has worked successfully in other crops (Verhalen,
Baker and McNew (1975), Redden and Jensen (1974), and
Matzinger and Wernsman (1968)) as well as in maize (Gardner,
1973). Lonnquist's modifications for ear-to-row breeding
in maize have made that system one of the most successful
(Webel and Lonnquist (1967)).

8. Quantitative genetic studies conducted in self-fertilizing
species have lead to the adoption of recurrent selection
methods for population improvement and to provide a greater
array of genotypes among which selection can be practiced
to develop new varieties or breeding material. In sorghum,
we have used genetic male sterile genes to convert a normally
self-pollinating species into an open-pollinated one and
are utilizing recurrent selection schemes in an attempt to
improve those populations.

9. Genotype-environment interaction as a component of the
deviation of a phenotype from its genotypic value has been
elucidated by the early work of Sprague and his co-workers

QUANTITATIVE GENETIC RESEARCH IN PLANTS

(Sprague and Federer (1951) and Rojas and Sprague (1952)) and by Comstock and Moll (1963). Measures of stability of genotypes have been provided by Finlay (1964) and by Eberhart and Russell (1966).

10. Finally, quantitative genetic information from populations has and will continue to play an important role in the plant breeders choice of base populations and sub-populations with which to work, as well as in his choice of breeding systems.

RESEARCH NEEDS

As we look to the future, there are a number of areas where quantitative genetics will play an important role in plant breeding, but where additional developments are needed.

1. An improvement is needed in methods for determining optimum effective population size for different recurrent selection programs. Many such programs have been conducted where the effective population size has been much too small. This leads to fixation of undesirable genes and limits ultimate progress possible.

2. The effects of linkage and linkage disequilibrium when populations are combined or created for recurrent selection programs, and when selections are recombined to form the next generation, need further study.

3. Methods for simultaneous selection for several traits need further investigation. Undesirable correlated responses have often occurred in single trait selection in plants. For populations to have commercial value, many traits must be improved simultaneously.

4. New approaches for studying epistasis and making maximum use of it in breeding programs are needed. Epistasis appears to be very important in heterosis observed in hybrids among inbred lines, in cultivars of self-pollinating

species, and in asexually reproduced plants, but we know relatively little about it.

5. Plant physiological processes and the chemical composition of plant parts are likely to be under multigenic control and will need to be studied using quantitative genetic techniques. Such traits will undoubtedly become a part of our selection criterion in the future as we develop better indices and better measurement techniques.

6. If cell culture techniques are developed to where haploid plants can easily be produced, quantitative geneticists will face a whole new set of problems and opportunities. Griffing has anticipated this and has already developed a selection theory for haploids.

7. Finally, in the years ahead, we will be faced with a variety of problems related to the shortage of fossil fuels as sources of energy. Fertilizer usage, irrigation practices, and tillage operations are bound to change, which in turn will change the plant environment and the plant breeder's goals. Alleles that were most favorable under near optimum conditions may no longer be the most favorable ones.

As we learn more about nitrogen fixing bacteria and their relationship to members of the grass family as well as to the legume family, we may need to develop breeding systems for the simultaneous selection of the crop species and the related bacteria species to maximize yields.

Although no new quantitative genetic principles are involved in dealing with these changes, quantitative genetics will play a major role. Changes in environment in its broadest sense are not new to plant breeders. Plant pathogens and insects and other pests are constantly undergoing evolution to form new races and new biotypes which become part of the plant's environment. This is what keeps plant breeders in business.

BIBLIOGRAPHY

(1) Cockerham, C. C. (1954) An extension of the concept of partitioning hereditary variance for analysis of covariances among relatives when epistasis is present. Genetics 39:859-882.

(2) Cockerham, C. C. (1963) Estimation of genetic variances. Statistical Genetics and Plant Breeding (W. D. Hanson & H. F. Robinson, Eds.). Washington, D. C., National Academy of Sciences-National Research Council Publ. 982, 53-94.

(3) Comstock, R. E. & Moll, R. H. (1963) Genotype-environment interaction. Statistical Genetics and Plant Breeding (W. H. Hanson & H. F. Robinson, Eds.). Washington, D.C.: National Academy of Science-National Research Council Publ. 982, 164-196.

(4) Comstock, R. E. & Robinson, H. F. (1948) The components of genetic variance in populations of biparental progenies and their use in estimating the average degree of dominance. Biometrics 4:254-266.

(5) Comstock, R. E. & Robinson, H. F. (1952) Estimation of average dominance of genes. Heterosis (J. W. Gowen, Ed.). Ames, Iowa: Iowa State College Press, 494-516.

(6) Comstock, R. E., Robinson, H. F. & Harvey, P. H. (1949) A procedure designed to make maximum use of both general and specific combining ability. Agron. Jour. 41:360-367.

(7) Eberhart, S. A., Debela, S. & Hallauer, A. R. (1973) Reciprocal recurrent selection in the BSSS and BSCB1 maize populations and half-sib selection in BSSS. Crop Sci. 13:451-456.

(8) Eberhart, S. A. and Russell, W. A. (1966) Stability parameters for comparing varieties. Crop Science 6:36-40.

(9) Eberhart, S. A. & Russell, W. A. (1975) Hybrid performance of selected maize lines from reciprocal recurrent and testcross selection programs. Crop Sci. 15:1-4.

(10) Empig, L. T., Gardner, C. O. & Compton, W. A. (1972) Theoretical gains for different population improvement procedures. Univ. of Nebr. Agr. Exp. Sta. Misc. Pub. 26 Revised.

(11) Falconer, D. S. (1960) *Introduction to Quantitative Genetics* New York: The Ronald Press.

(12) Finlay, K. W. & Wilkinson, G. N. (1963) The analysis of adaptation in a plant-breeding programme. Australian J. Agr. Res. 14:742-754.

(13) Fisher, R. A. (1918) The correlation between relatives on the supposition of Mendelian inheritance. Trans. Roy. Soc. Edinb. 52:399-433.

(14) Gardner, C. O. (1961) An evaluation of effects of mass selection and seed irradiation with thermal neutrons on yield of corn. Crop Sci. 1:241-245.

(15) Gardner, C. O. (1963) Estimates of genetic parameters in cross-fertilizing plants and their implications in plant breeding. *Statistical Genetics and Plant Breeding*, (W. D. Hanson & H. F. Robinson, Eds.). Washington, D.C.: National Academy of Sciences-National Research Council Publ. 982, 225-252.

(16) Gardner, C. O. (1973) Evaluation of mass selection and of seed irradiation with mass selection for population improvement in maize. Genetics 74:s88-s89.

(17) Griffing, B. (1956) Concept of general and specific combining ability in relation to diallel crossing systems. Australian J. Biol. Sci. 9:463-493.

(18) Haldane, J. B. S. (1924) A mathematical theory of natural and artificial selection. Parts I, Trans. Camb. Phil. Soc. 23:19-41.

(19) Hallauer, A. R. (1967) Development of single-cross hybrids from two-eared maize populations. Crop Sci. 7:192-195.

(20) Kempthorne, O. (1954) The correlations between relatives in a random mating population. Proc. Roy. Soc. Lond. B 143:103-113.

(21) Kempthorne, O. (1957) *An Introduction to Genetic Statistics*. New York: John Wiley and Sons.

(22) Kimura, M. (1957) Some problems of stochastic processes in genetics. Ann. Math. Stat. 28:882-901.

(23) Lonnquist, J. H. & Williams, N. E. (1967). Development of maize hybrids through selection among full-sib families. Crop Sci. 7:369-370.

(24) Mather, K. (1949) **Biometrical** Genetics. New York: Dover Publications.

(25) Mather, K. & Jinks, J. L. (1971) **Biometrical** Genetics. Ithaca, N.Y.: Cornell University Press.

(26) Matzinger, D. F. & Wernsman, E. A. (1968) Four cycles of mass selection in a synthetic variety of an autogamous species Nicotiana tabacum L. Crop Sci. 8:239-243.

(27) Redden, R. J. & Jensen, N. F. (1974) Mass selection and mating systems in cereals. Crop Sci. 14:345-350.

(28) Rojas, B. A. & Sprague, G. F. (1952) A comparison of variance components in corn yield trials. III. General and specific combining ability and their interactions with locations and years. Agron. J. 44:462-466.

(29) Schnell, F. W. (1961) Some general formulations of linkage effects in inbreeding. Genetics 46:947-957.

(30) Schnell, F. W. (1963) The covariance between relatives in the presence of linkage. **Statistical** Genetics and Plant Breeding (W. D. Hanson & H. F. Robinson, Eds.). Washington, D.C.: National Academy of Science-National Research Council Publ. 982, 468-483.

(31) Sprague, G. F. & Federer, W. T. (1951) A comparison of variance components in corn yield trials. II. Error, year x variety, location x variety and variety components. J. Amer. Soc. Agron. 43:535-541.

(32) Verhalen, L. M., Baker, J. L., & McNew, R. W. (1975) Gardner's grid system and plant selection efficiency in cotton. Crop Sci. 15:588-591.

(33) Webel, O. D. & Lonnquist, J. H. (1967) An evaluation of modified ear-to-row selection in a population of corn (Zea mays L.). Crop Sci. 7:651-655.

(34) Wright, S. (1935) The analysis of variance and the correlations between relatives with respect to deviations from an optimum. Jour. Genetics 30:243-256.

The general area: Setting the stage

Dewey L. Harris
USDA-ARS
PURDUE UNIVERSITY, WEST LAFAYETTE, IN 47906

The interrelationships between the different aspects of Quantitative Genetics, Statistical Genetics, Population Genetics and Applied Animal Breeding are reviewed. A key feature is the ability to predict the genetic response to alternative testing, selection, and mating schemes.

INTRODUCTION

In setting the stage for this International Conference on Quantitative Genetics, I will describe the interrelationships between several areas of activity relative to quantitative genetics. In particular, I will describe how these areas contribute and lead to the decisions and activities of a breeder who is developing improved breeding stock of some agriculturally important species. These interrelationships are illustrated in the schematic diagram of Figure 1. In this diagram we see how the four basic areas of statistics, basic genetics, economics, and the husbandry of the particular species lead through various activities and considerations. The two key aspects in this diagram from the point of view of the practicing animal breeder are (1) the selection objective and (2) the ability to predict response to alternative breeding schemes. The lack of definitions of economic selection objectives in a precise soundly based manner is one of the serious weaknesses of much animal breeding of the past. However, this is not among the concerns of the conference this week. However, the prediction of response to alternative schemes is very much of concern and, in fact, should be of primary importance to quantitative genetics. In the various sessions this week, we will undoubtedly examine situations in which this ability is good and will probably encounter other situations in which this ability is not so good. In situations where the reliability of predictions is not good, it is usually necessary to make some modification of theory. This modification of theory is generally in the direction of including more and more complex phenomena in the possible occurrences to be accounted for in the predicted responses.

In the development of our selection theory we are at a stage where there have been considerable experimental results reported that seem to be in contradiction to our theory. In my opinion, most of these are due to inadequate experimentation with inadequate replication to assess the magnitude of variability around the expected response to selection. However, there are some

FIG 1

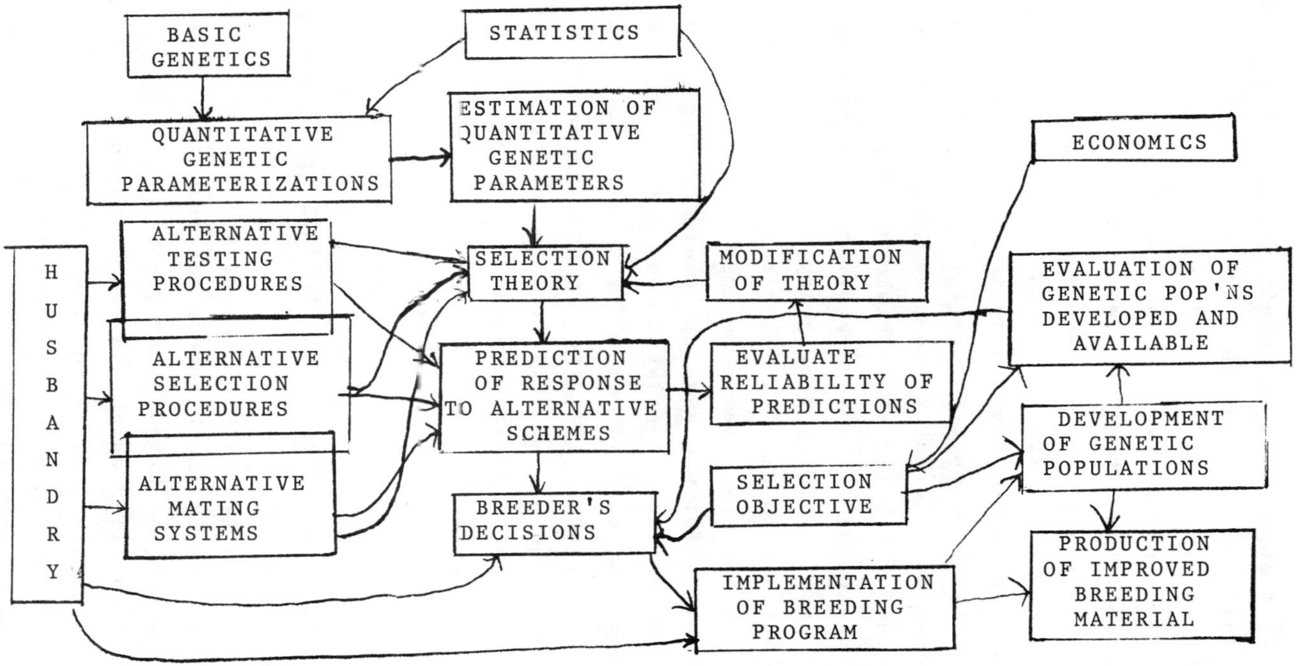

Schematic diagram showing interrelationships between different aspects of quantitative genetics and applied animal breeding.

observations suggesting the manifestation of phenomena in finite populations that are not accounted for in some theoretical developments relative to conceptual infinite populations. Undoubtedly we will hear more discussion of these phenomena during this week. The overriding question for the practicing animal breeder in this situation is going to be whether we can modify our selection theory in such a manner that he can bring these occurrences into his prediction of response to alternative schemes in order to provide a sound basis for the decisions which he has to make in designing a breeding program to be implemented. In particular the difficulty seems to be in his present ability to predict "long-term" response to alternative schemes. We are seemingly at the very unsatisfactory point where we are aware of the potential decline in response to selection over the long-term but we are limited in our ability to predict the magnitude of response to alternative schemes except that we know that the ultimate limit is a function of effective population size times intensity of selection (Robertson, 1960).

There are several characteristics of the population and of the design of the breeding program that influence the breeders expected response to selection under alternative schemes. These fall into three groupings as follows:
1. Characteristics of the breeding program that are under the breeders control as he implements the breeding program.
2. Quantities which can be estimated or evaluated that are indicative of the nature of the genetic influence upon the traits and characteristics of concern to the breeder.

SETTING THE STAGE

3. Those detailed characteristics of the genetic mechanism as it influences the quantitative traits that the breeder does not know and cannot estimate.

Let us first list those characteristics which are under the breeders' control as follows:
1. Test population sizes.
2. Intended selection proportions.
3. Dam to sire mating ratios and number of breeders.
4. Length of use of selected breeders.
5. Alternative selection criteria.

The characteristics which can be estimated and evaluated are as follows:
1. Initial covariance between relatives, between traits, and response first few generations of selection.
2. Measures of genetic relationship such as the probabilities of alikeness by descent for genes.
3. Inbreeding depression effects.
4. Results obtained from selection in other similar populations.

Those quantities which the breeder does not know and cannot estimate are as follows:
1. Number of segregating loci which influence traits of interest.
2. Distribution of gene frequencies.
3. Distribution of genotypic values.
4. Arrangement of loci on chromosomes and recombination values.

Of course, these latter characteristics are of fundamental importance in determining response to selection and other characteristics of interest to the breeder. However, those traits which are influenced by more than a very few genetic factors and are influenced by environmental contributions to the variability are usually termed "quantitative". This term "quantitative" usually implies an inability to determine the characteristics in the latter list. At times, major genes which influence quantitative traits have been discovered. This has usually occurred only in laboratory species, and those occurrences in practical breeding species have been responsible for only a limited portion of the total variation of interest to the breeder. The necessary complex genetic control over the complex physiological characteristics which are of interest to the breeder of agriculturally important species will continue to require that the knowledge of the details of the genetic control of these characteristics and traits be unavailable and unattainable.

Thus the breeder is likely to continue to need to characterize his populations by gross parameters that are indicative of expected genetic improvement when alternative selection schemes are used rather than to attempt to completely "break the genetic code" for the multitude of genetic influences upon the traits of concern. This does not mean that it will not be of interest in a research context to study the detailed genetic controls influencing some relatively simple yet quantitative characteristics of a laboratory species. However, these approaches will primarily help to reassure us that there are genetic loci and alleles underlying quantitative traits but are not likely to provide tools to be useful for the practicing animal breeder.

BIBLIOGRAPHY

Robertson, A. (1960). A theory of limits in artificial selection. Proc. Royal Soc., B, 153:234-249.

Introductory statement: Poultry

A. W. Nordskog
DEPARTMENT OF ANIMAL SCIENCE
IOWA STATE UNIVERSITY, AMES, IA 50011

Chickens have played an important part in both fundamental and applied aspects of genetics. In the early days of this century, Bateson and Punnett demonstrated Mendelian segregation in the inheritance of plumage color. Also one of the first cases of sex-linked inheritance involving the barring gene, was demonstrated by Bateson. Somewhat later R. A. Fisher used plumage color mutants in poultry to support his theory of the evolution of dominance.

Compared with the breeding of other economically important animals, poultry breeding has been the first to leave the farm, so to speak, to become part of a sophisticated breeding industry. On a commercial level, chickens have been the first to be commercially exploited by the application of inbreeding-hybridization technics, as earlier used in corn, as well as by methods of selective improvement using the principles of quantitative genetics. Thus, the poultry industry, compared to other animal industries, seems to have been the quickest to apply modern methods of genetic improvement, including the employment of formally trained geneticists to handle breeding technology plus the use of computers and other modern business methods.

In corn, essentially all commercial breeding is based on inbreeding and hybridization. In chickens, it is not nearly as clear that hybridization is the sole alternative to generating high level performance. Today, there are perhaps two commercial breeders in the U.S. using corn hybridizing methods but there are far more breeding companies that produce strain crosses from parental lines with low levels of inbreeding which have undergone selective improvement.

Looking at where we are today, laying flocks which average 260 or more eggs per year are not uncommon. Average production in the U.S. today on an annual basis is at the rather remarkable level of 230 eggs. The improvement in broilers and the development of our broiler meat industry is truly one of the remarkable accomplishments of the poultry industry. Today, the average U.S. citizen consumes 15 chickens a year, each averaging 2 3/4 pounds. The average family of 4 consumes 60 meat chickens, 2 turkeys, and uses all the eggs laid in a year by 4 hens - about 100 dozen. The improvement that has been made in growth and efficiency of turkeys has only been slightly less phenomenal than in the case of broilers. Turkeys will average one pound of gain per week over a 20 week period so that turkey toms when marketed at 20 weeks will weigh 20 pounds.

Thus, it seems that the poultry breeder has every right to be proud of his accomplishments and, accordingly, that quantitative genetics has played an important role in his accomplishments seems a most reasonable inference.

However, this success story is only one side of the coin. As quantitative geneticists we realize that phenotypic performance also has an environmental basis. We are certain that knowledge and application of disease control, better management and better feeding have played a very important role in the development of our broilers, turkeys and laying strains. We are, in fact, really

POULTRY

in the dark as to how much genetics has really contributed to the improvement that we observe. The high level improvement we as geneticists see in our egg-laying varieties of chickens, we like to think, is a result of effective selection. On the other hand, Dr. Hutt of Cornell University, long ago pointed out that ducks which seem not to have been subjected to much sophisticated selection pressure, as in the case of chickens, have been known to average 300 eggs per year with some individual records of over 360 eggs per year.

Although meat chickens and turkeys have seemed to respond to selection, more or less according to expectation from quantitative genetics theory, this seems not to hold for such reproductive traits as fertility and rate of lay. For the past 25 years a lively controversy has existed among poultry breeders as to whether we have made any improvement in rate of egg production, as a sole consequence of selection, even though genetic variance seems not to have become exhausted. True, we seem to have made improvement but we aren't sure whether such improvement is a result of discarding inferior stocks or if there has been real improvement within the breeding populations themselves.

According to quantitative genetics theory, traits which are not highly heritable should be improved by family selection. Theoretically, selecting on an index made up of the individual's record, the average of its sire's progeny and its dam's progeny, put together in some optimum combination, for example, Lush's combination selection index, will maximize genetic gains. Yet, to my knowledge this has never been proved by experimental test. Certainly, it has not been proved unequivocally in any other species, including laboratory animals.

Commercially, most of our successful breeders have proved to be, first of all, innovators and entrepreneurs, but almost never trained geneticists. A typical pattern of a successful poultry breeder is a person who, if he has college training, it is in some field other than breeding or genetics. After he becomes successful, only then can he afford the luxury of having scientists on his staff. Perhaps there may be some exceptions to this generalization. I like to think that formally trained geneticists contribute in some substantial way to the continuing success of the breeding business. Yet, we don't have overwhelming evidence that the application of quantitative genetics theory is critically important to the continuing success of a poultry breeding enterprise.

I raise the question as to whether the statistical training almost all quanitatively trained geneticists receive today may perhaps be of more value to him than his knowledge of quantitative genetics, per se. His statistical training alone permits him to design valid comparisons, to make more efficient field tests and to otherwise make efficient use of data.

With some confidence, we think, as teachers, that specific training in quantitative genetics contributes to genetic progress in breeding applications. Most importantly, quantitative genetics provides the theoretical basis for laying breeding plans. A good example, is the application of selection index theory, but even then I am not sure, how critically important this is. For example, if we had no genetics theory, a trained statistician could still use his knowledge of statistics and regression theory in a breeding operation. He could predict gains from fitting observations of response on mean selection intensities over generations. As I see it, this is equivalent to what Falconer calls realized heritability. That method is purely statistical and empirical; not based on quantitative genetics.

Let me conclude by stating that I think we are at the stage in the science of quantitative genetics as applied to breeding where we need to give more thought to experimental testing of the validity of certain assumptions: the relative efficiency of selecting on family averages versus individual records, genetic variance as related to population size, the conflict between artificial and natural selection, asymmetry of responses when we deal with bidirectional selection, efficiency of selection indexes when based on genetic information from static populations as compared with only phenotypic information, and finally, the testing of observed results of multiple-trait selection with theoretical expectation. Are physical incompatibilities between traits important but unknown parameters? Pure theory is clear enough. Its validity in application is far from clear. We don't seem to understand why observed results don't agree well with expectation, especially in the case of reproduction in poultry.

Perhaps our general problem is that quantitative genetics theory is based on static populations but our applications are to dynamic populations. Finally, in poultry, some researchers are returning again to searching for mendelian segregation of major gene effects as they may bear on important economic traits. There is now evidence, for example, that important differences in the cancer-like disease in chickens, leukosis, are controlled by single genetic loci.

II

*Implications of
Population Genetics
and Molecular Biology*

The relevance of molecular biology to plant and animal breeding

R. C. Lewontin
MUSEUM OF COMPARATIVE ZOOLOGY
HARVARD UNIVERSITY, CAMBRIDGE, MA 02138

Quantitative genetics, upon which modern plant and animal breeding is based, is an attempt to produce knowledge by a systematization of ignorance. It is rooted in the certainty that the phenotypes of organisms are the manifestations of the actions of genes in particular sequences of environment, genes that are organized into chromosomes that have a regular behavior during gametogenesis. But beyond those generalities we can see only dimly. How many genes influence a measurement? How many and what kinds of alleles are segregating at these gene loci? What are the allelic frequencies? What are the phenotypic effects of gene substitutions? What are the norms of reaction of the genotypes across the present array of environments or in new environments? What is the organization of the relevant part of the genome within and between chromosomes? How much recombination is there between these loci? How much inter-locus interaction is there in the effect on phenotype and of what kinds? The predictions of quantitative genetics depend upon the answers to these questions because the models of quantitative genetics have each of these questions built into them. To some extent predictions may be quite robust to differences in the actual values of the parameters in these questions. That is, observations on the phenotypic level may be quite sufficient to predict the outcomes of breeding schemes quite directly, without the necessity of knowing the detailed structure of the genome and its relation to phenotype. For example, Fisher's Fundamental Theorem is useful precisely because such a little knowledge of the actual genetic phenotypic situation goes such a long way. But, if we wish to look beyond the immediate rate of response to selection in the present set of environments, we really do require knowledge of two general sorts that is not available to us from the first and second moments of phenotypic distributions.

First, the predictions of quantitative genetics depend upon the <u>structure of the genome</u>. In particular, we need to know the numbers of loci, their linkage relations, their allelic frequency distribution and the mutational and recombinational sources of new variation. Second, we need to know the <u>relations between gene and organism</u>, how gene action, in particular environments, is translated into phenotype. The knowledge about these questions can come to us only by opening up the black box whose outer shape we have so far been describing, and seeing what the machinery inside really looks like. This is the task of molecular and developmental genetics and some general knowledge is already available to us from the recent activity in these fields. Our models of quantitative genetics must either take cognizance of these findings or else show that they are, in fact, irrelevant because of the robustness of our theory.

1. The Number of Genes is not Large

While higher organisms have enough DNA to specify from 100,000 to 1,000,000 proteins of average size, it appears that the actual number of cistrons does not exceed a few thousand. Thus, saturation lethal mapping at the fourth chromosome (Hochman, 1971) and the X chromosome (Judd, Shen and Kaufman, 1972) of <u>Drosophila melanogaster</u> make it appear that there is one cistron per salivary chromosome band, of which there are about 5,000 in this species. Whether

RELEVANCE OF MOLECULAR BIOLOGY

5,000 is a large or small number of total genes depends, of course, on the degree of interaction of various cistrons in influencing various traits. Nevertheless, it is apparent that either a given trait is strongly influenced by only a small number of genes, or else there is a high order of gene interaction among developmental systems. With 5,000 genes we cannot maintain a view that different parts of the organism are both independent genetically and each influenced by large numbers of gene loci.

2. Genes of Related Function are Closely Linked

Although there are known exceptions, it is often observed that controlling elements are closely linked to the genes they control and that cistrons controlling sequential steps in biochemical reactions are close to each other, and even sequential, on the chromosome. As a result, models of multi-gene phenotypic characters where it is assumed that the genes are independently segregating are suspect. This does not mean that independently segregating blocks of genes relevant to a character do not exist. There is experimental evidence that several chromosomes contribute genetic variance to one measurement, as for example, bristle number in Drosophila (Thoday, 1965), but it is likely that there are separate functional systems influencing quite separate morphogenetic processes that contribute to the total number of bristles. We must not make the mistake of assuming that a measurement arbitrarily chosen by quantitative geneticists necessarily corresponds to a unitary character from a genetic or natural selective standpoint. The number of cases in which structural genes and their controlling elements are closely juxtaposed and serially ordered forces us to consider this as a usual phenomenon that should be reflected in model-making. In particular, linkage cannot be regarded as a second-order nuisance to be ignored except in special cases.

3. Coordinate Repression and Induction are Common

The transcription and translation of genes into protein product are sensitive to materials in the environment by means of control circuits that affect more than one gene product simultaneously. Thus a single operon may, and often does, contain several structural genes that will be simultaneously switched off or on by the presence or absence of a single end-product of reaction. For quantitative genetics, this means that environmental correlations may be expected which cannot be broken by selection or remaking of the genotype. It has long been supposed that correlations between characters, especially genetic correlations, could be broken by carefully designed selection schemes and by searching for genetic variants with different phenotypic effects. While this is no doubt true for some cases, we should also expect environmental correlation that cannot be broken by genetic manipulation, correlation that arises out of the basic logic of the control circuitry of gene action.

4. Genetic Variation is Dense Along the Genome

Since the first systematic studies of genic variation of structural genes by Harris (1966) and Hubby and Lewontin (1966), an immense amount of information on genetic variation in enzymes has accumulated (see Powell, 1976, for a review). Plants and animals, vertebrates and invertebrates, domesticated and wild organisms, prokaryotes and eukaryotes, consistently show a very high function of these structural gene loci to be polymorphic. A rough modal value is that about one-third of structural gene loci in a species are polymorphic and the heterozygosity per individual is about 10 percent. Recent experiments of Singh, Lewontin and Felton (1976) and Coyne (1976) as well as more recent unpublished data from our laboratory show that these figures are considerable underestimates, at least for heterozygosity. While plant and animal breeders and experimental population geneticists have always known of the existence of genetic variation for virtually any measurement in virtually any noninbred and unselected population, this observed variance could have been the result of very few segregating genes. We now know that the density of segregating genes along the chromosome is very high. What we do not know is whether the structural genes for which this variation has been documented represent the genetic variations of shape, size, metabolic rate, etc. in which plant and animal breeders are interested. It may be that structural gene variation is irrelevant to crop and animal improvement and that it is controlling genes that are the object of artificial selection. It is not known how variable these genes are, nor whether there are specific interactions between allelic variants of structural genes and allelic variants of controlling genes, if they exist.

5.   Multiple Alleles are The Rule

In Drosophila a polymorphic locus will have a very large number of different alleles segregating. The work of Coyne, Felton, Lewontin and Singh, cited above, shows that a polymorphic locus in Drosophila typically has a dozen to three dozen alleles segregating at a locus. Models of quantitative genetics with two alleles, a positive and a negative modifier of some measurement, are completely inappropriate. Models ought to be recast with a large number of alleles having a spectrum of gene effects.

6.   Intracistronic Recombination is Common, if Not The Rule

The classical picture of genes as beads on a string, with recombination occurring between the beads, is wrong. Recombination occurs between DNA bases and as a result recombination can be a very powerful source of the production of allelic variation. Of course, recombination gives rise to new combinations of already existing alleles at different loci, as in the classical picture, but it also generates quite new allelic forms in heterozygotes that differ from each other by more than a single base substitution within a locus. This, in turn, means that the rate of appearance of new variance for phenotypic traits will be much higher in crossbred as opposed to inbred materials, not only as a consequence of re-shuffling of variation between loci, but because wholly new alleles will arise by intracistronic recombination. There is no reason to suppose that such new alleles will have properties that are in any way predictable from those of the old alleles. The effect of bringing two base substitutions into the same DNA strand may be to completely cancel out the effect of each or give a quite new function. At the level of DNA base substitutions, epistasis is the rule, not the exception.

7. Genetic Engineering Could be a Powerful Supplement to Conventional Breeding

The most recent discoveries in molecular genetics bring into the realm of reality a number of possibilities for genetic manipulation of plants and animals by totally new methods. These consist essentially of several techniques for the introduction of genic material between widely divergent organisms on a selective basis. The normal methods of plant and animal breeding not only require that two organisms be sexually compatible and produce fertile offspring, but they produce results of crossing that are nonspecific and require future generation of selection. The classical situation is that of disease resistance to a virus or fungus. The resistance may exist only outside of the cultivated species in which case a series of indirect crosses will be required to introduce the needed genes. Moreover, even if resistance factors are found within the species, but especially if they are not, such genes will occur in strains of poor yield, configuration, physiology, etc. The introduction of the resistance genes must then be followed by a laborious selection and backcrossing procedure to recover the yield qualities lost in the original cross.

Methods now exist both in theory and practice for the introduction of genic material <u>selectively</u> and from <u>widely divergent</u> species. These methods include viral transduction, plasmid formation and, more restrictively, genetic transformation. In the case of viral transduction, a virus with a restricted site of attachment would be useful, but not necessary, since each transducing particle carries only a very small part of the donor genome into the host. Thus, the amount of cotransduction of different genes would be small. Imagine, for example, growing a virus on a smut resistant variety, treating meiotic tissue with viral particles from the resistant donor, and then using very high levels of disease infection as a screen for transduced progeny. Depending upon the efficiency of transduction, one might recover one progeny plant in a thousand that was genetically resistant and with virtual certainty such a plant would have the rest of its genome undisturbed. Plasmid technology would work in a similar way, but would probably be much more efficient but less selective. While no effort up to the present has been devoted to the introduction of specific small sections of the genome as a method of crop or animal improvement, it seems likely that a considerable effort will be mounted in the near future in view of the immense advantage such specific gene transfer would have over the usual methods of breeding.

Aside from the mildly futuristic ramifications of specific gene transfer, the most direct general relevance of discoveries in molecular biology up to the present has been the much clearer picture it has given us of the actual structure of the genome in terms of linkage and control relations. What molecular biology has not yet really illuminated is the relation between genes, environment and organism. As yet we know so little about the mechanisms of morphogenesis that we cannot make any predictions about the phenotypes of different genotypes in different environments. For example, the long-standing question of whether it is better to select for a trait under optimal or stress conditions remains a matter for <u>ad hoc</u> empirical investigation on a case-by-case basis. Yet the solutions to problems of genotype-environment interaction become increasingly important as high technology and capital intensive agriculture make the conditions of agricultural production more specialized. Such questions cannot be answered by estimation of components of phenotypic variance because such estimations depend so strongly on the actual environments experienced exactly when there is a substantial genotype-environment interaction. Just as realistic models of the genome could not be derived from the essentially statistical observations of quantitative genetics, but had to await experimental studies of the genome itself, so a correct understanding of the relation between genotype and environment in producing phenotype, cannot come from the internal operation of the techniques of quantitative genetics, but will have to be provided exogenously by future progress in developmental biology.

BIBLIOGRAPHY

(1) Coyne, J. A. (1976). Lack of genic similarity between two sibling species of Drosophila as revealed by varied techniques. Genetics, November, 1976. (in press)

(2) Harris, H. (1966). Enzyme polymorphism in man. Proc. Roy. Soc. Ser. B. 164, 298-310.

(3) Hochman, B. (1971). Analysis of chromosome 4 in Drosophila melanogaster. II. Ethyl methanesulfonate induced lethals. Genetics 67, 235-252.

(4) Hubby, J. L. and Lewontin, R. C. (1966). A molecular approach to the study of genic heterozygosity in natural populations. I. The number of alleles at different loci in Drosophila pseudoobscura. Genetics 54, 577-594.

(5) Judd, B. H., Shen, M. W. and Kaufman, T. C. (1972). The anatomy and function of a segment of the X-chromosome of Drosophila melanogaster. Genetics 71, 139-156.

(6) Powell, J. R. (1976). Protein variations in natural populations of animals. Evolutionary Biology 8, 79-119.

(7) Singh, R. S., Lewontin, R. C. and Felton, A. (1976). Genetic heterogeneity within electrophoretic "alleles" of xanthine dehydrogenase in Drosophila pseudoobscura. Genetics, November, 1976. (in press)

(8) Thoday, J. M. (1965). Effects of selection for genetic diversity. Proc. XI Int. Congr. Genetics 3, 533-540.

Applications of meiotic drive in animal breeding and population control

Daniel L. Hartl
DEPARTMENT OF BIOLOGICAL SCIENCES
PURDUE UNIVERSITY, WEST LAFAYETTE, IN 47907

Meiotic drive is defined and distinguished from gametic selection and gamete competition. Known cases of meiotic drive are briefly reviewed, including accessory chromosomes, nonrandom disjunction, sex chromosomal meiotic drive, and autosomal meiotic drive. Three cases are discussed in detail: the sc^4sc^8 X chromosome in *Drosophila melanogaster*, t alleles in the house mouse, and *SD* in *D. melanogaster*. Detailed suggestions are presented concerning several ways and under what conditions alleles exhibiting meiotic drive might be found in agriculturally important plants and animals. Prospects for finding such alleles are quite favorable. The most important potential application of meiotic drive in animal breeding is in the genetic control of the sex ratio. Meiotic drive can also be applied in the control of populations of insect pests; control programs based on meiotic drive are presently feasible in fruit flies and mosquitoes.

INTRODUCTION

Twenty years have passed since Sandler and Novitski (1957) summarized the handful of cases of meiotic drive that were then known and predicted that more examples would be found, in a wider variety of species, as the genetic study of higher organisms became more precise and extensive. It therefore seems appropriate at this symposium to inquire whether their prediction has been validated.

The answer is yes, and the answer is no; yes because the number of known examples of meiotic drive has increased from a handful to more than twenty and because the number of species exhibiting the phenomenon has increased; no because meiotic drive is apparently an unusual phenomenon. Mendel's law of segregation is still safe. Nevertheless, the increase in the number of examples of meiotic drive has been impressive, so much so that we can now begin to consider how -- or if -- the phenomenon can be put to some practical use.

In this paper I want first to define meiotic drive and distinguish it from phenomena that are related but not identical. Then I will mention some specific examples of meiotic drive, avoiding details inasmuch as a recent review is available (Zimmering, Sandler and Nicoletti 1970). Three cases will be discussed in some detail because more recent information has become available and because their mechanisms prove to be interesting. I will then suggest some ways for plant and animal breeders to look for meiotic drive, both where and how. Finally, some potential practical uses of meiotic drive will be pointed out. Although I cannot envisage an immediate use of meiotic drive in traditional breeding programs, I do think the phenomenon has potential application in manipulation of the sex ratio in animals and in population control of various insect pests.

APPLICATIONS OF MEIOTIC DRIVE

DEFINITIONS

A heterozygote for alleles A_1 and A_2 is ordinarily expected to produce functional gametes carrying each of the alleles with a frequency of 50%. This is, of course, Mendelian segregation. Non-Mendelian segregation or meiotic drive (Sandler and Novitski 1957) or segregation distortion (Sandler, Hiraizumi and Sandler 1959) -- the terms have come to be used almost interchangeably -- refer to instances in which A_1/A_2 heterozygotes produce functional gametes in a ratio deviating from one-to-one because gametes carrying one of the alleles may fail to be formed or may fail to function properly. In meiotic drive, the deviation from a one-to-one ratio of gametic types is a property of the diploid, heterozygous genotype which produces the gametes and not a property of the genetic constitution of the gametes themselves.

Operationally, meiotic drive in a heterozygous parent leads to a deviation from the expected one-to-one genotypic distribution among the offspring in a backcross. Many other things can also lead to such a deviation, zygotic selection probably being the most common. When the deviation from a one-to-one distribution is large -- say, when the ratio recovered is *4:1* or greater -- then it is often relatively easy to determine from mortality tests or litter sizes whether there is sufficient zygotic mortality to account for the effect. When the deviation is small, however, then such experiments are impractical because enormous numbers of offspring may be required to establish whether the effect is real, let alone how many may be required to determine its cause.

Although small departures from one-to-one genotypic ratios are conventionally attributed to zygotic selection, many of them may actually result from meiotic drive. Indeed, Hanks (1965, 1969) and Hanks and Torgerson (1969) have studied several such cases in *Drosophila* in detail and have shown them to be due to meiotic drive. In any event, because experimental difficulties in dealing with small deviations from Mendelian expectation muddy the waters, it is still not clear how widespread small amounts of meiotic drive may be. There is one trick that is sometimes useful in testing for meiotic drive of an autosomal gene. Many (but not all) cases of meiotic drive occur in only one sex. Thus, if the cause of a deviation from expectation is meiotic drive, one will generally expect different recovery ratios in reciprocal crosses; if the cause is zygotic selection, then one generally expects the same recovery ratio in reciprocal crosses.

Meiotic drive has not only to be operationally distinguished from zygotic selection, it must also be distinguished from such forms of prezygotic selection as gametic selection and gamete competition. In gametic selection, the viability or functional ability of a gamete depends on its own genetic constitution; in gamete competition, the functional ability of a gamete depends on the types of gametes with which it is competing for fertilization. In both these cases, A_1-bearing and A_2-bearing gametes from an A_1/A_2 heterozygote will fail to give rise to zygotes in equal frequencies. Both gametic selection and gamete competition seem to be important and widespread phenomena in plants (see Mulcahy 1975).

Gametic selection and gamete competition share one important feature that distinguishes them from meiotic drive: the functional ability of a gamete does *not* depend on whether the genotype of the organism that produced the gamete was heterozygous or homozygous. Meiotic drive, as mentioned before, occurs only in heterozygous genotypes.

Although the terms meiotic drive and segregation distortion and non-Mendelian segregation have come to be used synonymously, meiotic drive, as originally defined, had a more narrow meaning. Sandler and Novitski (1957) coined the term meiotic drive to refer to instances of non-Mendelian segregation in which the deviation from one-to-one segregation "is a consequence of the mechanics of the meiotic divisions". This nuance has largely been dropped for the simple reason that, in most cases, it is impossible to determine operationally whether or not a deviation from Mendelian segregation is a consequence of the mechanics of meiosis.

ACCESSORY CHROMOSOMES

Accessory chromosomes (also called supernumerary chromosomes or B-chromosomes) should be considered along with meiotic drive. Accessory chromosomes are chromosomes that are not homologous, or only partly homologous, to chromosomes in the regular set, and which are found in some individuals but not in others of natural populations of plants and animals. In some cases, the majority of the population may carry one or more accessory chromosomes. Accessory chromosomes have accumulation mechanisms (not always meiotic) by which they can be maintained at high frequency even in the face of considerable adverse selection, and in this way they are similar to meiotic drive. As with all systems of meiotic drive, accessory chromosomes "break most of the laws of classical genetics" (Darlington 1956).

Accessory chromosomes were first described by Lutz (see Stevens 1908) in the insect genus *Diabrotica*. Numerous reports have appeared since then. White (1973) lists 98 species having accessory chromosomes, ranging all the way from platyhelminths to molluscs, crustacians, insects and vertebrates (including two cases in mammals). Accumulation mechanisms in animals are exceedingly diverse and include endomitotic reduplication of the accessory chromosome, mitotic nondisjunction in the embryo with the accessory chromosome being directed toward the germ cell precursors, mitotic nondisjunction in spermatogonia coupled with a greater proliferation rate of spermatogonia that carry accessory chromosomes, and, in the case of mealy bugs (Nur 1962), meiotic nondisjunction which consigns the accessory chromosome to the pole of the spermatocyte which is destined to form functional sperm (see White, 1973, for review).

Accessory chromosomes are even more widespread in plants than in animals. First discovered by Langley (1927) and Randolph (1928) in maize, accessory chromosomes are now known to occur thoughout the plant kingdom, although they tend to be rare in long-lived or polyploid species (Darlington 1956). Parker (see Brown 1960)

reports the occurrence of accessory chromosomes in 199 species of
88 genera in higher plants alone. Battaglia (1964) lists 163
cases in higher plants, including many commercially important
species and genera. Indeed, Darlington (1956) estimates that up
to 10% of all short-lived, diploid flowering plants may have
accessory chromosomes. Accumulation mechanisms in plants are as
bewildering as those in animals, often involving mitotic nondisjunction in the first or second post-meiotic mitosis or (in
sorghum) extra mitoses of pollen nuclei bearing accessory
chromosomes, with one of the accessory chromosome bearing nuclei
becoming the generative nucleus. In maize, there is preferential
fertilization by the pollen nucleus bearing accessory chromosomes
(see Battaglia, 1964, for review).

Accessory chromosomes are usually heterochromatic, and their
effects on fitness and performance traits are still controversial.
Although accumulations of 4 or more accessory chromosomes in rye,
and 10 or more in maize, reduce vigor and fertility, there are
reports in *Trillium* and *Achillea* that accessory chromosomes may
enhance seed fertility (see Müntzing 1958, Battaglia 1964). The
main argument, in essence, is whether a single accessory chromosome
in an organism has any appreciable effect on fitness or performance,
granted that large numbers are harmful. In modern terminology:
is a single accessory chromosome selectively neutral?

Although geographical clines in the frequency of accessory
chromosomes have been found (in rye, for example, increasing from
Europe to Asia), and although provocative correlations with
ecological parameters have been discovered, there are many
populations that do not fit the overall patterns, and the patterns
themselves may have arisen from founder-like effects rather than
selection (see Müntzing 1958). Indeed, a population geneticist
feels a sense of *déjà vu* in reading this literature because much
of it is precisely of the sort that characterizes contemporary
efforts to determine whether allozyme polymorphisms are selectively
neutral: I should say one feels a sobering sense of *déjà vu*, for
the issue involving accessory chromosomes has never actually been
resolved -- it simply went out of fashion.

NONRANDOM DISJUNCTION

In certain species, such as *Drosophila melanogaster*, meiotic drive can occur in females which are heterozygous for structurally different (or heteromorphic) homologous chromosomes (Novitski 1967). In *Drosophila* females, as in many other organisms, the four products of meiosis are arranged more or less in a linear chain, one of the two outer products of the chain being destined to form the functional egg nucleus. Nonrandom disjunction is associated with crossing over in the following manner: a single exchange in a bivalent composed of hetermorphic homologues will generate, at the following anaphase, two asymmetric dyads. The longer chromatid of each dyad will "drag" relative to the shorter one, and this will create an orientation of the centromeres such that, at anaphase II, the longer chromatids will proceed to the two inner nuclei in the chain of four, and the shorter chromatids will pass to the outer nuclei. Since one of the outer products forms the functional egg nucleus, the shorter chromosomes will be recovered preferentially.

Nonrandom disjunction occurs in *Drosophila* females which are heterozygous for hetermorphic homologues of any chromosome of the set, and the sorts of aberrations that can give rise to it include deficiencies, translocations, and attached-X and other compound-X chromosomes (Sandler and Novitski 1957, Novitski 1967). Nonrandom disjunction has been invoked in humans to account for a differential transmission, between males and females, of segregation products in translocation heterozygotes (Hamerton *et al*. 1961, Jacobs 1972). The phenomenon probably occurs quite widely in the animal kingdom, but rather special genetic conditions are required to detect it.

CONTROL OF THE SEX RATIO

The ability to control the sex ratio in such organisms as cattle and poultry would be of tremendous practical importance, but efforts so far have been largely failures. Attempts to shift the sex ratio in cattle by separating X-bearing and Y-bearing sperm by means of differential centrifugation, electrophoresis, or chromatography have resulted in sex ratios of 60:40, at best, whereas sex ratios of about 90:10 would be required for economic feasibility (Church 1974). Attempts to separate sperm based, not on their gross physical properties, but on biochemical properties resulting from haploid expression of the genes they carry may be even less successful, as there is considerable evidence against haploid gene expression in animal spermatozoa (Beatty 1970, 1975). On the other hand, in the mouse, certain embryonic cell surface antigens may show haploid expression in sperm (Gluecksohn-Waelsch and Erickson 1971; Artzt and Bennett 1975). In any event, because of the nature of sex determination in poultry, sperm separation by any means will not alter the sex ratio, as females are the heterogametic sex.

The sex ratio has also proved to be refractory to direct selection, in part because the sampling error of measurement almost completely obscures what real genetic variation there may be, and in part because the sex ratio seems not to be subject to much polygenic variation (Falconer 1954). Although Falconer (1954) in *Drosophila* and Weir (1953, 1960) in mice (selecting for blood pH and obtaining a correlated response of the sex ratio) were successful in altering the sex ratio by selection, much or all of the effect seems to be due to the action of one or a few major genes. Indeed, King's (1918) classic success in selecting for sex ratio in rats is now attributed to such a major gene in the initial population (Falconer 1954).

This, I believe, is precisely the most important point -- that the sex ratio is under the control of major genes, and many departures from a 1:1 sex ratio may be the result of meiotic drive of these major genes in the heterogametic sex (Zimmering, Sandler and Nicoletti 1970). Weir's (1960) deviant sex ratio in the mouse is transmitted by the male, and Bar-Anan and Robertson (1975) have found highly significant differences in sex ratio between progeny groups in Israeli-Frisian cattle, with significant corre-

APPLICATIONS OF MEIOTIC DRIVE

lations between sire and son and between half sibs suggesting that the differences in sex ratio have a genetic basis.

Many instances of meiotic drive of major genes affecting the sex ratio are known. The classic case is that of *sex ratio* in *Drosophila pseudoobscura* and related species (Sturtevant and Dobzhansky 1936, Jungen 1967), in which males carrying the *sex ratio* X chromosome produce almost exclusively female offspring. The mechanism of this phenomenon is now known to be the dysfunction of the Y-bearing sperm (Policansky and Ellison 1970, Hauschteck-Jungen and Maurer 1976). Other examples of meiotic drive affecting the sex ratio have been studied in *D. melanogaster* (Erickson 1965; Hanks 1965, 1969; Faulhaber 1967, Kataoka 1967; Hanks and Torgerson 1969), in the mosquito *Aedes aegypti* (Craig, Hickey and VandeHey 1960, Hickey and Craig 1966), in the housefly *Musca domestica* (Wagoner 1969), and in the butterfly *Danaus chrysippus* (Smith 1975).

In *D. melanogaster*, and perhaps in some other organisms as well, X chromosomes that carry large deficiencies of the basal heterochromatin (the X is acrocentric) are associated with high rates of nondisjunction and meiotic drive. In the classical case studied by Gershenson (1933), denoted the sc^4sc^8 X-chromosome, among offspring that result from regular disjunction ($oo^4oo^8 \leftrightarrow Y$), about 66% carry the sc^4sc^8 chromosome; among the approximately 20% (at 25°C) of offspring that result from nondisjunction ($sc^4sc^8 + Y \leftrightarrow 0$), about 90% arise from the *nullo*-XY gamete (see Peacock, Miklos and Goodchild 1975). Moreover, the degrees of nondisjunction and meiotic drive are correlated, higher levels of nondisjunction corresponding with higher levels of meiotic drive among the regular segregants. Peacock, Miklos and Goodchild (1975) interpret this to mean that there is an underlying variable -- they propose it to be the probability of pairing -- low values of which lead both to an increased probability of nondisjunction and an increased amount of meiotic drive.

Peacock, Miklos and Goodchild (1975) have studied spermiogenesis in sc^4sc^8/Y males with the electron microscope. Spermiogenesis in *D. melanogaster* occurs synchronously in syncytial-like cysts containing 64 spermatid nuclei. The spermatids elongate symmetrically, heads oriented toward the base of the testis, until they extend almost the entire length of the 2mm-long *Drosophila* testis. At this time a spindle-shaped cystic bulge forms in the head region of the cyst and begins to progress along the tails, disrupting the syncytium as it goes along, removing unneeded organelles, parts of the nuclear membrane, nucleoplasm, and cytoplasm, and leaving behind it 64 individualized sperm tails, each invested in its own membrane. (The process of spermiogenesis has been beautifully analyzed by Tokuyasu, Peacock and Hardy 1972a, 1972b).

Peacock, Miklos and Goodchild (1975) find that, in sc^4sc^8/Y males, the degree of meiotic drive is strongly correlated with the amount of failure of sperm tail individualization. (In *Drosophila*, sperm that fail in individualization are degraded.) Thus, the mechanism of meiotic drive in the case of sc^4sc^8 seems to be sperm dysfunction.

Chromosomes that behave like sc^4sc^8 are by no means rare in *D. melanogaster*. Lindsley and Sandler (1958) studied 13 X-chromosomes that had large deletions of the basal heterochromatin, and 6 of these showed nondisjunction and meiotic drive similar to that observed for sc^4sc^8. Peacock and Miklos (1973) have accounted for this with a so-called "pairing-dysfunction" model. On this hypothesis, pairing between the X and the Y chromosome depends on many different sites being present on both chromosomes. These sites are assumed to be genetically active, and the process of pairing is assumed to be required to switch them off during meiosis in order that normal spermatid development may ensue. (This sort of model was clearly foreshadowed by Sandler and Carpenter, 1972). In the case of sc^4sc^8 or other grossly deleted X chromosomes, pairing sites on the Y remain unpaired, thus

APPLICATIONS OF MEIOTIC DRIVE 73

genetically active, and this subsequently leads to dysfunction of the Y-bearing sperm or, in the case of nondisjunction, of the X+Y-bearing sperm.

This phenomenon is obviously of some importance if it can be demonstrated in other organisms, because it implies that certain deletions of the X or Y chromosome may be associated with meiotic drive. Such chromosomes could be used to genetically control the sex ratio.

AUTOSOMAL MEIOTIC DRIVE

Meiotic drive of autosomal genes is not as easy to detect as that of the sex chromosomes, because sex chromosomes are associated with easily recognized phenotypes. Detection of meiotic drive at an autosomal locus requires either that one of the alleles at the locus itself have some recognizable phenotypic effect, or that a recognizable phenotypic effect be associated with a linked locus. However, meiotic drive induced by linkage decreases linearly with the recombination fraction: if the proportion of functional A-bearing gametes from an AB/ab heterozygote is k, say, then the proportion of functional B-bearing gametes will be $k - (2k - 1)r$, where r is the recombination fraction.

Nevertheless, a large number of instances of autosomal meiotic drive are known. (Meiotic drive in plants is most conveniently classified as autosomal meiotic drive.) Examples in plants include three cases in maize: abnormal chromosome 10 (Rhoades 1942, Carlson 1969); "aberrant ratio" (symbol AR), which is associated with infection by barley stripe mosaic virus (Sprague and McKinney 1966); and a case involving a maize-Tripsacum hybrid studied by Maguire (1963). There is also a case in tobacco ("pollen killer" of Cameron and Moav (1957), also associated with interspecific hybridization); cases in wheat (Loegering and Sears 1963); rye (Müntzing 1968); *Arabidopsis* (Rédei 1965); and an astonishing series of examples found recently by Perkins and associates in natural populations of *Neurospora*

(Perkins, Turner and Barry 1976; Raju 1976; Turner and Perkins 1976).

In animals, autosomal meiotic drive is known in the house mouse (two cases: t-alleles (Chesley and Dunn 1936) and *low ratio* (Dunn and Bennett 1968)); in the deer mouse (affecting the transferrin locus (Canham, Birdsall and Cameron 1970)); in *Drosophila melanogaster* (*segregation distorter* (Sandler, Hiraizumi and Sandler 1959) and another case discovered by Minamori (1970) which, like AR, seems to be associated with some sort of virus infection); in the butterfly *Danaus chrysippus* (Smith 1976); and even, evidently, in humans (Alport's syndrome (Shaw and Glover 1961) and Holt-Oram syndrome (Gall *et al.* 1966)).

One of the most intriguing systems of meiotic drive is that represented by the t alleles in the house mouse. The T locus itself is located on chromosome 17 of the mouse genetic map, about 14 crossover units to the left of $H-2$, the major histocompatibility locus. Two kinds of mutations at the T locus are known. There are dominant T mutations which, when heterozygous, cause a short tail and which, when homozygous, are lethal. Then there is a large number of recessive t mutations (most derived from natural populations) which are distinguished phenotypically by the fact that T/t genotypes are tailless (Gluecksohn-Waelsch and Erickson 1970, Bennett and Dunn 1970, Artzt and Bennett 1975).

Recessive t alleles are homozygous lethal or semilethal, and the precise stage at which embryonic development fails is allele-specific (Gluecksohn-Waelsch and Erickson 1970). However, heterozygotes for two different t alleles, denoted t_i/t_j, are frequently viable. The t alleles also have effects on spermatogenesis: t_i/t_j males are usually sterile. Moreover, t alleles are characterized by a high degree of meiotic drive, $+/t$ or T/t males transmitting the t allele to 90% or more of their progeny.

The mechanism of segregation distortion of the t alleles is not known with certainty, but the proximity between the T locus

APPLICATIONS OF MEIOTIC DRIVE 75

and the H-2 locus has suggested that the system may somehow involve cell surface antigens (Glueckeohn-Waelsch and Erickson 1971). Indeed, an antigen specific to the T locus has now been discovered. It is biochemically very similar to an H-2 antigen (Vitetta et al. 1975), and it appears that the T locus represents an embryonic analogue of the adult major histocompatibility locus (H-2) (Artzt and Bennett 1975). Both systems operate as mediators of cell to cell recognition, the T system functioning in the embryo and the H-2 system functioning in the adult. H-2 antigens are present on all adult somatic cells, but not on sperm or on cells of the embryo before 7-9 days of development. Antigens coded by the T locus are absent from all adult somatic cells, but they are present on sperm and on early embryonic cells. Moreover, T-type antigens are present on the sperm of all mammals, including humans (Buc-Caron et al. 1974).

How the T antigens relate to segregation distortion is not known. There is some evidence for haploid gene expression of the antigens on sperm (Lyon, Glenister and Hawker 1972; Artzt and Bennett 1975), and, on this hypothesis, the distortion comes about because the female allows preferential fertilization by sperm that carry certain antigens. On the other hand, the evidence for haploid expression of the T locus is by no means conclusive. Indeed, the sterility of t_i/t_j males would militate against it. The meiotic drive could as well result from gene action in diploid cells in spermatogenesis.

In any case, the practical implications of this locus are considerable. By analogy with the T system, I think it would be extremely worthwhile for animal breeders to examine the segregation ratio, when heterozygous, of recessive lethals linked to major histocompatibility loci. It would also be worthwhile to look for segregation distortion, induced by linked loci, of certain histocompatibility alleles. An analogue of a t-bearing chromosome in, say, cattle would be a first step in genetically controlling the sex ratio. The next step would be to obtain a translocation between this chromosome and the X or the Y, or to place the t-analogue onto an existing translocation by crossing over.

The *segregation distorter* system in *Drosophila melanogaster*, a system which resembles the t system in many particulars (Braden et al. 1972, Hartl and Childress 1972), has also been extensively studied (review in Hartl and Hiraizumi 1976). Meiotic drive in this case is due to certain second chromosomes designated SD (Sandler, Hiraizumi and Sandler 1959). Like the t system, $SD/+$ males transmit the SD chromosome to 90% or more of their offspring, and SD_i/SD_j males are sterile (Hartl 1973). Unlike the t system, SD chromosomes do not seem to have effects on embryonic development.

The distortion caused by SD chromosomes is due to an interaction of two mutations spanning the centromeric region of chromosome 2 (Hartl 1974). The locus on the left arm is denoted Sd, that on the right arm is denoted Rsp. Operationally, the Sd^+ allele behaves as a recessive suppressor of the meiotic drive of the Rsp locus. That is, meiotic drive in favor of Rsp occurs in Rsp/Rsp^+ males unless the males are also homozygous for Sd^+. A molecular model for the allelic interaction between Sd and Rsp has been proposed by Hartl (1973).

In the case of SD, it is known that the mechanism of distortion in $SD/+$ males is the dysfunction of the $+$-bearing sperm (Hartl, Hiraizumi and Crow 1967; Nicoletti, Trippa and DeMarco 1967; Peacock, Tokuyasu and Hardy 1972). Recent evidence suggests that the dysfunction may come about as a consequence of failure of these sperm to undergo the normal transition from relatively lysine-rich somatic histones to relatively arginine-rich sperm histones, a transition that occurs normally in most animals (including mammals) and which is thought to facilitate condensation of DNA in the spermatid nucleus. The first evidence implicating the histone transition was that of Kettaneh and Hartl (1976), who found that spermatids in sterile SD_i/SD_j males fail to undergo the histone transition. More recently, Tokuyasu,

Peacock and Hardy (1977) have studied spermiogenesis in
heterozygous *SD*'s with the electron microscope and have found
that the earliest observable abnormality presaging sperm
dysfunction occurs in the chromatin in the latter half of the
period of nuclear condensation, corresponding in time with the
histone transition.

As in the case of *t* alleles, whether distortion in the case
of *SD* involves haploid gene expression is not known, though the
bulk of the evidence favors a diploid effect (see Hartl and
Hiraizumi 1976). Haploid gene expression is thought not to occur
during spermiogenesis in *Drosophila*, the best evidence for this
being the finding that sperm all but devoid of chromosomes can
still function in fertilization (Lindsley and Grell 1969). However, this evidence does not apply to genes for sperm histones,
as a nucleus without chromosomes hardly needs sperm histones to
aid in chromosome condensation.

In any event, the existence of *SD* makes worthwhile another
approach to finding cases of meiotic drive in agriculturally
important animals. One could perhaps find *SD*-like elements by
examining the segregation ratio, when heterozygous, of recessive
genes which, when homozygous, cause male sterility associated
with an absence of the histone transition.

PROSPECTS FOR DISCOVERING OTHER CASES OF MEIOTIC DRIVE

(The following discussion will be focused on non-Mendelian
segregation in the strictest sense and accessory chromosomes will
be excluded.) The examples of meiotic drive cited in this paper
represent a relatively complete list of known cases. There are
not very many -- twenty or so. On the other hand, it is worth
noting that at least one verified case is known in virtually every
species which has been genetically characterized to an extent
sufficient to allow the detection of meiotic drive. In such well-
studied species as maize and mice and *Drosophila*, several cases
are known. Therefore, although meiotic drive is not a common
phenomenon, I think the prospects of finding it in agriculturally
important species are quite favorable.

The question is, How does one go about searching for meiotic drive? Four methods suggest themselves. First, of course, one can directly observe segregation ratios of marker genes in appropriate crosses. This method is limited by the rather large sampling variance of segregation ratios and the consequent large number of offspring required to ascertain a significant segregation bias, except in cases in which the segregation bias is itself large.

Another approach, suggested earlier, is to search in the reverse direction: to look for physiological effects which, in other species, are associated with meiotic drive, and then to examine the segregational properties of the genes causing these effects. Thus, for example, one would examine the segregational behavior of sex chromosome deletions by analogy with sc^4sc^8, of recessive lethals linked to the major histocompatibility locus by analogy with t alleles, or of recessive male steriles associated with a defective histone transition by analogy with SD.

A third approach to finding meiotic drive is to examine segregation in offspring of interspecific hybrids. This approach is essentially limited to plants, but it has proven successful in maize (Maguire 1963), tobacco (Cameron and Moav 1957), and *Neurospora* (Perkins, Turner and Barry 1976). Most known systems of meiotic drive have been found segregating in natural populations, and natural populations containing segregation distorters also contain modifiers of distortion -- suppressors, or homologues of the distorting chromosome which are insensitive to distortion (*e.g.* Hartl 1975). Examining segregation in interspecific hybrids partially overcomes the modifying effects of the coadapted genetic background. Plant breeders should in any case be aware that alien chromosomes carrying disease resistance genes introduced into commercial varieties may not segregate in the expected Mendelian fashion.

Finally, a fourth method for finding meiotic drive is based on features that are common, but not universal, in known cases of meiotic drive. For instance, since meiotic drive is usually sex specific, differences in segregation between males and females would be suggestive of meiotic drive. Then, too, most cases of meiotic drive show ageing effects -- different degrees of distortion in individuals of different ages. Paternal age effects on the secondary sex ratio in humans have been reported (Novitski and Sandler 1956), as have paternal age effects on segregation in the *ABO* blood groups (Hiraizumi et al. 1973). Unfortunately, such ageing effects do not necessarily imply meiotic drive.

The next question is, Where should one look for meiotic drive? Since most cases of meiotic drive have been discovered in natural populations, natural populations seem to be the best place to look. However, the frequency of segregation distorters is typically quite low -- 1% in the case of *SD* (Hartl 1975) -- and sufficiently large samples of wild genomes have to be studied. One of the deficiencies of present methods of sampling from wild populations is that routine samples are not large enough to detect such rare but potentially very useful phenomena as meiotic drive. This is particularly troubling in light of the steady disappearance of local populations of wild relatives of cultivated plants and domestic animals.

Segregation distorters that are polymorphic in natural populations are likely to have harmful phenotypic effects in heterozygotes or homozygotes or both, as seen in the lethality of *t/t* mice and the male sterility of *SD/SD* in *Drosophila*. It is to be expected that a polymorphic allele which is favored by segregation will have harmful effects at some other stage of the life cycle, for the simple reason that, if there were not some balancing selective effect, the allele would not be polymorphic, but would have spread rapidly through the entire population and would be fixed. How often the fixation of a segregation distorter occurs is not known, but Sandler and Novitski (1957) have suggested that such an allele could be detected by examining segregation in the offspring of interpopulational hybrids.

MEIOTIC DRIVE IN POPULATION CONTROL

In addition to potential applications of meiotic drive in the genetic control of the sex ratio, one further practical application should be mentioned. This is the use of meiotic drive in population control. Although various schemes for the genetic control of insect pests have been considered (Smith and von Borstel 1972), meiotic drive has not been much discussed (but see Hartl 1975).

Lyttle (1977) has recently studied population cages of *Drosophila* and has shown that the release of males carrying a *Y-SD* translocation can cause the population to become extinct. Such males produce almost exclusively sons, and the frequency of the *Y-SD* translocation in the population cage gradually increases until the population is left without enough females to replenish itself.

However, the method does not work nearly so well if the population contains suppressors or other modifiers of distortion, as natural populations are known to carry (Hartl 1975, Hartl and Hartung 1975, Trippa and Loverre 1975). I believe that the problem of suppressors can be overcome by repeatedly overflooding the target population with males bearing the appropriate sorts of translocations. Although the proper release ratios have not been determined experimentally, the scale on which the overflooding would have to be performed would seem to be much less than that routinely employed in the male sterile technique of population control (Smith and von Borstel 1972).

Using this method, population control by means of meiotic drive is feasible at the present time in both *Drosophila melanogaster* and *Aedes aegypti*, as the appropriate meiotic drive systems are already available. The approach is in any case sufficiently promising and sufficiently immediate to warrant further experimental and field tests.

ACKNOWLEDGEMENT

This work was supported by NSF grant BMS74-19708 and NIH grant 7RGM21732. The author is recipient of Research Career Award GM0002301.

BIBLIOGRAPHY

Artzt, K. & Bennett, D. (1975). Analogies between embryonic (T/t) antigens and adult major histocompatibility (H-2) antigens. Nature 256, 545-547.

Bar-Anan, R. & Robertson, A. (1975). Variation in sex ratio between progeny groups in dairy cattle. Theor. Appl. Genetics 46, 63-65.

Battaglia, E. (1964). Cytogenetics of B-chromosomes. Caryologia 17, 245-299.

Beatty, R. A. (1970). The genetics of the mammalian gamete. Biol. Rev. 45, 73-119.

Beatty, R. A. (1975). Genetics of animal spermatozoa. *In* "Gamete Competition in Plants and Animals" (D. L. Mulcahy, Ed). North-Holland, Amsterdam. Pp. 61-68.

Bennett, D. & Dunn, L. C. (1970). Transmission ratio distorting genes on chromosome IX and their interactions. *In* "Proc. Symp. Immunogenetics of the H-2 System" (A. Lengerova, Ed.). Karger, Basel. Pp. 90-103.

Braden, A. W. H., Erickson, R. P., Gluecksohn-Waelsch, S., Hartl, D. L., Peacock, W. J. & Sandler, L. (1972). A comparison of effects and properties of segregation distorting alleles in the mouse (*t*) and in Drosophila (*SD*). *In* "Edinburgh Symposium on the Genetics of the Spermatozoon" (R. A. Beatty & S. Gluecksohn-Waelsch, Eds). Bogtrykkeriet Forum, Copenhagen. Pp. 310-312.

Brown, W. V. (1960). Supernumerary chromosomes in a population of *Tradescantia edwardsiana*. Southwestern Naturalist 5, 49-60.

Buc-Caron, M., Gachelin, G., Hofnung, M. & Jacob, F. (1974). Presence of a mouse embryonic antigen on human spermatozoa. Proc. Nat. Acad. Sci. USA 71, 1730-1733.

Cameron, D. R. & Moav, R. (1957). Inheritance in *Nicotiana tabacum*. XXVIII. Pollen killer, an alien genetic locus inducing abortion of microspores not carrying it. Genetics 42, 326-335.

Canham, R. P., Birdsall, D. A. & Cameron, D. G. (1970). Disturbed segregation at the transferrin locus of the deer mouse. Genet. Res., Camb. 16, 355-357.

Carlson, W. R. (1969). Factors affecting preferential fertilization in maize. Genetics 62, 543-554.

Chesley, P. & Dunn, L. C. (1936). The inheritance of taillessness (anury) in the house mouse. Genetics 21, 525-536.

Church, R. B. (1974). Molecular and reproductive biology in animal genetics. Genetics 78, 511-524.

Craig, G. B., Jr., Hickey, W. A. & VandeHey, R. C. (1960). An inherited male-producing factor in *Aedes aegypti*. Science 132, 1887-1889.

Darlington, C. D. (1956). Natural populations and the breakdown of classical genetics. Proc. Royal Soc., B., 145, 350-364.

Dunn, L. C. & Bennett, D. (1968). A new case of transmission ratio distortion in the house mouse. Proc. Nat. Acad. Sci. USA 61, 570-573.

Erickson, J. (1965). Meiotic drive in Drosophila involving chromosome breakage. Genetics 51, 555-571.

Falconer, D. S. (1954). Selection for sex ratio in mice and Drosophila. American Naturalist 88, 385-397.

Faulhaber, S. H. (1967). An abnormal sex ratio in *Drosophila simulans*. Genetics 56, 189-213.

Gall, J. C., Jr., Stern, A. M., Cohen, M. M., Adams, M. S. & Davidson, R. T. (1966). Holt-Oram syndrome: clinical and genetic study of a large family. Am. J. Hum. Genet. 18, 187-200.

Gershenson, S. (1933). Studies on the genetically inert region of the X-chromosome of Drosophila. I. Behavior of an X-chromosome deficient for a part of its inert region. J. Genet. 28, 297-313.

Gluecksohn-Waelsch, S. & Erickson, R. P. (1970). The T-locus of the mouse: Implications for mechanisms of development. Curr. Topics in Dev. Biol. 5, 281-316.

Gluecksohn-Waelsch, S. & Erickson, R. P. (1971). Cellular membranes: A possible link between H-2 and T-locus effects. In "Proc. Symp. Immunogenetics of the H-2 System" (A. Lengerova, Ed.). Karger, Basel. Pp. 120-122.

Hamerton, J. L., Cowie, V. A., Giannelli, F., Briggs, S. M. & Polani, P. E. (1961). Differential transmission of Down's Syndrome (Mongolism) through male and female translocation carriers. Lancet 2, 956-958.

Hanks, G. D. (1965). Are deviant sex ratios in normal strains of Drosophila caused by aberrant segregation? Genetics 52, 259-266.

Hanks, G. D. (1969). A deviant sex ratio in *Drosophila melanogaster*. Genetics 61, 595-606.

Hanks, G. D. & Torgerson, R. O. (1969). Aberrant segregation: A new case in *Drosophila melanogaster*. Can. J. Genet. Cytol. 11, 305-316.

Hartl, D. L., Hiraizumi, Y. & Crow, J. F. (1967). Evidence for sperm dysfunction as the mechanism of segregation distortion in *Drosophila melanogaster*. Proc. Nat. Acad. Sci. USA 58, 2240-2245.

Hartl, D. L. & Childress, D. (1972). Genetic studies of sperm function and utilization in *Drosophila melanogaster*. In "Edinburgh Symposium on the Genetics of the Spermatozoon" (R. A. Beatty & S. Gluecksohn Waelsch, Eds.). Bogtrykkeriet Forum, Copenhagen. Pp. 269-288.

Hartl, D. L. (1973). Complementation analysis of male fertility among the segregation distorter chromosomes in *Drosophila melanogaster*. Genetics 73, 613-629.

Hartl, D. L. (1974). Genetic dissection of segregation distortion. I. Suicide combinations of SD genes. Genetics 76, 477-486.

Hartl, D. L. (1975). Segregation distortion in natural and artificial populations of *Drosophila melanogaster*. In "Gamete Competition in Plants and Animals" (D. L. Mulcahy, Ed.). North-Holland, Amsterdam. Pp. 83-91.

Hartl, D. L. & Hartung, N. (1975). High frequency of one element of segregation distorter in natural populations of *Drosophila melanogaster*. Evolution 29, 512-518.

Hartl, D. L. & Hiraizumi, Y. (1976). Segregation distortion. *In* "Genetics and Biology of Drosophila", Vol. 1b (M. Ashburner and E. Novitski, Eds.). Academic Press, New York. Pp. 615-666.

Hauschteck-Jungen, E. & Maurer, B. (1976). Sperm dysfunction in *sex ratio* males of *Drosophila subobscura*. Genetica (in press).

Hickey, W. A. & Craig, G. B., Jr. (1966). Genetic distortion of sex ratio in a mosquito, *Aedes aegypti*. Genetics 53, 1177-1196.

Hiraizumi, Y., Spradlin, C. T., Ito, R. & Anderson, S. A. (1973). Birth-order dependent segregation frequency in the ABO blood groups of man. Am. J. Hum. Gen. 25, 277-286.

Jacobs, P. (1972). Chromosome abnormalities and fertility in man. *In* "Edinburgh Symposium on the Genetics of the Spermatozoon" (R. A. Beatty & S. Gluecksohn-Waelsch, Eds.). Bogtrykkeriet Forum, Copenhagen. Pp. 346-358.

Jungen, H. (1967). Abnormal sex ratio linked with inverted gene sequence in populations of *D. subobscura* from Tunisia. Droso. Inf. Ser. 42, 109.

Kataoka, Y. (1967). A genetic system modifying segregation distortion in a natural population of *Drosophila melanogaster* in Japan. Japan. J. Genetics 42, 327-337.

Kettaneh, N. P. & Hartl, D. L. (1976). Histone transition during spermiogenesis is absent in *segregation distorter* males of *Drosophila melanogaster*. Science (in press).

King, H. D. (1918). Studies on inbreeding. III. The effect of inbreeding with selection on the sex ratio of the albino rat. J. Exp. Zool. 27, 1-35.

Langley, A. E. (1927). Supernumerary chromosomes in *Zea mays*. J. Agr. Res. 35, 769-784.

Lindsley, D. L. & Sandler, L. (1958). The meiotic behavior of grossly deleted X chromosomes in *Drosophila melanogaster*. Genetics 43, 547-563.

Lindsley, D. L. & Grell, E. H. (1969). Spermiogenesis without chromosomes in *Drosophila melanogaster*. Genetics 61, suppl. 1, 69-78.

Loegering, W. Q. & Sears, E. R. (1963). Distorted inheritance of stem-rust resistance of timstein wheat caused by a pollen-killing gene. Can. J. Genet. Cytol. 5, 65-72.

Lyon, M. F., Glenister, P. H. & Hawker, S. G. (1972). Do the H-2 and T-loci of the mouse have a function in the haploid phase of sperm? Nature 240, 152-153.

Lyttle, T. W. (1977). Experimental population genetics of meiotic drive systems. I. Pseudo-Y chromosomal drive as a means of eliminating cage populations of *Drosophila melanogaster*. Genetics (in press).

Maguire, M. P. (1963). High frequency transmission of a Tripsacum chromosome in corn. Genetics 48, 1185-1194.

Minamori, S. (1970). Extrachromosomal element delta in *Drosophila melanogaster*. III. Induction of one-sided gamete recovery from male and female heterozygotes. Genetics 66, 505-515.

Mulcahy, D. L. (Ed.). (1975). *Gamete Competition in Plants and Animals*. North-Holland, Amsterdam.

Müntzing, A. (1958). A new category of chromosomes. Proc. Xth. Int. Genet. Congr. 1, 453-467.

Müntzing, A. (1968). A case of differential fertilization in inbred rye. Hereditas 59, 298-302.

Nicoletti, B., Trippa, G. & DeMarco, A. (1967). Reduced fertility in SD males and its bearing on segregation distortion in *Drosophila melanogaster*. Atti Acad. Naz. Lincei 43, 383-392.

Novitski, E. & Sandler, L. (1956). The relationship between parental age, birth order and the secondary sex ratio in humans. Ann. Hum. Genet. 21, 123-131.

Novitski, E. (1967). Nonrandom disjunction in Drosophila. Ann. Rev. Genetics 1, 71-86.

Nur, U. (1962). Population studies of supernumerary chromosomes in a mealy bug. Genetics 47, 1679-1690.

Peacock, W. J., Tokuyasu, K. T. & Hardy, R. W. (1972). Spermiogenesis and meiotic drive in *Drosophila*. In "Edinburgh Symposium on the Genetics of the Spermatozoon" (R. A. Beatty and S. Gluecksohn-Waelsch, Eds.). Bogtrykkeriet Forum, Copenhagen. Pp. 247-268.

Peacock, W. J. & Miklos, G. L. G. (1973). Meiotic drive in
   Drosophila: New interpretations of the segregation distorter
   and sex chromosome systems. Adv. in Genetics 17, 361-409.

Peacock, W. J., Miklos, G. L. G. & Goodchild, D. J. (1975). Sex
   chromosome meiotic drive systems in *Drosophila melanogaster*.
   I. Abnormal spermatid development in males with a
   heterochromatin-deficient X chromosome (sc^4sc^8). Genetics
   79, 613-634.

Perkins, D. D., Turner, B. C. & Barry, E. G. (1976). Strains
   of Neurospora collected from nature. Evolution 30, 281-313.

Policansky, D. & Ellison, J. (1970). "Sex ratio" in *Drosophila
   pseudoobscura*: spermiogenic failure. Science 169, 888-889.

Raju, N. B. (1976). Cytology and gene action of Spore Killer-2
   in Neurospora. Genetics 83, suppl. s59.

Randolph, L. F. (1928). Types of supernumerary chromosomes in
   maize. Anat. Rec. 41, 102 (abstr.).

Rédei, G. P. (1965). Non-mendelian megagametogenesis in
   Arabidopsis. Genetics 51, 857-872.

Rhoades, M. M. (1942). Preferential segregation in maize.
   Genetics 27, 395-407.

Sandler, L. & Novitski, E. (1957). Meiotic drive as an
   evolutionary force. American Naturalist 41, 105-110.

Sandler, L., Hiraizumi, Y. & Sandler, I. (1959). Meiotic drive
   in natural populations of *Drosophila melanogaster*. I.
   The cytogenetic basis of segregation-distortion. Genetics
   44, 233-250.

Sandler, L. & Carpenter, A. T. C. (1972). A note on the
   chromosomal site of action of SD in *Drosophila melanogaster*.
   In "Edinburgh Symposium on the Genetics of the Spermatozoon"
   (R. A. Beatty & S. Glueksohn-Waelsch, Eds.). Bogtrykkeriet
   Forum, Copenhagen. Pp. 233-246.

Shaw, R. F. & Glover, R. A. (1961). Abnormal segregation in
   hereditary renal disease with deafness. Am. J. Hum. Gen.
   13, 89-97.

Smith, D. A. S. (1975). All-female broods in the polymorphic
   butterfly *Danaus chrysippus* L. and their ecological
   significance. Heredity 34, 363-371.

Smith, D. A. S. (1976). Evidence for autosomal meiotic drive in the butterfly *Danaus chrysippus* L. Heredity 36, 139-142.

Smith, R. H. & von Borstel, R. C. (1972). Genetic control of insect populations. Science 178, 1164-1174.

Sprague, G. F. & McKinney, H. H. (1966). Aberrant ratio: an anomaly in maize associated with virus infection. Genetics 54, 1287-1296.

Stevens, N. M. (1908). The chromosomes in *Diabrotica vittata, Diabrotica soror* and *Diabrotica 12-punctata*: a contribution to the literature on heterochromosomes and sex determination. J. Exp. Zool. 5, 453-470.

Sturtevant, A. H. & Dobzhansky, Th. (1936). Geographical distribution and cytology of "sex-ratio" in *Drosophila pseudoobscura* and related species. Genetics 21, 473-490.

Tokuyasu, K. T., Peacock, W. J. & Hardy, R. W. (1972a). Dynamics of spermiogenesis in *Drosophila melanogaster*. I. Individualization process. Z. Zellforsch. 124, 479-506.

Tokuyasu, K. T., Peacock, W. J. & Hardy, R. W. (1972b). Dynamics of spermiogenesis in *Drosophila melanogaster*. II. Coiling process. Z. Zellforsch. 127, 492-525.

Tokuyasu, K. T., Peacock, W. J. & Hardy, R. W. (1977). Dynamics of spermiogenesis in *Drosophila melanogaster*. VII. Effects of *segregation distorter (SD)* chromosome. J. Ultrastruct. Res. (in press).

Turner, B. C. & Perkins, D. D. (1976). Spore killer genes in Neurospora. Genetics 83, suppl. s77.

Trippa, G. & Loverre, A. (1975). A factor on a wild third chromosome (III^{Ra}) that modifies the segregation distortion phenomenon in *Drosophila melanogaster*. Genet. Res. Camb. 26, 113-125.

Vitetta, E. S., Artzt, K., Bennett, D., Boyse, E. A. & Jacob, F. (1975). Structural similarities between a product of the T/t locus isolated from sperm and teratoma cells, and H-2 antigens isolated from splenocytes. Proc. Nat. Acad. Sci. USA 72, 3215-3219.

Wagoner, D. E. (1969). Presence of male determining factors found on three autosomes in the house fly, *Musca domestica*. Nature 223, 187-188.

Weir, J. A. (1953). Association of blood-pH with sex ratio in mice. J. Heredity 44, 133-138.

Weir, J. A. (1960). A sex ratio factor in the house mouse that is transmitted by the male. Genetics 45, 1539-1552.

White, M. J. D. (1973). 3rd ed. *Animal Cytology and Evolution*. Cambridge University Press, Cambridge.

Zimmering, S., Sandler, L. & Nicoletti, B. (1970). Mechanisms of meiotic drive. Ann. Rev. Genetics 4, 409-436.

On gene action

M. Gillois and *Cl. Chevalet*
LABORATOIRE DE GÉNÉTIQUE CELLULAIRE
CENTRE DE RECHERCHES DE TOULOUSE I.N.R.A.
B.P. 12, 31320, CASTANET-TOLOSAN, FRANCE

A. Micali
U.E.R. DE MATHÉMATIQUES
UNIVERSITÉ DES SCIENCES ET TECHNIQUES DU LANGUEDOC
PLACE EUGÈNE BATAILLON, 34060, MONTPELLIER CEDEX, FRANCE
and IMECC, UNIVERSIDADE ESTADUAL DE CAMPINAS
CAIXA POSTAL 1170, CEP 13100 CAMPINAS, SP, BRASIL

R. Baron
UNITÉ DE RECHERCHES BIOMÉTRIQUES
ECOLE NATIONALE SUPÉRIEURE AGRONOMIQUE
PLACE PIERRE VIALA, 34060, MONTPELLIER CEDEX, FRANCE

R. Costa
INSTITUTO DE MATEMÁTICA E ESTATÍSTICA
UNIVERSIDADE DE SÃO PAULO
CIDADE UNIVERSITÁRIA CAIXA POSTAL 20570
AGENCIA IQUATEMÍ, SÃO PAULO, SP, BRASIL

Two principal ideas are considered. The first is that genes code sequences, particularly polypeptide chains. The heterogeneity of two allele gene sequences ensures the variability of the stereo-specific mechanism and the variability of enzymatic activity. The

second idea is that genes do not act independently but according to
one or many ordered sequences.

In this model, phenotypes are expressed in terms of two parameter sets. The first set explains, in biophysical or biochemical terms, the information coded in the enzyme structure gene; the second interprets the initial conditions of the system.

STEREOSPECIFICITY AND BIOLOGICAL ACTIVITY

It seems time to revisit quantitative genetics. We consider two principal ideas. First, genes code sequences, particularly polypeptide chains. The heterogeneity of two allelic gene sequences ensures the variability of the stereospecific mechanisms, and the variability of enzymatic biological activity. Secondly, genes do not act independently, but according to one or more ordered sequences.

The main property of proteins (and of other sequential molecules: RNA, DNA, glucids, glycolipids) is their ability to develop specific interactions with other molecules. Thus, enzymes interact with their substrates, antibodies with antigens, and proteins with other molecules for building complicated multimolecular structures. Consequently, the catalytic, regulatory, structural or immunological functions of proteins come more or less directly from these specific interactions.

It is very important to recall a fundamental difference between noncovalent and covalent bonds. The former do not need catalysis, but depend strictly on the spatial structures of ligands and proteins. The latter need enzyme activity to be formed or broken.

We shall use the law of mass action to describe the quantitative results of the stereospecificity. The association or dissociation stereospecific constants, expressed in terms of absolute velocity and not of steady state, have a genetical variability. The enzymatic activity is always conditioned by realization of a multimolecular stereospecificity association. These general ideas preserve the fundamental hypothesis of Henry and Michaelis about the building of a complex enzyme-substrate, and introduce association or activity constants in terms of absolute value, which depend on genetic information and are independent of equilibrium conditions.

DETERMINISTIC MODEL

We consider first only biochemical associations involving the activity of one enzyme E associated with substrates A, B, C and D. The graph representing this enzymatic activity is the following one:

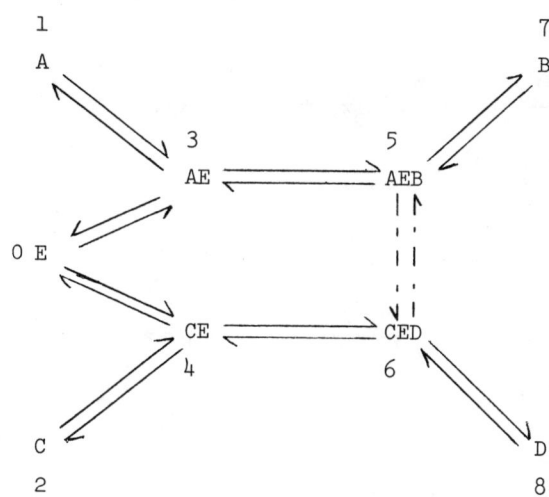

AE indicates the stereospecific association of molecules A and E, AEB that of A, E, B, etc. In AEB, our rule is that the association AE is assigned priority relative to the association AEB, so we can neglect such associations as EB (cf. Ricard, 1973).

Let $x_A(t)$ denote the concentration of substrate A at time t. We describe the kinetics during the time interval [t, t+h]. Thus we have the equation

$$x_A(t+h) = x_A(t) + \lambda' x_{AE}(t)h - \lambda x_A(t)x_E(t)h ,$$

where λ' is the dissociation constant of complex AE, so that $\lambda' x_{AE}(t)h$ is the amount of substrate A appearing during time interval [t,t+h] from dissociation of complex AE. Similarly, λ stands for a constant of association between molecules A and E. The constants

λ and λ' depend on the genetic nature of enzyme E.

The limit of the foregoing expression for $h \to 0$ gives the differential equation

$$\frac{d\, x_A}{dt} = \lambda' x_{AE}(t) - \lambda x_A(t) x_E(t) .$$

If we now replace letters by indices, the dynamical system may be written as follows:

$$\frac{dx_0}{dt} = \lambda' x_3 - \lambda x_0 x_1 + \rho' x_4 - \rho x_0 x_2 ,$$

$$\frac{dx_1}{dt} = \lambda' x_3 - \lambda x_0 x_1 ,$$

$$\frac{dx_2}{dt} = \rho' x_4 - \rho x_0 x_2 ,$$

$$\frac{dx_3}{dt} = -\lambda' x_3 + \lambda x_0 x_1 + \mu' x_5 - \mu x_3 x_7 ,$$

$$\frac{dx_4}{dt} = -\rho' x_4 + \rho x_0 x_2 + \nu' x_6 - \nu x_4 x_8 ,$$

$$\frac{dx_5}{dt} = -\mu' x_5 + \mu x_3 x_7 + \beta x_6 - \alpha x_5 ,$$

$$\frac{dx_6}{dt} = -\nu' x_6 + \nu x_4 x_8 - \beta x_6 + \alpha x_5 ,$$

$$\frac{dx_7}{dt} = \mu' x_5 - \mu x_3 x_7 ,$$

$$\frac{dx_8}{dt} = \nu' x_6 - \nu x_4 x_8 .$$

In these equations the coefficients ρ', μ' and ν' are dissociation coefficients like λ', and ρ, μ and ν are association coefficients like λ. Coefficients α and β concern only the high energy activity of the enzyme E after its association with both substrates A and B (or C and D).

We have the obvious relations $x_0 + x_3 + x_4 + x_5 + x_6 = b_0$, $x_1 + x_3 - x_7 = b_1$, $x_2 + x_4 - x_8 = b_2$ and $x_0 - x_1 - x_2 = b_3$, where

b_i (i=0,1,2,3) are real constants. As b_0 is the total quantity of enzyme we can guess that it is small with respect to the absolute values of b_1, b_2 and b_3. The following inequalities may then be stated: $b_3 < 0$, $b_1 + b_3 < 0$, $b_2 + b_3 < 0$, $b_1 + b_2 + b_3 < 0$. These are shown to be sufficient conditions for the dynamic system to have one equilibrium state. Approximate expressions for equilibrium values of x_1 and x_2 may be expressed as functions of environmental parameters b_1, b_2 and b_3, and of the ratio $g = \frac{a}{b}$, where $a = \alpha \frac{\mu}{\mu'} \frac{\lambda}{\lambda'}$ and $b = \beta \frac{\nu}{\nu'} \frac{\rho}{\rho'}$. These are

$$x_1 = \frac{g b_1 + b_2 + 2 b_3 + \sqrt{(g b_1 + b_2)^2 + 4 g b_3 (b_1 + b_2 + b_3)}}{2(g - 1)} + O(x_0),$$

and

$$x_2 = -\frac{g b_1 + b_2 + 2 g b_3 + \sqrt{(g b_1 + b_2)^2 + 4 g b_3 (b_1 + b_2 + b_3)}}{2(g - 1)} + O(x_0),$$

where x_0 ($0 < x_0 < b_0$) is the amount of free enzyme at equilibrium.

In the case where $a = b$, we have the formulas

$$x_1 = -\frac{b_3 (b_2 + b_3)}{b_1 + b_2 + 2 b_3} + O(x_0),$$

$$x_2 = -\frac{b_3 (b_1 + b_3)}{b_1 + b_2 + 2 b_3} + O(x_0).$$

We recall that x_1 and x_2 are, respectively, the quantities of substrates A and C at equilibrium.

CONCLUSION

We have studied the case when only one monomeric enzyme is involved. This is only relevant to a haploid, or diploid homozygote, case. Similar dynamic systems may be derived to deal with polymeric enzymes or polyploid zygotes with various alleles coding for the same enzyme.

The following results, which apply to the simplest case, are worth bringing to light.

1) Equilibria are expressed in terms of two parameter sets: The first set (a and b) explains, in biophysical or biochemical terms, the information coded in the enzyme structure gene. The second (b_1, b_2, b_3) interprets the initial conditions of the system, i.e., environment conditions.

2) Even in the simple case of a homozygote for the action of two homologous genes, the phenotypic expression is a nonlinear function of the genetic and environment parameters.

3) The genetic parameters ($a = \alpha \frac{\lambda}{\lambda'} \frac{\mu}{\mu'}$ and $b = \beta \frac{\rho}{\rho'} \frac{\nu}{\nu'}$) involve, on the one hand, the stereospecific capabilities of the genetically coded enzyme molecule and, on the other hand, its capabilities of exchanges of high energy bonds (α, β) which concern its strictly enzymatic properties. Therefore, we can understand the action of a mutation at the level of the phenotypic expression of a mutated gene, by some adjustment of the values of these parameters.

The ratio g takes into consideration not only all the genetic information of the coefficients α, β, λ, λ', ... but also the direction of the departure from the equilibrium $A + B \rightleftharpoons C + D$, and is identical to the coefficient $K = [C][D] / [A][B]$, in kinetics.

4) If we suppose that the parameters g_i of an enzyme class are not very different from a standard value g_r, each having the form $g_i = g_r + \omega_i$, then the equilibrium values $x_1(g_i)$ are the sum of two terms:

The first term, $x_1(g_r)$, is common to every enzyme of this class.

The second term is specific to an enzyme and is the deviation of a given gene from the standard phenotypic value $x_1(g_r)$.

For instance, if $g_r = 1$, we get

$$x_1 = - \frac{b_3(b_2+b_3)}{b_1+b_2+2b_3} + \frac{b_3(b_1+b_3)(b_2+b_3)(b_1+b_2+b_3)}{(b_1+b_2+2b_3)^3} \omega + O(x_0) + O(\omega^2 |b_3|),$$

$$x_2 = -\frac{b_3(b_1+b_3)}{b_1 + b_2 + 2b_3} - \frac{b_3(b_1+b_3)(b_2+b_3)(b_1+b_2+b_3)}{(b_1 + b_2 + 2b_3)^3} \omega + O(x_0) + O(\omega^2 |b_3|),$$

where $g = 1 + \omega$. These expressions allow us to put in a concrete form the idea of substitution effect of a gene; for two genes having the hereditary characteristics ω, ω', the effect of the substitution of one for the other is

$$\frac{b_3(b_1 + b_3)(b_2 + b_3)(b_1 + b_2 + b_3)}{(b_1 + b_2 + 2b_3)^3} (\omega - \omega') \text{ for } x_1.$$

This expression involves the action of the gene interacting with the environment.

We have to prove, in other situations, that analogous expressions describe the effects of dominance, of allosteric regulation, of feedback regulation and of strictly genic regulation.

ACKNOWLEDGMENT

The authors are particularly grateful to J. R. Sedcole, J. Lewis and E. Pollak for their willingness to correct an early English version.

BIBLIOGRAPHY

A more complete version of some aspects of this paper, including a stochastic approach to the problems outlined here, may be found in:

(1) Gillois, M., Chevalet, C., Micali, A., Baron, R., Costa, R. Biochemical and molecular approaches of quantitative genetics. Submitted to Theor. Pop. Biology.

For the mathematical developments about the dynamical system defined here, we refer the reader to:

(2) Gillois, M., Chevalet, C., Micali, A. Sur un système dynamique intéressant la Génétique, in preparation.

For general notions about enzymology, we refer to:

(3) Ricard, J. Cinétique et mécanismes d'action des enzymes. Vol. 1 et 2, Doin 1973.

APPENDIX: Open Systems

In the preceding discussion, only one pathway was considered, that leading from a molecular species, A, to another one, C, and the system we studied was closed. In fact, biochemical pathways are connected with each other and are regulated by such phenomena as feedback inhibition, allosteric regulations, genic control, repression, and so on. Therefore the next step in our approach is to deal with such a network as shown in the following figure.

In the figure, three enzymatic reactions are considered, leading from the species A_1 to the species C_3. Each reaction is assumed to involve stereospecific interactions analogous to the one we considered. Substrate C_1, the output of first reaction, is then the input A_2 of the second reaction. Complexes between A_i, E_i, C_i, as well as substrates B_i and D_i, have been omitted for clarity. The new symbols and interactions, G_1, G_2 and G_3, included in the graph stand for the <u>structural</u> genes coding for enzymes E_1, E_2, E_3. They may be repressed if a repressor R is associated with their initiator(s). Hence, R is assumed to be able to have some affinity with A_1, allowing the synthesis of genes G_1, G_2, G_3 if the amount of A_1 is large enough in the medium, and inhibiting it if the amount of A_1 is low.

On the other hand, a retroinhibition is considered through the ability of C_3 and E_1 to form some heterosteric association C_3E_1.

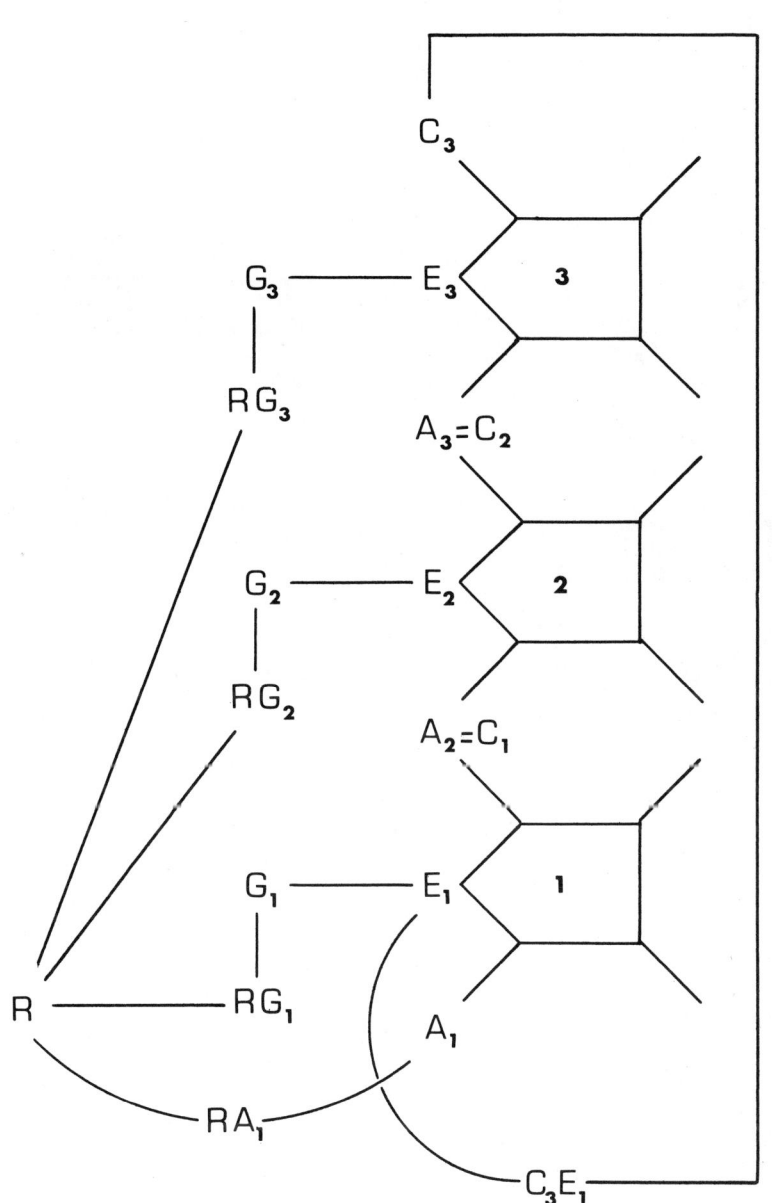

III
What Have We Learned from Laboratory Species?

Some results of the Edinburgh selection experiments with mice

D. S. Falconer
AGRICULTURAL RESEARCH COUNCIL, UNIT OF ANIMAL GENETICS
DEPARTMENT OF GENETICS, UNIVERSITY OF EDINBURGH

This paper is exclusively a review of work done in Edinburgh. To have included the Edinburgh work on Drosophila, or the immense amount of work on mice done in other laboratories, would have required a paper many times longer. Furthermore, the review is restricted to experiments on selection and mainly to two problems, namely asymmetrical responses and limits.

When the work on selection in mice started in Edinburgh in 1947 we knew so little about selection responses that whatever happened was bound to be interesting. It was interesting to know even that the character would respond to selection. Since then the accumulation of experience from selection experiments in many places has done for qualitative characters what electrophoresis has done for proteins, and we now know that virtually every measurable character is genetically variable in non-inbred populations.

In the introduction to the first paper on the Edinburgh work (Falconer, 1953) I posed some specific questions to be asked of a selection experiment at that time, and I would like now to consider briefly how far we have got toward answering them. The questions were:

(1) "How far are the predictions of 'heritability analyses' borne out in practice?"

(2) "Is the genetic situation symmetrical: that is to

say, does selection in opposite directions yield equal rates of progress?"

(3) "How long does the progress under selection continue, and are there limits to what can be achieved by selection?"

(4) "How accurately are the results of selection repeatable in different experiments?"

Validity of Predictions

On testing the validity of predictions we have done almost nothing in Edinburgh. The latest experiment (Falconer, 1973) was designed with this intention, but various things went wrong and it failed in this respect. To get sufficiently precise estimates of the heritability in the population before selection and of the realized heritability, very large numbers are needed which are not easily obtained with mice, and for really critical tests we shall perhaps have to be content with the work on Drosophila and Tribolium. I do not think it is very useful to compare the heritability estimated by offspring-parent regression in the base population with the heritability realized by selection. The question at issue is the more basic one of whether the offspring-parent regression is linear. We have not tried to test the linearity of regression directly. A non-linear regression would, however, be expected to result in different realized heritabilities in lines selected in opposite directions, and this asymmetry of response has been a feature of several experiments.

Asymmetrical Responses

The asymmetry of the response in the first experiment was a surprise. The selection was for body weight and the realized heritability was much less in the up-line than in the down-- about 20% vs 50%. The same happened with selection for litter size, where the realized heritability was 8% upwards and 25%

downwards (Falconer, 1971). Asymmetrical responses have been obtained in very many selection responses with mice and other organisms since then. The direction of the asymmetry is variable but it may perhaps be more usual to find a lower heritability in the 'up' line than in the 'down', when up means 'good' and down means 'bad'. The most important question is whether asymmetry and its direction can be predicted. Asymmetry of response occurs if the mean genotypic value at the start is not mid-way between the two limits; in other words if there is further to go in one direction than in the other. This can arise from unequal gene frequencies (e.g. + alleles being at higher frequencies than - alleles) or from directional dominance (+ alleles being dominant). If there is directional dominance the character will show inbreeding depression, and in a small enough population inbreeding depression will cause an asymmetrical response in the same direction as the directional dominance. Thus, characters that are subject to inbreeding depression might be expected to respond asymmetrically to selection, with upward selection being less effective than downward. There are probably now enough recorded two-way selection experiments to see how well, if at all, the prediction is borne out.

Hill's studies of the sampling variance of the realized heritability (Hill, 1971, 1972a, b) show that random drift may often be an important source of asymmetry in selection responses. Indeed, with populations of the size we have used in our experiments, it seems to me that random drift may well be a much more potent cause of asymmetry than any of the causes considered hitherto. Realized heritabilities are usually estimated from the regression of response on cumulated selection differential, and the standard error of the regression coefficient is taken to be the standard error of the realized heritability. The essence of Hill's work is to show that this is not a realistic standard error of the realized heritability. The heritability realized in any selection line is regarded as an estimate of the heritability in the base population. Different lines selected from the same base will, however, give different estimates because of random drift. The regression of response on cumulated selection differential in any one line estimates the response of that line, and the standard error of the regression is a valid measure of the precision of that estimate of response. But if the regression is to be regarded as an estimate of the heritability in the base population its standard error must be increased to include the drift variance between replicated lines. Hill has shown that the standard error of the realized heritability may easily be several times greater than the standard error of the regression. Consequently responses may be obtained which, on the basis of the regression and its standard error, are significantly asymmetrical, but which do not imply any significant asymmetry of the realized heritability. In other words the asymmetry of response may be real enough without there being any real genetic asymmetry underlying it. Our replicated selection (Falconer, 1973) illustrates the point well. What we found was, briefly, as follows.

There were six up-lines and six down-lines, each maintained by 8 pairs with minimal inbreeding. We first looked at the responses of the whole set of lines, as if it were one large experiment with 48 pairs in each generation. Up to generation 10 of selection the regressions of response on cumulated selection differentials were: 0.398 ± 0.020 upwards, and 0.328 ± 0.014 downwards. There was asymmetry, though in the opposite direction from what had been found before, and the difference in the regression coefficients was significant at $P \sim 0.01$. Next we looked at the replicate lines in each direction separately. The lines differed markedly in their regression coefficients. Among the up-lines the lowest regression was 0.251 and the highest 0.448, while among the down-lines the range was from 0.159 to 0.501. Furthermore these differences of response between lines selected in the same direction were significant ($P < 0.05$ among the up-lines, and $P < 0.001$ among the down-lines). The variation between the six lines selected in the same direction provide empirical estimates of the sampling variances of the realized heritabilities in the two directions. The empirical standard errors of the realized heritabilities were 0.031 in the up-lines and 0.046 in the down-lines, whereas the standard errors of the pooled regression coefficients were 0.020 and 0.014. Furthermore the asymmetry of the realized heritabilities is now not significant when assessed by the empirical standard errors ($P \sim 0.3$). Thus the replication of the selection has shown that the significant asymmetry of the responses was not evidence of asymmetry of the realized heritabilities. It is worth asking how large the asymmetry would have had to be for this experiment to have detected it as significant; a difference of less than 0.124 between the realized heritabilities upward and downward would not have been detected as significant at $P = 0.05$. This was a much larger experiment than any we had done before, with upward and downward selection each based on populations of size $N_e = 192$. It is discouraging to find how imprecise an experiment on this scale is for estimating the realized heritability.

There is another form of asymmetry of selection responses that I would like to comment on briefly, and that is the finding of qualitatively different responses in the two directions. This refers to characters that have two, or more, measurable components, which respond differently. Litter size is an example. The two components are the number of eggs ovulated and the proportion of eggs that are fertilized and survive to birth. When we selected up and down for litter size, we found that the two components responded differently. Increased litter size was brought about by increased ovulation rate, but decreased litter size was brought about by decreased embryonic survival which was a property of the mother, not of the embryos (Falconer, 1960, 1963). Ovulation rate itself provides another example (Land & Falconer, 1969). Ovulation rate is determined by the output of gonadotrophic hormone (FSH) and by the sensitivity of the ovary to FSH. We selected up and down for natural ovulation rate and found the up-line responded by increased FSH output, but the down-line responded by reduced ovarian sensitivity. In neither of these experiments was the selection replicated and we therefore have no means of knowing whether the asymmetrical responses of the components were chance effects of random drift. The responses of the components are, of course, correlated responses dependent on the genetic correlation between the component and the character selected, and correlated characters are notorious for their erratic responses as shown for example by Bohren, Hill and Robertson (1966). Until we know that the asymmetry of the components is not just random drift I see little point in trying to invent explanations.

Limits

The early experiments with Drosophila and mice proved that selective breeding could produce strains transcending the limits of variation in the original stock. It is difficult for us now to realise how surprising this was at that time, but it is in fact still surprising to students being introduced to current ideas about selection. About 50% of my students each year say they would expect the limit to be about two standard deviations from the mean.

When we started our experiments the questions to be asked were purely empirical ones: how far can we get, and how long will it take to get there? The answers were very consistent about the time taken--roughly 20 generations--but less so about the total response achieved. For weight, the difference between up and down lines in phenotypic standard deviations was roughly 5 to 10 σ_p, but for litter size it was much less-- about 1.5 σ_p. The development of a theory of limits by Robertson (1960), Hill and Robertson (1966) and Robertson (1970) now provides a basis for asking more precise questions. Our experiments on body weight have been analysed by Roberts (1966a). I shall summarize his results and add the results of the replicated selection for body weight and of the selection for litter size.

The limits cannot be predicted from observations on the base population, as explained by Robertson (1960), because they depend on the number of segregating genes, the magnitude of their effects, and their initial frequencies, none of which are known. So the observed limits cannot be tested for agreement with the theory. But some deductions can nevertheless be made from the observed limits. The additive genetic variance in the base population, which can be estimated from observed data, may be caused by many genes with small effects or few genes with large effects, or anything in between, with the relationship between number and effects being dependent on gene frequencies and dominance. The larger the number of genes the greater is the total response in relation to the initial variance. The total response actually attained is, however, limited by the effective population size N. There is a conflict between selection tending to fix favourable alleles and random drift tending to fix some unfavourable alleles, particularly when the favourable allele is at low frequency. With very large N, the favourable alleles at all loci will eventually be fixed, but with small N unfavourable alleles will be fixed by random drift at loci that have small effects, thus reducing the total response obtained. Robertson (1960) showed that the maximum possible response with a given population size is approximately $2Nih^2\sigma_P$, where $ih^2\sigma_P$ is the initial response per generation. If the actual response is much less than the theoretical maximum, most of the response must have been due to genes with effects large enough for selection to have overcome random drift, and if the selection were replicated, the favourable alleles at these loci would be fixed in most of the lines. Table 1 summarizes our results, showing the observed and theoretical maximum total responses in units of the phenotypic standard deviation. In every case the observed limit is clearly much less than the theoretical maximum. This shows that the genes responsible for the major part of the responses had effects large enough to give the favourable alleles a high probability of being fixed by the selection.

Corresponding to the theoretical limit outlined above, there is a theoretical maximum duration of the response. Since the approach to the limit is in principle asymptotic, the time taken to get to the limit itself is not a very meaningful concept, and the duration of the selection response is best expressed as the half-life, i.e., the number of generations taken to get half-way to the limit. Robertson (1960) has shown that when the maximum possible response is obtained the half-life will be between about N and 2N generations, depending on dominance and initial gene frequencies. Our results (Table 1) are consistent in showing half-lives of about $\frac{1}{2}N$. Since this is less than the theoretical maximum, the observed duration of the responses is consistent with the limits attained and the conclusions drawn from them.

The duration of the response can be used to calculate a number that is tenuously connected with the number of loci

Table 1: Selection limits and duration of responses

Strain	Character selected	Direction of selection	Total response[2] Observed (σ_P)	Total response[2] Max.expected (σ_P)	½-life of response[3]
N	Weight	Up	3.4	7.2	.6 N
		Down	5.6	15.9	.6 N
		Divergence	9.0	22.4	
CF	Growth	Up	2.0	7.4	.3 N
		Down	4.5	13.7	.5 N
		Divergence	6.5	20.8	
Q^1	Weight	Up	3.9	15.8	.2 N
		Down	3.6	9.6	.4 N
		Divergence	7.5	25.5	
J	Litter size	Up	1.2	2.3	.5 N
		Down	0.5	7.7	.5 N
		Divergence	1.7	8.3	

1. All replicates combined
2. In units of the total phenotypic standard deviation
3. N = Effective population number (not always known exactly)

contributing to the response. What this number, n, means is
this: if all the genes were unlinked, all had effects that
were additive and equal in magnitude, and all had initial
frequencies not too near the extremes, then n such loci
would account for the response. Roberts (1966a) worked out
the numbers for the size-selection experiments and obtained
numbers ranging very roughly from 10 to 20, and the effects
of these genes were roughly a difference between homozygotes of
0.5 and 1.5 phenotypic standard deviations. The litter-size
selection and the replicated selection for body weight yield
similar numbers.

Another aspect of limits about which we have some information is the genetic properties of lines after the limit has been reached. The questions here are whether all relevant loci have been fixed, and whether different lines have the same alleles fixed at all, or most, of the loci. We have evidence that some lines were not fixed at all loci affecting the character. The evidence, which comes from reversed selection and from inbreeding, is briefly as follows. (i) Roberts (1966b) studied one line selected up and one selected down for body size. The large line failed to respond to reversed selection, and so appeared to have been fixed, but the small line did respond to reversed selection and so was not fixed. (ii) The line selected for high litter size depressed on inbreeding (Falconer, 1971), and when the last four surviving inbreds were crossed a substantial gain over the original limit was obtained. Consequently recessive genes must have been segregating in the line at the limit. To account for the results it was necessary to postulate about 40 recessives with effects of about 0.5 phenotypic standard deviations at gene frequencies of about 0.2 in the line at the limit. (iii) Al-Murrani & Roberts (1974) applied the method of inbreeding and crossing to one of the replicated lines selected for high body weight. Though no improvement was gained by the crossing, the line proved not to be fixed because the inbred lines became differentiated and reversed selection with inbreeding was successful. The results could be accounted for by the segregation of about 10 recessives with effects of $2/3 \ \sigma_P$ at frequencies of about 0.2.

The segregation of recessive genes means, of course, that the final limit had not been reached, though response appeared to have ceased. How near to the real limit were these lines? Al-Murrani & Roberts (1974) calculated that elimination of all the unfavourable recessives would have yielded a gain of only 0.5 g, which is about 0.2 phenotypic standard deviations. In this case, therefore, the line was very close to its real limit. In the case of the high litter-size line, crossing the four best inbreds derived from it yielded a gain of $0.7 \ \sigma_P$ showing that the apparent limit was less than two-thirds of the way to the real limit.

It used to be widely believed that over-dominant loci would often be found as causes of residual variance in lines at their limits. This would be expected particularly of litter size, a character that depresses severely on inbreeding. We have good evidence, however, that overdominance is not an important factor in inbreeding depression or in residual variance at the limit. The evidence comes from inbreeding in an unselected population (Bowman & Falconer, 1960) and in the high litter-size line at its limit (Falconer, 1971). In both cases one or more lines survived to high levels of inbreeding without any reduction of their mean performance. If overdominant loci were an important source of variation it would be impossible to produce inbred lines with means equal to the original population.

The second question about the genetic properties at the limits is whether different lines have the same alleles fixed. If different alleles have been fixed at some loci, then

(a) crossing lines and reselecting should yield a renewed response, and (b) replicate lines should reach different levels at their limits. We have a little evidence on this question, but it is not at all critical. We have twice crossed lines at their limits and obtained responses to renewed selection (Falconer & King, 1953; Roberts, 1967). The lines crossed, however, had not been selected from the same base, so the differences between them at the limits could be attributed to different loci segregating in the base populations. In the replicated selection for body weight, the lines were selected from the same base, but the evidence is very incomplete. When the paper was written (Falconer, 1973), the up-lines looked as if they were approaching the same level at their limits. Subsequent generations, however, have shown that both the up- and down-lines show about as much between-line variation as do the controls. Crosses between them have not been made.

Repeatability of Responses

The last of my original questions about the repeatability of responses has been partially answered by the references already made to the replicated selection (Falconer, 1973). Having now seen how much variation there was between the replicates, I realize that all our experiments should have been done with replication, and I view with considerable skepticism any conclusions drawn from unreplicated selection. If resources are limited, as they always are, I think it better to have several smaller replicates rather than one larger population, because then one has an estimate of the error variance instead of having to rely on guessing or theory. The variance between our replicate lines was probably mainly due to random drift. We do not know, however, if lines sampled from the same base can differ appreciably in the amount of additive variance they contain. This is a problem still to be tackled.

Summary

The results of the Edinburgh experiments with mice concerning asymmetrical responses and selection limits are briefly reviewed. The most recent selection with replication of lines suggested that many cases of asymmetrical responses in unreplicated two-way selection experiments are due to random drift. Studies of total responses, duration of responses and properties of lines at their selection limits show that in all cases most of the response was due to genes with effects large enough to give the favourable alleles a high probability of being fixed by the selection. Some lines retained some residual genetic variation due to recessives at low frequencies, and were therefore not at their true limits though response appeared to have ceased.

BIBLIOGRAPHY

(1) Al-Murrani, W. K. & Roberts, R. C. (1974). Genetic variation in a line of mice selected to its limit for high body weight. Animal Production 19:273-289.

(2) Bohren, B. B., Hill, W. G. & Robertson, A. (1966). Some observations on asymmetrical correlated responses to selection. Genetical Research 7:44-57.

(3) Bowman, J. C. & Falconer, D. S. (1960). Inbreeding depression and heterosis of litter size in mice. Genetical Research 1:262-274.

(4) Falconer, D. S. (1953). Selection for large and small size in mice. Journal of Genetics 51:470-501.

(5) Falconer, D. S. (1960). The genetics of litter size in mice. Journal of Cellular and Comparative Physiology 56 (Suppl. 1):153-167.

(6) Falconer, D. S. (1963). Qualitatively different responses to selection in opposite directions. In Statistical Genetics and Plant Breeding Ed. Hanson, W. D. and Robinson, H. F. Publ. 982 National Academy of Sciences - National Research Council, Washington, D. C. pp.487-490.

(7) Falconer, D. S. (1971). Improvement of litter size in a strain of mice at a selection limit. Genetical Research 17:215-235.

(8) Falconer, D. S. (1973). Replicated selection for body weight in mice. Genetical Research 22:291-321.

(9) Falconer, D. S. & King, J. W. B. (1953). A study of selection limits in the mouse. Journal of Genetics 51:561-581.

(10) Hill, W. G. (1971). Design and efficiency of selection experiments for estimating genetic parameters. Biometrics 27:293-311.

(11) Hill, W. G. (1972a). Estimation of realised heritabilities from selection experiments. I. Divergent selection. Biometrics 28:747-765.

(12) Hill, W. G. (1972b). Estimation of realised heritabilities from selection experiments. II. Selection in one direction. Biometrics 28:767-780.

(13) Hill, W. G. & Robertson, A. (1966). The effect of linkage on limits to artificial selection. Genetical Research 8:269-294.

(14) Land, R. B. & Falconer, D. S. (1969). Genetic studies of ovulation rate in the mouse. Genetical Research 13:25-46.

(15) Roberts, R. C. (1966a). The limits to artificial selection for body weight in the mouse. I. The limits attained in earlier experiments. Genetical Research 8:347-360.

(16) Roberts, R. C. (1966b). The limits to artificial selection for body weight in the mouse. II. The genetic nature of the limits. Genetical Research 8:361-375.

(17) Roberts, R. C. (1967). The limits to artificial selection for body weight in the mouse. III. Selection from crosses between previously selected lines. Genetical Research 9:73-85.

(18) Robertson, A. (1960). A theory of limits in artificial selection. Proceedings of the Royal Society B 153:234-249.

(19) Robertson, A. (1970). A theory of limits in artificial selection with many linked loci, In Mathematical Topics in Population Genetics, ed. Kojima, K., Springer-Verlag, New York, pp. 246-288.

Antagonistic selection index results with mice

E. J. Eisen
ANIMAL SCIENCE DEPARTMENT
NORTH CAROLINA STATE UNIVERSITY, RALEIGH, NC 27607

INTRODUCTION

The theory of selection index was first proposed by Smith (1936) and Hazel (1943) as an objective method of selecting for a linear function of traits defined as the aggregate breeding value. The selection index procedure theoretically maximizes the correlation between the aggregate breeding value and the index and is never inferior to tandem selection or independent culling levels (Hazel and Lush, 1942; Young, 1961; Finney, 1962). Since its inception, selection index theory has been studied in detail and adapted to many special cases (Kempthorne and Nordskog, 1959; Tallis, 1962; Williams, 1962; Van Vleck, 1970; Cunningham et al., 1970). However, there is a lack of experimental data to test adequately selection index theory. Of particular interest in the present study are those situations where the selection goals defined by the selection index are antagonistic to the genetic correlation between the traits involved (Rutledge et al., 1973; Nordskog et al., 1974).

The present paper summarizes the results of two selection index experiments with mice that are categorized as forms of antagonistic selection. The first study involved a restricted selection index (Kempthorne and Nordskog, 1959), where a two-way selection experiment was designed to change postweaning weight gain while holding change in feed intake to zero. The second experiment was designed to select for a decrease in litter size and an increase in 6-week body weight in one line and vice versa in a second line. This problem was treated as an example of a desired gains index (Pesek and Baker, 1969). In each case the genetic correlation between the two traits involved in the index is positive.

MATERIALS AND METHODS

Mice used to form all the lines in the present study were obtained from the randombred ICR population (Eisen and Hanrahan, 1974).

Experiment 1: The restricted selection index procedure (Kempthorne and Nordskog, 1959) was adapted to maximize the change in postweaning gain from 3 to 6 weeks of age, while keeping genetic response in feed intake at zero. The genetic and phenotypic variance-covariance matrices needed to find the appropriate restricted index weights were obtained from a report on the same strain of mice used in the present study (Jara-Almonte and White, 1973). The restricted selection index was found to be $I_r = P_G - .067\ P_F$, where P_G and P_F are the phenotypic values for postweaning gain and total feed intake from 3 to 6 weeks of age. In a line denoted as I^+, selection was practiced for increased postweaning gain based on the restricted index, while selection was for decreased postweaning gain based on this index in a line designated as I^-. A randomly selected control line (C) was maintained contemporaneously with the selected lines. Ten generations of selection are reported herein.

Each of the three lines consisted ideally of ten males each mated to two females in every generation. Matings were made at random with the avoidance of half- and full-sib matings. Individual selection was practiced within each sex in the selected lines, with the reservation that a minimum of six sire families had to be represented among the selected individuals. The control line was maintained by randomly selecting one male from each sire family and one female from each full-sib family (Gowe et al., 1959).

Selected males and females were caged together for mating in the manner described above for 16 days beginning at 8 to 10 weeks of age. At the end of this period, dams were individually caged and checked daily until parturition. Litters were randomly standardized to eight pups at one day of age, with an attempt to equalize the sex ratio. Mice were weaned at 3 weeks of age and caged individually from 3 to 6 weeks to obtain feed intakes. Body weights were taken at 3 and 6 weeks of age. Further details on experimental procedures are given by Eisen (1977a).

Experiment 2: The objective was to use a selection index to select simultaneously, but in opposite directions, for genetic gains in litter size and 6-week body weight. How to determine the appropriate economic weights to realize this goal was not apparent. In lieu of establishing economic weights, desired genetic gains in each trait were determined first; then the desired gains index (Pesek and Baker, 1969; Baker, 1974; Yamada et al., 1975) was constructed. The absolute value of the desired genetic gain in litter size was taken to be the expected correlated response in litter size based on single-trait selection for 6-week body weight. In a similar manner, the absolute value of the desired genetic gain in 6-week body weight was calculated as the correlated response to single-trait selection for litter size. In this case, the expected response per generation was about equal for the two traits. Genetic and phenotypic parameter estimates used to construct the desired gains index were obtained from previously published data on the ICR strain (Hanrahan and Eisen, 1973, 1974). The calculated desired gains index was $I_d = .305\ P_W - .436\ P_L$, where P_W and P_L are the phenotypic values for 6-week body weight and litter size, respectively. The desired gains index lines were designated L^-W^+, selected for decreased litter size and increased 6-week body weight, and L^+W^-, selected for increased litter size and decreased 6-week body weight. Lines selected upward for litter size (L^+) or 6-week body weight (W^+) were also formed to obtain realized genetic parameters from single-trait selection, but details on their responses are presented elsewhere (Eisen, 1976a, 1977b). A randomly selected control line (K) was used to adjust for possible environmental trends. The present report includes 12 generations of index selection.

The intention was to have 20 parents of each sex represent each of the three lines each generation. Matings were made at random with the avoidance of sib matings. The control line was maintained by randomly selecting one male and one female from each full-sib family. Individual selection was conducted for females only in each selected line, while males were randomly selected. This procedure resulted in a relatively mild selection intensity but, at the same time, reduced the rate of inbreeding from what might be expected in selected populations of relatively small population size (Robertson, 1961). Justification for this approach was based on the desire to avoid the relatively high degree of inbreeding depression for litter size previously observed in the ICR strain (Eisen et al., 1973; Hauschka and Mirand, 1973).

Experimental protocol was similar to Experiment 1 with the following exceptions. Litters were randomly standardized to eight pups at 1 day of age, attempting to obtain two males and six females. At 3 weeks of age, mice were weaned and randomly assigned to cages containing four mice of like sex and line. Further details on this experiment, including correlated selection responses, will be given elsewhere (Eisen, 1977b).

Statistical analyses: The mean of 6-week body weight was additively adjusted for sex differences within each line and generation. Generation means for each trait were taken as a deviation from the appropriate control line mean. Genetic gain in each trait was estimated as the regression of the generation mean deviation on generation number. Realized heritability of the index was calculated as the index response regressed on the cumulative selection differential in index units (Falconer, 1960). Standard errors of the direct and correlated responses and of the realized heritability of the index were based on formulas presented by Hill (1972).

Since economic weights were not involved in arriving at either the restricted index or the desired gains index, it was appropriate to determine the realized heritability of the index in order to evaluate direct response from index selection, as opposed to the more typical criterion of the realized response in the aggregate breeding value. The heritability of the selection index was defined as

$$h_I^2 = \text{Var.}(\sum_i b_i A_i)/\text{Var.}(\sum_i b_i P_i) = \sigma_{A_I}^2/\sigma_I^2 ,$$

where, for the ith trait, b_i = the weighting factor in the index, A_i = the breeding value and P_i = the phenotypic value, σ_I^2 = the variance of the index and $\sigma_{A_I}^2$ = the variance of the sum of the breeding values each weighted by b_i.

In the case of the restricted selection index (I_r) involving two traits, the genetic regression of the restricted index on the unrestricted trait (U) has an expectation of

$$b_{g_{UI_r}} = 1 - r_{g_{UR}}^2 ,$$

which upon rearrangement yields the realized genetic correlation between the unrestricted trait (U) and the restricted trait (R)

$$r_{g_{UR}} = (1 - b_{g_{UI_r}})^{1/2} \quad \text{(Eisen, 1977a)}.$$

Realized selection index weights were calculated by a technique described by Dickerson et al. (1954). An attempt was made to find the realized heritabilities of and the genetic correlation between the component traits of each selection index (Harvey, 1972), but satisfactory solutions were not attained. This difficulty probably arose because of multicollinearity among variables defined in the design matrix.

RESULTS

Experiment 1: Summary information on population size, selection intensity and inbreeding is presented in Table I for the restricted selection index lines and the control line. The absolute cumulative selection differential was higher in the I^- line than in I^+. The realized selection intensity of -1.49 in I^- approached the expected value of -1.44. The lower absolute realized selection intensity of 1.22 in I^+ was explained by higher infertility and mortality rates in this line. However, a comparison of weighted versus unweighted cumulative selection differentials revealed no apparent effect of natural selection interfering with the selection criterion per se. Secondary selection intensities show that the overwhelming selection pressure was directed toward postweaning gain with little selection intensity being applied to feed intake, particularly in the I^+ line. The ratios of realized relative index weights for postweaning gain/feed intake in the restricted index lines were 1/-.077 and -1/.054 for I^+ and I^-, respectively. The absolute values of these ratios are in agreement with the ratio of realized relative index weights of 1/.067 derived from the genetic parameter estimates of Jara-Almonte and White (1973).

TABLE I

Experiment 1 - Restricted Index: Summary of Data on Population Size, Selection Intensity and Inbreeding in each Line

Item	I^+	I^-	C
Individuals/gen.	125.2	137.9	149.2
Sires/gen.	10.9	11.9	11.4
Dams/gen.	17.6	21.3	20.1
Cum. sel. diff. (index units)	25.55	-31.26	—
Sel. intensity (index)/gen.	1.22	-1.49	—
Sel. intensity (postweaning gain)/gen.	1.10	-1.43	—
Sel. intensity (feed intake)/gen.	.04	- .27	—
Cum. inbreeding coef. - $\overline{F}_{10}$ (%)	15.8	15.5	9.0
Effective size - N_e [a]	29.4	29.9	53.1

[a] Calculated from $\overline{F}_{10} = 1 - (1 - \Delta F)^{10}$, where $\Delta F = \frac{1}{2N_e}$.

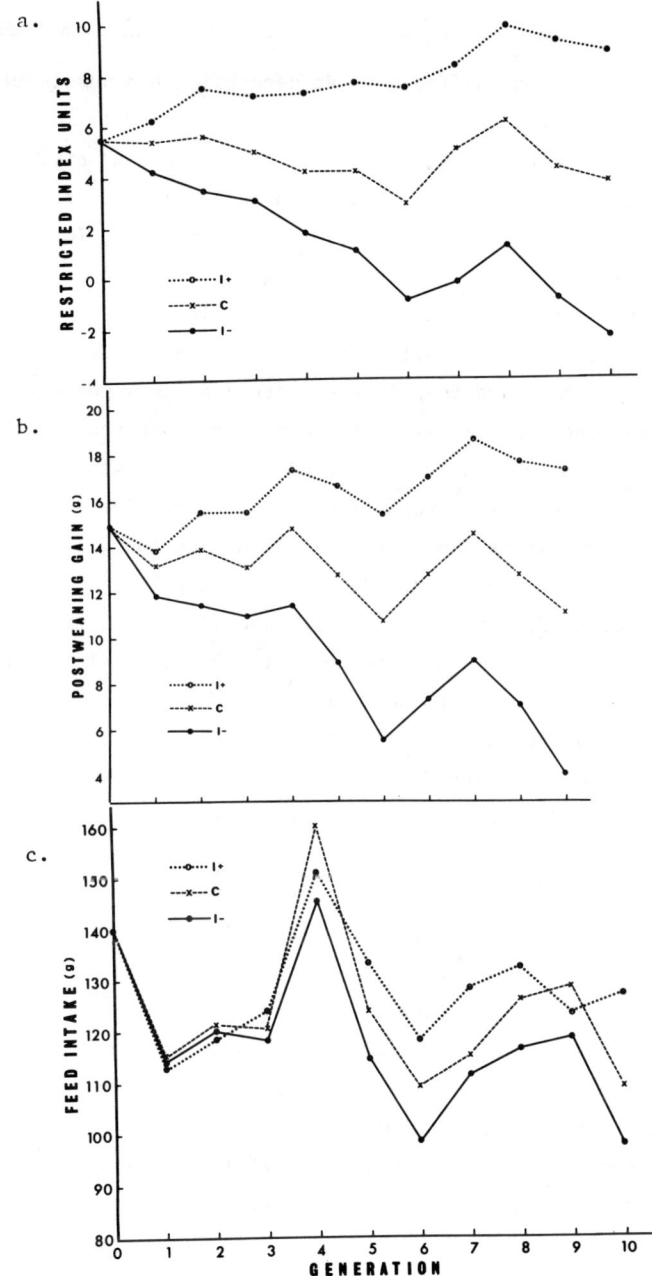

FIG. 1

Experiment 1: Generation means in the I^+, I^- and C lines.
a. restricted index, b. postweaning gain, c. feed intake.

The mean inbreeding coefficients at generation ten were similar for the two selected lines and clearly higher than the level of inbreeding in the control line (Table I). This result is explained by the pedigree control mating system which is designed to reduce the rate of inbreeding relative to complete random mating (Gowe et al., 1959) and also by the higher rate of inbreeding expected from selection in finite populations (Robertson, 1961).

Figures 1a,b,c give generation means of the restricted index units, postweaning gain and feed intake plotted against generation number for each line. No significant trends were observed for the control line. The responses in restricted index units were significant ($P<.01$) in both of the index lines (Table II). The apparent asymmetry of index response is readily explained by the greater selection intensity in the I^- line. Realized heritabilities of the restricted selection index were $.18 \pm .026$ and $.19 \pm .023$ in the I^+ and I^- lines, respectively, in agreement with the paternal half-sib estimate of $.19 \pm .09$ found in the C line.

The response in postweaning gain based on the restricted index was in the intended direction ($P<.01$) in both lines. The realized correlated response in postweaning gain per unit selection intensity in the upward (.46g) and downward (-.43g) directions was in agreement with prediction ($\pm.40$g). On the other hand, the response in feed intake was not significant in either line. From Figure 2, it can be seen that as postweaning gain responded to selection, little change in feed intake occurred in the first four generations of selection. However, as the response in postweaning gain progressed beyond approximately one absolute phenotypic standard deviation, genetic gain in feed intake tended to follow the genetic change in postweaning gain, although the difference was not significant. The realized genetic correlations between postweaning gain and feed intake were $.44 \pm .26$ and $.33 \pm .20$ in the I^+ and I^- lines, respectively. The pooled realized genetic correlation of $.38 \pm .17$ was not significantly different from the paternal half-sib estimate of $.61 \pm .19$.

TABLE II

Experiment 1 - Restricted Index: Regression Coefficients ± S.E.
of Responses in the Restricted Selection Index
and its Components on Generation Number

Response	I^+	I^-	C
Restricted index units	.46 ± .078**	−.58 ± .073**	−.11 ± .078
Postweaning gain (g)	.56 ± .085**	−.64 ± .093**	−.20 ± .119
Feed intake (g)	1.40 ± .69	−.98 ± .76	−1.46 ±1.42

**$P<.01$.

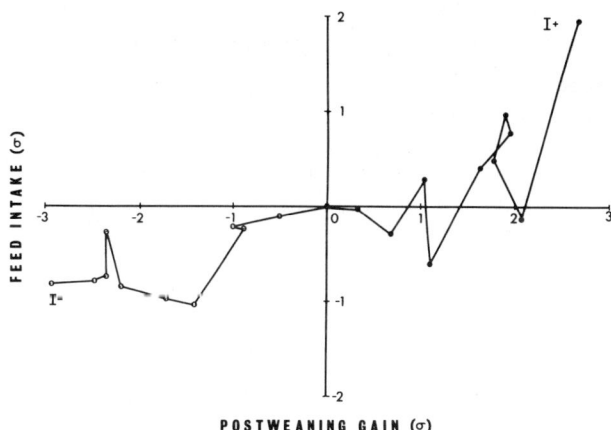

FIG. 2

Experiment 1: Generation responses in feed intake plotted against postweaning gain, both measured in phenotypic standard deviation units (σ).

Experiment 2: Data on selection differentials, population sizes and inbreeding coefficients are given in Table III. The cumulative selection differential for the desired gains index was higher in L^-W^+ compared to L^+W^-. This was due to a higher variance of index units, pooled within generations, in L^-W^+ ($\sigma^2_{I_d}$ = 1.59) than in L^+W^- ($\sigma^2_{I_d}$ = 1.36). Consequently, the standardized selection differentials were essentially equal. The secondary selection intensities were approximately twice as large for litter size as for 6-week body weight. The intended ratio of absolute index weights of litter size/6-week body weight was 1/.70. The realized ratios of index weights of -1/.80 and 1/-.76 in L^-W^+ and L^+W^-, respectively, suggest that slightly less relative weight was actually placed on litter size than was intended. The mean inbreeding coefficients were larger in the selected populations, as expected, but the difference was small due to the relatively mild selection intensity practiced.

TABLE III

Experiment 2 - Desired Gains Index: Summary of Data on Population Size, Selection Intensity and Inbreeding in each Line

Item	L^-W^+	L^+W^-	K
Males scored/gen.	63.8	48.4	54.4
Females scored/gen.	88.5	105.3	97.4
Sires/gen.	17.8	17.8	18.7
Dams/gen.	21.8	20.4	20.4
Cum. sel. diff. (index units)	9.39	-8.49	—
Sel. intensity (index)/gen.	.62	-.61	—
Sel. intensity (litter size)/gen.	-.52	.47	—
Sel. intensity (6-week body wt.)/gen.	.24	-.25	—
Cum. inbreeding coef. - $\bar{F}_{12}$ (%)	10.55	11.63	9.51
Effective size - N_e [a]	54.1	48.8	60.3

[a] Calculated from $\bar{F}_{12} = 1 - (1 - \Delta F)^{12}$, where $\Delta F = \dfrac{1}{2N_e}$.

Figures 3a,b,c present generation means of the desired gains index, litter size and 6-week body weight for the three lines. Regression coefficients of responses in the desired gains index and its components are given in Table IV. The control (K) line showed a decline (P<.05) in 6-week body weight and no significant change in the desired gains index or litter size. The downward trend in body weight observed in the K line may have been due to genetic drift, environmental trends or a combination of both. The realized heritabilities of the desired gains index were .19 ± .038 and .09 ± .030 in the L^-W^+ and L^+W^- lines, respectively, suggesting an asymmetric response. The realized heritability, calculated from divergence, was .14 ± .019. The heritability of the desired gains index estimated from base population statistics was .26 ± .12.

TABLE IV

Experiment 2 - Desired Gains Index: Regression Coefficients ± S.E. of Responses in the Desired Gains Selection Index and its Components on Generation Number

Response	L^-W^+	L^+W^-	K
Desired gains index units	.15 ± .03**	-.06 ± .02*	-.02 ± .02
Litter size	-.08 ± .03*	.12 ± .05*	-.05 ± .04
6-week body weight (g)	.39 ± .04**	.06 ± .09	-.15 ± .07*

*P<.05 , **P<.01.

The asymmetry of response in the desired gains index was reflected in the asymmetry of the correlated response in 6-week body weight. The L^-W^+ line evinced an increase (P<.01) in 6-week body weight of .39g per generation, whereas 6-week body weight in the L^+W^- line declined (P>.05) by only -.06g per generation. However, it is clear from Figure 2a, that after generation two, 6-week body weight in L^+W^- was consistently below the control line except in generation 10. Omitting generation 10 from the analysis yielded a regression of 6-week body weight on generation number of -.11g

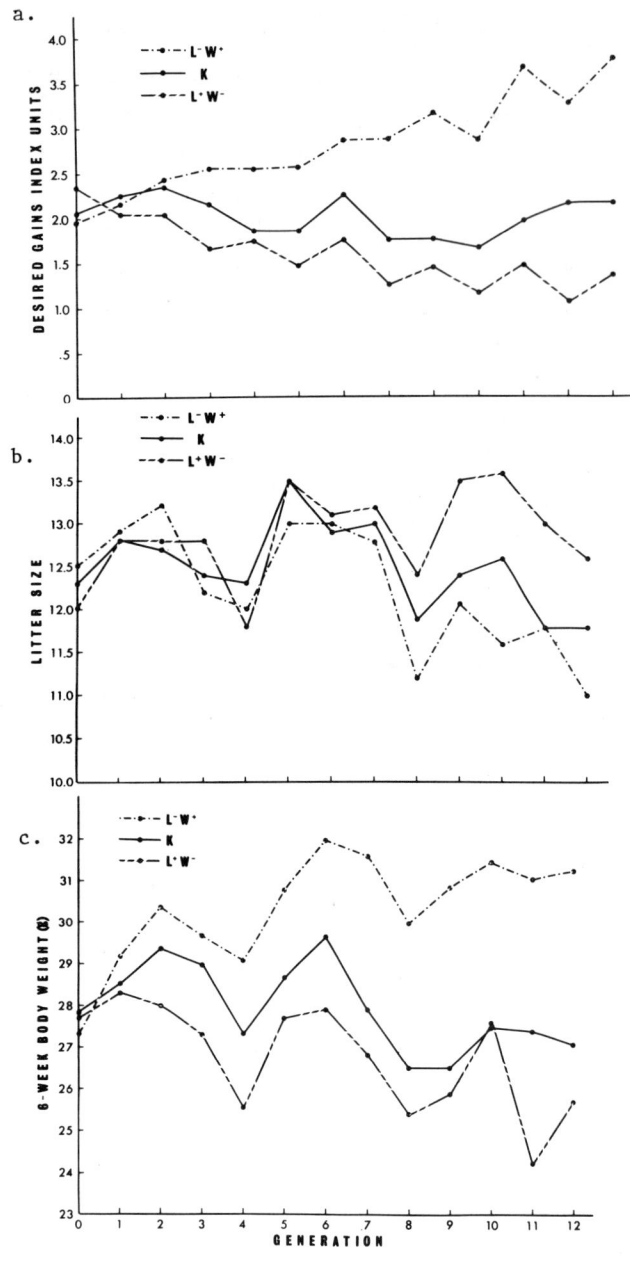

FIG. 3

Experiment 2: Generation means in the L^-W^+, L^+W^- and K lines. a. desired gains index, b. litter size, c. 6-week body weight.

(P>.05). This may still be an underestimate of the downward response in 6-week body weight in L^+W^- since the exact cause of the decline in this trait in the control line cannot be determined. Litter size showed significant (P<.05) responses in the intended direction in both index lines. However, divergence in litter size between L^+W^- and L^-W^+ did not become consistent until generation seven (Figure 3b). The absolute expected correlated responses in litter size and 6-week body weight per unit selection intensity were .38 and .42g, respectively. The observed correlated responses per unit selection intensity for litter size were -.13 and .19 in L^-W^+ and L^+W^-, respectively. Correspondingly, the correlated responses in 6-week body weight were .63g and -.10g (-.18g).

The graph of phenotypic standard deviation units in litter size plotted against 6-week body weight (Figure 4) shows some of the gyrations in the component responses of the index from another perspective. Although the desired gains index was expected to yield approximately equal genetic gains in each trait, this was not the case in L^-W^+, where the standardized response was greater in 6-week body weight than in litter size. The standardized responses appear to be approximately equal in L^+W^-.

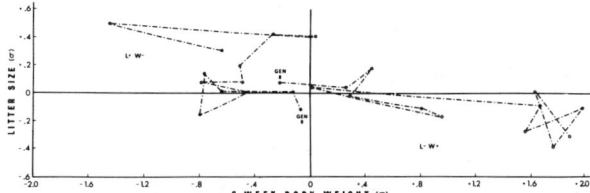

FIG. 4

Experiment 2: Generation responses in litter size plotted against 6-week body weight, both measured in phenotypic standard deviation units (σ).

DISCUSSION

Experiment 1: Results of two-way selection for the restricted index were at least partially in agreement with theoretical expectation. The realized responses in postweaning gain were in reasonable accord with prediction. The overall regressions of feed intake on generation number were not significantly different from zero. The first four generations of selection appeared to verify the biological validity of the restricted selection index since genetic gain in feed intake was essentially zero. Subsequently, however, feed intake tended to change in the same direction as postweaning gain.

The failure of a complete genetic restriction on change in feed intake may be the result of a biological incompatibility between the goals of the restricted selection index and the genetic correlation between the traits in the restricted index. Clearly, both postweaning gain and feed intake are partially under genetic control and are positively correlated genetically. Blaxter (1968) has emphasized that the chief component determining variation in growth rate, after dietary deficiencies are eliminated, is energy intake. Growth equations relating body weight to energy requirements above maintenance (Blaxter, 1968) are closely related to growth-age curves (Fitzhugh, 1976). In the present study, when feed intake was adjusted for estimated average metabolic body size during the 3-week test period, no genetic trends in feed intake were evident in either index line. Previous studies on single-trait selection designed to increase postweaning gain in mice have led to positive correlated responses in appetite as well as feed efficiency (see reviews by Eisen, 1974, 1976b; Sutherland et al., 1974). A major objective of the present experiment was to alter feed efficiency, but circumvent genetic change in appetite by holding genetic gain in feed intake to zero. The experimental results suggest that this goal was feasible up to a response in postweaning gain of about one absolute phenotypic standard deviation. The possibility also exists that after this point the genetic correlation between postweaning gain and feed intake was altered sufficiently to invalidate the restricted index. This hypothesis is extremely difficult to test experimentally in selection experiments involving relatively small population sizes. Moderately intense single-trait selection practiced for a duration of only ten generations has generally not yielded evidence of change in genetic parameters. However, it is conceivable that the restricted selection index is more sensitive to shifts in genetic parameters.

There is a paucity of restricted selection index experiments reported in the literature. Abplanalp et al. (1963) attempted to increase 8-week body weight of turkeys without increasing body weight at 24 weeks. Realized responses were in general agreement with prediction, although there was a tendency toward a negative response in 24-week body weight. Scheinberg et al. (1967) reported a restricted selection index study in Tribolium involving high larval weight, reduced developmental time and high pupal weight. In each of three treatments, selection was practiced for one of the above three criteria while restricting the other two. The attempted restrictions of larval weight and pupal weight resulted in negative realized responses in all cases and the intended restriction on developmental time was realized in only one case. Genetic gains in the unrestricted traits were generally lower than predicted, which was attributed to overestimated genetic parameters. Okada and Hardin (1967, 1970) selected for response in larval weight while restricting adult weight of Tribolium. Although the authors stated that adult weight was satisfactorily held constant, it would appear from their graphs that adult weight tended to decrease.

Experiment 2: Bi-directional selection using the desired gains index yielded significant index responses which were less than expected, particularly when selection was for increased litter size and decreased 6-week body weight. Selection for decreased litter size and increased 6-week body weight resulted in a greater selection response in the desired gains index and 6-week body weight than did selection in the opposite direction. Although genetic gains were predicted to be approximately equal for litter size and 6-week body weight, the latter trait responded to a greater degree in the L^-W^+ line. On the other hand, genetic gains in the two traits were nearly equal in the L^+W^- line.

Several factors could result in a realized response lower than what was predicted for the desired gains index or its components. The antagonism between the selection goals in the desired gains index and the positive genetic correlation between litter size and 6-week body weight may be a contributing factor. Single-trait selection for litter size and 6-week body weight, conducted concurrently with the desired gains index selection, yielded a realized genetic correlation of .60 ± .10 after ten generations (Eisen, 1976a). This compares with a genetic correlation estimated from daughter-dam records of .57 ± .12. Another factor which could have influenced response was the relatively low selection intensity applied in the index lines. This seems unlikely since the single trait selected lines have responded readily for increased litter size and increased 6-week body weight using similar selection intensities.

There are other examples of antagonistic index selection experiments where genetic gains were less than expected. Rutledge et al. (1973) found that an antagonistic selection index for body weight and tail length at 6 weeks of age in mice yielded responses which were less than predicted. The realized genetic correlation between these two traits was .33, based on single-trait selected lines. Nordskog et al. (1974) selected bi-directionally in chickens for an antagonistic index involving body weight and egg weight at 32 weeks of age. The realized genetic correlation between these two traits was .42 when pooling results of several single-trait selected lines. Realized responses and their standard errors in the index and its components were not given by Nordskog et al. (1974). The index response in the line selected for high body weight and low egg weight was considerably greater than the index response for low body weight and high egg weight, and most of the response in both lines appeared to be due to genetic gains in body weight. In a qualitative sense these results agree with the present findings where large body size - small litter size evinced a greater response than the opposite index criterion. Another case where antagonistic selection deviated from expectation involved selection for increased 12- to 21-day gain and decreased 51-day body weight in one line of mice and vice versa in a second line (Berger and Harvey, 1975). Response in 12- to 21-day gain based on the antagonistic index followed the same direction as the expected selection response for 51-day weight rather than the expected response for 12- to 21-day gain. In this case, genetic maternal effects may be involved since females having a relatively small adult body weight may provide an inferior preweaning nutritional environment for the pups. Knowledge of the additive maternal-direct genetic variance-covariance matrix could provide a selection index which might circumvent this difficulty (Van Vleck, 1970). The possibility also exists of an incompatibility between the goals of the selection index and the genetic control of growth (Berger and Harvey, 1975). The former explanation may be more logical since McCarthy (1971) has reported success in altering the growth curve by selecting for postweaning body weights using a selection index or independent culling. Comparisons of predicted with realized responses were not given. Two studies have demonstrated reasonable success in antagonistic selection using independent culling procedures: larval weight and pupal weight in _Tribolium_ (Bell and Burris, 1973) and coxal and sternopleural bristle numbers in _Drosophila melanogaster_ (Sheridan and Barker, 1974).

Genetic and environmental maternal effects contributing to discrepancies in index selection response should not be overlooked. In the present study, environmental maternal effects of the dam's litter size on the daughter's adult body weight were minimized by standardizing litter size soon after parturition (Eisen, 1970). The estimated direct-maternal genetic correlation between 6-week body weight and litter size was only -.09 (Hanrahan and Eisen, 1974), which suggests that this factor could not contribute markedly to discrepancies from predicted response.

Moll et al. (1975) used the desired gains index to select for increased yield and decreased ear height in maize, where the genetic correlation between the two characters was .56 from paternal half-sib estimates and .26 from realized responses to single-trait selection. The observed response in the index was actually greater than expected, but there were large discrepancies between observed and expected responses in yield and ear height. A nonlinear relationship was found between the two traits which the authors suggest may have contributed to these discrepancies. Nonlinear relationships were not observed between litter size and 6-week body weight in the present study.

General conclusions: Experimental results with two antagonistic selection indexes in mice support the general validity of index selection theory. Many deviations from expectation were noted, which confirm previous investigations with an array of traits and organisms. In some cases, differences between realized and expected responses to antagonistic selection could be explained by poorly estimated base population parameters (Scheinberg et al., 1967; Bell and Burris, 1973). Harris (1963) has emphasized the need for accurate estimates of genetic parameters to construct selection indexes.

The present theory of quantitative genetics is adequate for predicting short-term direct responses in single-trait selection experiments. Hill (1972) and Bohren (1975) have reviewed the factors contributing to variation in response, including genetic drift, natural selection, inbreeding depression and genotype x environment interactions. It is questionable whether present theory is adequate for reliably predicting correlated responses of individual traits due to multi-trait selection, even if correct estimates of base population parameters are used to construct the index. Bohren et al. (1966) demonstrated that the genetic covariance between two traits is subject to asymmetry due to change in gene frequency and hence asymmetry in correlated responses would be prevalent. They concluded that prediction of short-term correlated responses would be valid over fewer cycles of selection than it would for direct response. The influence of multi-trait selection on the genetic covariance between the traits probably could cause similar difficulties in predicting the correlated responses in component traits of the selection index. The problem of prediction may be more complex for the selection index than for single-trait selection since many constraints can be placed on the selection index and its component traits.

The antagonistic selection index further complicates the situation since physiological or anatomical incompatibilities between traits can be introduced after a few generations of selection. This can take the form of a nonlinear genetic relationship between the traits. If maternal genetic effects are important and are not taken into account in the index, this can also cause deviations from expected response in component traits.

ACKNOWLEDGMENTS

Paper No. 5102 of the Journal Series of the North Carolina Agricultural Experiment Station, Raleigh.

The competent technical assistance of Ms. B. J. Edwards is greatly appreciated.

BIBLIOGRAPHY

(1) Abplanalp, H., Ogasawara, F. X. & Asmundson, V. S. (1962). Influence of selection for body weight at different ages on growth of turkeys. Brit. Poul. Sci. 4, 71-82.

(2) Baker, R. J. (1974). Selection indexes without economic weights for animal breeding. Can. J. Anim. Sci. 54, 1-8.

(3) Bell, A. E. & Burris, M. J. (1973). Simultaneous selection for two correlated traits in Tribolium. Genet. Res. 21, 29-46.

(4) Berger, P. J. & Harvey, W. R. (1975). Realized genetic parameters from index selection. J. Anim. Sci. 40, 38-47.

(5) Blaxter, K. L. (1968). The effect of the dietary energy supply on growth. Growth and Development of Mammals (G. A. Lodge & G. E. Lamming, Eds.). New York: Plenum Press.

(6) Bohren, B. B. (1975). Designing artificial selection experiments for specific objectives. Genetics 80, 205-220.

(7) Bohren, B. B., Hill, W. G. & Robertson, A. (1966). Some observations on asymmetrical correlated responses to selection. Genet. Res. 7, 44-57.

(8) Cunningham, E. P., Moen, R. A. & Gjedrem, T. (1970). Restriction of selection indexes. Biometrics 26, 67-74.

(9) Dickerson, G. E., Blunn, S. T., Chapman, A. B., Kottman, R. M., Krider, J. L., Warwick, E. J., Whatley, J. A., Jr., Baker, M. L., Lush, J. L. & Winters, L. M. (1954). Evaluation of selection in developing inbred lines of swine. Mo. Agr. Exp. Sta. Res. Bull. 551.

(10) Eisen, E. J. (1970). Maternal effects on litter size in mice. Can. J. Genet. Cytol. 12, 209-216.

(11) Eisen, E. J. (1974). The laboratory mouse as a mammalian model for the genetics of growth. First World Congress on Genetics Applied to Livestock Production, Vol. 1, 467-492. Madrid, Spain.

(12) Eisen, E. J. (1975). Population size and selection intensity effects on long-term selection response in mice. Genetics 79, 305-323.

(13) Eisen, E. J. (1976a). Mass selection for litter size and body weight in mice. J. Anim. Sci. 43, 215 (Abstr.).

(14) Eisen, E. J. (1976b). Results of growth curve analyses in mice and rats. J. Anim. Sci. 42, 1008-1023.

(15) Eisen, E. J. (1977a). Restricted selection index: an approach to selection for feed efficiency. J. Anim. Sci. (Submitted).

(16) Eisen, E. J. (1977b). Single-trait and antagonistic index selection for litter size and body weight in mice. Genetics (In preparation).

(17) Falconer, D. S. (1960). An Introduction to Quantitative Genetics. New York: The Ronald Press Co.

(18) Finney, D. J. (1962). Genetic gains under three methods of selection. Genet. Res. 3, 417-423.

(19) Fitzhugh, H. A., Jr. (1976). Analysis of growth curves and strategies for altering their shape. J. Anim. Sci. 42, 1036-1051.

(20) Gowe, R. S., Robertson, A. & Latter, B. D. H. (1959). Environment and poultry breeding problems. 5. The design of poultry control strains. Poultry Sci. 38, 462-471.

(21) Hanrahan, J. P. & Eisen, E. J. (1973). Sexual dimorphism and direct and maternal genetic effects on body weight in mice. Theoret. Appl. Genetics 43, 39-45.

(22) Hanrahan, J. P. & Eisen, E. J. (1974). Genetic variation in litter size and 12-day weight in mice and their relationships with post-weaning growth. Anim. Prod. 19, 13-23.

(23) Harris, D. L. (1963). Influence of errors of parameter estimation upon index selection. Statistical Genetics and Plant Breeding (W. D. Hanson & H. F. Robinson, Eds.). Publ. 982, N.A.S.-N.R.C., Washington, D. C.

(24) Harvey, W. R. (1972). Direct and indirect response in two-trait selection experiments in mice. Proc. 21st Annual Session, National Breeders Round Table. Kansas City, Mo.

(25) Hauschka, T. S. & Mirand, E. A. (1973). The "Breeder: Ha (ICR)" Swiss mouse, a multipurpose stock selected for fecundity. Roswell Park Memorial Institute 75th Anniversary Volume. New York: Alan R. Liss.

(26) Hazel, L. N. (1943). The genetic basis for constructing selection indexes. Genetics 28, 476-490.

(27) Hazel, L. N. & Lush, J. L. (1942). The efficiency of three methods of selection. J. Heredity 33, 393-399.

(28) Hill, W. G. (1972). Estimation of realised heritabilities from selection experiments. II. Selection in one direction. Biometrics 28, 767-780.

(29) Jara-Almonte, M. & White, J. M. (1973). Genetic relationships among milk yield, growth, feed intake and efficiency in laboratory mice. J. Anim. Sci. 37, 410-416.

(30) Kempthorne, O. & Nordskog, A. W. (1959). Restricted selection indices. Biometrics 15, 10-19.

(31) McCarthy, J. C. (1971). Effects of different methods of selection for weight on the growth curve in mice. Xth International Conference on Animal Production.

(32) Moll, R. H., Stuber, C. W. & Hanson, W. D. (1975). Correlated responses and responses to index selection involving yield and ear height of maize. Crop Sci. 15, 243-248.

(33) Nordskog, A. W., Tolman, H. S., Casey, W. D. & Lin, C. Y. (1974). Selection in small populations of chickens. Poultry Sci. 53, 1188-1219.

(34) Okada, I. & Hardin, R. T. (1967). An experimental examination of restricted selection index, using Tribolium castaneum. I. The results of two-way selection. Genetics 57, 227-236.

(35) Okada, I. & Hardin, R. T. (1970). An experimental examination of restricted selection index, using Tribolium castaneum. II. The results of long-term one-way selection. Genetics 64, 533-539.

(36) Pesek, J. & Baker, R. J. (1969). Desired improvement in relation to selection indexes. Can. J. Plant. Sci. 49, 803-804.

(37) Robertson, A. (1961). Inbreeding in artificial selection programmes. Genet. Res. 2, 189-194.

(38) Rutledge, J. J., Eisen, E. J. & Legates, J. E. (1973). An experimental evaluation of genetic correlation. Genetics 75, 709-726.

(39) Scheinberg, E., Bell, A. E. & Anderson, V. L. (1967). Genetic gain in populations of Tribolium castaneum under uni-stage tandem selection and under restricted selection indices. Genetics 55, 69-90.

(40) Sheridan, A. K. & Barker, J. S. F. (1974). Two-trait selection and the genetic correlation. I. Prediction of responses in single trait and in two-trait selection. Aust. J. Biol. Sci. 27, 75-88.

(41) Smith, H. F. (1936). A discriminant function for plant selection. Ann. Eugenics 7, 240-250.

(42) Sutherland, T. M., Biondini, P. E. & Ward, G. M. (1974). Selection for growth rate, feed efficiency and body composition in mice. Genetics 78, 525-540.

(43) Tallis, G. M. (1962). A selection index for optimum genotype. Biometrics 18, 120-122.

(44) Van Vleck, L. D. (1970). Index selection for direct and maternal genetic components of economic traits. Biometrics 26, 477-483.

(45) Williams, J. S. (1962). The evaluation of a selection index. Biometrics 18, 375-393.

(46) Yamada, Y., Yokouchi, K. & Nishida, A. (1975). Selection index when genetic gains of individual traits are of primary concern. Japanese J. Genetics 50, 33-41.

(47) Young, S. S. Y. (1961). A further examination of the relative efficiency of three methods of selection for genetic gains under less-restricted conditions. Genet. Res. 2, 106-121.

Effects of specific genes

J. M. Thoday
DEPARTMENT OF GENETICS
UNIVERSITY OF CAMBRIDGE, DOWNING STREET, CAMBRIDGE
CB2 3EH

In their textbook of Genetics, published in 1940, Sturtevant and Beadle, writing about continuous variation discuss Emerson and East's results with ear-length in maize as follows:-

Numerous attempts have been made to develop methods of estimating the number of gene-pairs involved in crosses such as this. All such attempts, however, must yield unsatisfactory results because of one circumstance: there is no way of determining the relative phenotypic differences conditioned by the various gene-pairs involved; these may range from fairly large ones down to those which are at the limit of the sensitivity of the measuring system used. The slight differences will play little part in the results unless they are numerous, and will be practically impossible to estimate numerically – *yet there is no reason to expect them to constitute a class sharply distinct from the gene differences with larger effects. In practice, most attempts to solve this problem have started with the arbitrary and improbable assumption of equal effects of all gene differences concerned.* (Italics mine)

Since that time Biometrical approaches have developed substantially and, partly in association with these, partly in parallel, there have been two other developments. One involves attempts to use more nearly Mendelian methods to study some aspects of the genetic architecture of relevant characters, most particularly in Drosophila melanogaster, in the hope of complementing the information revealed by more strictly biometrical studies. The other involves speculation concerning the nature of the genetic differences concerned in 'polygenic' systems.

Of these latter speculations I will mention only four, three of which involve some assumption that the genes of so-called polygenic systems are of different kind from those familiar in the study of discontinuous variation, and which therefore go against the view expressed in the above quotation.

One of these was put forward by Mather (1944) on the basis of his demonstration that Y chromosome variation could influence polygenic variables in Drosophila. Clearly he then considered the possibility that polygenic systems might involve heterochromatin. This was an attractive speculation at the time, for it suggested a function for so-called inert heterochromatin, and threw up the interesting possibility, which Mather discussed, that variation of heterochromatin might form a pool of rather unspecific genetic variation from which by recombination and selection new genes of more specific effect might be evolved. This still remains an interesting proposition, but the evidence we have cannot be interpreted as demonstrating an exclusively heterochromatin location of "polygenic loci".

Pandey (1972) provides another example of such speculations in his arguments that polygenes are regulatory genes. Some may perhaps be, but I know of no good evidence. Though polygenic variations may provide the fine adjustment determining the effects of major segregations (Mather, 1949) this does not imply that they involve regulatory genes.

EFFECTS OF SPECIFIC GENES 143

Turning to the classification of genes given by Grant (1975) we clearly see how confused the position can become. Grant makes a serious attempt to classify gene systems "on the basis of functional relationships between the component genes" into multiple gene systems whose individual members produce the same growth substances or similar ones and have cumulative effects, their end product being a quantitative character, and serial gene systems where the component genes control different steps in a developmental sequence. Then he further separates, while saying he does not know where to draw the line, multiple genes and polygenes. He goes on to discuss the possibility that polygenic systems may involve supernumerary DNA, heterochromatin, repetitive DNA, all in a context which implies a belief that polygenes, multiple genes and serial genes must involve different kinds of genetic loci.

These three speculations contrast with the fourth, that polygenes are isoalleles, a speculation that at least has the merit that it concentrates on the nature of the allelic differences, but, like all the rest, it lacks sufficient supporting evidence, and also errs in the assumption that polygenes must all be of one kind. It should be stressed that Mather (1949, p.21) was one of those who expressed the view that isoalleles may play some part in the production of polygenic variation, a clear indication that he did not think that all polygenes were necessarily alike.

Opinions concerning the importance of these speculations about the nature of polygenes may vary. Biometricians, as such, very much involved as they are in analysis of the total relevant variance, among other things in order to aid applied geneticists, need not necessarily worry over much about the molecular biology or developmental genetics of the variable involved. However, I find it difficult to believe they are uninterested in, or that their techniques might not be improved by, knowledge of the fundamental nature of the genetic differences they handle.

And I think it self-evident that such knowledge will only be obtainable if we can handle individual loci sufficiently well to study the specific effects of specific segregations. Much work of this kind has been done. Payne (1918), Sturtevant (1918) and Wright (see Wright, 1952) did pioneer work showing some of the underlying individual segregations could be detected. Mather and his colleagues (see particularly Breese and Mather, 1957) did a great deal, though I think it fair to say that their aim was rather to validate the underlying assumptions of biometrical genetics than to determine how the relevant genes act in development, and this I think is one of the critical things we need to know. I and my colleagues have taken Mather's approach a little further in two ways: first by adding our technique of progeny testing of a particular kind which both allows the multiplication of particular chromosomes so that estimates of their effects can be made more precise, and places those chromosomes on increasingly homogeneous backgrounds (Thoday, 1961); second by studying the ways in which the resulting chromosomes affect the continuous variable under study.

I wish now to consider a few of our results, and what they tell us about polygenic systems, polygenic loci as Thompson and I (1974) have named them or QTLs (quantitative trait loci) which is I think a better term used by Gelderman (1975).

Before I begin, however, I will give you a definition which is the only definition I consider unconfusing. A polygenic system is a system of loci at which segregation contributes to the genetic variance associated with variation of a continuous variable. Any locus at which segregation contributes to that variance is a polygenic or quantitative trait locus.

It is the effects of allelic differences at such loci we wish to understand. Given such understanding, it may be possible to deduce what kinds of loci can be involved. It seems a priori

EFFECTS OF SPECIFIC GENES

probable that all possible kinds of loci as they might be classified by molecular geneticists can be involved, and that the distinctive feature of polygenic systems is not a matter of kinds of loci but rather is the magnitude of effects on the variable of allelic substitutions relative to other sources of variance. This implies, of course, that if we can control the other sources of variance well enough we should be able to produce a situation in which a single gene-pair that is a component of a polygenic system can be made to give discontinuous variation.

The techniques that can be used to get at specific allele substitutions therefore depend on means of reducing the magnitude of other sources of variance. The first need is to minimise the misclassification arising from environmental variance. This we have done by using progeny testing devices, most particularly by serial progeny testing of single chromosomes held against markers in male Drosophila thus permitting the classification of chromosome means even though individuals cannot be classified. (Recent information (Hiraizumi, 1971: Sochacka and Woodruff, 1976) shows that care must be taken to ensure that one is not using chromosomes that induce male recombination.) The second requirement is to minimise background genetic variance arising from other chromosomes, which in our hands is also ensured by serial backcrossing to some standard inbred marker stock. Swank and Bailey's (1973) and Oliviero's (1974) use of recombinant inbred strains involve similar principles. Finally it is necessary to separate parts of the chromosome to reduce the number of allele-pairs segregating. This we do by generating numbers of recombinants with a marker chromosome, if necessary getting rid of the markers by a further round of recombination with a standard wild-type chromosome. We have been able in this way to isolate pieces of chromosome of the order of 2 map units in length (sometimes less) containing genes that influence the character under study.

The principles will be found in Thoday,(1961).

The most critical information we have obtained concerns sterno-pleural chaeta number, and modifiers of wing vein mutants in <u>Drosophila melanogaster</u>.

Sternopleural Chaeta Number

Our first and most striking results were obtained in our analyses of the sternopleural chaeta number selection lines of Thoday and Boam (1961).

Figure 1 summarises the selection results and the alleles found to distinguish the selected lines from inbred Oregon which was used as a standard. 5 loci were involved and accounted for over 80% of the genetic difference under study. There were strong interactions between the left hand chromosome III locus and those on chromosomes I and II, but there is little interaction with the right hand locus on chromosome III. Details will be found in Thoday, Gibson and Spickett (1964) and Spickett and Thoday (1966). See also Thoday,(1967a,b,c)

Figure 2 provides the summary of the critical studies done by Spickett (1963) on the effects of the allelic substitutions at the three autosomal loci together with those at a fourth locus that affected fly weight but not chaeta number.

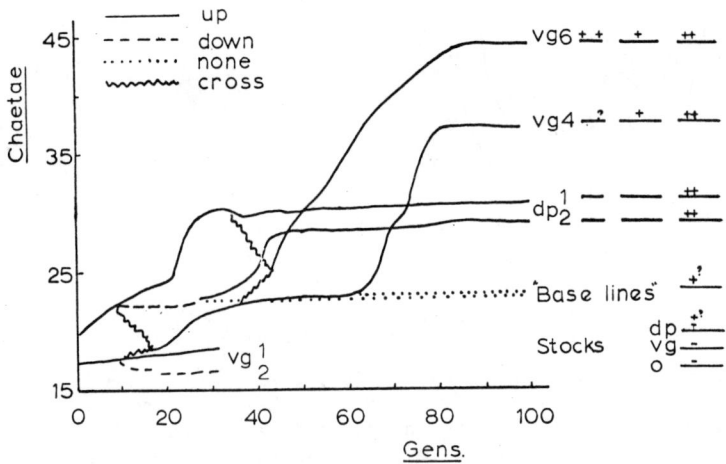

FIG. 1

The responses to selection and the genes located in the various lines of Thoday and Boam. The selection used is indicated in the top left corner. The alleles found to distinguish the lines from Oregon inbred stocks on the right, chromosomes I, II and III to be read from left to right. The 'stocks' provided the ancestors of all the lines. (From Thoday, 1973).

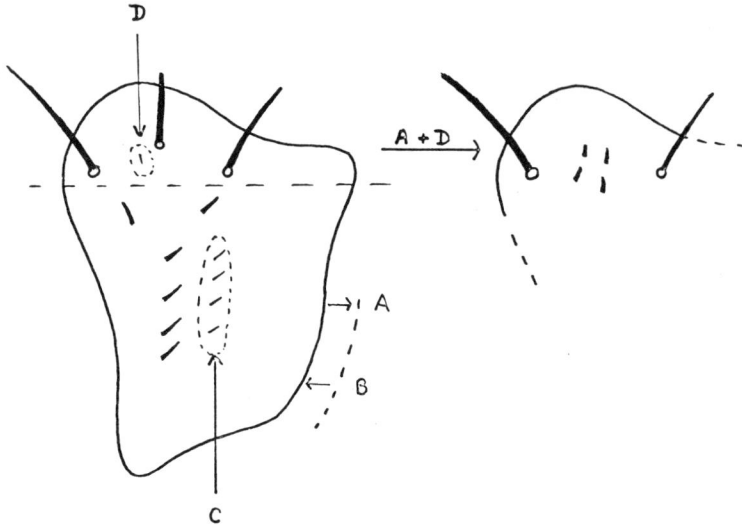

FIG. 2

The effect of four genes isolated from the high sternopleural chaeta number line vg4 from Fig.1. The left hand figure shows the pattern of hairs on the sternopleurite of wild-type Oregon inbred and, enclosed in dotted lines, the effects of the individual alleles isolated from vg4. The right hand figure shows the effect of the genes A and D together.

A, the left hand chromosome III locus, increases cell number and hair number in all regions, including other parts of the fly.

B, a locus not shown in Fig.1, decreases cell size, compensating for the fly size effect of A without influencing hair number.

C, the right hand locus on chromosome III, has a local effect on hair number tending to give a double row of microchaetes.

D, on chromosome II, has a very local effect adding one hair near the middle macrochaete. With A it replaces this macrochaete by several microchaetes. It works by delaying differentiation of the middle macrochaete initial.

EFFECTS OF SPECIFIC GENES 149

These findings lead to four striking conclusions. First, most of the difference of mean chaeta number produced by over 80 generations of selection is explicable in terms of alleles at only 5 loci. Second, the different loci affect chaeta number in specifically different ways which seems incompatible with the concept that polygenic variation involves "multiple gene systems whose individual members produce the same growth substances" (Grant, 1975). Neither am I sure that it is compatible with the view that "The polygene .... is one of a system whose parts are apparently interchangeable in development" (Mather, 1949, p.17). Third, the example shows that in principle there is an additional way of approaching specific components of polygenic systems which is by analysing the character into simple components. Fourth, the example shows that the type of analysis of polygenic differences we have used can throw light on the nature of complex character differences such as continuous variables must be, and guide character analysis such as applied breeders may need in dealing with complex variables such as yield.

Some of the same conclusions apply to Spickett's findings on fly weight where two loci were involved, one affecting cell number (and chaeta number), the other cell size. Our knowledge of the different morphological consequences of cell number changes and cell size changes clearly indicates that these cannot be regarded as in any way equivalent loci except for their equivalent effects on fly size.

The gene pairs referred to above showed partial dominance. We have located others affecting sternopleural chaeta number in analyses of disruptive selection experiments. They include gene pairs with complete dominance (Wolstenholme and Thoday, 1963), a pair with classical maternal inheritance (Gibson and Thoday, 1963), and two pairs that show a lethal position effect interaction though they are 20 map units apart (Gibson and Thoday, 1962a,b).

Biometrical geneticists may find it puzzling that so large a part of the variance in these experiments could be attributed to so few loci, especially since in some cases the differences studied were established through many generations of continuous response to selection.

I believe that at least a partial answer to this problem lies in consideration of fitness effects. Dinsley and Thoday (1961) published the results of a clearcut experiment that throws light on this. They had the good fortune to pick up a partial back mutant, vg^{p^D}, in a line homozygous for vg that was under selection for sternopleural chaeta number. This back mutant greatly improved the fertility and viability of the flies. They were able to show

1. that the vg^{p^D}/vg segregation had no effect on chaeta number
2. that the loss of selection advance for chaeta number under relaxed selection was less rapid in vg^{p^D}/vg^{p^D} lines than in vg/vg lines
3. that selection both for low and for high chaeta number could proceed further before a line plateaued in vg^{p^D}/vg^{p^D} lines than in vg/vg lines
4. that the vg^{p^D}/vg^{p^D} lines reached plateaus approximately when the productivity of these lines had fallen to the productivity level of vg/vg lines at plateau.

It follows that the reduced fitness commonly observed as a result of intense selection, some of which (though not all: Breese and Mather, 1960) must be consequent on pleiotropic effects on viability of the alleles directly selected as affecting chaeta number, can be compensated for by other alleles affecting viability that have no direct influence on the character under artificial selection.

Two things further follow. First, the rate of response to selection will depend not only on the primary response of chaeta number genes but also on secondary natural selection for viability

genes that compensate for the low fitness effects of the genes
directly selected. Second, these viability genes which will
affect the genetic variance of such a character as chaeta number
and which therefore are the subject of concern to biometrical
geneticists, are not genes that could be described as chaeta
number genes from the point of view of developmental, biochem-
cal or molecular geneticists.

Wing Vein Modifiers

To turn to our work on modifiers of expression of wing vein
mutants I would like just to highlight a few results. First,
we have shown that, though some of the genes concerned are vein
specific or specifically affect a particular part of the vein,
many of the genetic modifications of one mutant are effective on
other mutants (Thompson, 1973). Second, the modifiers selected
as enhancers of, say, veinlet, can themselves produce vein gaps
in non-veinlet flies. Their action can be independent of the
mutant at the major-gene locus. Third, such modifiers are
segregating in wild populations. Fourth, such modifiers may
act additively as well as interacting with the mutant (Thompson,
1975a).

Fig.3 illustrates this latter point which once again shows
specificity in the action of genes of continuous variation. It
shows that by itself an allele that was selected to enhance
veinlet will produce gaps in vein L4 in a region in which unenhanced
veinlet does not itself produce a recognisable effect. The
allele must in the first place have been selected because, in the
mid region of the vein, ve and it must act in combination to
raise the probability of vein absence or else because an allele
at another locus which enhances the effect of ve in region 3
also interacts with the allele that gives gaps in region 2. Elsewhere ve and the region 2 allele are additive in effect in a more
conventional sense.

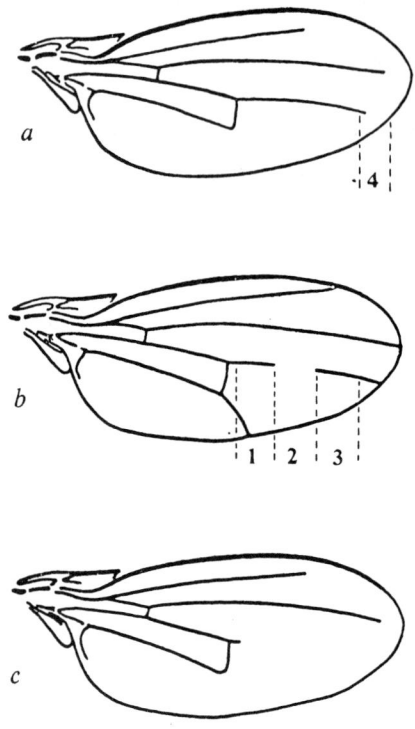

FIG. 3

Wings of Drosophila melanogaster from a reduced expression ve/ve line (a) and an enhanced expression ve/ve line (c) showing regions of the fourth longitudinal (L4) vein affected by different components of the polygenic system. One difference between ve reduced expression (a) and ve increased expression (c) selection lines is accounted for by only one locus (located on the right arm of chromosome II). This eliminates vein from the middle of the L4 vein (b, region 2). The other polygenic loci influence the expression of this second chromosome polygene by enlarging the gap formed in the middle of the L4 (regions 1 and 3). The same gene or genes may act on both regions 1 and 3. The ve mutant itself does not form vein at the tip (region 4), even in reduced expression lines. Thus the phenotype of the increased expression ve line is simply the sum of the background effect and the ve phenotype. (From Thompson, 1975a). (By his and the Editor of Nature's permission)

One further experiment with vein modification provides a specific test of the question whether the substitution of isoalleles at a relevant locus (ci) modifies ve expression as one might expect. ve and ci are reciprocally modified in the same direction by backgrounds selected to enhance or reduce the expression of either. Substitution of chromosome IVs with ci isoalleles ci^{+3} and ci^{+5} had no effect on the expression of ve (Thompson, 1975b). This does not mean that isoalleles cannot be components of polygenic systems. Scharloo (1970) and Boyer, Parris and Milkman (1973) have indeed produced evidence suggesting involvement of isoalleles in their systems. But Thompson's evidence shows that apparently relevant isoallelic differences do not necessarily do the trick.

Conclusion

Before concluding I should point to a danger in our thinking in this area, for we often fall into typological traps. We mainly study continuous variation by biometrical techniques and, essentially because the techniques for their study differ, the genes have been given different labels - polygenes versus major genes. Then we tend to discuss the differences of these kinds of genes in terms of different kinds of loci, perhaps forgetting that the only evidence we have relates to differences between alleles, that evidence about loci can only be obtained by critical location through recombination tests, and that to prove that an allelic difference affecting a continuous variable is not at a locus that can give mutants of major effect is very difficult.

Then a further step may happen. Someone, by special techniques, manages to locate some of the allele-pairs contributing to a difference in a continuous variable, and people react by concluding that the located genes no longer fall into the category of members of polygenic systems.

A splendid example of this error is provided by the following quote from Serra (1966, p.357) referring to our work

"Besides polygenes, major loci have been found to control the number and arrangement of sternopleural chaetae."

This need not have been said of our work for the mutant $\underline{Sp}$ which increases sternopleural chaeta number and is fully classifiable except at lower temperatures has been known since 1923 (Lindsley and Grell, 1967). Our alleles are not so classifiable except after sufficient replication on controlled backgrounds so that we classify chromosome means not individual flies.

But the quotation illustrates an attitude I have found to be quite common, so that I think it important to stress that if you start by analysing the variance of, or a difference involving, a polygenic character, no matter to what degree some allele-pair you abstract proves to be capable of giving discontinuity in the new situation you create, that allele-pair is still part of the polygenic system you began with.

Of course the allele-pairs you study this way must be a non-random sample of greater than average effect. Of course they may be complexes that might be broken down into components by more fine structure analysis (see Thoday, 1973; McMillan and Robertson, 1976), which also applies to 'major' genes. But it is only this biased sample of polygenes of larger effect that can give us some more precise information concerning the nature of polygenes.

Their properties do not conform to the definitions of polygenes I have quoted above. Nor does the structure of polygenic systems conform closely to the underlying assumptions of biometrical genetics; as will be clear if we consider the following conclusions, which for convenience I express in terms of chaeta number. Many of them can be reached for other variables, such as coat-colour variation in guinea-pigs (Wright, 1935), flowering time and ear-length in wheat (Wehran and Allard, 1965; Law, 1967), or wing-vein modifiers in $\underline{Drosophila}$ (see above) to mention but a few characters.

EFFECTS OF SPECIFIC GENES

The conclusions are:-

1. Serious attempts to identify the allele-pairs with direct and interactive effects on a continuous variable such as sternopleural chaeta number in <u>Drosophila melanogaster</u> explain a very large part of the genetic differences produced by artificial selection in terms of a very few loci. The polygenic system involves allele-pairs of large as well as relatively small quantitative effect.

2. Each locus affects the variable in developmentally different ways. The loci might in principle be complex but their effects are specific and there is no reason to consider these loci as different in kind from those familiar in the study of discontinuous variables. It is much more plausible to suppose that allelic differences at any kind of locus can play a part in polygenic systems.

3. Biometrical techniques will detect segregation at these loci and also at the loci directly responsible for the remaining 15-20% of the genetic differences of chaeta number. In addition they may detect effects on the genetic variance of segregation at loci which are not sternopleural chaeta number genes in the sense of having effects on the developmental processes influencing chaeta number, but which do influence the fitness of flies with unusual combinations of chaeta number genes and hence affect the genetic variance of chaeta number. These fitness gene-pairs are chaeta number genes to the biometrician but not to the developmental geneticist. They may have a profound effect on selection-responses and may explain part of the difference in number of loci estimated by biometrical approaches and the number we find.

4. It would be useful if, when material is suitable, biometricians included the possibility of using segregating markers affecting discontinuous variables in their theoretical basis for methods of partitioning variance, attempting

to abstract from the rest of the variance those parts associated with segregation at these marker loci. Attempts to do this such as that of Gelderman (1975) and others are much to be welcomed.

BIBLIOGRAPHY

Boyer, B.J., Parris, D.L. and Milkman, R. (1975). The cross veinless polygenes in an Iowa population. *Genetics 75*, 169-179

Breese, E.L. and Mather, K. (1957). The organization of polygenic activity within a chromosome in Drosophila. I. Hair characters. *Heredity, Lond. 11*, 373-395

Breese, E.L. and Mather, K. (1960). The organization of polygenic activity within a chromosome of Drosophila. II. Viability. *Heredity, Lond. 14*, 375-400

Dinsley, M. and Thoday, J.M. (1961). Fitness and artificial selection. *Heredity 16*, 113-121

Gelderman, H. (1975). Investigations on inheritance of quantitative characters in animals by gene markers. I. Method. *Theor.Appl.Genet. 46*, 319-330

Gibson, J.B. and Thoday, J.M. (1962a). Effects of disruptive selection. VI. A second chromosome polymorphism. *Heredity 17*, 1-26

Gibson, J.B. and Thoday, J.M. (1962b). An apparent 20 map-unit position effect. *Nature 196*, 661-662

Gibson, J.B. and Thoday, J.M. (1963). Maternal inheritance of a sternopleural chaeta number difference in Drosophila melanogaster. *D.I.S. 37*, 81-82

Grant, V. (1975). *Genetics of flowering plants*. Columbia University Press

Hiraizumi, Y. (1971). Spontaneous recombination in Drosophila melanogaster. *Proc.Nat.Acad.Sci. 68*, 268-270

Law, C.N. (1966). The location of genetic factors affecting a quantitative character in wheat. *Genetics, 53*, 487-498

Lindsley, D.L. and Grell, E.M. (1967). Genetic variations in Drosophila melanogaster. *Carnegie Institution of Washington Publ.No.627*

McMillan, I. and Robertson, A. (1974). The power of methods for the detection of major genes affecting quantitative characters. *Heredity 32*, 349-356

Mather, K. (1944). The genetical activity of heterochromatin.
   Proc.Roy.Soc.(B) 132, 308-332

Mather, K. (1949). *Biometrical Genetics.* Methuen

Oliviero, A. (1974). Genetic factors in the control of drug
   effects on the behaviour of mice. *The Genetics of Behaviour.*
   J.N.F. Van Abeelen (ed.)   North Holland/American Elsevier

Pandey, K.K. (1972). Origin of genetic variation: regulation
   of genetic recombination in higher organisms: a theory.
   Theor.Appl.Genet. 42, 250-261

Payne, F. (1918). An experiment to test the nature of variation
   on which selection acts. *Indiana University Studies 5*, 1-45

Scharloo, W. (1970). Stabilising and disruptive selection on a
   mutant character in Drosophila. II. Polymorphism caused by
   a genetical switch mechanism. *Genetics 65*, 681-691

Serra, J.A. (1966). *Modern Genetics, Vol.2.* Academic Press

Sochacka, J.H.M. and Woodruff, R.C. (1976). Induction of male
   recombination in Drosophila melanogaster by injection of
   extracts of flies showing male recombination. *Nature 262*,
   287-289

Spickett, S.G. (1963). Genetic and developmental studies of a
   quantitative character. *Nature 199*, 870-873

Spickett, S.G. and Thoday, J.M. (1966). Regular responses to
   selection. 3. Interaction between located polygenes. *Genet.
   Res. 7*, 96-121

Sturtevant, A.H. (1918). An analysis of the effects of selection.
   Carnegie Inst.Wash.Publ.264

Sturtevant, A.H. and Beadle, G.W. (1940). *An Introduction to
   Genetics.* Sanders

Swank, R.T. and Bailey, D.W. (1973). Recombinant inbred lines:
   Value in the genetic analysis of biochemical variation.
   Science 181, 1249-1251

Thoday, J.M. (1961). Location of polygenes. *Nature 191*, 368-370

Thoday, J.M. (1967a). Summing up: uses of genetics in physiological studies. *Mems.Soc.End.No.15*, pp.297-311

Thoday, J.M. (1967b). Genes in the study of continuous variation.
   Ciencia e Cultura 19, 54-63

Thoday, J.M. (1967c). New insights into continuous variation. *Proc.3rd.Int.Cong.Human Gen.* pp.339-350. Johns Hopkins, Baltimore

Thoday, J.M. (1973). The origin of genes found in selected lines. *Atti della Academia delli Scienze dell' Instituto di Bologna Anno 26,1*, pp.15-26

Thoday, J.M. and Boam, T.B. (1961). Regular responses to selection. I. Description of responses. *Genet.Res. 2*, 161-176

Thoday, J.M., Gibson, J.B. and Spickett, S.G. (1964). Regular responses to selection. II. Recombination and accelerated response. *Genet.Res. 5*, 1-19

Thompson, J.N. Jr. (1973). General and specific effects of modifiers of mutant expression. *Genet.Res. 22*, 211-215

Thompson, J.N. Jr. (1975a). Quantitative variation and gene number. *Nature 258*, 665-668

Thompson, J.N. Jr. (1975b). A test of the influence of iso-allelic variation on a quantitative character. *Heredity 35*, 401-406

Thompson, J.N. Jr. and Thoday, J.M. (1974). A definition and standard nomenclature for "Polygenic Loci". *Heredity 33*, 430-437

Wehran, C. and Allard, R.W. (1965). The detection and measurement of the effects of individual genes involved in the inheritance of a quantitative character in wheat. *Genetics 51*, 109-119

Wolstenholme, D.R. and Thoday, J.M. (1963). Effects of disruptive selection. VII. A third chromosome polymorphism. *Heredity 18*, 413-431

Wright, S. (1952). The genetics of quantitative variability. *Quantitative Inheritance*. E.C.R. Reeve and C.H. Waddington (eds) London HMSO

A genetic linkage study of a quantitative trait in *Drosophila melanogaster*

S. D. Jayakar, L. Della Croce, M. Scacchi and G. Guazzotti

LABORATORIO DI GENETICA BIOCHIMICA ED EVOLUZIONISTICA
C.N.R., PAVIA, ITALY
INSTITUTO DI GENETICA, UNIVERSITY OF PAVIA, ITALY

ABSTRACT

An attempt has been made to locate loci which influence the length of development from oviposition to eclosion of the imago in Drosophila melanogaster. Analysis of data presented suggests the presence near the taxi locus on the third chromosome of a locus or loci which have such an effect. The method presented should provide a useful tool in the study of quantitative variation.

INTRODUCTION

Despite extensive experimentation with artificial selection for quantitative traits in several organisms, few serious attempts been made to look systematically for single loci with detectable effects on continuously varying characters. Not surprisingly, such attempts

have been made on Drosophila melanogaster. Thoday and
his collaborators were successful in locating single
loci influencing the number of sternopleural bristle
number in D. melanogaster (see Thoday, 1961, and
Thoday, 1966, for an account of this work and the
relevant bibliography).

A formal experimental approach to the problem and
a statistical methodology were suggested by Jayakar
(1970) following Thoday's general approach. A major
advantage of Jayakar's method is that one need not use
selected lines - individuals selected at random from any
population can be used. Much the same genetical model,
but with a different statistical approach, was put
forward by Haseman and Elston (1973). These two different methods of analysis are complementary in that
they are probably suitable for different types of data.
For example, Jayakar's methods are more suitable for
data on Drosophila and other organisms where large
sibships are easily obtainable, whereas Hasemann and
Elston's methods are possibly more suitable for human
data where sibships are of limited size.

The basic approach suggested by Jayakar consists
(1) of producing heterozygotes for a chromosome with
the chosen marker and a chromosome lacking the marker;
(2) examining the progeny of crosses of those heterozygotes either with similar heterozygotes or with
homozygotes for the mutant. Though both types of
crosses can be used, the backcross is by far the more
efficient type of cross for this analysis. Using the
progeny data from either type of cross, one can test
for the presence of linkage between the marker locus
and a locus influencing a quantitative trait. By using
a battery of marker loci, one could then map the genes

influencing any given trait. In order to test whether such an analysis could fruitfully be employed for locating 'quantitative genes', we have studied the length of development from oviposition to the emergence of the imago in laboratory stocks of Drosophila melanogaster.

EXPERIMENTAL DESIGN AND METHODS

The scheme of the experimental and the control crosses in shown in Fig. 1.

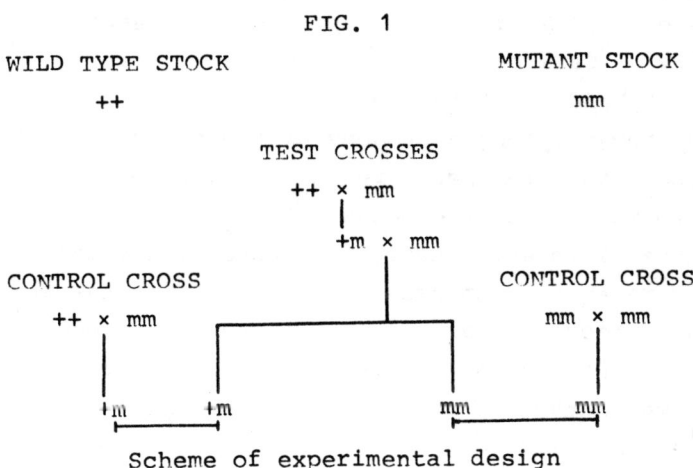

Scheme of experimental design

Since in Drosophila there is no crossing-over in the male, the two types of backcross, one with the male heterozygous and the other with the female heterozygous, have different outcomes. In previous pilot experiments, we had used the former type of back-cross with one mutant per autosome (Della Croce, Guazzotti, and Jayakar, 1975). The results of these experiments indicated that chromosome III was probably the most influential in the control of the length of development, though there may be some influence of second chromosome loci too. Thus as a starting point, we began looking for loci affecting this trait on the third chromosome, and the results and the analyses of these experiments will be described in this paper. The wild type stock used throughout this work is the Canton stock maintained at the Istituto di Genetica of the University of Pavia. The mutant stocks, sepia, scarlet, spineless, ebony, and taxi were also from the same laboratory, while the mutant stocks rosy and glass were obtained from the Department of Genetics, University of California, Davis. For each mutant, a number of hybrids with the Canton wild type stock were made and from the female progeny of these crosses, 15 unselected virgin females were individually backcrossed to males from the mutant stock. At the same time, since different genotypes in general develop at different rates 4 F_1 crosses and 4 paired matings from the mutant stock were also set up to serve as controls for the two segregating classes among the backcross progeny. The method of measuring the length of development was straightforward. For each paired mating, eggs were collected at a fixed hour each day, and were placed in groups of 25 eggs into food vials. Putting a constant number of eggs into the vials was in order to avoid any differential crowding effects.

The eggs were collected in the initial period of egg laying immediately after mating. As many groups of 25 eggs as possible were collected from each day's eggs until the required number of eggs for any mating was reached; the parents were eliminated. Usually, all the eggs were collected within 3 or 4 days of mating. The vials in which the eggs were placed were kept at a constant temperature and observed every 24 hours and any imagine that were found to have emerged were removed. The period of development was therefore measured in days. The variation in this period was from 7 to 12 days, but the large majority of individuals emerged at 8 or 9 days. A total of 150 eggs (6 vials) were set up for each backcross, whereas 200 eggs (8 vials) were collected from the control progeny.

Depending on the season of the year and on the mutant, not all the crosses set up gave progeny, and varying proportions of the eggs collected went through to emergence.

STATISTICAL ANALYSIS

Table I presents a modified version of Table 1a from Jayakar (1970). We see from this Table that in the presence of linkage between the marker on the one hand and locus with some effect on the quantitative trait on the other, we expect a difference in the mean value of the trait between the two segregating classes within some of the backcross sibships but not in others. Thus, a test for the presence of linkage involves the detection of such differences. On the average, however, the differences are zero.

One way of measuring the difference between the means of the trait in the two classes would be the statistic suggested by an analysis of variance, namely

$$(2T - V_1 - V_2),$$

TABLE I

	CROSS		PROB.	PROGENY					
				A_1A_1	M_1M_1 A_1A_2	A_2A_2	A_1A_1	M_1M_2 A_1A_2	A_2A_2
1.	M_1M_1 A_1A_1	M_1M_2 A_1A_1	a_1^4	$\frac{1}{2}$			$\frac{1}{2}$		
2.	A_1A_1	A_1A_2	$a_1^3 a_2$	$\frac{1}{2}(1-r)$	$\frac{1}{2}r$		$\frac{1}{2}r$	$\frac{1}{2}(1-r)$	
3.	A_1A_1	A_2A_1	$a_1^3 a_2$	$\frac{1}{2}r$	$\frac{1}{2}(1-r)$		$\frac{1}{2}(1-r)$	$\frac{1}{2}r$	
4.	$\left.\begin{array}{l} A_1A_1 \\ A_2A_2 \end{array}\right\{$	$\begin{array}{l} A_2A_2 \\ A_1A_1 \end{array}$	$2a_1^2 a_2^2$		$\frac{1}{2}$			$\frac{1}{2}$	
5.	$\left.\begin{array}{l} A_1A_2 \\ A_2A_1 \end{array}\right\{$	$\begin{array}{l} A_1A_1 \\ A_1A_1 \end{array}$	$2a_1^3 a_2$	$\frac{1}{4}$	$\frac{1}{4}$		$\frac{1}{4}$	$\frac{1}{4}$	
6.	$\left.\begin{array}{l} A_1A_2 \\ A_2A_1 \end{array}\right\{$	$\begin{array}{l} A_1A_2 \\ A_1A_2 \end{array}$	$2a_1^2 a_2^2$	$\frac{1}{4}(1-r)$	$\frac{1}{4}$	$\frac{1}{4}r$	$\frac{1}{4}r$	$\frac{1}{4}$	$\frac{1}{4}(1-r)$
7.	$\left.\begin{array}{l} A_1A_2 \\ A_2A_1 \end{array}\right\{$	$\begin{array}{l} A_2A_1 \\ A_2A_1 \end{array}$	$2a_1^2 a_2^2$	$\frac{1}{4}r$	$\frac{1}{4}$	$\frac{1}{4}(1-r)$	$\frac{1}{4}(1-r)$	$\frac{1}{4}$	$\frac{1}{4}r$
8.	$\left.\begin{array}{l} A_1A_2 \\ A_2A_1 \end{array}\right\{$	$\begin{array}{l} A_2A_2 \\ A_2A_2 \end{array}$	$2a_1 a_2^3$		$\frac{1}{4}$	$\frac{1}{4}$		$\frac{1}{4}$	$\frac{1}{4}$
9.	A_2A_2	A_1A_2	$a_1 a_2^3$		$\frac{1}{2}(1-r)$	$\frac{1}{2}r$		$\frac{1}{2}r$	$\frac{1}{2}(1-r)$
10.	A_2A_2	A_2A_1	$a_1 a_2^3$		$\frac{1}{2}r$	$\frac{1}{2}(1-r)$		$\frac{1}{2}(1-r)$	$\frac{1}{2}r$
11.	A_2A_2	A_2A_2	a_2^4			$\frac{1}{2}$			$\frac{1}{2}$

where T is the total variance in the sibship, and V_1 and V_2 are the variances within the two segregating classes taken separately. The variances V_1 and V_2 can be easily estimated. The estimate of T however has to be corrected for the difference in the values of the trait in control groups when the number of individuals in the two classes are not the same. If the numbers of individuals and the means corresponding to the two classes are n_1, n_2, X_1, and X_2 respectively, then T can be estimated as

$$p_1 V_1 + p_2 V_2 + p_1 p_2 (X_1 - c_1 + c_2)^2,$$

where c_1 and c_2 are the means of the respective control groups and

$$p_1 = n_1/(n_1+n_2), \qquad p_2 = n_2/(n_1+n_2).$$

If there is a recombination frequency of $\underline{r}$ between the marker locus and the locus influencing the quantitative trait, the expected value of the quantity $(2T - V_1 - V_2)$ over all the sibships is $k \cdot (1-2r)^2$, where k is a constant depending on the parameters of the model, in other words on the magnitude of the effect of the locus, the dominance effect, and the gene frequency of the variants at the locus. An unbiased estimate of the quantity $(2T - V_1 - V_2)$ will have the same expected value. This expected value is zero for $r=1/2$ and is maximum for $r=0$, and within this range it varies parabolically with $\underline{r}$. We can therefore use this statistic, i.e. the estimate of $(2T - V_1 - V_2)$ as an index of linkage between the marker and a gene for the trait. We will refer to this index as I_1. Due to systematic difference between the two sexes in their speed of development, the analysis was also carried out separately for the two sexes.

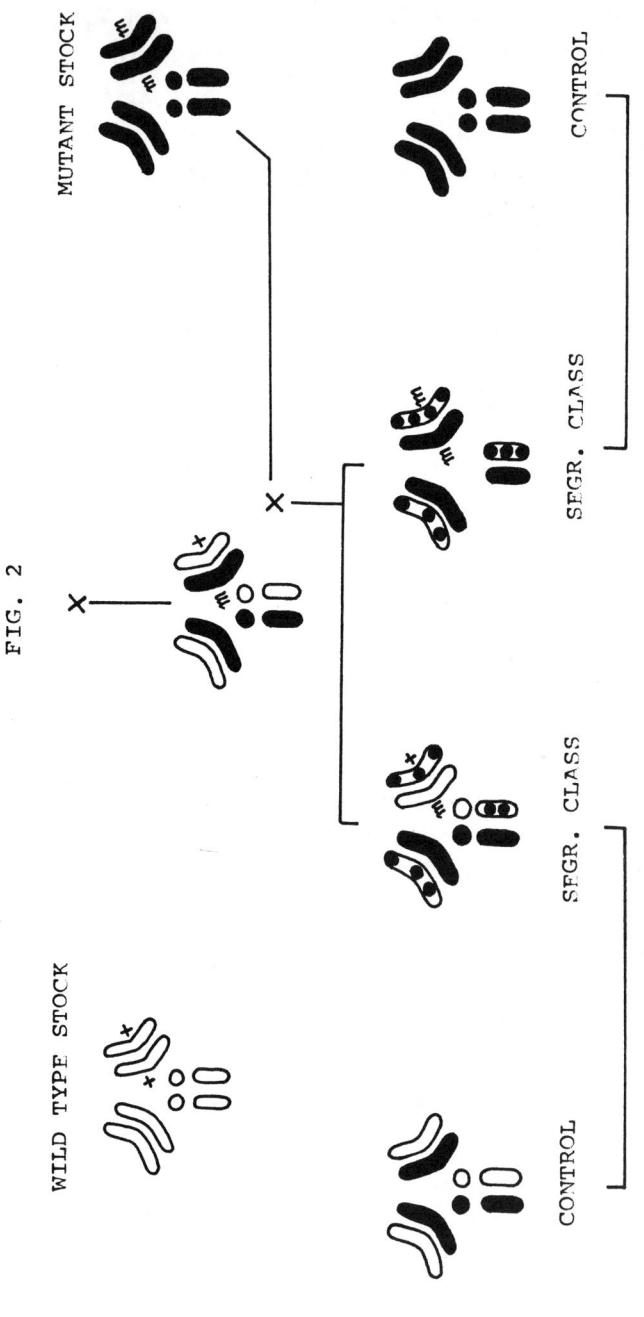

Diagram showing genetic difference between the four classes

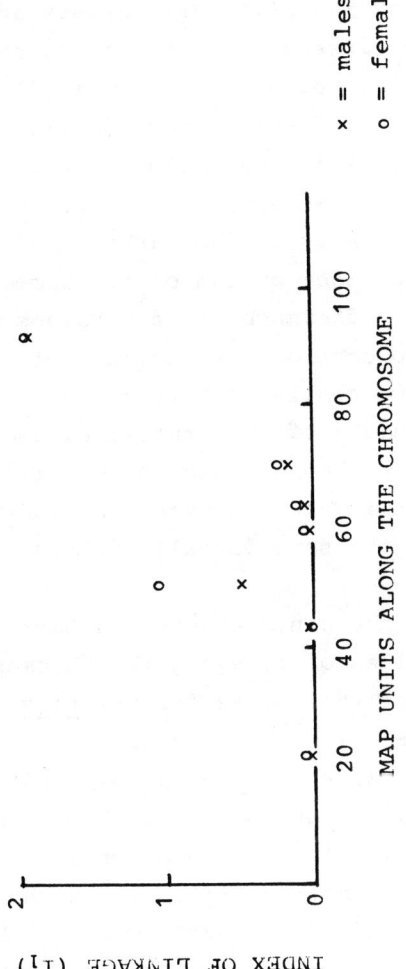

FIG. 3

Length of development in Drosophila melanogaster. Linkage with seven third chromosome markers.

Fig. 3 shows the values of I_1, the index we have constructed, for the different markers and the positions of these markers along the third chromosome.

There is however a difficulty involved in the use of I_1 as an index of the linkage we wish to detect. This difficulty is diagrammatically explained in Fig. 2. We see that in the absence of a locus influencing the trait in the vicinity of the marker, but in the presence of such loci on the other chromosomes, there could be large differences in the values of the trait between the two control classes but not within the classes which we really wish to compare. This would produce high values of the statistic, and consequently indicate the presence of linkage when it is absent. In the absence of linkage, however, the index would be expected to be the same for all markers on the chromosome.

Fig. 3 on the contrary shows a pattern of variation of the values of I_1 along the chromosome. With the exception of the values for the rosy locus, which will be discussed later, the values show an increase of the index towards the taxi locus, indicating the presence of a locus or loci near this marker which influences the length of development.

One can use another statistic which removes the possible source of error involved in I_1. From Table I, one can calculate the expected value of the variance among sibships of the differences between the means of the segregating classes, i.e.

$$\text{var}(X_1 - X_2).$$

This turns out to be again a function of $(1-2r)^2$, of the form

$$K_1 + K_2 \cdot (1-2r)^2,$$

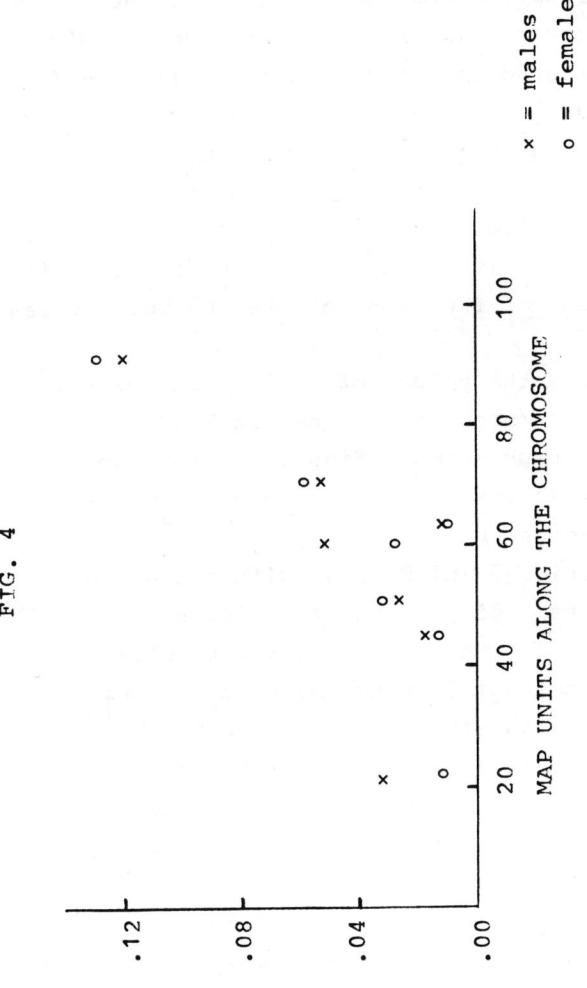

FIG. 4

Length of development in Drosophila melanogaster. Linkage with seven third chromosome markers.

where K_1 and K_2 are as before functions of the parameters of the model. Thus the estimate of this variance can be used as another index of linkage (I_2), the expected value of this index being again maximum for r=0, but going to a non-zero minimum value for r=1/2, again with a parabolic variation in r.

That this statistic removes the error to which I_1 could be subject is evident from the fact that it does not make use of the means of the control crosses at all.

Fig. 4 shows the values of I_2 for each marker locus. Here, even more clearly than in Fig. 3, for both sexes the graph shows a high value for the taxi locus and lower values as we move away to the left from the taxi locus.

Thus both Fig. 3 and Fig. 4 (with the exception of the points corresponding to the rosy locus in Fig. 3) tell the same story, that there is a chromosomal region, nearer to taxi than to any of the other loci tested, which has an effect on the length of development. Unfortunately, we have not been able to test with a marker to the right of taxi. Thus we do not know whether the region is nearer to taxi or to the chromosomal extremity. The anomalous results for the rosy locus for I_1 could either be due to the difficulties associated with this index, as discussed before, or due to the fact that for this particular locus the control crosses, due to technical difficulties, were not run exactly at the same time as the backcrosses but a few days later. On the other hand it may indicate the presence of genes near the rosy locus which have not been detected by the other index. The rosy mutant is also more difficult to

distinguish from the wild type. The experimental
results suggest the region of chromosome III near <u>taxi</u>
as being the "single" most important region in its
contribution to the genetic variance for the length of
development in the stocks examined, but it is not
possible from the available data to estimate what
fraction of the total genetic variance in controls.
This will be possible, when similar analyses have been
carried out on all the chromosomes.

DISCUSSION

In 1961, Thoday stated that he had published
"with the view of encouraging others to approach their
own problems in quantitative genetics with the
possibility in mind of locating polygenes". He does
not seem to have been very successful. It is not clear
to me whether the lack in the follow up of Thoday's
suggestions is due to any doubt concerning the
possibilities of applying the methods he suggested, or
whether there was just not enough interest in the
subject. I hope this presentation will contribute
further towards encouraging linkage and other research
on 'polygenes'. It is intended purely as a methodological contribution. The experimental work involved
is no more laborious or complicated than any other
approach to quantitative genetics and should prove
fruitful.

The value of this type of work lies not only in
its importance to formal genetics and the location of
these genes. As Thoday (1966) also stressed, the
identification of genes which affect a trait would help
in studying the physiology of the trait in much the
same way as mutants at single loci. This would almost

certainly lead to a better understanding of the nature of the trait itself.

The methods suggested here are suitable for the study not only of morphological and physiological characters but variation of a continuous nature also in behavioural traits (e.g. locomotor activity) and in biochemical characteristics (e.g. the activity of a particular enzyme). We are making a beginning in this laboratory on both these types of variables.

Evidently, such methods would be useful in the study of quantitative traits in any laboratory animal. Given sufficient data, there is no reason why they should not work for other organisms too. However, on the basis of some rather simple calculations, Jayakar and Matessi (1971) found that application of these methods in man would require enormous amounts of data. Possibly, with a more efficient statistical approach and/or with a better knowledge of human polymorphic loci and their map positions, the quantity of data necessary could be reduced to some extent.

BIBLIOGRAPHY

(1) Della Croce, L., Guazzotti, G. & Jayakar, S.D. (1975). Genetic studies on the length of development in *Drosophila melanogaster*. *Atti Ass. Genet. It., 20,* 79-81.

(2) Haseman, J.K. & Elston, R.C. (1972). The investigation of linkage between a quantitative trait and a marker locus. *Behavior Genetics, 2,* 3-19.

(3) Jayakar, S.D. (1970). On the detection of linkage between a locus influencing a quantitative character and a marker locus. *Biometrics, 26,* 451-464.

(4) Jayakar, S.D. & Matessi, C. (1971). Probability of locating a gene influencing a quantitative character in man. *Excerpta Medica, 233,* 95.

(5) Thoday, J.M. (1961). Location of polygenes. *Nature, 191,* 368-370.

(6) Thoday, J.M. (1966). New insights into continuous variation. Proc. III International Congress of Human Genetics, 339-350.

Selection experiments in tribolium designed to look at gene action issues

F. D. Enfield
DEPARTMENT OF GENETICS AND CELL BIOLOGY
UNIVERSITY OF MINNESOTA, ST. PAUL, MN 55108

Since the 1970 publication of the Franklin and Lewontin paper entitled "Is the gene the unit of selection?" a number of papers (for example Kojima, 1971; Clegg et al., 1974) and the book of Lewontin (1974) have suggested the existing theory based on single gene models leads to inaccurate predictions of the effects of selection. While these publications are dealing largely with evolutionary theory in natural populations it is obvious that if the natural populations theory has serious flaws then the prediction theory for expected change as a result of artificial selection for quantitative traits also has limitations.

I would like to take just a brief amount of time to spell out what I consider the main issues to be for those who are challenging the theory and then try to relate some of my own experimental work to these issues. Let me first point out that I am far more critical of the experimental works in quantitative genetics than I am of the theory. One of the areas in which I find myself in strong agreement with the writing of Lewontin (1974) is his criticism of much of the experimental work in population genetics in terms of empirical insufficiency. As he points out far too many of the experiments were not designed with sufficient statistical power to adequately test the posed hypotheses. The same is true in much of the quantitative genetic literature. The thing that makes this most disheartening is that probably no other area of biology has professed as much knowledgeability in experimental statistics as population and quantitative genetics. Given the nature of the field, part of our problem has undoubtly been a function of inadequate person power, funds, and time. Yet despite this, we have been willing to clutter the literature with results that are either (1) meaningless because of huge standard errors or (2) suspect because of no standard errors. As a result of this, before we too quickly decide to abandon existing theory we need to look seriously at the strength of the experimental data on which any decisions of inadequacy in the theory should, at least in part, be based.

If our existing theory is either inadequate or inaccurate, I assume we are saying that one or more of the basic assumptions in the model building is either (1) too important biologically not to be included as part of the model or (2) when it is included it is handled in an invalid way. I make a slight distinction between inadequacy and inaccuracy of theory although the difference may be too subtle to worry about. Let me try to illustrate with an example. There was a time when much of the theoretical treatment of the effects of selection assumed either no linkage or linkage equilibrium. Biologically we would consider this an unrealistic assumption and this to me then means that the theory is _inadequate_ until we have evaluated the effects of these assumptions. Quantitative geneticists probably still differ considerably in their view of whether ignoring linkage has led to _inaccuracy_ in the ability of the theory to predict the effects of selection. Our models should have biological realism but some of the properties, even though real, may sometimes be ignored without seriously altering the outcome. Apparent inadequacies in the model need not necessarily lead to inaccuracy in the prediction theory.

Much of the current disenchantment with the existing theory seems to center on the role of linkage, epistasis and the interaction between the two. Certainly the results from computer simulation models that have involved tight linkage of genes with multiplicative, heterotic effects represent a very restrictive highly specialized model which may or may not be biologically realistic. Under these conditions prediction of the ultimate fate of the genome from single or two locus models may lead to quite erroneous conclusions. The same is true for Dr. Wright's multiple adaptive peaks model where the selective plateau is a function of the particular array of allelic frequencies of the interacting genes. We are plagued with the question of whether these systems are of real biological importance in the same way we are still asking the question of how important is single locus heterosis. The extent to which we see a failure in the present theory, I think, is largely a function of how important we consider these issues to be.

Rather than attempt to make any review of the experimental literature to ascertain whether the theory and experimental results are in harmony, I'm going to use the time to present some of the more interesting aspects of my own data that bear on the first question listed under the aims of this conference; i.e., "What is good and what is bad about the current theory of quantitative genetics?"

Many of you are aware that I have been involved with a long term selection experiment for pupa weight in the flour beetle that was designed initially to investigate the question of whether any appreciable fraction of the genetic variation for this trait could be attributed to overdominant genes. Time prevents me from going into detail concerning many of the specifics of the experiment but they can be found in the literature in a number of places (Enfield, *et al*., 1966, Enfield,

1972, Enfield, 1974). The main selection experiment has now progressed through 109 generations.

The base populations for all the lines I will talk about were established from the cross of two highly inbred lines. We had several reasons for doing this. First, for interpreting some of the experimental results we thought it would be advantageous to be able to start with allelic frequencies of .5 for those genes that were segregating. Secondly, we were avoiding the complication of multiple alleles in interpreting the data. Thirdly, we had hoped that the inbreeding process would have served as a screen to minimize reproductive fitness problems. In essence we were trying to get our base populations to conform as closely as possible to the assumptions often found in some of the theoretical derivations. In retrospect it turns out we were being too idealistic on a couple of counts. First, our experience over the last 15 years has been that no matter how long and hard we inbreed in our experimental material, we always have far more variation maintained than the inbreeding theory would predict. Probably directly related to this, we have been confronted with real reproductive problems in advanced generations despite the inbreeding process suggesting the natural selection was probably important in maintaining variability within the initial inbred lines.

Throughout the course of the experiment selection has always been for pupa weight. We were interested in a trait that was moderately heritable and showed some heterosis. In addition to pupa weight, data on two reproductive parameters have been collected; i.e., percent sterility and number of live progeny per fertile mating. Complete pedigree data are available on all individuals. Effective population size based on calculated coefficients of inbreeding has been approximately 100 in all lines.

I will first summarize some of the statistical estimates of population parameters obtained in the early generations to serve as a base for making comparisons as the experiment has progressed.

Table 1 gives the pooled estimates of heritability for the first thirteen generations of the experiment along with the realized heritability for the same period.

TABLE I

Estimated and Realized Heritabilities for
Pupa Weight (Generations 0-12)

Method of analysis	Population	Male data	Female data
Realized heritability	S_1	.37±.03	
	S_2	.34±.03	
Components of variance $4S/(S+D+W)$	S_1	.50±.10	.23±.10
	C_1	.28±.09	.31±.10
	S_2	.31±.08	.28±.09
	C_2	.37±.09	.26±.09
	Pooled[+]	.36±.04	.27±.05
Sire-male offspring regression	Pooled	.34±.05	
Sire-female offspring regression			.36±.05

[+]Pooled estimate was obtained by weighting the four population estimates by the inverse of the variance.

These results provide a check on the predictability of short term response to selection in a population with no previous history of selection. There is nothing particularly profound in these results except for the remarkably close observed and predicted values which certainly gives credence to the theory under these conditions. I should point out that S_1 and S_2 are replicate populations selected for heavy pupa weight, C_1 and C_2 are stabilizing selection populations where individuals closest to the median of the half-sib family are selected. I will have more to say about the C-populations a bit later.

Figure 1 summarizes the pattern of response for the S and

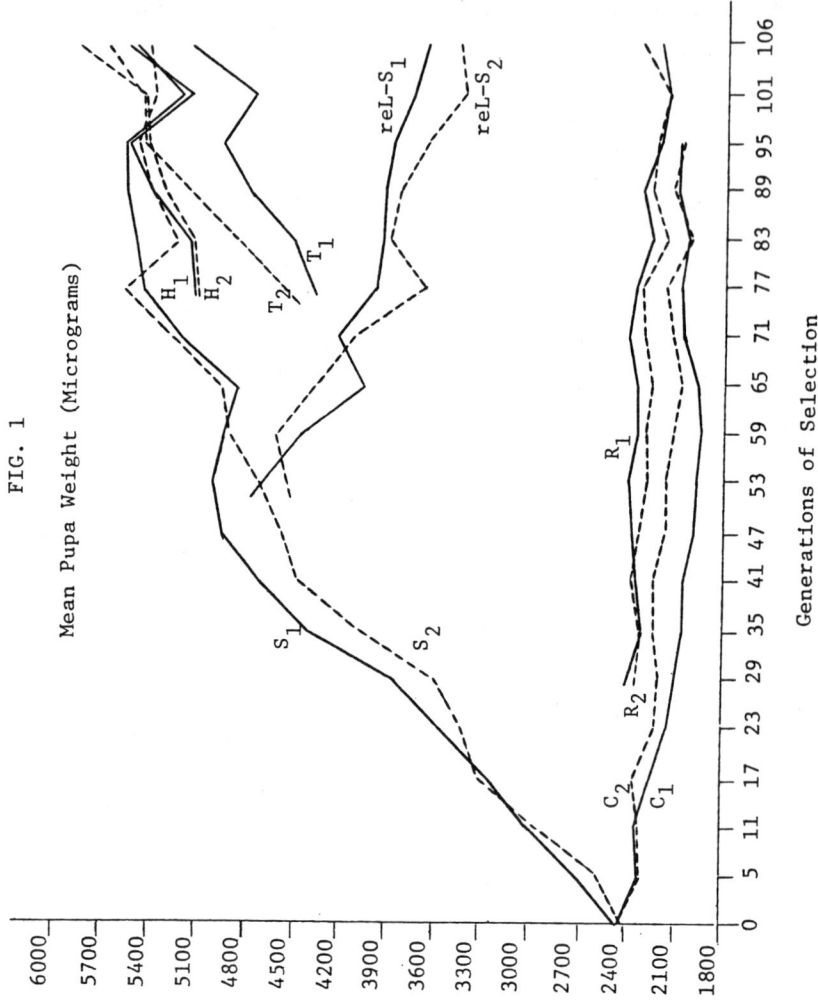

FIG. 1

Mean Pupa Weight (Micrograms)

Generations of Selection

C-populations and four additional populations started in later
generations. The R-populations are random bred, random selected
controls and the rel-S are relaxed selection lines derived from
the S-populations. The total response in the S-populations has
been about 3000 micrograms starting from a base of about 2450
micrograms. This represents a change of approximately 26
genetic standard deviations and 15 phenotypic standard deviations
in terms of base population estimates. There has been a very
slight but nonsignificant response in both S-populations since
generation 75. The relaxed S-populations have shown a rapid and
continuing decline since their initiation in generation 52.

The T-populations were established from sampling in the rel-S
populations after 22 generations of relaxed selection. The H-
populations originated from a cross of the S_1 and S_2 populations.
Selection in both the H and T-populations has been for heavy pupa
weight. The H-populations have now surpassed the S-populations in
both replicates indicating genetic differences in the S-populations
in advanced generations despite their common origin. One of the
T-populations, T_2, has now surpassed S_2 and is still responding to
selection. This supports the argument that at least part of the
strong negative correlation between pupa weight and fitness was
due to linkage effects as well as pleiotropy.

The long continuing response to selection in these experiments
is in direct contrast to nearly all such experiments reported prior
to the initiation of this experiment where the common pattern was
to reach plateaus in 30 generations or less. I interpret this
striking difference to be largely a function of the large
effective population sizes employed in our work. This has been a
critical aspect of this project since we have been attempting to
maintain any variation for overdominant genes which might exist
while at the same time fixing additive and dominant genes.

Perhaps the most serious questions concerning the validity of
the theoretical prediction equations for expected response to
selection are associated with populations with long selection

histories. Table II gives the estimates of heritability in the two
S-populations for a 12 generation period in the late generations
when there was no longer any response to selection despite an
effective selection differential.

TABLE II

Advanced Generation Estimates of Heritability for
Pupa Weight in S-Populations (Generations 84-95)

Population	Male data	Female data
S_1	.24±.09	.32±.09
S_2	.10±.09	.22±.09
Pooled	.17±.06	.27±.06

Estimates are pooled from components of variance and parent-
offspring regression analyses.

Does this then represent an inadequacy in the theory? To the
extent that we do not have adequate prediction equations that
allow us to take into account the joint effects of artificial and
natural selection, I think it does. We know that opposing
natural selection is a strong force at this stage but I know of no
way to assess whether it accounts for the complete lack of response.

One of the most interesting pieces of data that we have that
serves as an experimental check on selection theory is the comparison
of the C and R-populations over the course of the experiment. In
selection for an intermediate optimum for a trait somewhat correlated with fitness, selection should lead toward fixation for all
genes affecting the trait (Robertson, 1956, Wright, 1935) and, of
course, as a function of this lead to a reduction in additive
genetic variance. On the other hand, Waddington (1957) has argued
that selection for an intermediate may be selecting for genes that
increase the buffering capacity of the genotype. Environmental
variance may be reduced in this way. The genetic variation that

SELECTION EXPERIMENTS IN TRIBOLIUM

is affected may be genes that affect other than the character in question. The results from 95 generations of stabilizing selection are given in Table III.

TABLE III

Variances and Heritabilities for Pupa Weight
in C and R-Populations

Population	Parameter	Pooled Generation means	Regression of parameter on time
C	σ_p^2	36543±498	− 148± 31
	σ_g^2	10896±350	− 74± 31
	h^2	.29±.01	−.001±.001
R	σ_p^2	43854±188	23± 24
	σ_g^2	11108±349	11± 53
	h^2	.25±.01	000±.001

Note that there is a significant reduction in additive genetic variance and phenotypic variance. The reduction in the environmental fraction of the total phenotypic variance is about the same magnitude as the reduction in additive genetic variance supporting at least in part the Waddington proposal. A reduction of about 50 micrograms per generation in additive genetic variance would be expected in both the C and R-populations as a result of drift assuming initial allelic frequenices of .5. The data certainly does not provide evidence that stabilizing selection will be a very strong force in changing either the heritability or additive genetic variance for the trait. This would argue that if the Wright, Robertson theories are correct we must be dealing with a very large number of genes with very small effect. The most striking effect of the stabilizing selection was its effect on both measures of reproductive performance. These results are summarized in Table IV. Certainly these results suggest that

TABLE IV

Reproductive Data in C and R-Populations

Population	Number of Generations	Mean number of progeny per fertile mating	% Sterile matings
C_1	96	28.7±.4	21.8±1.0
R_1	68	22.1±.3	29.8±1.3
C_2	96	27.1±.4	16.9±1.1
R_2	68	23.7±.2	34.0±1.4

selection against extreme deviates for pupa weight has a beneficial effect on reproductive performance. This is consistent with the observations from directional selection where we observe significant changes in measures of fitness once a population has been selected beyond its normal range. However, within the normal range it appears that it is also reproductive advantageous to reduce the variation in the metric trait. There must be some equilibrium in nature, however, beyond which it is no longer advantageous to reduce variation for the metric trait or we would not expect our artificial selection to be effective.

Eariler in my presentation I emphasized the fact that if multiple peak epistasis is an important genetic property of a species then our existing single gene theory has limited predictive value unless we know the allelic frequencies for all interacting genes and their mode of interaction. The question of the reality and widespread importance of multiple peak systems has interested me for a long time. The presence or absence of multiple peak epistasis and overdominance are two of the most significant issues relative to gene action in the decision making on alternative

SELECTION EXPERIMENTS IN TRIBOLIUM

breeding and selection procedures for improvement of quantitative traits. We have some recent data that has been stimulating enough for us to embark on a new series of experiments designed to look at the question of the importance of gene interactions on the ultimate plateau attained for pupa weight selection. Preliminary indications that the S_1 and S_2-populations may be approaching different plateaus as a result of multiple peak epistasis have been provided by experiments designed to look at heterosis resulting from a cross of the two populations. In these investigations, S_1 and S_2 males were each crossed to both S_1 and S_2 females. The mean pupa weights for this experiment are shown in Table V. When the $S_1 \times S_2$ and $S_2 \times S_1$ means are compared with

TABLE V

Heterosis for Pupa Weight in Long Term Selected Populations

Population	Male data	Female data
$S_1 \times S_1$	5090±20	5234±23
$S_2 \times S_2$	5095±20	5339±25
$S_1 \times S_2$	4724±19	4840±23
$S_2 \times S_1$	5017±19	4855±22
Percent heterosis	−4.4	−8.3

the $S_1 \times S_1$ and $S_2 \times S_2$ means, a pattern of negative heterosis is apparent. In contrast, the original inbred lines showed positive heterosis when crossed (Table VI) indicating that, on average, the genes in the initial population were showing some directional dominance for heavier weight.

A change from positive to negative heterosis could arise if recessive genes were segregating at low allelic frequencies in the initial population and after a long period of selection were the major source of variation remaining. This would be unlikely,

TABLE VI

Heterosis in Crosses of Inbred Lines to Produce Base Populations

Population	Male data	Female data
CSI-5	1900±43	2071±31
CSI-10	2377±53	2514±82
5x10	2504±27	2558±59
10x5	2333±53	2471±33
Percent heterosis	13.0	9.7

however, since the populations were established by crossing two highly inbred lines. With this background, we would not expect a change in the <u>direction of heterosis</u> after the long selection history unless either (1) epistasis is important or (2) the pleiotropic effects of genes for pupa weight and fitness are such that natural selection favors the genes showing dominance for smaller size, thus keeping them segregating in the population, while artificial selection has been able to fix the genes where dominance for heavier weight is observed. While we cannot completely rule out the second possibility, we do observe some inbreeding depression in each of the S-populations in advanced generations indicating that <u>within</u> each S-population there are still some genes segregating which on the average show directional dominance for heavier weight.

To argue that the populations differ in epistatic gene complexes requires that either the population sizes were small enough so that drift enabled selection for different gene combinations in the two populations <u>or</u> the numbers of genes involved is so large that the magnitude of drift need not be very great to allow some differentiation. The fact that progress from selection continued for a minimum of 75 generations supports the hypothesis that a large number of genes are affecting the trait and could provide the opportunity for a tremendous number of nonallelic gene interactions.

A logical follow-up to the heterosis study might have been to determine whether a further breakdown would have been observed in the F_2. We have chosen not to do this, however, since it is apparent that we would not be able to distinguish an F_2 breakdown due to epistasis from a decline from natural selection resulting from the strong negative correlation between high pupa weight and fitness that exists in the populations. Rather we have chosen to initiate a set of replicated populations where the major difference among the populations is initial allelic frequencies. This will be accomplished by crossing and backcrossing inbred lines derived from our own selected populations and another selected population provided by Dr. Earl Bell at Purdue University. The basic experimental plan is to select for increased weight until plateaus are reached. It is likely that the rate of response will be different in the populations since we know with or without epistasis that the response at any point of time is a function of allelic frequencies. Thus the plateau limit becomes the critical test for epistasis. We have chosen to derive lines from populations with long selection histories to avoid the tedious process of eliminating the background variation of additive genes. Obviously, replication and population size are important experimental considerations for this experiment.

In summary, my own view of the status of the field is not one of pessimism that says our existing theory has no relevance to the real world. This is not to say that there remains no work for the theoreticians. The fact that many of the questions remain unanswered that were first posed by pioneers in the field emphasizes that it may be adequate experimentation that is lacking. I do not see the breakthroughs on the horizon via molecular genetics techniques that some of you may envision. This view is in part a function of my own biases concerning the nature of quantitative genetic variation. Basically, I still view many of the traits which are important from the plant and animal breeding point of view to be affected by large numbers of genes with small effects. If this basic premise is in fact correct, then there still remains plenty of opportunity to move forward via our traditional, but hopefully somewhat more enlightened approaches.

ACKNOWLEDGMENT

Supported by NSF grants G-1238, GB-5987 and USPHS grant GM-16074.

BIBLIOGRAPHY

(1) Clegg, M.T., R.W. Allard, and A.L. Kahler (1972). Is the gene the unit of selection? Evidence from two experimental plant populations. PNAS 69:2474.

(2) Enfield, F.D. (1972). Patterns of response to 70 generations of selection for pupa weight in Tribolium. Proc. 21st Animal Breeders Roundtable. pp. 52-69.

(3) Enfield, F.D. (1974). Recurrent selection and response plateaus. Proc. 1st World Congress Animal. Genetics 1:365.

(4) Enfield, F.D., R.E. Comstock, and O. Braskerud (1966). Selection for pupa weight in Tribolium castaneum. I. Parameters in base populations. Genetics 54:123.

(5) Franklin, I., and R.C. Lewontin (1970). Is the gene the unit of selection? Genetics 65:707.

(6) Kojima, Ken-ichi (1971). Is there a constant fitness value for a given genotype? No! Evolution 25:281.

(7) Lewontin, R.C. (1974). The Genetic Basis of Evolutionary Change. Columbia Univ. Press, New York.

(8) Robertson, A. (1956). The effect of selection against extreme deviants based on deviations or on homozygosis. J. Genet. 54:236.

(9) Waddington, C.H. (1957). The Strategy of the Genes. George Allen and Unwin Ltd. London.

(10) Wright, S. (1935). The analyses of variance and correlation between relatives with respect to deviations from an optimum. J. Genet. 30:243.

Multiple-trait selection experiments: Current status, problem areas and experimental approaches

P. Jeffrey Berger
DEPARTMENT OF ANIMAL SCIENCE
IOWA STATE UNIVERSITY, AMES, IA 50011

INTRODUCTION

In species of economic importance, the value of a potential replacement individual usually is a function of several quantitative traits. Those traits that determine aggregate breeding value may differ in their degree of genetic determination and may be either positively or negatively correlated, genetically and phenotypically.

Selection among individuals usually involves many traits. The process of selection, both voluntary and involuntary, must ultimately reduce to the classification of individuals into one of two categories, the selected or the culled. To determine the relative role of each trait and the role of an entire set of traits in selection, it is necessary to consider the interrelationships among those traits considered in the selection process.

Selection experiments have been used to study the effectiveness of selection and to estimate "realized" genetic parameters. Such estimates are used to estimate the genetic parameters in a large random-mating population. The theory of 'correlated responses' is well known and has been reviewed by Falconer (1960a). In this theory, the genetic correlation between the two traits plays an important part and determines the predicted pattern of direct and indirect responses found in different experiments. Hill (1971, 1972) discussed several methods of estimating genetic parameters and the variances of such estimates from single-trait and double-selection experiments. Berger and Harvey (1975) gave procedures for the estimation of realized heritabilities and genetic correlations from populations selected simultaneously for two traits.

Estimates of realized genetic correlations from selection experiments have been characteristically unstable. Falconer (1960b) selected mice for growth on high and low planes of nutrition and observed the correlated responses on the alternate nutritional level. The realized genetic correlations were equal for the first four generations of selection (0.67, 0.65) but were markedly different for generations 5 to 13 (1.25, -0.02). This discordance of the pattern of correlated responses from expectation was attributed to changes in the basic parameters, due to the selection applied, and to large changes in the phenotypic standard deviations.

Asymmetry of the realized genetic correlations also was observed by Bell and McNary (1963), who selected *Tribolium castaneum* for increased pupa weight in both a wet and a dry environment, and by Yamada and Bell (1963) where selection was for increased and decreased 13-day larval weight under good and poor nutritional levels. Siegel (1962) found a realized genetic correlation of 0.55 when selection was for body weight and 0.45 when selection was for breast angle in poultry. In another selection experiment with poultry, Nordskog and Festing (1962) observed asymmetry of the realized genetic correlation between body and egg weight when either the direction of selection or the trait being selected was considered. In both poultry studies, the asymmetry was attributed to differing genetic variances or heritabilities for the two traits.

The purpose of this paper will be to present selection index procedures that may be used to estimate realized genetic parameters under a multiple trait objective and to describe their application to an experiment designed to evaluate the efficiency of index selection.

EXPERIMENTAL PROCEDURE

Tribolium castaneum were selected for pupa weight, as a measure of growth, and family size, as a measure of reproduction. The basic design of the experiment is described in Table I. Mass selection in each line was based on the designated criterion. Selection for family size in the family size line and the index line was accomplished by mating all females from the previous generation and retaining families from the highest ranking parents on the designated criterion. Males were chosen strictly at random in these two lines. The index was constructed to allow an equal contribution by both traits to the response in aggregate genotype.

TABLE I

Design of Experiment

| Line | Selection Criterion | Matings/Gen. | | Individ. |
		Males	Females	
$PW^+FS^°$	Pupa Weight	18	54	324
$PW^°FS^+$	Family Size	54	162	324
Index	I = 0.0024PW + 0.094FS	54	162	324
Control	Random	54	54	324

All populations were cultured on standard medium (whole wheat flour enriched with 5% dried brewer's yeast) in an environmental chamber at $38 \pm 1°$ C and $70 \pm 5\%$ relative humidity. The base population originated from 24-hour egg lays taken from females obtained from a stock with a pearl eye mutant gene maintained in the laboratory. The stock is maintained by transferring 200 unsexed pupa every 30 days to new media. Individuals obtained from 24 hour egg lays of this random mating population were divided equally into the four lines. Families were cultured for 19 days in a 3/4 oz. creamer containing 2 g. medium. Three males and three females were randomly chosen from each family at 19 days, weighed individually, and placed in separate bottles. Matings in each generation were randomly divided into three sets, with matings and egg lays started across a 3-day period to distribute the work.

SELECTION INDEX PRINCIPLES

The criterion upon which the selection decision is made arises from a condensation of many factors. A linear function of total productivity might be represented as

$$I = \sum_{i=1}^{n} \alpha_i X_i + Z \quad (1)$$

where I is the ultimate measure upon which selection is practiced, the α_i describe the attention given to the ith trait, X_i is the phenotypic measure of the ith trait and Z is the collection of remaining factors involved in selection.

In the absence of knowing I, it is possible to construct an index

$$I^* = \sum_{i=1}^{n} a_i X_i \quad (2)$$

such that the r_{II^*} is maximized. The relative attention given in the intended index (α_i) would be identical to the relative attention in the actual index (a_i) if Z were independent of the X_i's. Allaire and Henderson (1966) and others have shown that to maximize this correlation the weights of the index I^* are obtained by solution of the following equations, where s_i is the observed selection differential for the ith trait in standard deviation units and $r_{P_1 P_2}$ is the phenotypic correlation:

$$\begin{bmatrix} 1.0 & r_{P_1P_2} & \cdots & r_{P_1P_n} \\ \vdots & \vdots & & \vdots \\ r_{P_nP_1} & r_{P_nP_2} & & 1.0 \end{bmatrix} \begin{bmatrix} a_1 \\ \vdots \\ a_n \end{bmatrix} = \begin{bmatrix} s_1 \\ \vdots \\ s_n \end{bmatrix} \quad (3)$$

This technique was first used by Dickerson et al. (1954) to examine selection in retrospect for traits in swine.

With selection based on (2), it can be shown that

$$E(\Delta G_i) = \text{Cov}(G_i, I) \frac{(I-u)}{\sigma_I^2} .$$

Dividing both sides of the equation by the phenotypic standard deviation for the ith trait and evaluating the covariance term

$$E(\Delta G_i / \sigma_{P_i}) = a_i g_i^2 + a_j r_{G_i G_j} g_i g_j ,$$

where

$$g_i = \frac{\sigma_{G_i}}{\sigma_{P_i}} .$$

Thus from each experimental line selected simultaneously for two traits, there are two equations for expected genetic progress and three unknown genetic parameters. Equations from two experimental lines may be combined as follows:

$$\begin{bmatrix} a_{11} & a_{21} & 0 \\ 0 & a_{11} & a_{21} \\ a_{12} & a_{22} & 0 \\ 0 & a_{12} & a_{22} \end{bmatrix} \begin{bmatrix} g_1^2 \\ g_1 r_{G_1 G_2} g_2 \\ g_2^2 \end{bmatrix} = \begin{bmatrix} \Delta G_{11} \frac{\sigma_I^2}{s_I} \\ \Delta G_{21} \frac{\sigma_I^2}{s_I} \\ \Delta G_{12} \frac{\sigma_I^2}{s_I} \\ \Delta G_{22} \frac{\sigma_I^2}{s_I} \end{bmatrix} \quad (4)$$

where a_{ij} is the attention given to the ith trait in the jth line, and the ΔG_{ij}'s are the estimates of genetic response in standard deviation units. By considering these equations as

$$X\beta = Y ,$$

then estimates of the realized genetic parameters may be obtained from the solution of the least squares equations

$$\hat{\beta} = (X'X)^{-1}X'Y .$$

It has been shown (Berger and Harvey, 1975) that the ratio, σ_I^2/s_I, equals one for each experimental line and therefore does not affect the solution for the realized genetic parameters. However, the a_i's from the index in retrospect are proportional to the α_i's of the intended index. Because the intended index may be different from one experimental line to another, this proportionality constant will be different for each experimental line. This suggests that the magnitude of the a_i's from one experimental line to another will influence the accuracy of the estimates combining lines.

RESULTS

The weight given each trait through selection is shown in Table II. To facilitate comparison, the actual attention achieved in selection for pupa weight has been expressed relative to the attention to family size. A positive sign indicates positive attention in selection, and a negative sign indicates negative attention in selection. The sign designating attention in selection does not necessarily follow the direction of the selection differential for individual traits. This is due to the correlation among traits, although for the two traits considered here, the attention in selection did follow the direction of the selection differentials.

TABLE II

Relative Attention in Selection for Pupa Weight (a_1') and Family Size (a_2')[a]

	Line					
	PW$^+$	FS$^\circ$	PW$^\circ$	FS$^+$	I = b_1 PW + b_2 FS	
Gen	a_1'	a_2'	a_1'	a_2'	a_1'	a_2'
1	0.9	1.0	0.2	1.0	0.6	1.0
2	58.6	1.0	0.2	1.0	0.7	1.0
3	193.4	1.0	0.2	1.0	0.5	1.0
4	3.7	-1.0	0.3	1.0	0.5	1.0
5	2.8	1.0	0.2	1.0	0.5	1.0
6	86.5	1.0	0.2	1.0	1.2	1.0
7	6.6	1.0	0.1	1.0	0.9	1.0
8	4.9	-1.0	0.1	1.0	0.3	1.0
9	2.6	-1.0	2.2	1.0	18.2	1.0
10	65.6	1.0	18.2	1.0	0.5	1.0
11	94.9	-1.0	0.0	1.0	0.6	1.0
12	10.6	-1.0	0.1	1.0	0.6	1.0
13	2.1	-1.0	-0.0	1.0	0.4	1.0
14	2.8	1.0	0.2	1.0	0.7	1.0
15	3.9	-1.0	0.0	1.0	0.6	1.0
16	4.4	-1.0	-0.0	1.0	0.4	1.0
Ave.	12.1	-1.0	0.1	1.0	0.6	1.0

[a] a_1' and a_2' were obtained by expressing a_1 and a_2 proportional to the absolute value of a_2.

As expected for single trait selection, pupa weight received considerably more attention than family size in the pupa-weight line. There was considerable variation in the attention given to pupa weight from generation to generation. In later generations of the experiment, the attention to family size became negative. In the family-size line, there was less variability in the attention given to family size, and pupa weight received positive attention throughout the experiment. Selection on the index, in the index line, was effective in giving positive attention to both traits and more attention to pupa weight than was achieved in the family-size line.

TABLE III

Relative Attention in Selection of Males and Females for Pupa Weight (a_1) and for Family Size (a_2)

Sex	PW+ a_1	FS° a_2	PW° a_1	FS+ a_2	$I = b_1$ PW + b_2 FS a_1	a_2
Males	18.2	1.0	-18.4	-1.0	19.6	1.0
Females	6.4	-1.0	0.3	1.0	1.1	1.0

The attention to pupa weight and to family size was different in the selection of males than females (Table III). Despite the random selection of males in the family-size and index lines, pupa weight received about the same relative attention to family size in all three lines. However, negative attention was given to pupa weight in selection of males in the family-size line. Pupa weight of females received positive attention, and for both sexes combined, the overall attention for the experiment was positive.

The responses of pupa weight in Figure 1 are expressed as deviations from the control population. Direct response of pupa weight in the pupa-weight line was 136 ± 3 ug per generation. This was followed by the intermediate response of 35 ± 2 ug in the index line. The indirect response in pupa weight in the family-size line was 9 ± 2 ug per generation.

Figure 2 shows the responses in family size expressed as deviations from the control population. The correlated response in family size, in the pupa-weight line, resulted in the loss of 1.11 ± 0.09 individuals per generation. This continued loss resulted in the eventual extinction of this line by generation 16 because of this loss in fertility. Response in family size increased 0.34 ± 0.12 and 0.27 ± 0.11 individuals per generation in the family-size and index lines, respectively.

MULTIPLE TRAIT SELECTION EXPERIMENTS

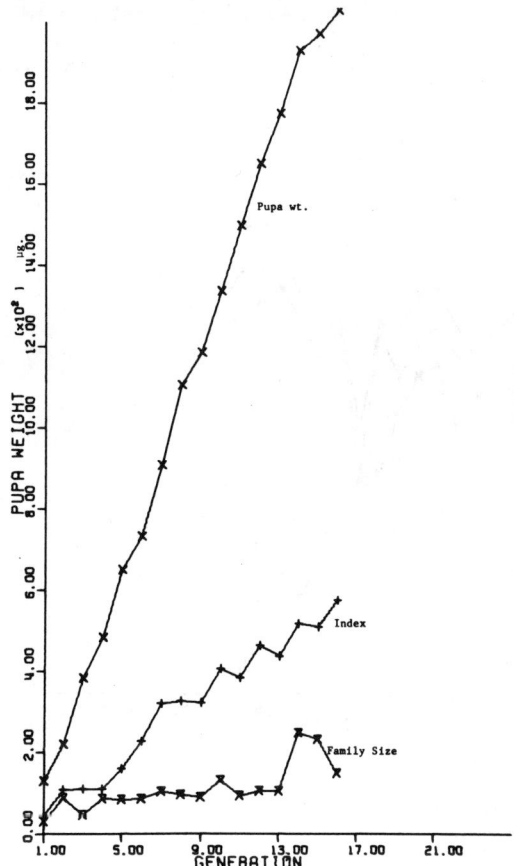

FIGURE 1. Response in Pupa Weight

Estimates of the realized genetic parameters in Table IV were obtained by the procedure described earlier. Estimates describing selection for increased growth were obtained by combining the pupa weight and index line. The family size and index line describe selection for improved reproduction. In general, the realized heritability estimates for both traits were lower selecting for increased reproduction than those selecting for increased growth. The differences between the

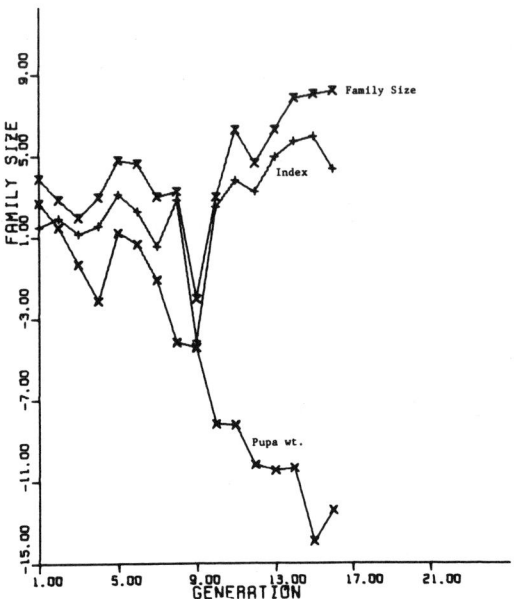

FIGURE 2. Response in Family Size

realized heritabilities and base population estimates (Table V) could be due to sampling error, although there was an indication that selection for increased growth yielded a higher direct response, while selection for increased reproduction yielded less direct response than expected.

Realized genetic correlations agreed least with expectations. The realized genetic correlation was larger selecting for increased growth and positive selecting for increased reproduction. This could be due to sampling error or further evidence supporting conclusions from other multiple trait experiments (Berger and

TABLE IV

Realized Heritability and Genetic Correlation Estimates[a]

Selection for High Pupa Weight		Selection for High Family Size	
Pupa Weight	Family Size	Pupa Weight	Family Size
0.36	−.43	0.33	0.03
	0.11		0.06

[a]Heritabilities are on the diagonal and the genetic correlation above the diagonal.

Harvey, 1975; Rutledge et al., 1973). Under the antagonistic selection practiced in these experiments, the realized genetic correlation was overestimated, suggesting that the pleiotropic effects of genes may be more powerful in retarding response in aggregate genotype than current theory would allow. On the other hand, some experiments have shown no change in the genetic correlation between two selected traits. For example, Bell and Burris (1973) concluded that either initial upward bias, a gradual decline in the genetic variances, or both accounted for predicted changes consistently greater than those observed.

TABLE V

Base Population Parameter Estimates[a]

Trait	Pupa Wt.	Family Size
Pupa Wt.	0.33 ± 0.04	−.17
Family Size	−.05	0.10 ± 0.05

[a]Heritabilities are on the diagonal, genetic correlation above the diagonal and phenotypic correlation below the diagonal.

There are several difficulties with the procedure used. It was not possible to obtain standard errors for the realized genetic parameters. On the other hand, it is impossible to determine the effect of antagonistic selection on indirect responses with single trait selection. In a recent Monte Carlo simulation study, Bruns and Harvey (1976) found the accuracy of realized heritability estimates depended on the attention in selection. More accurate estimates were obtained for the trait receiving the most attention in selection. In this experiment this would correspond to the estimates for pupa weight under high pupa weight selection and family size under high family size selection. In addition, Bruns and Harvey (1976) further concluded that the reduction in genetic and phenotypic variances and covariances, due to the selection of parents, biased realized heritabilities downward and the genetic correlation upward.

The generality of these results is uncertain. They do raise some question concerning the conclusions of some theoretical studies comparing the efficiency of selection for production and reproduction. In comparisons between the efficiency of selecting males and females on different indices with the efficiency of selecting both sexes on the same index, Moav and Hill (1966) and Smith (1964) concluded that the most efficient method of selection was one of splitting the original line into separate sire and dam lines, each selected on a specialized index. The correlated response of reduced fertility in the pupa weight line would suggest that it might be difficult to maintain a specialized sire line selected entirely on productivity. The index line demonstrates that index selection for both productivity and reproductivity is possible and may be used in developing specialized dam lines.

CONCLUSION

The seeming conflict between selection index principles and the results from selection experiments can be explained by two types of errors made in the development and utilization of multiple trait selection schemes. The first is that the true parameter estimates are often unknown, and the second is the failure to account for important correlated traits. The major benefit to come from multiple trait selection experiments is to identify the errors in parameter specification and the way these may change with selection. Once these errors are identified, the additional pertinent information must be considered.

ACKNOWLEDGMENTS

Journal Paper J-8651 of the Iowa Agriculture and Home Economics Experiment Station, Ames, Ia. Project 1053 as a collaborator under the North Central Regional Project NC-2, Improvement of Dairy Cattle Through Breeding.

BIBLIOGRAPHY

[1] Allaire, F. R. & Henderson, C. R. (1966). Selection practiced among dairy cows. II. Total production over a sequence of lactations. J. Dairy Sci. 49, 1435-1440.

[2] Bell, A. E. & Burris, M. J. (1973). Simultaneous selection for two correlated traits in Tribolium. Genet. Res., Camb. 21, 29-46.

[3] Bell, A. E. & McNary, H. W. (1963). Genetic correlation and asymmetry of the correlated response from selection for increased bodyweight of Tribolium in two environments. Proc. XI Int. Congr. Genet., 256.

[4] Berger, P. J. & Harvey, W. R. (1975). Realized genetic parameters from index selection in mice. J. Anim. Sci. 40, 38-47.

[5] Bruns, E. & Harvey, W. R. (1976). Effects of varying selection intensity for two traits on estimation of realized genetic parameters. J. Anim. Sci. 42, 291-298.

[6] Dickerson, G. E., Blunn, C. T., Chapman, A. G., Kottman, R. M., Krider, J. L., Warwick, E. J., Whatley, J. A., Jr., Baker, M. L., Lush, J. L., & Winters, L. M. (1954). Evaluation of developing inbred lines of swine. Mo. Agric. Exp. Stn. Res. Bull. 551.

[7] Falconer, D. S. (1960a). Introduction to Quantitative Genetics. Edinburgh: Oliver and Boyd.

[8] Falconer, D. S. (1960b). Selection of mice for growth on high and low planes of nutrition. Genet. Res. 1: 91-113.

[9] Hill, W. G. (1971). Design and efficiency of selection experiments for estimating genetic parameters. Biometrics 27: 293-311.

[10] Hill, W. G. (1972). Estimation of realized heritabilities from selection experiments. I. Divergent selection. Biometrics 28: 747-765.

[11] Moav, R. & Hill, W. G. (1966). Specialized sire and dam lines. IV. Selection within lines. Anim. Prod. 8, 375-390.

[12] Nordskog, A. W. & Festing, M. (1962). Selection and correlated responses in the fowl. Proc. XII Worlds Poult. Congr. 25-29.

[13] Rutledge, J. J., Eisen, E. J. & Legates, J. E. (1973). An experimental evaluation of genetic correlation. Genetics 75, 709-726.

[14] Siegel, P. B. (1962). A double selection experiment for body weight and breast angle at eight weeks of age in chickens. Genetics 47: 1313-1319.

[15] Smith, C. (1964). The use of specialized sire and dam lines in selection for meat production. Anim. Prod. 6, 337-344.

[16] Yamada, Y. & Bell, A. E. (1963). Selection for 13-day larval growth in Tribolium under two nutritional levels. Proc. XI Int. Congr. Genet., 256.

IV
Mathematical and Statistical Genetics

Selection with many loci and possible relations to quantitative genetics

Samuel Karlin
DEPARTMENT OF MATHEMATICS
STANFORD UNIVERSITY, STANFORD, CA 94301

ABSTRACT

One approach to the study of quantitative inheritance is through the analysis of multilocus systems. A "general multi-locus heterozygote advantage (or disadvantage)" mechanism has the following formulation. Consider an n-locus trait involving two possible alleles at each locus where the selection expression is a function of cumulative heterozygosity. Accordingly, let γ_i = the relative fitness of a genotype if i among its loci are heterozygous. Independent of the selection regime characterized by the coefficients $\{\gamma_i\}_0^n$ and the recombination process, a central polymorphism $\underline{x}^*$ always exists. Stability conditions of $\underline{x}^*$ for various forms of weak and strong epistasis and/or for cases of tight and loose linkage are indicated. Some interpretations of the results pertaining to chromosomal frequency data are extrapolated. Possible implications for problems of quantitative inheritance are also discussed.

Supported in part by NIH Grant GM 10452-13 and NSF Grant MCS75-23608.

INTRODUCTION

At this date a proper model describing the effects of separate genes on a continuously varying character in relation to selection, mating pattern and recombination forces is not available. The attempts to incorporate the genetic mechanism have relied on assumptions of additive gene contributions and global linkage equilibrium (usually tacitly postulated) which are basically inconsistent with the operation of differential selection effects.

There is a point of view suggesting that in many cases of polygenic and quantitative traits most of the variability (over 80%) can be ascribed to a few major loci (the major locus hypothesis) interacting with the environment. This does not preclude the existence of numerous genes contributing a minor part of the variability. In these cases the study of polygenic inheritance reduces to the analysis of suitable multilocus systems. Limits of multi-locus systems, ipso facto, approximate polygenic traits. Accordingly, apart from its independent value, the study of selection and recombination in the multilocus context is germaine in securing a deeper understanding of the dynamics of quantitative inheritance and gene interaction. In this paper we will report several results pertaining to a class of n loci selection models that have relevance with regard to evolutionary genetics and concomitantly in discerning characteristics of quantitative inheritance.

Our discussion divides into three sections. Section 1 offers a number of practical and natural ways by which to parameterize the recombination process. (With n loci there can exist up to $2^{n-1} - 1$ independent recombination frequencies.) The fact that selection for polygenic characters acts primarily on phenotypes, whereas segregation involves genotypes, points up the necessity for dealing jointly with an array of genotypes and phenotypes in a frame-

SELECTION WITH MANY LOCI

work that better accounts for the multifarious levels and forms of recombination and epistasis. In Section 2 we single out a class of selection regimes determined by an appropriate genotypic-phenotypic association. Section 3 describes some results, interpretations and implications bearing on the dynamic and equilibrium gamete frequency behavior. Further discussion and some speculations are presented in the concluding section.

§1. Some Representations of an n loci Recombination Mechanism

Consider a trait of a large diploid population determined at n loci with m_k possible alleles available at locus k. Unless stated otherwise, we stipulate reproduction by random mating. Let $i_0^{(k)}, i_1^{(k)}$ designate the alleles at the kth locus in gametes 0 and 1, respectively. Then chromosome (gamete - we use both terms interchangeably) types are described by n-tuples as

(1) $\quad \underline{i}_0 = (i_0^{(1)}, i_0^{(2)}, \ldots, i_0^{(n)}) \quad , \quad \underline{i}_1 = (i_1^{(1)}, i_1^{(2)}, \ldots, i_1^{(n)})$.

A typical genotype $\underline{g}$ composed of the two gametes (1) is displayed in the form

(2) $\quad g\left(\dfrac{\underline{i}_0}{\underline{i}_1}\right) = \begin{pmatrix} i_0^{(1)}, \ldots, i_0^{(n)} \\ i_1^{(1)}, \ldots, i_1^{(n)} \end{pmatrix}$

such that its allelic makeup at locus k consists of $i_0^{(k)}$ and $i_1^{(k)}$.

The <u>recombination frequencies</u> are summarized by the array of nonnegative quantities

(3) $\quad\quad\quad\quad\quad\quad R(\underline{\varepsilon})$,

where the n-tuples $\underline{\varepsilon} = (\varepsilon_1, \varepsilon_2, \ldots, \varepsilon_n)$ are generated from independent choices of $\varepsilon_i = 0$ or 1, $i = 1, 2, \ldots, n$, and these numbers are to be interpreted as follows.

The recombination event associated with $\underline{\varepsilon} = (\varepsilon_1, \varepsilon_2, \ldots, \varepsilon_n)$ means that at the positions (loci) where $\varepsilon_\nu = 1$, an exchange of genetic material occurs and subsequent Mendelian segregation produces the recombinant gametes

$$(i^{(1)}_{\varepsilon_1}, i^{(2)}_{\varepsilon_2}, \ldots, i^{(n)}_{\varepsilon_n}) \quad \text{and} \quad (i^{(1)}_{1-\varepsilon_1}, i^{(2)}_{1-\varepsilon_2}, \ldots, i^{n}_{1-\varepsilon_n}) \ .$$

Thus, the gametic output from segregation of genotype (2) comprises $(i^{(1)}_{\varepsilon_1}, i^{(2)}_{\varepsilon_2}, \ldots, i^{(n)}_{\varepsilon_n})$ and $(i^{(1)}_{1-\varepsilon_1}, i^{(2)}_{1-\varepsilon_2}, \ldots, i^{n}_{1-\varepsilon_n})$ with frequency $R(\underline{\varepsilon}) = R(\varepsilon_1, \varepsilon_2, \ldots, \varepsilon_n)$. The recombination frequencies are stipulated to obey the intrinsic relations $R(\underline{\varepsilon}) = R(\varepsilon_1, \varepsilon_2, \ldots, \varepsilon_n) = R(1-\varepsilon_1, 1-\varepsilon_2, \ldots, 1-\varepsilon_n) = R(\underline{1-\varepsilon})$ expressing the fact that two parental gametes contribute in a symmetrical manner in the segregation process; also $\sum_{\underline{\varepsilon}} R(\underline{\varepsilon}) = 1$ holds where the sum extends over all $\underline{\varepsilon} = (\varepsilon_1, \varepsilon_2, \ldots, \varepsilon_n)$, $\varepsilon_i = 0$ or 1; $i = 1, 2, \ldots, n$. The $R(\underline{\varepsilon})$ with $\underline{\varepsilon}$ varying cover all mutually exclusive recombination events.

Example 1. (Two loci, $n = 2$.) If r is the frequency of recombination between the two loci, then $R(0,0) = R(1,1) = \frac{1-r}{2}$, $R(0,1) = R(1,0) = \frac{r}{2}$.

Example 2. (Three loci, $n = 3$.) Let r be the frequency of the event of recombination between loci 1 and 2 joined with no recombination between loci 2 and 3. Let the recombination frequency between loci 2 and 3 with no recombination between loci 1 and 2 be s. Finally, let t be the frequency of simultaneous recombination between loci 1 and 2 and loci 2 and 3. With these definitions

$$R(0,0,0) = R(1,1,1) = \frac{1-r-s-t}{2}, \quad R(0,0,1) = R(1,1,0) = \frac{s}{2}$$

$$R(0,1,1) = R(1,0,0) = \frac{r}{2}, \quad R(0,1,0) = R(1,0,1) = \frac{t}{2}$$

A different parameterization of the recombination rates is more commonly used. Consider the case of 3 loci labeled A, B and C. Suppose r_1 is the crossover probability between the $\{A\}$ locus and the $\{B\}$ locus, that between the $\{B\}$ locus and

SELECTION WITH MANY LOCI

the {C} locus r_2, and that between the {A} and {C} loci r_3. The parameters in the two notations are connected by the relations $r_1 = r+t$, $r_2 = s+t$, $r_3 = r+s$.

The case of <u>no interference</u> is characterized by the equation $t = (r+t)(s+t)$ or equivalently by the relation $r_3 = r_1(1-r_2) + r_2(1-r_1) = r_1 + r_2 - 2r_1 r_2$. The situation where recombination between two adjacent positions precludes recombination in the next segment corresponds to the stipulation $t = 0$.

We single out next in the context of the general n-locus model a number of recombination frequency distributions that have intrinsic relevance.

Absolute Linkage (No Recombination)

(4) $\quad R(\underline{0}) = R(\underline{1}) = \frac{1}{2}$, $R(\underline{\varepsilon}) = 0$ for $\underline{\varepsilon} \neq \underline{0}$ or $\underline{1}$

where $\underline{0} = (0,0,\ldots,0)$ (reflects no exchange of genetic material) and $\underline{1} = (1,1,\ldots,1)$ signifying total exchange which is effectively equivalent to no exchange. We designate the <u>no recombination</u> case of (4) by $\mathscr{R}^0$.

Free Recombination: $\mathscr{R}^f$.

Here

(5) $\quad R(\underline{\varepsilon}) = \frac{1}{2^n}$ independent of $\underline{\varepsilon}$.

Where the individual loci are all located on different chromosomes then (5) applies.

The ordering of more or less recombination in the case of two loci is unambiguous since a single real recombination parameter is involved. In the situations of three or more loci, the recombination process is characterized by a vector of rates and consequently two recombination frequency prescriptions $\mathscr{R} = \{R(\underline{\varepsilon})\}$ and $\mathscr{R}' = \{R'(\underline{\varepsilon})\}$ are not always comparable. However, any partial ordering relation among natural recombination distributions certainly compels the view of $\mathscr{R}^0$ as a minimal recombination array whereas $\mathscr{R}^{(f)}$ should correspond to a maximal recombination distribution. The problem of delimiting what is "more recombination" in a multilocus framework and in delineating a natural partial ordering among recombination distributions will be elaborated in a separate publication.

Loosely Linked Blocks of Tightly Linked Genes

An appealing hybrid extension of (4) and (5) ensues when the n loci system divides intrinsically into clusters of loci consisting of $\ell_1, \ell_2, \ldots, \ell_p$ loci, $\sum_{i=1}^{p} \ell_i = n$ in the manner that the loci within each cluster are relatively tightly linked, where loose linkage or free recombination operates between the gene clusters.

Recombination Arrays Reflecting Specific Physical Characteristics of Loci

Suppose the loci are endowed with a physical arrangement along a single chromosome. We can consider independent probabilities of breaks between successive positions (the no interference postulate). Let r_i be the probability of a break between loci i and $i+1$. Then, for example, $R(\overbrace{1,1,\ldots,1}^{k},0,\ldots,0) = r_1 r_2 \cdots r_{k-1}(1-r_k)(1-r_k)\cdots(1-r_{n-1})$. Generally, $R(\underline{\varepsilon}) = \prod_{i=1}^{n-1} r_i^{|\varepsilon_i - \varepsilon_{i+1}|}(1-r_i)^{1-|\varepsilon_i - \varepsilon_{i+1}|}$.

It is possible to generate a hierarchy of recombination distributions that take account of the natural physical ordering in the loci points through notions of renewal processes, order statistics, and counter processes; (several such constructions in terms of renewal processes were studied by Owens; e.g., consult the book of Bailey (1961) for elaborations on this theme and other references.

§2. The Nature of the Selection Regime

The study of linkage and selection in multilocus multiallele systems has concentrated on four main categories of models delineated as follows.[*]

(i) Selection engendered by nonepistatic effects over loci;

---

[*] For a review of the corresponding two-locus versions, see Karlin (1975).

SELECTION WITH MANY LOCI

(ii) Phenotypic selection values based on certain classes of phenotypic-genotypic associations;

(iii) Genotypic fitness arrays invariant under natural group transformations;

(iv) Hybrid versions of the selection regimes of (i) to (iii).

An extensive discussion of nonepistatic models will appear in Karlin (1977)(1978) and Karlin and Liberman (1977). A notable contribution in this vein is the work of Roux (1974). In this paper we concentrate on some cases of category (ii) and defer the discussion of (iii) and (iv).

Phenotypic Model based on Cumulative Heterozygosity Involving n loci (A Generalized Heterozygosity Mechanism).

Consider an n loci trait involving <u>two</u> possible alleles at each of the different loci. There are $n+1$ quantities γ_i, $i = 0, 1, 2, \ldots, n$, with the following interpretation:

(6) γ_i = the relative fitness of a genotype if i among its loci are heterozygous.

For $n = 2$ this phenotypic selection relationship produces the classical Lewontin-Kojima symmetric viability regime (1960); see also Kojima and Lewontin (1970). For $n = 3$ involving alleles $\{A,a\}$, $\{B,b\}$, $\{C,c\}$ at the respective loci the fitness matrix takes the form

	ABC	ABc	AbC	Abc	aBC	aBc	abC	abc
ABC	γ_0	γ_1	γ_1	γ_2	γ_1	γ_2	γ_2	γ_3
ABc	γ_1	γ_0	γ_2	γ_1	γ_2	γ_1	γ_3	γ_2
AbC	γ_1	γ_2	γ_0	γ_1	γ_2	γ_3	γ_1	γ_2
Abc	γ_2	γ_1	γ_1	γ_0	γ_3	γ_2	γ_2	γ_1
aBC	γ_1	γ_2	γ_2	γ_3	γ_0	γ_1	γ_1	γ_2
aBc	γ_2	γ_1	γ_3	γ_2	γ_1	γ_0	γ_2	γ_1
abC	γ_2	γ_3	γ_1	γ_2	γ_1	γ_2	γ_0	γ_1
abc	γ_3	γ_2	γ_2	γ_1	γ_2	γ_1	γ_1	γ_0

(6a)

Some facets of the above 3 loci selection model were studied by Feldman, Franklin and Thomson (1974). They refer to this case as the completely symmetric three-locus model. Our emphasis is that the model of (6a) fits the framework of a phenotypic fitness expression determined by cumulative heterozygosity of a genotype. Based on this principle, a number of extensions and insights are readily forthcoming as reported in this and subsequent works.

A genotypic-phenotypic association of wide scope and interest assigns a fitness value $\varphi(g)$ to a genotype g such that

(7) $$\varphi(g) = \gamma(k_g, \ell_g, m_g)$$

where k_g = cumulative number of heterozygous loci of the genotype g ;
ℓ_g = cumulative value of favorable alleles in g ;
m_g = cumulative number of homozygous loci in g carrying only favorable alleles (or equivalently the number of recessive loci).

SELECTION WITH MANY LOCI

With different numbers of alleles at separate loci more flexibility in (7) is available.

The influences of the values of k_g, ℓ_g and m_g on fitness may be counteractive and/or reinforcing. To sort out the equilibrium and dynamic possibilities in their dependence on the effects of favorable alleles, on the locations and extent of heterozygous loci and on the distribution of recessive genes coupled to linkage relationships is an important aspect in understanding multilocus selection balance and in interpreting observations of gamete frequency data.

§3. Results and Interpretations

Throughout the remainder of this article we concentrate on the phenotypic selection model of (6) having two possible alleles A_i and a_i at locus i, $i = 1,2,\ldots,n$. With reference to the more general fitness model implicit in (7) we refer to Karlin (1978). The selection regime is characterized by an array $\{\gamma_i\}_{i=0}^{n}$ where γ_i is the fitness of any genotype carrying i heterozygous loci.

A gamete type can be denoted by an n-tuple $\underline{\xi} = (\xi_1, \xi_2, \ldots, \xi_n)$ where each $\xi_i = 0$ or 1 (1 signifying the presence of A_i and 0 stands for a_i). Henceforth, we arrange the gamete types in lexicographic order. Thus, for 3-loci, the ordering is $(1,1,1)$, $(1,1,0)$, $(1,0,1)$, $(1,0,0)$, $(0,1,1)$, $(0,1,0)$, $(0,0,1)$, $(0,0,0)$.

A genotype is composed of two gametes (say $\underline{\xi}$ and $\underline{\xi}'$) and will be put in the notation $g(\underline{\xi},\underline{\xi}') = \underline{\xi}/\underline{\xi}'$. With each genotype $g = \underline{\xi}/\underline{\xi}'$ we can associate $\underline{\eta} = (\eta_1, \eta_2, \ldots, \eta_n)$ where $\eta_i = 1$ or 0 accordingly as g is heterozygous or homozygous at locus i, respectively.

Almost all the results we describe can be adapted to the case where the fitness $\gamma(\underline{\eta})$ may depend both on the extent of heterozygosity and the locations of the heterozygous loci. Moreover, there are natural analogs allowing multiple alleles at the separate loci.

The strong symmetry with respect to loci and allelic effects entails that the gamete frequency vector

(8) $$\underline{x}^*(\underline{\xi}) = \frac{1}{2^n} \quad \text{independent of } \underline{\xi}$$

is an equilibrium extant for any recombination distribution. We refer to (8) as the <u>central polymorphism</u>. The conditions for local stability of $\underline{x}^*$, depending on the fitness regime $\{\gamma_i\}_0^n$, are delineated later. Note that the central equilibrium manifests zero linkage associations of all orders (2nd and higher). We will later consider the relevance of other equilibrium possibilities involving a partial set of gametes displaying intermediate measures of complementarity in their gametic configurations.

<u>Stability of the central equilibrium for different recombination distributions</u> $\mathscr{R} = \{R(\underline{\varepsilon})\}$ <u>and fitness prescriptions</u> $\{\gamma_i\}_0^n$. We start with a general result.

<u>Result I</u>. <u>Consider a selection regime</u> $\{\gamma_i\}_0^n$. <u>Where the central equilibrium</u> $\underline{x}^*$ <u>is stable in the case of no recombination</u> $\mathscr{R}^{(0)}$, <u>see (4)) then</u> $\underline{x}^*$ <u>is the unique globally stable equilibrium persisting for any recombination distribution</u>.

The foregoing rigorous finding is concordant with the dichotomy on the nature of polymorphism enunciated in Karlin (1975).

The following two results give precise conditions assuring stability of the central equilibrium for tight and loose linkage, respectively.

<u>Result II</u>. <u>The exact conditions for stability of the central equilibrium</u> $\underline{x}^*$ <u>under absolute linkage are that the following inequalities prevail</u>:

(9) $$\sum_{k=0}^{n} \gamma_k \sum_{j=0}^{\ell} (-1)^j \binom{\ell}{j}\binom{n-\ell}{k-j} < 0 \quad , \quad \ell = 1, 2, \ldots, n \ .$$

SELECTION WITH MANY LOCI   217

Result III.

(α) The central polymorphism is stable in the presence of free recombination $\mathcal{R}^{(f)}$, (see (5)) if and only if

(10) $$\sum_{k=0}^{n} \gamma_k \left[2\binom{n-1}{k} - \binom{n}{k} \right] < 0 .$$

(β) Where $\gamma_{n-k} \geq \gamma_k$ for all $0 \leq k \leq [\frac{n}{2}]$ entailing strict inequality for some k then (10) holds. In particular, if the fitness values increase with increasing heterozygosity, viz. $\gamma_0 < \gamma_1 < \ldots < \gamma_n$, then the central equilibrium is always locally stable for free recombination and likely for moderate to loose linkage.

Proofs of Results II and III are set forth in Karlin (1978). Consonant with Result I, the relations (9) imply the validity of (10), but not conversely. Thus, the central equilibrium is more likely stable in circumstances of loose linkage then with tight linkage.

If (9) does not hold then for moderate to loose linkage the coexistence of several stable polymorphic equilibria is feasible.

The validity of (9) is assured if the fitness values correspond to the ordinates of a strongly concave function with abscissa equally spaced as depicted

FIG.(11a)

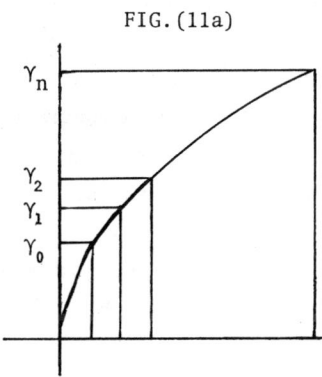

Qualitatively, the fitness regime implied by (11a) has the property that there is a marginal decrease in fitness contribution with each additional heterozygous locus. Equivalently, a few heterozygous loci can provide the bulk of fitness and entail substantial polymorphism for all the other loci involved in the trait. On the other hand, if the fitness array $\{\gamma_k\}_0^n$ constitute a convex sequence, i.e., $2\gamma_k \leq \gamma_{k+1} + \gamma_{k-1}$ holds, as described

FIG.(11b)

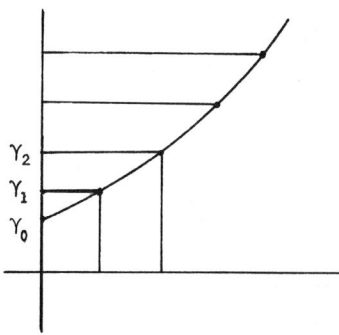

then the central polymorphism is <u>never locally stable</u> for small recombination rates although stable in line with Result III where the loci are mutually loosely linked. We discuss the implications of these facts in the next section. Where $\{\gamma_k\}_0^n$ constitutes a convex sequence then (11b) assuredly applies as for $\gamma_k = \left(\frac{k+1}{n+1}\right)^\alpha$, $k = 0,1,2,\ldots,n$, $\alpha > 1$. In contrast for $0 < \alpha < 1$ the shape of (11a) prevails and the central equilibrium is globally stable relative to any population state having all gamete types initially present.

We can inquire as to the nature of the stable equilibrium configurations occuring for tight linkage when the central point (8) is not stable.

SELECTION WITH MANY LOCI

Definition 1.

(α) A pair of chromosomes $\underline{c}^{(1)} = (\xi_1^{(1)}, \ldots, \xi_n^{(1)})$ and $\underline{c}^{(2)} = (\xi_1^{(2)}, \ldots, \xi_n^{(2)})$ are called complementary if $\xi_i^{(1)} + \xi_i^{(2)} = 1$ for all i.

(β) An equilibrium for the phenotypic selection array $\{\gamma_i\}_0^n$ with <u>no recombination</u> $(\mathscr{R}^{(0)})$ is called an <u>allelic polymorphism</u> if every allele at every locus is involved with positive frequency in at least one of the gamete types. A complete <u>chromosomal or gametic polymorphism</u> has <u>every</u> gamete type represented with positive frequency. Of course, $\underline{x}^*$ is the unique chromosomal polymorphism available in the case of absolute linkage.

The theory of small parameters (Karlin and McGregor (1972)) tells us that with small recombination rates (tight linkage) a stable allelic polymorphism translates into a bona fide stable chromosomal polymorphism located near the corresponding equilibrium state existing for absolute linkage.

Result IV. A pair of <u>complementary chromosomes</u> (necessarily occurring with equal frequency because of the natural symmetry implicit to definition (1)) <u>is stable for absolute linkage if</u>

(12) $\quad \gamma_n > \gamma_0 \quad$ and $\quad \gamma_n + \gamma_0 > \gamma_k + \gamma_{n-k} \quad$ <u>for</u> $\quad k = 1, 2, \ldots, n-1$.

The condition (12) entails that for the regular selection regimes of the form

(13) $$\gamma_k = \left(\frac{k+1}{n+1}\right)^\alpha, \quad k = 0, 1, \ldots, n,$$

then with $\alpha > 1$ the complementary chromosomal pairs comprise stable equilibria. These are the only stable possibilities for $\alpha > n - 1$. For $1 < \alpha < n - 1$, in addition to the equilibria composed from complementary pairs there exist other classes of stable equilibria with an intermediate number of chromosomal types; in other words, high complementarity is not an exclusive stable state for tight linkage even with such regular fitness arrays as in (13), $\alpha > 1$.

It is of interest to describe the equilibrium possibilities for the regular fitness pattern $\gamma_i = e^{\lambda i}$, $i = 0,1,\ldots,n$, $\lambda > 0$. The fitness model then reduces to a symmetric multiplicative non-epistatic selection array. Here, for tight linkage only complementary pair stable equilibrium configurations appear (Karlin (1977)) in accord with the simulation studies of Franklin and Lewontin (1970). With free recombination, consistent with Result III, the central equilibrium is exclusively stable.

Consider next the case where fitness reflects a threshold effect, to wit,

(14) $\mathscr{S}_k : \gamma_0 = \gamma_1 = \cdots = \gamma_{k-1} = 0$, $\gamma_k = \gamma_{k+1} = \cdots = \gamma_n = 1$

or more generally the fitness sequence corresponding to an S-shaped curve. With free recombination result III part (β) affirms that the central equilibrium $\underline{x}^*$ is stable. For tight linkage it is easy to check that a threshold selection pattern as in (14), with $k \geq 2$, cannot have the central equilibrium stable, and the equilibria sets consisting of complementary pairs are also not stable. Here only allelic polymorphic outcomes appear again involving a partial set (but not a small number) of gametes.

It is illuminating to describe the possibilities in the associated models embracing $n = 3$ and 4 loci. Consider first three loci with allele alternatives A-a, B-b and C-c at locus 1, 2 and 3, respectively. The fitness regimes is given by $\{\gamma_i\}_{i=0}^3$. Complementary chromosomal pairs (e.g., {ABC,abc} {abC,ABc}, etc.) are stable with tight linkage if and only if $\gamma_0 + \gamma_3 > \gamma_1 + \gamma_2$ and $\gamma_0 < \gamma_2$ (cf. Result IV). Stable symmetric sets involving 4 chromosomal types, e.g., {ABC,Abc,aBc,abC} = $\mathscr{F}$ occur if the conditions $\gamma_0 < \gamma_2$ and $\gamma_0 + 3\gamma_2 > 3\gamma_1 + \gamma_3$ hold. There can also exist nonsymmetric stable boundary equilibria (consult Karlin and Liberman (1976) for details). In the presence

SELECTION WITH MANY LOCI

of a threshold fitness array, viz. $\gamma_0 = 0$, $\gamma_1 = \gamma_2 = \gamma_3 = 1$ or $\gamma_0 = \gamma_1 = 0$, $\gamma_2 = \gamma_3 = 1$, then only the symmetric equilibrium configurations (like $\mathscr{F}$) are stable with tight linkage.

With 4 loci and tight linkage where the central equilibrium is not stable then stable boundary equilibria involving sets of 2, 4 and 8 chromosomal types can occur. Some representative combinations, that are stable under appropriate conditions include complementary pairs — {ABCD,abcd}; partial symmetric chromosomal sets — {ABCD,aBcd,abCd,AbcD} ; intermediate sets — {ABCD, ABcd, AbCd,AbcD,aBCd,aBcD,abCD,abcd}, etc.

§4. Discussion.

Multilocus data in natural populations is accumulating rapidly. The theory has been mostly confined to two loci (consult Karlin (1975) for a recent review) whereas the analytical base pertaining to many loci systems is as yet preliminary, although recent progress with new methods and concepts appear to be promising. The objectives of multilocus theory coupled to statistical analysis of field and laboratory observation and experiment strives to gain qualitative insights and clues for identifying blocks of genes related epistatically and/or chromosomally linked, to establish connections between single or multiple enzyme performances, to ascertain the extent that certain allelic and loci variants go together and their ecological and environmental correlates. The relevance of multilocus analysis in the medical sciences are familiar. Actually, the literature abounds with studies on disease associations between biochemical genes, marker genes, blood and serum typings, etc. With regard to objectives of quantitative inheritance, the possibilities of relating biochemical loci information in order to improve animal and plant breeding programs is mostly at its genesis and undoubtedly worth more effort.

We have focused in this paper on the selection regime of a

trait determined at n loci where the fitness value depends on the cumulative level and locations of the heterozygous loci. The model can be regarded as representing a multilocus "generalized heterozygote (advantageous, intermediate or disadvantageous) mechanism". This fitness scheme reflects epistatic selection and can depart to a lesser or greater extent from the classical models of additive or multiplicative nonepistatic loci effects. For this phenotypic model we reported several results on expectations for gene frequency arrays (that is, stable equilibrium configurations) in the presence of tight as against loose linkage. Our findings are robust with respect to perturbations maintaining fitness expression dependent largely on cumulative heterozygosity, but allowing other minor fitness accruements. We surmise that the selection recombination inferences uncovered here may be suggestive of the behavior for a number of important classes of multilocus selection relationships.

1. Where the fitness regime $\{\gamma_k\}_0^n$ (γ_k = the fitness of a genotype encompassing k heterozygous loci) has $\gamma_0 > \gamma_1 > \ldots > \gamma_n$ expressing a case of directional selection favoring the homozygotes, then independent of the recombination process, pure fixation is the only tenable outcome. A problem of greater interest and relevance, and certainly more recondite, concerns the nature of the gene frequency realizations resulting under the assumption

(15) $$\gamma_0 < \gamma_1 < \ldots < \gamma_n$$

where increasing heterozygosity tends to enhance the genomic fitness value. Result III tells us that with free recombination the central polymorphism (8) is persistently stable. The outcome for absolute linkage (and, therefore, for small positive recombination) is much more sensitive to the order and nature of increase in the sequence (15) (compare especially results II and III). Some possible implications based on these results are as follows.

(i) Consider the situation where the gene frequency data on a trait manifests a large number of segregating gamete types each occurring with reasonable frequency and consistently observed in different population areas. Suppose further that it is known, on the basis of other sources, that the genes are essentially tightly linked, then result III suggests that the fitness regime $\{\gamma_i\}_{i=0}^n$ relates more to the shape (11a), such that a few genes contribute most of the realizable fitness. In this situation, concomitantly, a host of interacting genes are kept segregating entailing much polymorphism. The HLA system in man (or the equivalent H_2 in mice) approximately conforms to the frequency pattern described above. Our theory proposes in this case that a few loci may be decisive to fitness and carry along due to epistatic or linkage connections numerous other loci substantially polymorphic.

In plant breeding, it is contended by some that where a large number of loci are adapted to their natural microenvironment and one is seeking to improve a particular trait then a few loci can be relevant and the selection expression may conform to (11b).

(ii) Suppose, unlike the situation of paragraph (i), a partial set of gamete types are mostly observed typically consisting of complementary chromosomal configurations or larger partial sets of chromosomal types with trace representations of the remaining chromosomal possibilities. Suppose further that in different sampled populations of the same trait analogous frequency data is noted with the principal gametes sometimes different from locality to locality. Where these genes are recognized to be relatively tightly linked then results I and II suggest that probably _many_ genes are indispensable to achieve large fitness or adaptability.

Put in another way, if only partial sets of gametes are around, which may vary spatially or temporally, then from our theory it is likely that the genes at hand are mostly tightly linked and many good genes are essential to adaptation. On the other side, when only a few good genes suffice to guarantee high fitness then many gamete types are likely segregating in approximate linkage equilibrium.

The assertion of result III (part β) is worth emphasis here. This affirms that where epistasis is rendered through aggregate heterozygosity value such that fitness increases with the number of heterozygous loci, and if these genes are mutually loosely linked, then significant polymorphism is expected with many correlated segregating loci.

2. There is a presumption prevalent in the literature concerning the interpretation of multilocus data asserting that observed linkage equilibrium of all orders implies weak selection differentials among genotypes. The phenotypic model discussed in this paper and many other selection regimes (both nonepistatic and epistatic) entail stable equilibrium configurations with no discernable associations (i.e., displaying global linkage equilibrium) of all orders although there is incontrovertible selection operating. There is also a common adage, stated often by experimental population geneticists, that only the alternatives (i) high linkage disequilibrium of all orders as against (ii) small linkage disequilibrium of all orders are viable competitors reflecting linkage selection balance. The phenotypic selection regimes even those induced by regular patterns; e.g. (13), already entail a variety of stable equilibrium arrays exhibiting mixed patterns of low and high measures of associations of various orders; (for more on this see Karlin (1977), on the experimental and observational front, e.g., Baker (1975), Mukai, Watanabe and Yamaguchi (1974)).

3. With an increasing number of loci involved (n increasing) maintaining a common aggregate range of fitness values standardized

to vary from 0 to 1 , over the n loci, we find a diminished
extent of recombination is needed in order to achieve significant
polymorphism. In the same vein, the influence of multiple
allelism also diminishes the amount of recombination necessary
to achieve a prescribed level of polymorphism. The precise
quantifications of these findings will be elaborated elsewhere.

BIBLIOGRAPHY

Bailey, N.T.J. (1961), Introduction to the Mathematical Theory of Genetic Linkage, Oxford Univ. Press, Oxford, England.

Baker, W. K. (1975), "Linkage disequilibrium over space and time in natural populations of Drosophila montana", PNAS 72: 4095-99.

Feldman, M. W., Franklin, I. R. and Thomson, G. (1974), "Selection in complex genetic systems: I. The symmetric equilibrium of the three locus symmetric viability model", Genetics 75: 135-162.

Franklin, I. R. and Lewontin, R. C. (1970), "Is the gene the unit of selection?", Genetics 65: 701-734.

Karlin, S. (1975), "General two-locus selection models: Some objectives, results and interpretations", Theor. Pop. Bio. 7: 364-398.

Karlin, S. (1977), "Theoretical aspects of multilocus selection balance I." Math. Assoc. of Amer., Studies in Mathematical Biology (S. Levin ed.).

Karlin, S. (1978), Mathematical Population Genetics, Acad. Press, N.Y. (to appear).

Karlin, S. and Liberman, U. (1976), "A phenotypic symmetric selection model for three loci, two alleles. The case of tight linkage", (to appear in Theor. Pop. Biol. 10, December).

Karlin, S. and Liberman, U. (1977), "Analysis and representations of some nonepistatis selection models" (to appear).

Kojima, K. and Lewontin, R. C. (1970), "Evolutionary significance of linkage and epistasis" in Topics in Math. Genetics (K. Kojima, ed.), Springer-Verlag, N.Y.

Lewontin, R. C. and Kojima, K. (1960), "The evolutionary dynamics of complex polymorphism", Evolution 14: 458-472.

Mukai, T.T.K., Watanabe, E. and Yamaguchi, O. (1974), "The genetic structure of natural populations of Drosophila melagonaster XII.", Genetics 77: 771-793.

Roux, C. Z. (1974), "Genetic loads: I. Hardy-Weinberg equilibria in random mating populations", Theor. Pop. Biol. 5: 394-416.

Multivariate normal genetic models with a finite number of loci

Joseph Felsenstein
DEPARTMENT OF GENETICS
UNIVERSITY OF WASHINGTON, SEATTLE, WA 98195

ABSTRACT

A genetic model due to Russell Lande is described. The model assumes a finite number of loci at each of which there are an infinite number of alleles whose effects on the phenotype are normally distributed. Analytic and numerical results using this model depend on the allele effects remaining multivariate normally distributed. This is almost never exactly true, but may often be a good approximation. Numerical results for several kinds of natural selection are discussed. A model involving overdominance is presented which seems to exhibit the Franklin-Lewontin crystallization effect. A model is presented of the maintenance of genetic variation by a cline along which there is linear change in the optimum phenotype under optimizing selection. The equilibrium has been found analytically for the case of an infinite cline. Remarkably, there is no linkage disequilibrium maintained at equilibrium in this case.

INTRODUCTION

The use of normal distributions for quantitative characters goes back to the Biometricians, and is most visible in selection theory (e.g. Cochran, 1951). The last twenty years of work on the theory of polygenic inheritance has concentrated on phenomena for which normal distributions are inadequate, such as selection limits and effects of linkage (for which the reader should consult the papers by Robertson and Karlin in this volume). Lately there has been a revival of interest in the normal theory, in connection with the application of quantitative genetics to evolutionary ecology and to the study of cultural evolution. Slatkin (1970) and Bulmer (1971a, b, 1972, 1973) have examined the effect of normalizing (optimizing) selection on genetic variability. Roughgarden (1972, 1974a, b), Bulmer (1974a, b), and Slatkin and Lande (1976) have examined frequency-dependent selection and the evolution of niche width. Lande (1976) has considered the rate of evolutionary change induced by a moving optimum when there is optimizing selection. Bulmer (1971c, d) has considered the maintenance of genetic variability by geographic variation. Cavalli-Sforza and Feldman (1976) have considered the equilibrium between mutation and normalizing selection. Feldman (this volume) incorporates phenotypic transmission (cultural inheritance) into this scheme.

The use of normal distributions in all of these papers is motivated by the assumption that they are good approximations to a large and relevant class of schemes for the genetic determination of quantitative characters, in particular models with many loci, each of small effect, and with the phenotype the sum of the effects of the individual loci. There has been almost no investigation of the accuracy of this approximation (but see Wiorkowski, 1972). One would like to have genetic models in which a normal distribution of phenotypes is achieved and maintained exactly. This has frequently been achieved when only a single locus is considered. Kimura (1965), in the first of the modern wave of normal-distribution papers, considered the equilibrium between optimizing selection and mutation. Eshel (1971, 1972, 1973) has considered the rate of evolution in models where there is a distribution of changes in fitness after mutation. His results are general, including normals as a special case. Bodmer and Cavalli-Sforza (1972) have simulated such a model in a finite population with normal distributions of mutation effect. Latter (1970, 1972) has simulated finite populations with optimizing selection and normal distribution of mutational effect. Finite populations will depart from normal distribution of the phenotype, so that analytical results are usually not available. The exception is the diffusion approximations to the directional selection case developed by Guess (1974).

When we assume more than a single locus, we must somehow incorporate the effects of linkage into the mathematics. Latter (1969) has simulated directional selection in a model with a finite number of loci each with infinitely many alleles falling in a normal distribution of allelic effects. Bulmer (1971a) has investigated the generation of linkage disequilibrium in a model with infinitely many loosely-linked loci, finding that there is a finite effect of disequilibrium, which can be incorporated into the normal-distribution approximately even though we are not keeping track of individual loci. In a later (1974b) paper, he approximated the effects of linkage by an analysis which is approximate unless the pairwise recombination fractions among all loci are the same. Cavalli-Sforza and Feldman (1976) have, in effect, ignored linkage disequilibrium when computing the equilibrium between mutation and normalizing selection.

If one set up a model involving a finite number of loci, each with a normal distribution of allele effects, it would be natural to ask under what conditions the joint distribution of allele effects at all loci would remain multivariate normal. This model has been posed, and the question answered, in a beautiful paper by Lande (1975). In the present paper I will present and slightly extend Lande's results.

THE MULTIVARIATE NORMAL MODEL - SELECTION

We consider an infinitely large diploid population with discrete generations. The life cycle of the organism will consist of random mating, followed by natural selection among the diploid zygotes, followed by production of gametes by meiosis:

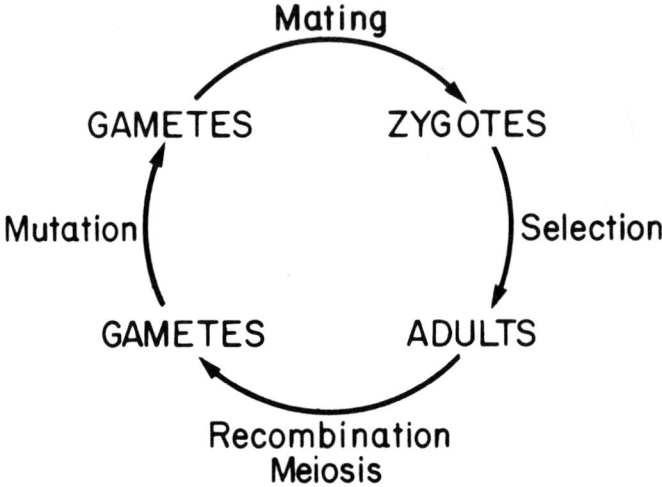

There are n loci. At each locus there are an infinite number of possible alleles. With each allele we associate a real number x, which is to be interpreted as the contribution of that allele to the phenotype. At each locus the values x have a distribution, which we take to be a normal distribution. Note that gene frequencies do not exist in this scheme. A gamete can be represented by a vector of n real numbers, which are the alleles at loci 1 through n: $(x_1, x_2, \ldots x_n)$. Let us begin by assuming that, at generation t, the gametes in the gamete pool are in a multivariate normal distribution with mean vector m and covariance matrix V. We wish to know whether the gametes of the next generation will still be in a multivariate normal distribution. We must determine what happens at each of four stages: formation of zygotes, selection, meiosis, and mutation.

(a) Formation of Zygotes

Each zygote will be composed of two gametes. We can number the genes in the maternal gamete 1 through n, and those in the paternal gamete n+1 through 2n. Thus we can represent an individual by a vector $\underline{x}$ of length 2n, whose distribution we wish to know. We are assuming that there are no differences between the sexes, so that the maternal and paternal gametes each have distribution $N(\underline{m}, \underline{V})$. Since random mating of individuals is equivalent (in this case) to random combination of gametes, the two halves of the vector $\underline{x}$ each are drawn from $N(\underline{m}, \underline{V})$ independently. There is (at this stage in the life cycle) no covariance between x_i and x_j if $i \leq n < j$. Clearly the distribution of $\underline{x}$ is multivariate normal.

$$\underline{x} \sim N\left(\begin{bmatrix} \underline{m} \\ \underline{m} \end{bmatrix}, \begin{bmatrix} \underline{V} & \underline{0} \\ \underline{0} & \underline{V} \end{bmatrix} \right). \qquad (1)$$

$\underline{m}$ being a column vector, and $\underline{0}$ being the n x n matrix of zeros. In what follows, I will use $\underline{V}$ sometimes for the 2n x 2n covariance matrix of genotypes, and sometimes for the n x n covariance matrix of gametes. It should be clear from the context which is intended.

Note that if, in the initial generation, each gene had a normal distribution independently of each other gene, we would be in a state which could only be called linkage equilibrium. In that state $\underline{V}$ would be a diagonal matrix. We shall consider this to be the definition of linkage equilibrium in the present context.

(b) Selection

If the fitness of an individual is the function $w(\underline{x})$ of its genotype vector, and if we take $f(\underline{x})$ as the density function of the random variable $\underline{x}$, then after selection, the survivors will be found to have the density

$$g(\underline{x}) = f(\underline{x}) w(\underline{x}) / \int_{R^n} f(\underline{x}) w(\underline{x}) \, d\underline{x}. \qquad (2)$$

The denominator of the right side of (2) is the mean fitness $\bar{W}$ of individuals in the original distribution.

We wish to find a fitness function which will leave the survivors still in a multivariate normal distribution. The particular form we will use is

$$w(\underline{x}) = \exp[\ \underline{c}'\ \underline{x} - \tfrac{1}{2}\ (\underline{x} - \underline{a})'\ \underline{B}\ (\underline{x} - \underline{a})\ ], \qquad (3)$$

where $\underline{c}$ and $\underline{a}$ are vectors of length 2n, and $\underline{B}$ is a symmetric 2n x 2n matrix. As we shall see shortly, this encompasses a number of interesting types of selection. One can readily show, using equation (2), that if $g(\underline{x})$ is to be a multivariate normal density, $w(\underline{x})$ must be of the form (3), with the trivial exception that it might be further multiplied by a constant. The converse is not true: for some $\underline{a}$, $\underline{B}$, and $\underline{c}$, $g(\underline{x})$ is not a multivariate normal density -- in fact, is not a density at all.

If we specify $f(\underline{x})$ and $g(\underline{x})$ as:

$$f(\underline{x}) = (2\pi)^{-n}\ |\ \underline{V}_1\ |^{-\tfrac{1}{2}}\ \exp[\ -\tfrac{1}{2}(\underline{x}-\underline{m}_1)'\ \underline{V}_1^{-1}\ (\underline{x}-\underline{m}_1)\ ] \qquad (4)$$

and

$$g(\underline{x}) = (2\pi)^{-n}\ |\ \underline{V}_2\ |^{-\tfrac{1}{2}}\ \exp[\ -\tfrac{1}{2}(\underline{x}-\underline{m}_2)'\ \underline{V}_2^{-1}\ (\underline{x}-\underline{m}_2)\ ], \qquad (5)$$

and keep in mind that the denominator of (2) is a constant, by equating like terms in (2) we can compute that:

$$\underline{V}_2^{-1} = \underline{V}_1^{-1} + \underline{B} \qquad (6)$$

and

$$\underline{V}_2^{-1}\ \underline{m}_2 = \underline{V}_1^{-1}\ \underline{m}_1 + \underline{B}\ \underline{a} + \underline{c}. \qquad (7)$$

These expressions can be used to compute the effect of selection on the means and covariances of the distribution of $\underline{x}$. If $\underline{V}_2$ is not a positive definite matrix, then the integral in the denominator of (2) diverges, and no density function $g(\underline{x})$ exists.

(c) Recombination

If we have n loci, there will be 2^n different recombination classes possible. Suppose that the k-th of these has probability

R_k. It might, for example, consist of the maternally-derived genes at loci 1 through 10, the paternally-derived genes at loci 11 and 12, and the maternally-derived genes at loci 13 through 16. The 16-tuple corresponding to this recombinant gamete would be $(x_1, \ldots, x_{10}, x_{27}, x_{28}, x_{13}, \ldots, x_{16})$. If we consider all the gametes in the population formed by this particular type of recombination, these will follow a multivariate normal distribution, since they are a subset of the 2n variables which are themselves multivariate normal. The mean vector of this particular recombinant class is simply the appropriate subset (ordered as above) of the mean vector $\underline{m}_2$ following selection. The covariance matrix of this recombinant class can be obtained by taking the appropriate subset of rows and the same subset of columns of $\underline{V}_2$.

In all of the cases which we shall consider, there is a particular symmetry property of the selection. Selection does not discriminate, in these cases, between which gene is maternally- and which paternally-derived. So for the vector $\underline{c}$,

$$c_i = c_{i+n} \tag{8a}$$

and for the matrix $\underline{B}$,

$$b_{ij} = b_{i+n,j+n} = b_{i,j+n} = b_{i+n,j} \tag{8b}$$

It is not difficult to show that this implies that if $\underline{V}_1$ (before selection) is of the form

$$\begin{bmatrix} \underline{V}_{11} & \underline{V}_{12} \\ \underline{V}_{12} & \underline{V}_{11} \end{bmatrix},$$

$\underline{V}_2$ will also show the same symmetry properties. It will also be true that since before selection $m_i = m_{i+n}$ (by random mating), this will continue to hold after selection.

The upshot of all of this is that if we consider all 2^n of the recombinant classes, (i) all of them have the same mean vector, and (ii) all of them have the same variances v_{ii}. This would seem to offer hope that all might follow the same multivariate normal distribution. But these hopes are dashed when

we consider the covariances. Consider the case of two loci
(n = 2). Suppose that

$$\underline{V}_2 = \begin{bmatrix} v_{11} & v_{12} & v_{13} & v_{14} \\ v_{12} & v_{22} & v_{14} & v_{13} \\ \hline v_{13} & v_{14} & v_{11} & v_{12} \\ v_{14} & v_{13} & v_{12} & v_{22} \end{bmatrix} \qquad (9)$$

There are four possible recombinant classes, but in the symmetrical cases which we consider, these reduce to two: recombinants and nonrecombinants. Each class of gametes is multivariate normal with the same mean vector, but different covariance matrices. A fraction (1-r) of the gametes have covariance matrix

$$\begin{bmatrix} v_{11} & v_{12} \\ v_{12} & v_{22} \end{bmatrix},$$

while r of them have covariances:

$$\begin{bmatrix} v_{11} & v_{14} \\ v_{14} & v_{22} \end{bmatrix}.$$

Thus the distribution of gametes is a mixture of two bivariate normals, with the same means and variances but with different covariances.

Under what conditions will such mixtures be themselves multivariate normal? We can investigate this by considering the characteristic function of the mixture. The characteristic function of a multivariate normal distribution with mean vector $\underline{m}$ and covariance matrix $\underline{V}$ is the complex-valued function

$$F(\underline{\theta}) = E[\exp(i\,\underline{\theta}'\,\underline{x})] = \exp(i\,\underline{\theta}'\,\underline{m} - \tfrac{1}{2}\,\underline{\theta}'\,\underline{V}\,\underline{\theta}), \qquad (10)$$

$\underline{\theta}$ being a vector $(\theta_1, \ldots, \theta_n)$. The characteristic function of the mixture of gamete subpopulations is

$$G(\underline{\theta}) = \sum_r R_r \exp(i\,\underline{\theta}'\,\underline{m}^{(r)} - \tfrac{1}{2}\,\underline{\theta}'\,\underline{V}^{(r)}\,\underline{\theta}) \qquad (11)$$

where $\underline{m}^{(r)}$ and $\underline{V}^{(r)}$ are respectively the mean vector and covar-

MULTIVARIATE NORMAL GENETIC MODELS

iance matrix of the r-th recombination class. Since distributions are uniquely determined by their characteristic functions, the distribution of gametes will be multivariate normal if and only if a mean vector $\underline{m}$ and covariance matrix $\underline{V}$ exist such that (11) is of form (10). In fact, the mean vectors of all of the recombinant classes are the same, so $\underline{m}^{(r)} = \underline{m}$ for all r. Since all the recombination classes have the same means, the overall covariance matrix is simply the mean of the covariance matrices $\underline{V}^{(r)}$.

The question of multivariate normality of the gamete population thus reduces to whether

$$\exp(\sum_r R_r \underline{\theta}' \underline{V}^{(r)} \underline{\theta}) = \sum_r R_r \exp(\underline{\theta}' \underline{V}^{(r)} \underline{\theta}). \tag{12}$$

for $\underline{\text{all}}$ vectors $\underline{\theta}$. It is not difficult to show that this almost never holds true. Consider only those $\underline{\theta}$ in which the j-th and k-th elements are the only nonzero ones, i.e. $\underline{\theta} = (0, 0, \ldots, 0, \theta_j, 0, \ldots, 0, \theta_k, 0, \ldots, 0)$. If (12) holds for all $\underline{\theta}$, it must hold for these in particular. For all values of θ_j and θ_k, it would have to be true that

$$\exp[\sum_r R_r (\theta_j^2 v_{jj}^{(r)} + \theta_k^2 v_{kk}^{(r)} - 2\theta_j \theta_k v_{jk}^{(r)})]$$

$$= \sum_r R_r \exp(\theta_j^2 v_{jj}^{(r)} + \theta_k^2 v_{kk}^{(r)} - 2\theta_j \theta_k v_{jk}^{(r)}). \tag{13}$$

All of the $v_{jj}^{(r)}$ and $v_{kk}^{(r)}$ are equal (respectively) to v_{jj} and v_{kk}. $\exp(x)$ is a convex function, so by Jensen's Inequality (13) is true only if $-2\theta_j \theta_k v_{jk}^{(r)}$ is equal for all those r for which R_r is nonzero. But $v_{jk}^{(r)}$ can have only two possible values. It was selected from either v_{jk} or $v_{j,k+n}$ in the original covariance matrix of diploid zygotes, $\underline{V}_2$.

We have the following result: a necessary condition for the population of gametes be multivariate normally distributed is that either (a) there is only one recombination class (i.e., no recombination at all), or (b) that $v_{jk} = v_{j,k+n}$ in $\underline{V}_2$ (after selection). It follows that recombination generally causes the distribution of gametes to depart from multivariate normality. In assuming that normality continues to apply, we are making an approximation. It will be accurate if all of the $\underline{V}^{(r)}$ are nearly equal, but as far as I know its accuracy has not been seriously investigated.

To find the parameters of the multivariate normal approximation to the distribution of gametes is not difficult. We need not compute all of the R_r and all of the $\underline{V}^{(r)}$. No computation of the mean vector is necessary, since under our symmetry conditions, it does not change. The gamete mean vector is simply the first n (or the last n) elements of the diploid genotype mean vector. An element of the new gamete covariance matrix, v_{jk}', is a weighted average of the $v_{jk}^{(r)}$. But the latter are each either v_{jk} or $v_{j,k+n}$, depending on whether recombination class $\underline{i}$ is nonrecombinant or recombinant for loci j and k. Thus we need only know the recombination fraction between loci j and k, r_{jk}, and can compute the gamete covariances as

$$v_{jk}' = (1 - r_{jk}) v_{jk} + r_{jk} v_{j,k+n}. \qquad (14)$$

The approximation we are using is to take as the distribution of gametes the multivariate normal distribution implied by the means and covariances of the gametes.

The essential conclusions of this section will be found in Lande's paper.

(d) Mutation

We may also wish to have mutation occur among the gametes. We shall assume that to each gene x_i is added an independent change in the identity of the allele. The change, due to mutation, is assumed to have mean zero and variance u_i at the i-th locus. If all of these mutational events are independent, clearly the gametes after mutation will each be the sum of two vectors, $\underline{x} + \underline{y}$, where $\underline{x}$ is drawn from the pre-mutation gamete distribution and $\underline{y}$ is the change due to mutation. Since both are multivariate

MULTIVARIATE NORMAL GENETIC MODELS

normal (the former by our recombination approximation) so is their sum. This type of mutation at the gamete stage leaves the mean vector unchanged, but adds to the covariance matrix a diagonal matrix (here called $\underline{U}$).

To carry a population through random mating, selection, recombination and mutation, we start with the mean vector $\underline{m}$ and covariance matrix $\underline{V}$ of the gametes. We then:

(i) Create the mean vector and covariance matrix of the diploid genotypes as shown in (1).

(ii) Compute the means and covariances in the diploid survivors of selection using (6) and (7).

(iii) Compute the covariance matrix of gametes after recombination, using (14). This is an n x n matrix. The mean vector of gametes is simply the first n elements of the mean vector of diploid genotypes (after selection).

(iv) Compute the covariance matrix after mutation by adding $\underline{U}$ to the matrix (i.e., $v_{ii}' = v_{ii} + u_i$ for all i).

RESULTS WITH VARIOUS KINDS OF SELECTION

Multiplicative Selection

If we restrict attention to types of selection which are within the framework of equation (3), the simplest case is when $\underline{a} = \underline{0}$, $\underline{B} = \underline{0}$, but $\underline{c}$ is nonzero. In this case, fitness is a product of fitnesses at the individual loci, with no dominance within loci. If $c_i = c_{i+n}$, then the fitness function is

$$w(\underline{x}) = \exp[\sum_i c_i(x_i + x_{i+n})] = \prod_{i=1}^{n} \exp[c_i(x_i + x_{i+n})] \quad (15)$$

It is easy to see what happens in such a case. Under selection, (6) shows that there is no change in the covariance matrix $\underline{V}_1$. However, by equations (1) and (14), recombination will cause all but the diagonal elements of $\underline{V}_1$ to become zero, provided that there is some recombination between all pairs of loci. If there is no further mutation, equation (7) shows that $\underline{m}$ will ultimately be changing by the constant amount $\underline{V}_1 \underline{c}$ each generation. If there is mutation, the diagonal elements of $\underline{V}_{-1}$ will increase linearly, and thus the amount of change of $\underline{m}$ will increase continually once $\underline{V}$ is essentially diagonal. We thus have no asymptote or equilibrium of $\underline{m}$, and no recombination effects, except possibly effects due to its breakdown of initial linkage disequilibrium.

Optimizing Selection

Suppose that selection favors phenotypes close to some optimum value A. If the genotypic contribution to the phenotype is simply the sum of allele effects $G = \Sigma x_i$, and if the environmental contribution E is normally distributed with variance V_E and is independent of G, then the fitness of $\underline{x}$ is

$$w(\underline{x}) = \int (2\pi V_E)^{-1/2} \exp[-E^2/(2V_E)] \, W(G + E) \, dE. \quad (16)$$

If the logarithm of fitness declines quadratically with deviation from the optimum, we can find a positive constant V_S such that

$$W(P) = \exp[- (P - A)^2/(2V_S)], \quad (17)$$

which when substituted into (16) gives

$$w(\underline{x}) = (1 + V_E/V_S)^{-1/2} \exp [- (G - A)^2/(2V_S + 2V_E)] \quad (18)$$

Since $G = \Sigma x_i = \underline{1}'\underline{x}$, this is (except for the leading constant), of form (3) with $\underline{c} = 0$,

$$\underline{a} = [A/(2n)] \, \underline{1} \quad (19a)$$

and

$$\underline{B} = [1/(V_S + V_E)] \, \underline{J}, \quad (19b)$$

where $\underline{J} = \underline{1}\,\underline{1}'$ is a matrix all of whose elements are ones.

This type of selection, balanced against mutation, has been extensively examined by Lande (1975). An equilibrium is reached in which the mean phenotype is at the optimum, genetic variance being maintained by a balance between mutation and selection. At this equilibrium there is negative (repulsion) linkage disequilibrium between different loci. The reader should consult Lande's paper for details, including some interesting comments on the opportunity for divergence of populations by random

MULTIVARIATE NORMAL GENETIC MODELS

genetic drift under this kind of selection-mutation balance.

Although investigation of the convergence of this case to the equilibrium has been incomplete, I have done numerical iterations of the mean vector and covariance matrix for a number of different parameter values, and have always found them to approach the equilibrium given by Lande. An interesting feature of Lande's solution is that the total genetic variance of the character does not depend on the pattern or amount of recombination. In the case where all mutation rates are equal, Lande's equation reduces to

$$V_G = 2n[u(V_S + V_E + n^2 u)]^{1/2} + 2nu^{1/2} \qquad (20)$$

A change in the recombination fractions redistributes this variance among the variances v_{ii} and the covariances v_{ij}, but does not change the total $V_G = \underline{1}'V\underline{1}$. If the mutation rates are made to approach zero, the equilibrium genetic variance also approaches zero, as can be seen in equation (20).

Frequency-Dependent Selection

Bulmer (1974) has presented a model of selection in which genetic variance is maintained at equilibrium, not by mutation, but by selection, based on the assumption that genotypes whose phenotypes are similar compete more than those whose phenotypes are more distant. Although Bulmer's specific fitness function is not of the form of equation (3), an analogous function can be developed which is. Fitness as a function of phenotype is taken to be

$$W(P) = [\int e^{-C(Q-P)^2} f(Q) \, dQ \,]^{-D}, \qquad (21)$$

where $f(Q)dQ$ is the density function of the phenotype immediately before this selection acts. There is not sufficient space here to show the derivation, but this leads to a fitness function of form (3), with $\underline{c} = 0$, $\underline{a} = \underline{m}$ (the current mean vector), and

$$\underline{B} = [\,-1/(2/[CD] + \underline{1}'V\underline{1}/D - V_E)]\,\underline{J}, \qquad (22)$$

provided that the quantity within the outer brackets is negative.

Note that the presence of $\underline{1}'\underline{V}\underline{1}$, the current diploid genetic variance, in (22) ensures that selection will weaken as the genetic variance increases.

If this selection acts alone, no equilibrium is reached. Not only does $V_G = \underline{1}'\underline{V}\underline{1}$ increase continually, but this increase accelerates, resulting in such a strong selection favoring the population extremes that after a finite number of iterations the distribution of $\underline{x}$ no longer exists, its covariance matrix no longer being positive definite.

We can achieve an equilibrium by assuming that this frequency-dependent selection occurs immediately before the optimizing selection given in equations (19). It is readily demonstrated that this compound selection is of form (3), with $\underline{B}$ the sum of (22) and (19b):

$$\underline{B} = [\ 1/(V_S + V_E) - 1/(2/[CD] + \underline{1}'\underline{V}\underline{1}/D - V_E)]\ \underline{J}. \qquad (23)$$

Equilibrium is reached when

$$\underline{1}'\underline{V}\underline{1} = V_G = D\ (V_S + 2V_E) - 2/C. \qquad (24)$$

When V_G exceeds this value, the net selection reduces V_G, and when it is smaller than this value V_G increases. The result is an equilibrium at this value. The most interesting feature of this equilibrium is that, when it is achieved, $\underline{B} = \underline{0}$. By the same argument as in the case of multiplicative selection, there must then be no linkage disequilibrium at equilibrium. The result is that $\underline{V}$ is then diagonal, its trace being V_G. The individual v_{ii} may be any values which add to V_G. At equilibrium, it will also be true that the mean phenotype equals the optimum phenotype.

Overdominance

The models used so far have had no non-additive variance in the phenotype. We could have nonadditive variance and still be within the framework of (3) if the logarithm of fitness were a quadratic form in the x_i. If we wished to have no epistasis, but to have dominance, then we could take

$$w(\underline{x}) = \exp[\sum_{i=1}^{n} (s_{1i} x_i^2 + 2s_{2i} x_i x_{i+n} + s_{1i} x_{i+n}^2)]$$

$$= \prod_{i=1}^{n} \exp(s_{1i} x_i^2 + 2s_{2i} x_i x_{i+n} + s_{1i} x_{i+n}^2). \quad (25)$$

I have investigated numerically two symmetric special cases of (28). They are the cases (i) $s_{1i} = -s_{2i} = s$, and (ii) $s_{1i} = 0$ and $s_{2i} = -s$. These may be thought of as being cases of overdominance, in the sense that the more different the two alleles at locus i in an individual are, the higher its fitness. In these cases, $\underline{a} = \underline{0}$, $\underline{c} = \underline{0}$, and the matrix $\underline{B}$ is given by:

in case (i) $\quad b_{ii} = -b_{i,i+n} = -b_{i+n,i} = b_{i+n,i+n} = -s$, (26a)

and in case (ii) $\quad b_{i,i+n} = b_{i+n,i} = s$, (26b)

all other elements b_{ij} being zero in both cases.

Computer iterations of equations (1), (6), (7), and (18) have been carried out for various choices of the r_{ij}, s, and the initial $\underline{V}$. In all cases I have tried, both of these overdominant models have behaved similarly, so that no further reference will be made to the distinction between them. For simplicity of computation, many runs where made for the case where $r_{ij} = r$ for all i, j such that $i \neq j$. This does not correspond to a realizable genetic map unless $r = 0$ or $r = 1/2$, but it is much easier to compute (if some minor algebra is done) and usually behaves like more realistic cases.

When one starts with no linkage disequilibrium, this sort of selection does not generate any (this can also be proven algebraically). This is not surprising. Since we are selecting for ever-more extreme alleles at each locus, we would expect the v_{ii} to increase without limit, which it does. What is less obvious, but not counterintuitive, is that when some critical value of v_{ii} is reached, the next generation's v_{ii} is negative, which actually means that v_{ii} has become infinite in a finite amount of time, and a probability density of $\underline{x}$ no longer exists.

One surprising result does emerge from the numerical runs. If we start with a small amount of covariance (linkage disequilibrium), either positive or negative, between any set of loci, an unusual pattern is seen. At first the covariance declines as expected. But as the v_{ii} approach their point of explosion, the v_{ij} not only begin to increase, but increase faster than the v_{ii}. This means that the correlation between non-alleles is increasing. This phenomenon occurs earlier in the process the smaller the r_{ij}. I believe it to be the result of the Franklin-Lewontin (1970) crystallization effect, found by them in the two allele case. Since, unlike the two-allele case, we are not approaching an equilibrium, this belief is rather hard to test. It does at least merit further investigation.

A Geographic Cline Model

Suppose that, in the absence of mutation, variation is maintained by having optimizing selection, with a cline in the position of the optimum. In particular, we suppose a one-dimensional geographic continuum of infinite length, along which the optimum phenotype changes linearly. Suppose that S measures the slope of the line of optimum phenotype. Let the strength of optimizing selection be V_S everywhere. Suppose that adults migrate after the action of optimizing selection, with the individuals at each point being drawn from a normal distribution of previous positions, with mean displacement zero and variance of displacement V_M. Bulmer (1971c, d) has considered some similar cases.

Let us call some point position 0, and let the optimum phenotype be Sy at position y. We seek an equilibrium, assuming that the mean genotype vector at y after selection is $[y/(2n)]\underline{1}$, so that the mean phenotype at each point is at the local optimum. The distribution of genotype vectors at y after migration has some characteristic function $E[\exp(i\underline{\theta}'\underline{x})]$. Since the migration is equivalent to mixing distributions, we take the expectation

MULTIVARIATE NORMAL GENETIC MODELS 243

to run over all parental positions z and within each such position over all genotypic vectors $\underline{x}$:

$$E(e^{i\underline{\theta}'\underline{x}}) = E_z E_x (e^{i\underline{\theta}'\underline{x}}) = E_z [\exp(i\underline{\theta}'[z/(2n)] \underline{1} - \tfrac{1}{2}\underline{\theta}'\underline{V}\underline{\theta})] \quad (27)$$

and since z is normally distributed,

$$E(e^{i\underline{\theta}'\underline{x}}) = \exp (i\underline{\theta}'[y/(2n)] \underline{1} - \tfrac{1}{2}\underline{\theta}'(\underline{V} + \underline{V}_B) \underline{\theta}) \quad (28)$$

which shows that after migration, $\underline{x}$ is still multivariate normal, with unchanged mean, but with a between-localities covariance matrix added to $\underline{V}$. This $\underline{V}_B$ is readily computed to be

$$\underline{V}_B = E[S(z-y) \underline{1}\,\underline{1}' S(z-y)] = [S^2 V_M/(2n)^2] \underline{J} \quad (29)$$

The net effect of optimizing selection followed by migration is thus the genotype covariance matrix

$$\underline{V}_2 = (\underline{V}_1^{-1} + [1/(V_S + V_E)] \underline{J})^{-1} + [S^2 V_M/(2n)^2] \underline{J}. \quad (30)$$

Now we try to find a diagonal matrix $\underline{V}$ such that the result of (30) is the same diagonal $\underline{V}_1$. If there is one such, then neither recombination nor mating will further alter it, so that it will be an equilibrium value of $\underline{V}$. We have already established that the mean vector $[Sy/(2n)] \underline{1}$ is unchanged by all of these operations. We will then have found an equilibrium. Since matrices of the form $a\underline{I} + b\underline{J}$ are easily inverted, we can find the required $\underline{V}$ in straightforward fashion (the details are omitted for lack of space). It has

$$v_{ii} = v = \frac{1}{2n} [1/2 \ S^2 V_M + 1/2 \ (S^4 V_M^2 + 4 [V_S + V_E]S^2 V_M)^{1/2}]. \quad (31)$$

The genetic variance at equilibrium will be $V_G = 2nv$, for which some typical values are:

$(V_S + V_E)/(S^2 V_M)$	$V_G/(S^2 V_M)$
0.01	1.0099
0.1	1.0916
0.2	1.1708
0.5	1.3660
1	1.6180
2	2

5	2.7913
10	3.7016
100	10.5125
1000	32.1267

Note that when selection is strong compared to the amount of migration, $V_G \cong S^2 V_M$. When selection is weak, $V_G \cong S\sqrt{V_M V_S}$.

The feature of the equilibrium (31) which is of interest is that it applies irrespective of the amount or pattern of linkage, and it involves no linkage disequilibrium. This is perhaps surprising, since both migration and optimizing selection tend to produce linkage disequilibrium. But they produce disequilibrium of opposite sign, and apparently in this model these exactly cancel. Slatkin (1977) has made a general analysis of clines in optimizing selection for a model with infinitely many loci. The present result indicates that his results may generalize to cases with finitely many linked loci.

While the above anlysis neither proves uniqueness nor stability of the equilibrium, I have not observed convergence to any other equilibrium in numerical runs of this model.

ACKNOWLEDGMENTS

This work has been supported by ERDA contract AT(45-1)2225 TA 5 with the University of Washington.

BIBLIOGRAPHY

[1] Bodmer, W. F. & Cavalli-Sforza, L. L. (1972). Variation in fitness and molecular evolution. *Proc. Sixth Berkeley Symp. Math. Statist. & Prob.*, Vol. V, 225-275.

[2] Bulmer, M. G. (1971a). The effect of selection on genetic variability. *Amer. Naturalist 105*, 201-211.

[3] Bulmer, M. G. (1971b). The stability of equilibria under selection. *Heredity 27*, 157-162.

[4] Bulmer, M. G. (1971c). Stable equilibria under the two-island model. *Heredity 27*, 321-330.

[5] Bulmer, M. G. (1971d). Stable equilibria under the migration matrix model. *Heredity 27*, 419-430.

[6] Bulmer, M. G. (1972). The genetic variability of polygenic characters under optimizing selection, mutation, and drift. *Genet. Res.* **19**, 17-25.

[7] Bulmer, M. G. (1973). The maintenance of the genetic variability of polygenic characters by heterozygous advantage. *Genet. Res.* **22**, 9-12.

[8] Bulmer, M. G. (1974a). Density-dependent selection and character displacement. *Amer. Naturalist* **108**, 45-58.

[9] Bulmer, M. G. (1974b). Linkage disequilibrium and genetic variability. *Genet. Res.* **23**, 281-289.

[10] Cavalli-Sforza, L. L., and Feldman, M. W. (1976). Evolution of continuous variation: direct approach through joint distribution of genotypes and phenotypes. *Proc. Natl. Acad. Sci. USA* **73**, 1689-1692.

[11] Cochran, W. G. (1951). Improvement by means of selection. *Proc. Second Berkeley Symp. Math. Statist. & Prob.*, 449-470.

[12] Eshel, I. (1971). On evolution in a population with an infinite number of types. *Theoret. Pop. Biol.* **2**, 209-236.

[13] Eshel, I. (1972). Evolution processes with continuity of types. *Adv. Appl. Prob.* **4**, 475-507.

[14] Eshel, I. (1973). Evolution in diploid populations with continuity of types. *Adv. Appl. Prob.* **5**, 55-65.

[15] Franklin, I. and Lewontin, R. C. (1970). Is the gene the unit of selection? *Genetics* **65**, 707-734.

[16] Guess, H. A. (1974). Evolution in finite populations with infinitely many types. *Theoret. Pop. Biol.* **5**, 417-430.

[17] Kimura, M. (1965). A stochastic model concerning the maintenance of genetic variability in quantitative characters. *Proc. Natl. Acad. Sci. USA* **54**, 731-736.

[18] Lande, R. (1975). The maintenance of genetic variability by mutation in a polygenic character with linked loci. *Genet. Res.* **26**, 221-235.

[19] Lande, R. (1976). Natural selection and random genetic drift in phenotypic evolution. *Evolution* **30**, 314-334.

[20] Latter, B. D. H. (1970). Selection in finite populations with multiple alleles. II. Centripetal selection, mutation and isoallelic variation. *Genetics 66*, 165-186.

[21] Latter, B. D. H. (1972). Selection in finite populations with multiple alleles. III. Genetic divergence with centripetal selection and mutation. *Genetics 70*, 475-490.

[22] Roughgarden, J. (1972). Evolution of niche width. *Amer. Naturalist 106*, 683-718.

[23] Roughgarden, J. (1974a). The fundamental and realized niche of a solitary population. *Amer. Naturalist 108*, 232-235.

[24] Roughgarden, J. (1974b). Niche width: biogeographic patterns among *Anolis* lizard populations. *Amer. Naturalist 108*, 429-441.

[25] Slatkin, M. (1970). Selection and polygenic characters. *Proc. Natl. Acad. Sci. USA 66*, 87-93.

[26] Slatkin, M. and Lande, R. (1976). Niche width in a fluctuating environment - density independent model. *Amer. Naturalist 110*, 31-55.

[27] Slatkin, M. (1977). Spatial patterns in the distribution of polygenic characters. *Submitted for publication.*

[28] Wiorkowski, J. (1972). Some statistical methods for the study of quantitative genetic traits. *Unpublished Ph.D. Dissertation, University of Chicago.*

Two-locus theory in quantitative genetics

B. S. Weir and C. Clark Cockerham
DEPARTMENT OF STATISTICS
NORTH CAROLINA STATE UNIVERSITY, RALEIGH, NC 27607

INTRODUCTION

The consideration of the effects of two or more loci on quantitative traits goes back at least as far as Fisher's 1918 paper. Fisher introduced the concept of epistasis whereby genes at different loci act in a dependent fashion. Apart from epistasis there are the complications of linkage and linkage disequilibrium for pairs of loci, and a comprehensive theory should also take account of inbreeding. It is only recently (Gallais, 1974) that all these factors have been considered simultaneously, although not in full generality.

Between 1918 and 1974 there were a series of papers dealing with some of these factors and evaluating some or all of the mean and variance of a quantitative trait or the covariance (or correlation) between the genotypic values of relatives. Reviews of the literature up to that time have been provided by Kempthorne (1955, 1957) and Gallais (1974). Some of the key references are marked with asterisks in the Bibliography.

Paper No. 5120 of the Journal Series of the North Carolina Agricultural Experiment Station, Raleigh, North Carolina. This investigation was supported in part by NIH Research Grant No. GM 11546 from the National Institute of General Medical Sciences.

One of the main difficulties with two-locus work in quantitative genetics is that expressions quickly become unwieldy. A systematic procedure is needed to avoid missing terms and such procedures are heavily dependent on the set of parameters chosen to reflect the effects of inbreeding, linkage and linkage disequilibrium. Parameters can be chosen to allow means, variances and covariances to be expressed as linear combinations of additive, dominance and epistatic effects in a reference population.

For genes at one locus, a complete set of parameters was given by Gillois (1964) and Harris (1964), and methods for calculating an equivalent set were given by Cockerham (1971). Denniston (1967) and Gallais (1970) introduced a set of two-locus parameters which allowed covariances to be found, although those of Denniston were restricted to non-inbred relatives. Both of these sets of measures were for eight genes, four at each of two loci.

When we introduced our four-gene descent measures (Cockerham and Weir, 1973) we remarked that they could be used to provide means and variances for a two-locus quantitative model of gene effects. Such expressions are derived here, and the problem of determining the covariance of relatives is discussed.

While the expressions, especially that for the mean, do offer some insight into the effects of the various dependencies, they also illustrate our belief that exact treatments of completely general situations are unlikely to be of great use in quantitative genetics. An adequate approximate treatment is needed.

ONE-LOCUS THEORY

A brief review of one-locus theory will be of considerable help in introducing our approach for the two-locus case. For completely independent genes of course, one-locus theory can be extended directly to multi-locus situations.

For locus A the genotypic value of individuals with genotype $A_i A_k$ will be written as G_k^i and is the sum of a mean, additive effects for each of the two A genes and a dominance effect for the

TWO-LOCUS THEORY IN QUANTITATIVE GENETICS

interaction of these genes.

$$G_k^i = \mu_0 + a_i + a_k + d_{ik}^a \quad , \quad G_k^i = G_i^k \quad .$$

The superscript a on the dominance term emphasizes that the A locus is being considered.

The mean and effects may be defined for any population, although with difficulty in some cases. We use an infinite random-mating reference population which has alleles A_i with frequency p_i. Then the usual least squares values for the mean and effects are

$$\mu_0 = \sum_{i,k} p_i p_k G_k^i = G_{..}^{..} \quad , \quad a_i = \sum_k p_k G_k^i - G_{..}^{..} = G_{.}^i - G_{..}^{..}$$

$$d_{ik}^a = G_k^i - G_{.}^i - G_k^{.} + G_{..}^{..}$$

so that
$$\sum_i p_i a_i = 0, \quad \sum_i p_i d_{ik}^a = 0, \quad i \neq k \quad . \tag{1}$$

Means

In any generation after the initial one, the mean genotypic value is obtained as the expectation of the value of any member of that generation. Expectation, written as $\mathscr{E}$, refers to the averaging over all possible replicate populations of the same size and mating system. We write the mean as μ_F to indicate that such populations may be inbred. If a random member of the population has genotype $A_i A_k$, then

$$\mu_F = \mathscr{E} G_k^i = \mathscr{E}(\mu_0 + a_i + a_k + d_{ik}^a) = \mu_0 + \mathscr{E} d_{ik}^a \quad .$$

The additive effects have zero expectation from relation (1), but for the dominance effects we need to take account of the fact that genes A_i and A_k may be identical by descent. With probability $1 - F_1$ the two genes are not identical by descent and so have descended from genes on distinct initial gametes. The probability that a random pair of initial gametes carry genes A_i, A_k is taken to be $\theta_k^i = p_i p_k$. With probability F_1 an individual carries genes which are identical by descent, and which must have descended from a gene on one initial gamete. This single ancestral gene is A_i with probability $\theta_{.}^i = p_i$ so that

$$\mathcal{S} \ d_{ik}^a = (1 - F_1) \sum_{i,k} p_i p_k \ d_{ik}^a + F_1 \sum_i p_i d_{ii}^a \ .$$

From (1) the first sum here is zero, and we recover the well-known result

$$\mu_F = \mu_0 + F_1 \sum_i p_i d_{ii}^a \ . \tag{2}$$

If the mean of a completely inbred population, $F_1 = 1$, is written as μ_1 then (2) can be expressed as

$$\mu_F = \mu_0 + F_1 (\mu_1 - \mu_0) \ ,$$

as given by Wright (1951). With inbreeding then, the change in the population mean is a linear function of the inbreeding coefficient and involves dominance effects.

By using the inbreeding coefficient we were able to express a property, the mean, of a population in terms of properties of an initial population. The inbreeding coefficient summarizes the progress of the population since the formation of the initial population, while the frequencies θ_k^i summarize the structure of this initial population. We assumed here that $\theta_k^i = p_i p_k$, with the p_i the same as in the reference population for which effects are defined. This assumption is not necessary and we have previously (Cockerham and Weir, 1973) discussed other situations. Specific sets of initial gametes or even specific initial individuals may be accommodated. For specific gametes θ_k^i is modified, and for specific individuals $1 - F_1$ is extended to distinguish the cases of two gametes between or within individuals.

Another aspect of initial populations deserves mention. We are often interested in the cross population $\mathcal{J}_1$ from parental populations Π_I and Π_{II}. This situation has been considered by several authors previously. Stuber and Cockerham (1966) used the terms "dually defined" and "uniquely defined" to refer to the two cases when gene effects are defined for genes from Π_1 and Π_2 separately or for the combined set respectively.

In the dual case G_k^i now refers to individuals receiving a gamete carrying A_i from Π_I and a gamete carrying a_k from Π_{II}, and we

write
$$G_k^i = \mu_C + a^i + a_k + d_k^{ai}.$$

We must also distinguish between p^i and p_k, the frequencies of A_i and A_k in populations Π_I and Π_{II} respectively. If $\mathcal{J}_1$ is formed by the random union of gametes from the two parental populations, then the least squares values for the mean and effects are

$$\mu_C = \sum_{i,k} p^i p_k G_k^i = G_{\cdot}^{\cdot}, \quad a^i = \sum_k p_k G_k^i - G_{\cdot}^{\cdot} = G_{\cdot}^i - G_{\cdot}^{\cdot}$$

$$a_k = \sum_i p^i G_k^i - G_{\cdot}^{\cdot} = G_k^{\cdot} - G_{\cdot}^{\cdot}, \quad d_k^{ai} = G_k^i - G_{\cdot}^i - G_k^{\cdot} + G_{\cdot}^{\cdot}.$$

The possible relations between μ_C, the mean of $\mathcal{J}_1$, and the means of Π_I, Π_{II} have been discussed by Kempthorne (1957). The dual formulation shows that even though the same alleles may be contained in both populations, they have different effects unless the two parental populations are identical.

For the $\mathcal{J}_2$ and subsequent generations, when inbreeding is a possibility, we employ the unique definitions. Care must be taken of course to account for any relatedness of populations Π_I and Π_{II}.

<u>Variances</u>

In formulating variances we need the expectations of squares of effects. It is helpful to write out the linear model for G_k^i as headings in a two-way table, and identify the classes with different expectations.

	μ_0	a_i	a_k	d_{ik}^a
μ_0	1	2	2	3
a_i	2	4	5	6
a_k	2	5	4	6
d_{ik}^a	3	6	6	7

There are seven classes, the first of which is just μ_0^2 and the second of which has zero expectation. The expectations of four of the remaining five classes depend on the identity by descent of the two genes A_i, A_k. These five expectations are now shown.

Class	Expectations
3	$\mu_0 F_1 \sum_i p_i d_{ii}^a$
4	$\sum_i p_i a_i^2$
5	$F_1 \sum_i p_i a_i^2 + (1 - F_1) \sum_{i,k} p_i p_k a_i a_k$
6	$F_1 \sum_i p_i a_i d_{ii}^a + (1 - F_1) \sum_{i,k} p_i p_k a_i d_{ik}^a$
7	$F_1 \sum_i p_i (d_{ii}^a)^2 + (1 - F_1) \sum_{i,k} p_i p_k (d_{ik}^a)^2$

Recognizing that the second sums for classes 5 and 6 are zero, and defining the familiar additive and dominance variance components as

$$\sigma_A^2 = 2\sigma_a^2 = 2\sum_i p_i a_i^2 \;, \quad \sigma_D^2 = \sum_{i,k} p_i p_k (d_{ik}^a)^2$$

we can now find $\mathcal{E}(G_k^i)^2$ by collecting terms over classes 1-7. If we then subtract μ_F^2 from (2) we have the variance:

$$\sigma_{G_F}^2 = (1 + F_1) \sigma_A^2 + (1 - F_1) \sigma_D^2 + 4F_1 \sum_i p_i a_i d_{ii}^a$$
$$+ F_1 \sum_i p_i (d_{ii}^a)^2 - (F_1)^2 (\sum_i p_i d_{ii}^a)^2. \quad (3)$$

Setting $F_1 = 1$ or 0 gives the variances $\sigma_{G_1}^2$ or $\sigma_{G_0}^2$ for completely inbred or non-inbred populations respectively and (3) can be rewritten in the form given by Wright (1951):

$$\sigma_{G_F}^2 = (1 - F_1) \sigma_{G_0}^2 + F_1 \sigma_{G_1}^2 + F_1 (1 - F_1) (\mu_1 - \mu_0)^2 \;.$$

The form of equation (3) is helpful in showing how both additive and dominance effects are involved in the way inbreeding changes population variances. Without dominance, complete inbreeding will increase genetic variance to twice the non-inbred value, but with dominance nothing can be said. It would be possible for example for inbreeding to decrease variance.

Covariance Between Relatives

To find the covariance between the genotypic values G_k^i of individual X: $A_i A_k$ and G_ℓ^j of individual Y: $A_j A_\ell$ we have six classes of expectations:

	μ_0	a_i	a_k	d_{ik}^a
μ_0	1	2	2	3
a_j	2	4	4	5
a_ℓ	2	4	4	5
$d_{j\ell}^a$	3	5	5	6

As in the variance situation, the first class has expectation μ_0^2, the second has zero expectation and the third has expectation $\mu_0 F_1 \sum_i p_i d_{ii}^a$. The remaining three classes require account to be taken of the identity relations between two, three or four of the four genes within X and Y. A complete set of four-gene measures was given by Gillois (1964) and Harris (1964), and we use the equivalent set of Cockerham (1971). Recall that the mean and variance required only the two-gene measure F_1.

For individuals W and Z with genes $A_1 A_2$ and $A_3 A_4$ respectively at the A locus, there are fifteen possible identity states. Generally we do not wish to distinguish between the two genes within an individual and so we can work with just nine measures. One measure, not needed for the expectations, is for the case where no subset of the four genes are identical by descent. Two further measures are the inbreeding coefficients F_{1W}, F_{1Z} for individuals W, Z. The remaining six measures are now defined in terms of the genes which are identical by descent in each case. An equivalence sign $\equiv$ denotes identity by descent, and nothing is implied about the identity of genes not connected by an equivalence sign.

Identity Measure	Definition
$4\theta_{WZ}$	$\text{prob}(A_1 \equiv A_3) + \text{prob}(A_1 \equiv A_4) + \text{prob}(A_2 \equiv A_3) + \text{prob}(A_2 \equiv A_4)$
$2\gamma_{W\ddot{Z}}$	$\text{prob}(A_1 \equiv A_2 \equiv A_3) + \text{prob}(A_1 \equiv A_2 \equiv A_4)$
$2\gamma_{\ddot{W}Z}$	$\text{prob}(A_1 \equiv A_3 \equiv A_4) + \text{prob}(A_2 \equiv A_3 \equiv A_4)$
$\delta_{\ddot{W}.\ddot{Z}}$	$\text{prob}(A_1 \equiv A_2 \text{ and } A_3 \equiv A_4)$
$2\delta_{\ddot{W}+\ddot{Z}}$	$\text{prob}(A_1 \equiv A_3 \text{ and } A_2 \equiv A_4) + \text{prob}(A_1 \equiv A_4 \text{ and } A_2 \equiv A_3)$
$\delta_{\ddot{W}\ddot{Z}}$	$\text{prob}(A_1 \equiv A_2 \equiv A_3 \equiv A_4)$

We recognize θ_{WZ} as the coancestry coefficient of individuals W and Z.

These identity measures are now applied directly to give the expectations for the remaining three classes needed in the covariance of G_k^i and G_ℓ^j. Only non-zero terms are shown.

Class	Expectations
4	$4\theta_{XY} \sum_i p_i a_i^2$
5	$2(\gamma_{\ddot{X}Y} + \gamma_{X\ddot{Y}}) \sum_i p_i a_i d_{ii}^a$
6	$\delta_{\ddot{X}.\ddot{Y}} (\sum_i p_i d_{ii}^a)^2 + \delta_{\ddot{X}\ddot{Y}} \sum_i p_i (d_{ii}^a)^2$ $+ 2\delta_{\ddot{X}+\ddot{Y}} \sum_{i,k} p_i p_k (d_{ik}^a)^2$

Collecting terms over classes 1-6, and subtracting the product $\mathscr{E} G_k^i \mathscr{E} G_\ell^j$ leads to the desired covariance

$$C_{XY} = 2\theta_{XY} \sigma_A^2 + 2\delta_{\ddot{X}+\ddot{Y}} \sigma_D^2 + 2(\gamma_{\ddot{X}Y} + \gamma_{X\ddot{Y}}) \sum_i p_i a_i d_{ii}^a$$
$$+ \delta_{\ddot{X}\ddot{Y}} \sum_i p_i (d_{ii}^a)^2 + (\delta_{\ddot{X}.\ddot{Y}} - F_{1X} F_{1Y})(\sum_i p_i d_{ii}^a)^2 . \quad (4)$$

As a check on this result we see that it reduces to (3) when X and Y are the same individual. The reduction uses

$2\theta_{XX} = 1 + F_{1X}$, $2\delta_{X+X} = 1 - F_{1X}$, $\gamma_{XX} = F_{1X}$, $\delta_{XX}^{\cdots} = F_{1X}$, $\delta_{X.X} = 0$.

If there is no dominance, equation (4) shows that the covariance of relatives

$$C_{XY} = 2\theta_{XY}\,\sigma_A^2$$

depends explicitly only on the degree of relatedness of the relatives. If neither relative is inbred

$$C_{XY} = 2\theta_{XY}\,\sigma_A^2 + 2\delta_{X+Y}\,\sigma_D^2$$

involves only the additive and dominance variance components. In other situations, however, the use of covariances in estimating variance components must take proper account of the three sums involving d_{ii}^a terms.

There is great simplification for the case of only two alleles at a locus. Explicit expressions for the effects and variance components are

$$a_1 = \frac{-p_2}{p_1}\,a_2 = p_2[p_1(G_1^1 - G_2^1) + p_2(G_2^1 - G_2^2)]$$

$$\frac{p_1}{p_2}d_1^1 = \frac{p_2}{p_1}d_2^2 = -d_2^1 = p_1 p_2 (G_1^1 + G_2^2 - 2G_2^1)$$

$$\sigma_A^2 = 2p_1 p_2 [p_1(G_1^1 - G_2^1) + p_2(G_2^1 - G_2^2)]^2,$$

$$\sigma_D^2 = p_1^2 p_2^2 (G_1^1 + G_2^2 - 2G_2^1)^2.$$

For two equally frequent alleles, $p_1 = p_2 = 0.5$, equations (3) and (4) reduce to

$$\sigma_G^2 = (1 + F_1)[\sigma_A^2 + (1 - F_1)\,\sigma_D^2]$$

$$C_{XY} = 2\theta_{XY}\sigma_A^2 + (2\delta_{X+Y} + \delta_{XY}^{\cdots} + \delta_{X.Y} - F_{1X}F_{1Y})\sigma_D^2.$$

Two general comments are in order at this point. In the first place, just as discussed for the mean, the treatment of variances

and covariances can be extended to include other types of initial
population. Equations (3), (4) were based on the simple assumption that the initial population was constituted from the random
pairing of gametes drawn at random from the infinite random-mating reference population. Secondly, the expressions for means,
variances and covariances involve various identity measures. Procedures are well established (e.g. Cockerham, 1971) for the evaluation of these measures in any generation of an inbreeding system.

TWO-LOCUS THEORY

We have seen that one-locus theory is concerned with the dependencies between the actions or frequencies of allelic genes
caused by dominance or inbreeding. To these features we must add
epistasis, linkage and linkage disequilibrium when two-locus models
are constructed.

The general two-locus model for the genotypic value of individuals with genotype $A_i B_j / A_k B_\ell$ formed by the union of gametes
$A_i B_j$ and $A_k B_\ell$ is written as

$$G_{k\ell}^{ij} = \mu_0 + a_i + a_k + b_j + b_\ell + ab_{ij} + ab_{i\ell} + ab_{kj} + ab_{k\ell}$$
$$+ d_{ik}^a + d_{j\ell}^b + ad_{ij\ell}^b + ad_{kj\ell}^b + bd_{jik}^a + bd_{\ell ik}^a + d^a d_{ikj\ell}^b$$

where the effects are defined for an infinite random-mating reference population in linkage equilibrium. Additive or dominance
effects at the B locus are written as b or d^b respectively. The
epistatic effects are written as ab (additive by additive), ad^b
or bd^a (additive by dominance), and $d^a d^b$ (dominance by dominance).
Least squares values of the effects may be found as in the one-locus
case.

Means and variances are sought for a population initiated
from the random union of gametes $A_i B_j$ in frequency $(p_i q_j + \mathcal{D}_{ij})$,
where p_i and q_j are allelic frequencies and $\mathcal{D}_{ij}$ the initial linkage
disequilibria. When expectations are taken of $G_{k\ell}^{ij}$ or $(G_{k\ell}^{ij})^2$, we
need to take notice of which genes have descended from the same
initial gamete so that the effects of such initial linkage

disequilibria may be incorporated. We use our two-locus descent measures for this purpose.

A pair of genes are said to be equivalent by descent if they both descend from genes on one initial gamete. Allelic genes which are equivalent by descent are also identical by descent. For a random individual with genes $A_1 B_1 / A_2 B_2$ in the population of interest there are six pairs of genes, and fifteen possible arrangements on gametes of the initial genes from which A_1, B_1, A_2, B_2 are descended. The number of arrangements reduces to nine when each locus is assumed to be equally inbred. The probabilities, or descent measures, of these nine arrangements sum to one and we choose to work with an equivalent set of eight summary measures. These eight measures, and the corresponding set of equivalent genes, are as follows.

Descent Measure	Equivalent Genes	Descent Measure	Equivalent Genes
F_1	$A_1 \equiv A_2$ or $B_1 \equiv B_2$	F_{11}	$A_1 \equiv A_2$ and $B_1 \equiv B_2$
F^1	$A_1 \equiv B_1$ or $A_2 \equiv B_2$	F^{11}	$A_1 \equiv B_1$ and $A_2 \equiv B_2$
$_1F$	$A_1 \equiv B_2$ or $A_2 \equiv B_1$	$_{11}F$	$A_1 \equiv B_2$ and $A_2 \equiv B_1$
$_1F^1_1$	$A_1 \equiv B_1 \equiv B_2$ or $A_2 \equiv B_1 \equiv B_2$ or $A_1 \equiv A_2 \equiv B_1$ or $A_1 \equiv A_2 \equiv B_2$	F^{11}_{11}	$A_1 \equiv B_1 \equiv A_2 \equiv B_2$

In each of these definitions, nothing is implied about the equivalence by descent of genes not mentioned. The measures hold for an arbitrary number of alleles at a locus and allow the treatment of populations descending from a random sample or a specific set of initial gametes. They can be modified to accommodate specific initial individuals. We have also established general procedures for evaluating the measures in any situation, and have given detailed results for selfing, sib mating, mixed self and random mating, and random mating in finite monoecious populations (Cockerham and Weir, 1973, Weir and Cockerham, 1973, 1974).

Means

The descent measures allow genes on a random member of a

population to be related to genes on initial gametes. This treatment is the essence of the expectation process. Frequencies of initial gametes are needed and the notation used is as follows. Two initial gametes carry genes A_iB_j and A_kB_ℓ respectively with probability $\theta^{ij}_{k\ell}$. Three or four initial gametes carry genes A_iB_j, A_k, B_ℓ or A_i, B_j, A_k, B_ℓ with probabilities $\theta^{ij}_{k|\ell}$ or $\theta^{i|j}_{k|\ell}$ respectively. Appropriate sums, denoted by dots, give frequencies for one, two or three genes and these initial frequencies were discussed by Cockerham and Weir (1973).

The expectations of the mean, the additive and dominance effects were given in the one-locus treatment. For the expectations of the remaining effects in $G^{ij}_{k\ell}$ we make use of the fact that effects where any one of the genes is independent of the others have zero expectations (Gallais, 1974). Specifically, each of the following sums is zero:

$$\sum_i p_i ab_{ij}, \quad \sum_j q_j ab_{ij}, \quad \sum_i p_i ad^b_{ij\ell}, \quad \sum_j q_j ad^b_{ij\ell}, \quad \sum_i p_i ad^b_{ijj},$$

$$\sum_i p_i d^a d^b_{ikj\ell}, \quad \sum_j q_j d^a d^b_{ikj\ell}, \quad \sum_i p_i d^a d^b_{ikjj}, \quad \sum_j q_j d^a d^b_{iij\ell}.$$

Terms leading only to such sums are omitted in the following list of expectations.

Effect   Expectation

additive by additive

ab_{ij} $F^1 \sum_{i,j} \theta^{ij}_{..} ab_{ij} = F^1 \sum_{i,j} \mathcal{D}_{ij} ab_{ij}$

$ab_{i\ell}$ $_1F \sum_{i,j} \theta^{ij}_{..} ab_{ij} = {}_1F \sum_{i,j} \mathcal{D}_{ij} ab_{ij}$

additive by dominance

$ad^b_{ij\ell}$ $_1F^1_1 \sum_{i,j} \theta^{ij}_{..} ad^b_{ijj} = {}_1F^1_1 \sum_{i,j} \mathcal{D}_{ij} ad^b_{ijj}$

bd^a_{jik} $_1F^1_1 \sum_{i,j} \theta^{ij}_{..} bd^a_{jii} = {}_1F^1_1 \sum_{i,j} \mathcal{D}_{ij} bd^a_{jii}$

dominance by dominance

$d^a d^b_{ikj\ell}$ $F^{11}_{11} \sum_{i,j} \theta^{ij}_{..} d^a d^b_{iijj}$

                         $+ (F_{11} - F^{11}_{11}) \sum_{i,j} \theta^{i\cdot}_{\cdot j} d^a d^b_{iijj}$

$$+ (F^{11} - F_{11}^{11}) \sum_{i,j,k,\ell} \theta_{k\ell}^{ij} d_{ikj\ell}^a d_{ikj\ell}^b$$

$$+ (_{11}F - F_{11}^{11}) \sum_{i,j,k,\ell} \theta_{kj}^{i\ell} d^a d_{ikj\ell}^b$$

$$= F_{11} \sum_{i,j} p_i q_j d^a d_{iijj}^b$$

$$+ F_{11}^{11} \sum_{i,j} \mathcal{D}_{ij} d^a d_{iijj}^b$$

$$+ (F^{11} + {}_{11}F - 2F_{11}^{11}) \sum_{i,j,k,\ell} \mathcal{D}_{ij} \mathcal{D}_{k\ell} d^a d_{ikj\ell}^b.$$

Collecting the expectations of all sixteen terms in $\delta\, G_{k\ell}^{ij}$ gives

$$\mu_F = \mu_0 + F_1 (\sum_i p_i d_{ii}^a + \sum_j q_j d_{jj}^b) + F_{11} \sum_{i,j} p_i q_j d^a d_{iijj}^b$$

$$+ 2(F^1 + {}_1F) \sum_{i,j} ab_{ij} \mathcal{D}_{ij} + 2{}_1F^1 \sum_{i,j} (ad_{ijj}^b + bd_{jii}^a) \mathcal{D}_{ij}$$

$$+ F_{11}^{11} \sum_{i,j} d^a d_{iijj}^b \mathcal{D}_{ij} + (F^{11} + {}_{11}F - 2F_{11}^{11}) \sum_{i,j,k,\ell} d^a d_{ikj\ell}^b \mathcal{D}_{ij} \mathcal{D}_{k\ell}. \quad (5)$$

This result is equivalent to that given by Gallais (1974).

For the non-inbred initial population, $F^1 = F^{11} = 1$ and all other descent measures are zero so that the mean becomes

$$\mu_I = \mu_0 + 2 \sum_{i,j} ab_{ij} \mathcal{D}_{ij} + \sum_{i,j,k,\ell} d^a d_{ikj\ell}^b \mathcal{D}_{ij} \mathcal{D}_{k\ell}$$

which reflects the effects of initial linkage disequilibrium. For completely inbred populations we know (Cockerham and Weir, 1973) that

$$F_1 = F_{11} = 1, \quad F^1 = {}_1F = {}_1F_1^1 = F^{11} = {}_{11}F = F_{11}^{11}$$

and the common value of the last six measures is written as $F_{(\infty)}^1$. Equation (5) then gives

$$\mu_1 = \mu_0 + \sum_i p_i d_{ii}^a + \sum_j q_j d_{jj}^b + \sum_{i,j} p_i q_j d^a d_{iijj}^b$$

$$+ F_{(\infty)}^1 \sum_{i,j} (4ab_{ij} + 2ad_{ijj}^b + 2bd_{jii}^a + d^a d_{iijj}^b) \mathcal{D}_{ij}.$$

Evidently there is no longer a simple relation between μ_0, μ_1 and μ_F.

In the absence of linkage disequilibrium, the mean is seen to involve only dominance and dominance by dominance effects. The result for this special case was given by van Aarde (1974), who then equated F_{11} and F_1^2 in the case of no linkage to duplicate the result of Kempthorne (1957). In general however, we know that $F_1 \geq F_{11} \geq (F_1)^2$ with the last equality holding in the case of no linkage only when all matings are specified and all members of a generation have the same pedigree. Previously (Weir and Cockerham, 1969) we discussed the identity disequilibrium

$$\eta_{11} = F_{11} - (F_1)^2$$

and showed that, although it is quite small, it is still non-zero for no linkage in finite random-mating populations.

Variances

The first step in the evaluation of $\mathcal{E}(G_{k\ell}^{ij})^2$ is to identify the distinct classes of expectations of squares and products of effects, as was done in the one-locus situation. Recognizing symmetry between loci, there are 34 classes, the first 7 of which were treated in that previous case. The classes are now listed and the expectation process illustrated for classes 23 and 24.

Class	Typical element	No. of terms in class	Class	Typical element	No. of terms in class
1	μ_0^2	1	7	$(d_{ik}^a)^2$	2
2	$\mu_0 a_i$	8	8	$\mu_0 ab_{ij}$	8
3	$\mu_0 d_{ik}^a$	4	9	$\mu_0 ad_{ij\ell}^b$	8
4	a_i^2	4	10	$\mu_0 d_{ik}^a d_{j\ell}^b$	2
5	$a_i a_k$	4	11	$a_i b_j$	8
6	$a_i d_{ik}^a$	8	12	$a_i ab_{ij}$	16

TWO-LOCUS THEORY IN QUANTITATIVE GENETICS

Class	Typical element	No. of terms in class	Class	Typical element	No. of terms in class
13	$a_i ab_{kj}$	16	24	$ab_{ij} ad^b_{kj\ell}$	16
14	ab^2_{ij}	4	25	$d^a_{ik} ad^b_{ij\ell}$	8
15	$ab_{ij} ab_{ik}$	8	26	$d^a_{ik} bd^a_{jik}$	8
16	$ab_{ij} ab_{k\ell}$	4	27	$(ad^b_{ij\ell})^2$	4
17	$a_i d^b_{j\ell}$	8	28	$ad^b_{ij\ell} ad^b_{kj\ell}$	4
18	$ab_{ij} d^a_{ik}$	16	29	$ad^b_{ij\ell} bd^a_{jik}$	8
19	$d^a_{ik} d^b_{j\ell}$	2	30	$a_i d^a d^b_{ikj\ell}$	8
20	$a_i ad^b_{ij\ell}$	8	31	$ab_{ij} d^a d^b_{ikj\ell}$	8
21	$a_i ad^b_{kj\ell}$	8	32	$d^a_{ik} d^a d^b_{ikj\ell}$	4
22	$a_i bd^a_{jik}$	16	33	$ad^b_{ij\ell} d^a d^b_{ikj\ell}$	8
23	$ab_{ij} ad^b_{ij\ell}$	16	34	$(d^a d^b_{ikj\ell})^2$	1

$$\mathscr{E}(ab_{ij} ad^b_{ij\ell}) = {}_1F^1_1 \sum_{i,j} \theta^{ij}_{..} ab_{ij} ad^h_{ijj} + (F_1 - {}_1F^1_1) \sum_{i,j} \theta^{i}_{.j} ab_{ij} ad^h_{ijj}$$

$$+ (F^1 + {}_1F - 2{}_1F^1_1)/2 \sum_{i,j,\ell} \theta^{ij}_{.\ell} ab_{i\ell} ad^b_{ij\ell}$$

$$\mathscr{E}(ab_{ij} ad^b_{kj\ell}) = F^{11}_{11} \sum_{i,j} \theta^{ij}_{..} ab_{ij} ad^b_{ijj} + (F_{11} - F^{11}_{11}) \sum_{i,j} \theta^{i}_{.j} ab_{ij} ad^b_{ijj}$$

$$+ ({}_1F^1_1 - F^{11}_{11}) \sum_{i,j,\ell} \theta^{ij}_{.\ell} ab_{i\ell} ad^b_{ij\ell}$$

$$+ (F^{11} - F^{11}_{11}) \sum_{i,j,k,\ell} \theta^{ij}_{k\ell} ab_{ij} ad^b_{kj\ell}$$

$$+ ({}_{11}F - F^{11}_{11}) \sum_{i,j,k,\ell} \theta^{i\ell}_{kj} ab_{i\ell} ad^b_{kj\ell}.$$

Simplification of the sum of all these expectations follows the introduction of the following familiar variance components:

$$\sigma_A^2 = 2\sigma_a^2 + 2\sigma_b^2 = 2\sum_i p_i a_i^2 + 2\sum_j q_j b_j^2, \quad \sigma_{AA}^2 = 4\sigma_{ab}^2 = 4\sum_{i,j} p_i q_j ab_{ij}^2$$

$$\sigma_D^2 = \sigma_{d^a}^2 + \sigma_{d^b}^2 = \sum_{i,k} p_i p_k (d_{ik}^a)^2 + \sum_{j,\ell} q_j q_\ell (d_{j\ell}^b)^2$$

$$\sigma_{AD}^2 = 2\sigma_{ad^b}^2 + 2\sigma_{bd^a}^2 = 2\sum_{i,j,\ell} p_i q_j q_\ell (ad_{ij\ell}^b)^2 + 2\sum_{i,j,k} p_i q_j p_k (bd_{jik}^a)^2$$

$$\sigma_{DD}^2 = \sigma_{d^a d^b}^2 = \sum_{i,j,k,\ell} p_i q_j p_k q_\ell (d^a d^b_{ikj\ell})^2$$

When the square of the mean is subtracted from $\mathcal{E}(G_{k\ell}^{ij})^2$ we obtain the general variance formula displayed in equation (6) on the next two pages. Some features of this unwieldy result can be noted.

The coefficients of the five variance components in the first line of (6) almost show the kind of pattern often suggested in the literature. The coefficients of the epistatic components differ from the products of the corresponding additive and dominance components by identity disequilibrium:

$$\sigma_{G_F}^2 = (1 + F_1) \sigma_A^2 + (1 - F_1) \sigma_D^2 + (1 + F_1)^2 \sigma_{AA}^2$$
$$+ (1 + F_1)(1 - F_1) \sigma_{AD}^2 + (1 - F_1)^2 \sigma_{DD}^2$$
$$+ \eta_{11} (\sigma_{AA}^2 - \sigma_{AD}^2 + \sigma_{DD}^2) + \ldots$$

The identity disequilibrium again features in the variance of genes which are unlinked, for which there is no linkage disequilibrium, and for which there are no epistatic effects. We find from equation (6) that

$$\sigma_{G_F}^2 = \sigma_{G_F}^2 \text{ (locus A only)} + \sigma_{G_F}^2 \text{ (locus B only)}$$
$$+ 2\eta_{11} \sum_i p_i d_{ii}^a \sum_i q_i d_{jj}^b \ .$$

TWO-LOCUS THEORY IN QUANTITATIVE GENETICS

The first two, one-locus, variances are as given in equation (3) and the last term may be regarded as the covariance between the effects at the two loci. There is still a dependence between the loci, caused by the inbreeding system, depending on dominance at each locus and the amount of identity disequilibrium.

$$\sigma^2_{G_F} = (1+F_1)\sigma^2_A + (1-F_1)\sigma^2_D + (1+2F_1+F_{11})\sigma^2_{AA} + (1-F_{11})\sigma^2_{AD}$$

$$+ (1-2F_1+F_{11})\sigma^2_{DD} + F_1 [\sum_i p_i(4a_i + d^a_{ii})d^a_{ii} + \sum_j q_j(4b_j + d^b_{jj})d^b_{jj}]$$

$$+ 2F_{11}(\sum_i p_i d^a_{ii})(\sum_j q_j d^b_{jj}) + 2(F_1 + F_{11})\sum_{i,j} p_i q_j [(ad^b_{ijj})^2$$

$$+ (bd^a_{jii})^2 + 4ab_{ij}(ad^b_{ijj} + bd^a_{jii})] + 4F_{11} \sum_{i,j} p_i q_j (d^a_{ii} ad^b_{ijj}$$

$$+ d^b_{jj} bd^a_{jii} + 2ad^b_{ijj} bd^a_{jii}) + F_{11} \sum_{i,j} p_i q_j [4(a_i + b_j) + 8ab_{ij}$$

$$+ 2(d^a_{ii} + d^b_{jj}) + 4(ad^b_{ijj} + bd^a_{jii}) + d^a d^b_{iijj}] d^a d^b_{iijj}$$

$$+ 4(F_1 + F_{11}) \sum_{i,j} p_i q_j (a_i ad^b_{ijj} + b_j bd^a_{jii})$$

$$+ (F_1 - F_{11}) \sum_{i,j,k} p_i q_j p_k (2d^a_{ik} + 4bd^a_{jik} + d^a d^b_{ikjj}) d^a d^b_{ikjj}$$

$$+ (F_1 - F_{11}) \sum_{i,j,\ell} p_i q_j q_\ell (2d^b_{j\ell} + 4ad^b_{ij\ell} + d^a d^b_{iij\ell}) d^a d^b_{iij\ell}$$

$$+ 4(F^1 + {}_1F) \sum_{i,j} a_i b_j \mathscr{D}_{ij} + 4(F^1 + {}_1F + 2{}_1F^1_1) \sum_{i,j} (a_i + b_j) ab_{ij} \mathscr{D}_{ij}$$

$$+ 2(F^1 + {}_1F + 4{}_1F^1_1 + 2F^{11}_{11}) \sum_{i,j} (ab_{ij})^2 \mathscr{D}_{ij} + 2F^{11}_{11} \sum_{i,j} d^a_{ii} d^b_{jj} \mathscr{D}_{ij}$$

$$+ 4{}_1F^1_1 \sum_{i,j} [(a_i + 2ab_{ij})d^b_{jj} + (b_j + 2ab_{ij})d^a_{ii}] \mathscr{D}_{ij}$$

$$+ 2({}_1F^1_1 + F^{11}_{11}) \sum_{i,j} [(2b_j + 4ab_{ij} + bd^a_{jii})bd^a_{jii}$$

$$+ (2a_i + 4ab_{ij} + ad^b_{ijj})ad^b_{ijj}] \mathscr{D}_{ij}$$

$$+ 4{}_1F^1_1 \sum_{i,j} [(2a_i + d^a_{ii})bd^a_{jii} + (2b_j + d^b_{jj})ad^b_{ijj}] \mathscr{D}_{ij}$$

$$+ 2(F^1 + {}_1F - 2{}_1F^1_1) \sum_{i,j,k} p_k [(2a_k + d^a_{ik})bd^a_{jik} + 2ab_{kj}d^a_{ik}] \mathscr{D}_{ij}$$

$$+ 2(F^1 + {}_1F - 2{}_1F^1_1) \sum_{i,j,\ell} q_\ell [(2b_\ell + d^b_{j\ell})ad^b_{ij\ell} + 2ab_{i\ell}d^b_{j\ell}] \mathscr{D}_{ij}$$

$$+ 8({}_1F^1_1 - F^{11}_{11}) \sum_{i,j,k} p_k(d^a_{ik} + bd^a_{jik})ad^b_{kjj}\vartheta_{ij}$$

$$+ 8({}_1F^1_1 - F^{11}_{11}) \sum_{i,j,\ell} q_\ell(d^b_{j\ell} + ad^b_{ij\ell})bd^a_{\ell ii}\vartheta_{ij}$$

$$+ 4F^{11}_{11} \sum_{i,j} (d^a_{ii}ad^b_{ijj} + 2ad^b_{ijj}bd^a_{jii} + d^b_{jj}bd^a_{jii})\vartheta_{ij}$$

$$+ 2(F^1 + {}_1F - 2F^{11}_{11}) \sum_{i,j,k} p_k(2ab_{kj} + bd^a_{jik})bd^a_{jik}\vartheta_{ij}$$

$$+ 2(F^1 + {}_1F - 2F^{11}_{11}) \sum_{i,j,\ell} q_\ell(2ab_{i\ell} + ad^b_{ij\ell})ad^b_{ij\ell}\vartheta_{ij}$$

$$+ F^{11}_{11} \sum_{i,j} [4(a_i + b_j) + 8ab_{ij} + 2(d^a_{ii} + d^b_{jj}) + 4(ad^b_{ijj} + bd^a_{jii})$$

$$+ d^ad^b_{iijj}]d^ad^b_{iijj}\vartheta_{ij}$$

$$+ 2({}_1F^1_1 - F^{11}_{11}) \sum_{i,j,k} p_k(2a_k + 4ab_{kj} + d^a_{ik} + 2ad^b_{kjj} + 2bd^a_{jik}$$

$$+ d^ad^b_{ikjj})d^ad^b_{ikjj}\vartheta_{ij}$$

$$+ 2({}_1F^1_1 - F^{11}_{11}) \sum_{i,j,\ell} q_\ell(2b_\ell + 4ab_{i\ell} + d^b_{j\ell} + 2bd^a_{\ell ii} + 2ad^b_{ij\ell}$$

$$+ d^ad^b_{iij\ell})d^ad^b_{iij\ell}\vartheta_{ij}$$

$$+ 2(4F^{11}_{11} - F^{11} - {}_{11}F - 4_1F^1_1 + F^1 + {}_1F) \sum_{i,j,k,\ell} p_k q_\ell(4ab_{k\ell}$$

$$+ 2ad^b_{kj\ell} + 2bd^a_{\ell ik} + d^ad^b_{ikj\ell})d^ad^b_{ikj\ell}\vartheta_{ij}$$

$$+ (F^{11} + {}_{11}F - 2F^{11}_{11}) \sum_{i,j,k,\ell} [4a_i ad^b_{kj\ell} + 4b_j bd^a_{\ell ik} + 4ab_{ij}ab_{k\ell}$$

$$+ 8ab_{ij}(ad^b_{kj\ell} + bd^a_{\ell ik}) + 4d^a_{ik}ad^b_{ij\ell} + 4d^b_{j\ell}bd^a_{jik} + 8ad^b_{ij\ell}bd^a_{jik}$$

$$+ 2ad^b_{ij\ell}ad^b_{kj\ell} + 2bd^a_{jik}bd^a_{\ell ik} + (4a_i + 4b_j + 8ab_{ij} + 2d^a_{ik} + 2d^b_{j\ell}$$

$$+ 2d^a_{ik}d^b_{j\ell} + 4ad^b_{ij\ell} + 4bd^a_{jik} + d^ad^b_{ikj\ell})d^ad^b_{ikj\ell}]\vartheta_{ij}\vartheta_{k\ell}$$

$$- (\mu_F - \mu_0)^2 . \qquad (6)$$

Covariances

Expectations of terms in the product $G_{k\ell}^{ij} G_{pq}^{mn}$ for relatives X: $A_i B_j / A_k B_\ell$ and Y: $A_m B_n / A_p B_q$ require eight-gene descent measures. Such measures would give the probabilities of all the ways in which eight genes shared by two individuals could be located on one to eight initial gametes. In the one-locus case we found many more four-gene measures than there were two-gene measures, so in this two-locus case we expect many more eight-gene measures than there are four-gene measures. There is great reduction in the number of eight-gene measures however when symmetry between the loci is assumed and if the two individuals X, Y are in the same generation of a regular pedigree or monoecious random-mating system. This latter case allows the arrangement of the four gametes among individuals X, Y to be ignored.

Regardless of the number of measures, however, we know that the covariance of $G_{k\ell}^{ij}$ and G_{pq}^{mn} will contain exactly the same terms involving effects, linkage disequilibria and gene frequencies as does the variance expression (6). The difference will be in the descent measure coefficients. The variance has coefficients which are measures defined for genes in one individual while the covariance has coefficients which are measures defined for genes in two individuals. The same relationship is apparent between equations (3) and (4).

We have not identified or evaluated the measures for two individuals. Gallais (1974) has identified one set, but gave no general indication of how to evaluate components of the set. By restricting epistasis to additive by additive he could list an expression for covariance. (This expression contains two omissions: $E\beta_{i_1 i_1} \zeta_{\overline{j_1 k_2}}$, $E\beta_{i_1 * j_1} \zeta_{j_1 k_2 *}$ in his notation). The terms he gives reduce to the appropriate terms in equation (6) when the two individuals are the same.

The task is made considerably easier if linkage equilibrium is assumed. It is then not necessary to consider the equivalence

by descent of non-allelic genes, as the initial frequency of any
set of such genes is just the product of the corresponding
allelic frequencies regardless of on how many gametes these
initial non-allelic genes are located. We can work entirely
with identity measures. For one individual there are 2 identity
measures, F_1, F_{11}. For two individuals there are 9 identity
measures at each locus, and so 81 at two loci. However we do not
need the measure for complete non-identity, and we generally
assume that each locus is equally inbred so that we need to work
with 44 identity measures. In this case of linkage equilibrium,
the two-locus variance has 33 terms, the coefficients of which
are combinations of 2 identity measures and the two-locus covariance
will have the same 33 terms but with coefficients that are
combinations of 44 identity measures.

Denniston (1967, 1975) and van Aarde (1963, 1975) both made
the further assumption that individuals are not inbred. Identity
by descent is still possible for genes in different individuals
(relatives) however. The two-locus variance and covariance in
such cases of linkage equilibrium and no inbreeding involve only
the five variance components. Both van Aarde and Denniston discussed
the evaluation of the various measures of "between individual"
identity by descent.

DISCUSSION

We have sought to demonstrate the use of our two-locus descent
measures in finding two-locus means, variances and covariances.
Both positive and negative features emerge.

On the positive side we believe that the measures, for means
and variances, provide coefficients that are capable of immediate
interpretation. The two-locus inbreeding coefficient F_{11} for example
is an obvious extension of the one-locus coefficient F_1.
The expressions also explicitly demonstrate the effects of linkage
disequilibrium on means and variances, and initial allelic frequencies
enter explicitly. The effects of linkage and inbreeding

are accounted for by the descent measures. A variety of assumptions about the initial population can be made. Of some importance we feel is that procedures have been established (Cockerham and Weir, 1973), and illustrated (Weir and Cockerham, 1973, 1974), for evaluating these measures in any situation. The effects of mating system, linkage and time are thus accommodated and demonstrated.

Balancing these factors are two negative ones. We do not have a detailed theory of eight-gene measures, and hesitate to establish what would necessarily be a complex machinery. The complexity of the expressions, such as equation (6), is the second negative feature. As it stands, the result is of little use. The most promising approach would seem to be to find simple approximations for descent measures, maybe in terms of the one-locus inbreeding coefficient as has been suggested recently by Gillois (this conference).

BIBLIOGRAPHY

*Cockerham, C. C. (1954). An extension of the concept of partitioning hereditary variance for analysis of covariances among relatives when epistasis is present. Genetics 39:859-882.

*Cockerham, C. C. (1956). Effects of linkage on the covariances between relatives. Genetics 41:138-141.

Cockerham, C. C. (1971). Higher order probability functions of identity of alleles by descent. Genetics 69: 235-246.

Cockerham, C. C., Weir, B. S. (1973). Descent measures for two loci with some applications. Theor. Pop. Biol. 4: 300-330.

*Denniston, C. (1967). Probability and genetic relationship. Unpublished thesis, University of Wisconsin, Madison.

*Denniston, C. (1975). Probability and genetic relationship: two loci. Annals of Human Genetics 39:89-104.

*Fisher, R. A. (1918). The correlation between relatives on the supposition of Mendelian inheritance. Trans. Roy. Soc. Edinburgh 52:399-433.

*Gallais, A. (1970). Covariances entre apparentés quelconques avec linkage et épistasie. I - Expression générale. Ann. Génét. Sél. Anim. 2:281-310.

*Gallais, A. (1974). Covariances between arbitrary relatives with linkage and epistasis in the case of linkage disequilibrium. Biometrics 30:429-446.

Gillois, M. (1964). La relation d'identité en génétique. Thèse Fac. Sci., Universite de Paris, Paris.

*Gillois, M. (1966). Note sur la variance et la covariance génotypiques entre apparantés. Ann. Inst. Henri Poincaré 2:349-352.

*Harris, D. L. (1964). Genotypic covariances between inbred relatives. Genetics 50:1319-1348.

*Kempthorne, O. (1954). The correlation between relatives in a random mating population. Proc. Roy. Soc. London B 143:103-113.

*Kempthorne, O. (1955). The theoretical values of correlations between relatives in random mating populations. Genetics 40:153-167.

*Kempthorne, O. (1957). An introduction to genetic statistics. Wiley, New York.

Malécot, G. (1948). Les mathématiques de l'hérédité. Masson, Paris.

*Schnell, F. W. (1963). The covariance between relatives in the presence of linkage. In Statistical Genetics and Plant Breeding. Natl. Acad. Sci., Natl. Res. Council. Publication 982:468-483.

Stuber, C. W., Cockerham, C. C. (1966). Gene effects and variances in hybrid populations. Genetics 54:1279-1286.

*van Aarde, I. M. R. (1963). Covariances of relatives in random mating populations with linkage. Unpublished thesis, Iowa State University, Ames.

*van Aarde, I. M. R. (1974). The effect of linkage on the mean value of inbreds derived from a random mating population. Genetics 78:1245-1249.

*van Aarde, I. M. R. (1975). The covariance of relatives derived from a random mating population. Theor. Pop. Biol. 8:166-183.

Weir, B. S., Cockerham, C. C. (1969). Group inbreeding with two linked loci. Genetics 63:711-742.

Weir, B. S., Cockerham, C. C. (1973). Mixed self and random mating at two loci. Genetical Research 21:247-262.

Weir, B. S., Cockerham, C. C. (1974). Behavior of pairs of loci in finite monoecious populations. Theor. Pop. Biol. 6:323-354.

*Wright, S. (1935). The analysis of variance and the correlations between relatives with respect to deviations from an optimum. J. Genetics 30:243-256.

Wright, S. (1951). The genetical structure of populations. Annals of Eugenics 15:323-354.

Estimation of genotypic variance components with dominance in small consanguineous populations

Cl. Chevalet and M. Gillois

LABORATOIRE DE GÉNÉTIQUE CELLULAIRE
CENTRE DE RECHERCHES DE TOULOUSE, I.N.R.A.
B.P. 12, 31320, CASTANET-TOLOSAN, FRANCE

ABSTRACT

Dominance interactions, together with small effective size, are generally not dealt with in applied quantitative genetics. Identity by descent between genes provides a good understanding of these phenomena in random mating populations. Results about computation of identity coefficients and derivation of genotypic covariances between inbred relatives are reviewed. New components of variance that arise with inbreeding are involved. Their estimation may be drawn either from several inbred lines, or from a single inbred line. The first method relies on an analysis of variance among inbred lines and among F_1 and F_2 crosses between them; it is useable in practice for plants or for laboratory animals. The second method is available if the quantitative character is affected by many loci. It is based on special statistical devices combined with suitable mating designs and applies to those cases in which a single inbred line is available, as in farm animal breeding.

INTRODUCTION

Dominance interactions, together with restricted effective size, are generally not dealt with in applied quantitative genetics. Indeed, no commonly used statistical method is satisfactory in such cases, although the genetic understanding of these phenomena has been thoroughly investigated in the last decade. In the field of statistical genetics, developments should rely on the general expressions of genotypic variances and covariances. Such expressions, taking both dominance and inbreeding into account, have been first established by Gillois (1964) and Harris (1964). They introduce two generalizations, when compared to previous results: (1) It is shown that coefficients of inbreeding and of kinship are not sufficient to describe correlations. Identity situations involving three and four genes, and their probabilities are required. (2) Fisher's partition of the genetic variance into additive and dominance variance does not give a sufficient description of a quantitative character when there is inbreeding. Two or three other components were thus defined.

The basic tool in the derivations is, in fact, the identity situation and its probability, the identity coefficient. Generalizing the coefficient of kinship, identity coefficients convey many notions in population genetics, as well as in quantitative genetics. They provide a thorough insight into such topics as population structures, genetic distances, expressions of genic frequency moments in evolving populations, equilibrium structures under mutation or migration. This is why the first part of this paper is devoted to a summary of identity situations and rules of calculation from the pedigree or from the mating scheme. Subsequent parts deal with applications in quantitative genetics.

RELATIONSHIPS BETWEEN GENES, IDENTITY SITUATIONS, IDENTITY COEFFICIENTS

1.1 Definitions

In previous papers, Gillois (1964, 1965, 1966, 1967) has

GENOTYPIC VARIANCE COMPONENTS IN SMALL POPULATIONS   273

developed a logical analysis of the basic concepts used in mathematical and quantitative genetics. The set of genes present in a population is given a structure through the different relationships that can be defined between genes; two genes may be <u>homologous</u> if they share the same locus, <u>isoactive</u> or "alike in state" if they are homologous and have the same effect on phenotype, <u>identical</u> or "alike by descent" if they derive without mutation from a common ancestral gene.

These relationships divide the set of genes in a population into equivalence classes. Any partition into identity classes is an <u>identity situation</u> for the population (Gillois, 1964, 1965). Given initial conditions and a rule for matings, probabilities of such identity situations can be defined, and are referred to as "<u>identity coefficients</u>." Malécot's (1948, 1969b) inbreeding and kinship coefficients (f and ϕ) are identity coefficients involving only two genes, either in a single zygote, or in two zygotes.

1.2. Characterization of Identity Situations

Any set of N homologous genes is split into identity classes by the equivalence identity relationship. Any such partition is an identity situation of "order" N. Given two genes G_1 and G_2, they may be either identical ($G_1 = G_2$), or nonidentical ($G_1 \circ G_2$). For three genes G_1, G_2 and G_3, the following five situations can be found:

$(G_1 = G_2 = G_3)$, or $(G_1 = G_2) \circ (G_3)$, or $(G_1) \circ (G_2 = G_3)$,
or $(G_1 = G_3) \circ (G_2)$, or $(G_1) \circ (G_2) \circ (G_3)$

Between four genes G_1, G_2, G_3 and G_4, fifteen situations are possible (Table 1 from Gillois, 1964) and referred to by symbols S_1 to S_{15}.

To deal with situations involving any number of genes, the following rule may be used. A situation between N genes, given in a fixed order of enumeration $(G_1, G_2, \ldots, G_N)$, will be represented by a string of N integers, $I_1, I_2, \ldots, I_N$, so that two genes G_i and G_j are associated with the same integer $I_i = I_j$ if and only if these genes are identical. Any value which appears in the string I_i can be identified with an identity class, as in Bouffette (1966).

TABLE I

Identity Situations Between Four Genes

	Identity Situations	Numerical Notation $G_1\ G_2\ G_3\ G_4$				Probability
S_1	$(G_1 = G_2 = G_3 = G_4)$	1	1	1	1	δ_1
S_2	$(G_1 = G_2 = G_3) \circ (G_4)$	1	1	1	2	δ_2
S_3	$(G_1 = G_2 = G_4) \circ (G_3)$	1	1	2	1	δ_3
S_4	$(G_1 = G_3 = G_4) \circ (G_2)$	1	2	1	1	δ_4
S_5	$(G_2 = G_3 = G_4) \circ (G_1)$	1	2	2	2	δ_5
S_6	$(G_1 = G_2) \circ (G_3 = G_4)$	1	1	2	2	δ_6
S_7	$(G_1 = G_2) \circ (G_3) \circ (G_4)$	1	1	2	3	δ_7
S_8	$(G_1) \circ (G_2) \circ (G_3 = G_4)$	1	2	3	3	δ_8
S_9	$(G_1 = G_3) \circ (G_2 = G_4)$	1	2	1	2	δ_9
S_{10}	$(G_1 = G_3) \circ (G_2) \circ (G_4)$	1	2	1	3	δ_{10}
S_{11}	$(G_2 = G_4) \circ (G_1) \circ (G_3)$	1	2	3	2	δ_{11}
S_{12}	$(G_1 = G_4) \circ (G_2 = G_3)$	1	2	2	1	δ_{12}
S_{13}	$(G_1 = G_4) \circ (G_2) \circ (G_3)$	1	2	3	1	δ_{13}
S_{14}	$(G_2 = G_3) \circ (G_1) \circ (G_4)$	1	2	2	3	δ_{14}
S_{15}	$(G_1) \circ (G_2) \circ (G_3) \circ (G_4)$	1	2	3	4	δ_{15}

For instance, situation S_7 between genes G_1, G_2, G_3 and G_4, previously written as $(G_1 = G_2) \circ (G_3) \circ (G_4)$, can be now written as 1.1.2.3, or 5.5.1.3, or x.x.y.z, provided $x \neq y \neq z \neq x$.

Two sequences may convey the same identity situation, and a comparison may be made if, for a fixed order in gene enumeration, strings are converted to the equivalent one in which the n-th appearing identity class is referred to by the integer n.

1.3 Computation of Identity Coefficients from a Pedigree

There are two main approaches used to compute inbreeding coefficients. The original one, due to Wright (1922) and Malécot (1948) relies on the enumeration of independent ancestral chains between zygotes. An alternative approach is based on the relationships between kinship coefficients of Malécot (1941). Thus, if I and J are the parents of M, one has for any N that

$$\phi_{M,N} = \frac{1}{2} \phi_{I,N} + \frac{1}{2} \phi_{J,N}$$

Both approaches can be extended to the computation of identity coefficients. The first one has been carried out by Gillois (1964, 1966a), according to an analysis of the algebraic properties of zygotic networks. A zygotic network can be seen as an ordered set, through the ancestral relationship. Also, a zygotic network may convey several mutually exclusive gametic networks in which two genes are linked by an arrow if one of them derives from the other. In the gametic set, those networks provide a partition into independent connected parts, which are trees made up of identical genes. In those conditions, identity situations are completely defined. Any general identity situation is made up of several identity classes, each of which cannot be obtained, except as the basis of some gametic tree. General computation rules have been given by Gillois that allow the systematic identification of zygotic subsets giving rise to gametic trees. Such rules were used in computer programming (Nadot (1971), Nadot and

Vaysseix (1973)).

A direct extension of the second approach would be to describe a set of zygotes, or of genes, by probabilities that two, three or four genes are in all possible identity situations. Although Cockerham (1971) proposed this as a solution, it seems quite unrealistic, as in a set of N zygotes the number of necessary parameters would be

$$4\binom{2N}{4} + \binom{2N}{3} + \binom{2N}{2} \sim \frac{8}{3} N^4 .$$

On the contrary, manipulation of high order situations has been proved to provide an approximate, but tractable, way to undertake the practical computation of identity coefficients (Chevalet, 1971a). Notations of identity situations, as shown in Section 1.2, are very convenient if a new zygote is to be introduced. Let a set of zygotes $(Z_1, Z_2, \ldots, Z_N)$ be in a situation described by the string

$$\{I_1 I_2 \cdots I_{2i-1} I_{2i} \cdots I_{2j-1} I_{2j} \cdots I_{2N-1} I_{2N}\} .$$

If an $(N + 1)$-th zygote has Z_i and Z_j as parents $(i, j \leq N)$, then its genes will be characterized by their identity classes

$$I_{2i-1} I_{2j-1}, \text{ or } I_{2i-1} I_{2j}, \text{ or } I_{2i} I_{2j-1}, \text{ or } I_{2i} I_{2j} ,$$

each issue having the common probability 1/4 (for autosomal genes). This rule leads to the explicit construction of all possible issues of the segregation process of genes allowed by one pedigree. It is usable provided zygotes are put in such an order that the N-th appearing zygote has parents of lower ranks i and j. If one parent is unknown and assumed to be unrelated to any other individual, the gene it gives will be referred to by some new larger integer in the strings. Such an order is provided by dates of birth. After the introduction of one individual of rank N, and the exclusion of parents who have no offspring of higher rank, several identity situations may be confounded into a single one. Computation for large pedigrees requires the use of computers and a Monte Carlo approach, as the number of identity situations is much too large

GENOTYPIC VARIANCE COMPONENTS IN SMALL POPULATIONS

to enable exact calculation. Initial ancestors, whose parents are unknown, are assumed to be unrelated and are referred to by a string, $w_o = (1, 2, 3, \ldots, 2n-1, 2n)$, for n zygotes. Then the segregation process is simulated $\underline{S}$ times, providing a set of $\underline{S}$-independent and equally possible states $w_1, w_2, \ldots, w_S$, for any subset in the pedigree. For a subset of four genes, for example, this provides numbers $S_1, S_2, \ldots, S_{15}$ adding up to $\underline{S}$, so that $\frac{S_j}{S}$ appears as the frequency with which situation S_j occurred. This frequency is an approximate value of the identity coefficient δ_j.

1.4 Mean Values of Identity Coefficients in Random Mating Populations

For mean inbreeding and kinship coefficients, Malécot (1948) gave the following equations:

$$f_t = (1-u)^2 \phi_{t-1}$$

$$\phi_t = (1-u)^2 \frac{1}{2N_e} (1+f_{t-1}) + (1-u)^2 (1-\frac{1}{N_e}) \phi_{t-1}$$

where u is a beta mutation rate per generation (Gillois, 1966), and N_e is the effective size. The change of ϕ_t with t is defined approximately by

$$\phi_t \simeq \frac{1 - 2u}{1 + 4 N_e u} \{1 - (1 - 2u - \frac{1}{2N_e})^t\}$$

When situations involving four genes are dealt with, all situations defined between four genes taken from two, three and four diploid zygotes must be considered. To write down all the equations, it is convenient to introduce also identity situations between three genes from two and three zygotes. In total, Gillois (1964) had to derive 26 equations between two generations. The main qualitative results of the computations is that, when time is expressed by the parameter $F = 1 - \exp(-t/2N_e)$, then the value of any identity coefficient turns out to be approximately independent of N_e, and to depend only on the quantity $4 N_e u$. So ϕ or f are well approximated by F if N_e is large. In the same way, it appears that, as N_e increases, the curve describing the evolution of a coefficient tends to an asymptotic curve, relevant for large N_e values. Theoretical consideration of the largest eigen-values of the system of equations, and empirical observations of the computed curves for large values of N_e (100, 200 and 500) are sufficient to derive the equations of these asymptotic curves. Without mutation one obtains:

$$\delta_1 = 1 - \frac{9}{5}(1-F) + (1-F)^3 - \frac{1}{5}(1-F)^6$$

$$\delta_2 + \delta_3 + \delta_4 + \delta_5 = \frac{6}{5}(1-F) - 2(1-F)^3 + \frac{4}{5}(1-F)^6$$

$$\delta_6 = \frac{1}{5}(1-F) - \frac{1}{3}(1-F)^3 + \frac{2}{15}(1-F)^6$$

$$\delta_7 + \delta_8 = \frac{2}{3}(1-F)^3 - \frac{2}{3}(1-F)^6$$

$$\delta_9 + \delta_{12} = \frac{2}{5}(1-F) - \frac{2}{3}(1-F)^3 + \frac{4}{15}(1-F)^6$$

$$\delta_{10} + \delta_{11} + \delta_{13} + \delta_{14} = \frac{4}{3}(1-F)^3 - \frac{4}{3}(1-F)^6$$

$$\delta_{15} = (1-F)^6$$

(Chevalet, Gillois and Nassar, 1976).

These approximate expressions for large N_e values and no mutation give rise to the approximate expressions of the moments of orders three and four of the gene frequency distribution in a random mating population of finite size, as given by Crow and Kimura (1970, pp. 333-337). More generally, it is known that identity coefficients of order k are related to the moments of order k of gene frequency in a random mating population (Malécot, 1969a).

From results with N_e values of 2, 4, 8, 16, 256, it is seen that approximations are very good as soon as N_e is larger than 16. The absolute value of the difference between the actual coefficient and the predicted value for large effective size is always found to be less than 0.01. For smaller populations, the influence of sex ratio is important.

GENOTYPIC VARIANCE COMPONENTS IN SMALL POPULATIONS 279

IDENTITY BETWEEN GENES AND GENOTYPIC CORRELATIONS

In the following, nonidentity of genes is assumed to imply independence between their allelic states. It is known, however, that it is restricted to pure random mating. Consideration of assortative mating or selection would necessitate the introduction of more general expressions for the joint probabilities of genic and genotypic states (Gillois, 1966c, 1966d, 1967a, 1967b), Gillois, et. al. (1969a, 1969b).

2.1 Conditioned Genotypic Frequencies When Identical Genes are Involved

In a panmictic infinite group, where the Hardy-Weinberg law is assumed, all genes at a locus have the same and independent probabilities of carrying one given allele. All mating types have their frequencies derived directly from the Hardy-Weinberg law. On the other hand, in a line of restricted size, inbreeding occurs, and genes may have a positive probability of being identical by descent. Hence, the Hardy-Weinberg law of association of genes into genotypes must be generalized in the following way. Let q_i^x be the frequency of allele A_i^x at locus x in some panmictic group. In a line derived from this group, the probability that two genes g_1 and g_2 at locus x carry alleles A_i^x and A_j^x depends on the identity situation realized between these two genes. If they are identical ($g_1 = g_2$), genes g_1 and g_2 will be in state $A_i^x A_i^x$ with probability q_i^x, and in states $A_i^x A_j^x$ ($i \neq j$) with zero probabilities. If, on the contrary, they are not identical ($g_1 \circ g_2$), independence between genes may be assumed and genes g_1 and g_2 will be in state $A_i^x A_j^x$ with probability $q_i^x q_j^x$, according to the Hardy-Weinberg law. Given the four genes g_1, g_2, g_3, g_4 of two diploid zygotes, the 15 possible identity situations between them must be considered, and each one gives rise to one conditioned distribution of genotypes. For instance, if situation S_2 is realized at locus x, i.e., $(g_1 = g_2 = g_3) \circ (g_4)$, genotype

$A_i{}^x A_i{}^x A_i{}^x A_j{}^x$ will appear with probability $q_i{}^x q_j{}^x$, other genotypes like $A_i{}^x A_j{}^x A_i{}^x A_j{}^x$ ($j \neq i$) with zero probabilities.

2.2 Various Expressions of Covariances between Inbred Relatives

Up to this point, expectations were not taken with respect to those various possible identity situations. Taking these expectations introduce the probabilities of identity situations, i.e., identity coefficients, and leads to the known expressions of genotypic variances and covariances between inbred relatives (Gillois, 1964, 1966a, 1966b), (Harris, 1964). The reason is the following. At a given locus (x), in a given line derived from some original panmictic group, one and only one identity situation may be realized for a set of genes. If the four genes of two individuals are considered, only one situation, from S_1 to S_{15}, is realized. Hence, the probabilities of these situations, i.e., identity coefficients δ_1 to δ_{15} computed for these individuals, carry no useful information about these two individuals in the line they are picked from. In other words, identity coefficients are related to the pedigree of the line, but not to some one line having this pedigree.

It follows that formulae involving identity coefficients are strictly related to some pedigree; the experimental interpretation of them necessitates that several lines are derived from the same population and according to some one common pedigree. Experiments with plants can be designed in this way from Gillois-Harris expressions.

$$E(Z_I) = f_I E_1 (D) .$$

$$VAR(Z_I) = (1 + f_I)V_A + (1 - f_I)V_D + f_I V'_D + f_I C'_{AD}$$
$$+ f_I (1 - f_I)D^2$$

$$\text{COV}(Z_I Z_J) = 2\phi_{IJ} V_A + (\delta_9 + \delta_{12})V_D$$

$$+ \delta_1 V_D' + (\delta_1 + \frac{1}{4}(\delta_2 + \delta_3 + \delta_4 + \delta_5))C_{AD}'$$

$$+ (\delta_1 + \delta_6 - f_I f_J)D^2 .$$

The new parameters $E_1(D)$, V_D', C_{AD}' and D^2 are:

$$E_1(D) = \sum_x \sum_i q_i^x d_{ii}^x ,$$

$$V_D' = \sum_x [\sum_i q_i^x (d_{ii}^x)^2 - (\sum_i q_i^x d_{ii}^x)^2] ,$$

$$C_{AD}' = 4 \sum_x \sum_i q_i^x a_i^x d_{ii}^x ,$$

$$D^2 = \sum_x (\sum_i q_i^x d_{ii}^x)^2 ,$$

where a_i^x, the additive effect of allele A_i^x of locus x, and d_{ii}^x, the dominance deviation of homozygotes $A_i^x A_i^x$, are defined in the original panmictic population. Thus one has

$$\sum_i q_i^x a_i^x = \sum_i \sum_j q_i^x q_j^x d_{ij}^x = \sum_i \sum_j q_i^x q_j^x a_i^x d_{ij}^x = 0$$

and

$$V_A = 2 \sum_x \sum_i q_i^x (a_i^x)^2 ,$$

$$V_D = \sum_x \sum_i \sum_j q_i^x q_j^x (d_{ij}^x)^2 .$$

These formulae apply to any character, whatever the number of loci and the relative magnitude of their contributions may be. Two slightly different interpretations are possible. First, the formula can apply to a set of N pairs $I_i J_j$ drawn from N lines and showing the same pedigree, and the theoretical covariance is related to the experimental covariance of these data. But, secondly, it is also possible to consider the set of these N lines, and provided they have evolved in the same way, with the same genetic effective size, the coefficients may be seen as mean coefficients in the population. From this point of view, VAR appears to be the total genotypic variance over all the lines, COV is interpretated as the variance of line means, and VAR-COV as the mean of the variances within lines. This remark provides both the solution of a problem previously unsolved, and an easy way to estimate the new genetic variance components (Section 3).

Conversely, it is clear that these formulae have no meaning when attention is concentrated upon one single inbred line. However, if many segregating loci are involved in the building of a character, an identity coefficient δ_k may be related, in a single line, to the frequency with which the identity situation S_k occurs at these different loci. Averaging the contributions of many loci to genotypic variances and covariances may give rise to alternative formulae where identity coefficients take this meaning. Expressions are thus approximate. Detailed conditions for this operation to be valid are given by Chevalet (1971b). They can be summarized in a few words: (i) The number of loci is large; (ii) the ratio of the greatest to the smallest contribution is small with respect to the number of loci. Those are in fact classical hypotheses in quantitative genetics, which explain, in particular, the roughly normal distribution of inherited metric characters. It must be noticed that the preceding conditions do not allow a general use of the approximate formulae; no major gene must contribute, and many genes must still segregate. The limiting expressions, when ϕ, f and δ_1 tend to 1, are meaningless. These

GENOTYPIC VARIANCE COMPONENTS IN SMALL POPULATIONS

approximations are:

$$\text{COV}(Z_I Z_J) \simeq 2\phi_{IJ} V_A + (\delta_9 + \delta_{12}) V_D$$
$$+ \delta_1 V_D' + (\delta_1 + \frac{1}{4}(\delta_2 + \delta_3 + \delta_4 + \delta_5)) C_{AD}'$$

$$\text{VAR}(Z_I) \simeq (1 + f_I) V_A + (1 - f_I) V_D + f_I V_D' + f_I C_{AD}'$$

$$E(Z_I) \simeq f_I E_I(D)$$

It is found that component D^2 disappears in these formulae, because means are related to the actual mean genotypic value achieved in the line, and not to some overall mean value achieved among many lines. From them, experiments can be designed, which are relevant to animal populations, especially to farm animals for which replication of pedigrees is impossible (Section 4).

PARTITIONING THE GENETIC VARIANCES IN A SUBDIVIDED POPULATION

3.1 Expected Results with Dominance

Use of Gillois' expressions for genotypic variances and covariances, where identity coefficients are taken as average ones in homologous isolated lines derived from a base population, provides a generalization of the results of Wright (1951, 1952). Wright showed that, for an additive character whose initial genetic variance is V_o, the total variance among lines is $V_t = (1 + F)V_o$, at a time when the inbreeding coefficient is F. At the same time the average within line variance is $V_w = (1 - F)V_o$ and the variance of the line means is $V_b = 2F V_o$.

Deriving the gene frequency moments of orders 3 and 4, Crow and Kimura obtained a similar decomposition for a quantitative character due to one locus with dominance. Introduction of identity coefficients and of the five variance components V_A, V_D, V_D', C_{AD}' and D^2 of a multi-locus character allows an extension of this result. Furthermore, if the genetic effective size N_e of the lines is large enough ($N_e > 16$; see Section 1.4), mean identity coefficients can be related to F. The decomposition comes out to be

$$V_t = (1+F)V_A + (1-F)V_D + F V_D' + F C_{AD}' + F(1-F) D^2$$

$$V_B = 2F V_A + \frac{2}{3}K V_D + H V_D' + G C_{AD}' + (H + \frac{1}{3}K - F^2)D^2$$

$$V_W = (1-F)V_A + (1-F-\frac{2}{3}K)V_D + (F-H)V_D' + (F-G)C_{AD}' + (F-H - \frac{1}{3}K)D^2$$

where

F = the probability that, at the same locus, two gametes taken at random carry identical genes

$$\simeq 1 - \exp(-t/2 N_e)$$

G = the probability that three gametes taken at random carry identical genes

$$\simeq 1 - \frac{3}{2}(1-F) + \frac{1}{2}(1-F)^3$$

H = the probability that four gametes taken at random carry four identical genes

$$\simeq 1 - \frac{9}{5}(1-F) + (1-F)^3 - \frac{1}{5}(1-F)^6$$

K = the probability that four distinct gametes are distributed among two independent groups, each made up of two identical genes

$$\simeq \frac{3}{5}(1-F) - (1-F)^3 + \frac{2}{5}(1-F)^6$$

(according to Chevalet, Gillois and Nassar, 1976). The changes in these variances with time can therefore be studied as soon as relative values for five genetic variance components are given. Moreover, the redistribution of genetic variances can be calculated from F_1 and F_2 populations derived from inbred lines. From discussion of possible changes by means of numerical illustrations, two principal phenomena are expected from those results: (i) preservation or temporary increase in the mean genetic variance within lines is possible in cases more general than those that are already known; (ii) the genetic variance in F_1 populations obtained by interbreeding inbred lines may be considerably lower than the mean genetic variance in parental populations. Thus the reduction in variance experimentally observed in the F_1 generation must not always be attributed entirely to a reduction in the environmental variance (Chevalet and Gillois, 1976).

GENOTYPIC VARIANCE COMPONENTS IN SMALL POPULATIONS 285

3.2 Genetical Interpretation of an Analysis of Variance

Using the preceding results, estimates of the genetic components of variance may be derived from an experiment in which several independent inbred lines have been reared, together with all possible F_1 crosses and, if possible, with F_2 crosses derived from each F_1. Then the expected genetic variances within and between inbred parental lines, and F_1 and F_2 crosses, may be related to the statistics available in a diallel scheme. For simplicity, only the case where lines have reached fixation will be outlined here. If $_0V_e$, $_1V_e$ and $_2V_e$ stand for environmental variances in parental F_1 and F_2 populations, we get the following expressions:

a) In parental lines:
- within line variance $= {}_0V_e$,
- variance of line means $= 2V_A + V'_D + C'_{AD}$,

b) In F_1 crosses:
- within line variance $= {}_1V_e$,
- variance of main effects $= \frac{1}{2} V_A$,
- interaction variance $= V_D$,

c) In F_2 crosses:
- within line variance $= {}_2V_e + \{\frac{1}{2} V_A + \frac{1}{4} V'_D + \frac{1}{4} C'_{AD}\}$
  $+ \{\frac{1}{8} V'_D + \frac{1}{4} V_D + \frac{1}{4} D^2\}$,
- variance of main effects $= \frac{1}{2} V_A + \frac{1}{16} V'_D + \frac{1}{8} C'_{AD}$,
- interaction variance $= \frac{1}{4} V_D$.

Such an experiment may thus provide estimates of the unknown components V'_D and C'_{AD}, and maybe also of D^2 if environmental variance $_2V_e$ in F_2 crosses is available. Those results are useful for autogamous species in which fixation is easily realized. General results, related to inbred populations where the inbreeding coefficient is F, are obtained by introducing general coefficients G, H and K.

ESTIMATION OF GENETIC VARIANCE COMPONENTS WITH DATA
FROM A SINGLE INBRED LINE

4.1 The Statistical Problem

Closed populations with a small genetic effective size are commonly found in the field of animal genetics, not only in poultry, but also in pigs, rabbits and cattle. For them, classical statistical methods are generally used, whatever the deviations from the underlying genetic theory might be. Such deviations may be illustrated by one poultry line where kinship coefficients were calculated. Whereas current theory takes into consideration only three correlations (between full sibs, with $\phi_{FS} = \frac{1}{4}$; between half sibs, with $\phi_{HS} = \frac{1}{8}$; and between unrelated individuals, with $\phi_U = 0$), calculations proved these three kinds of obvious relationships could not be really distinguished from each other. Kinship coefficients were indeed found in three overlapping intervals (Chevalet, 1974): $0.43 < \phi_{FS} < 0.55$; $0.33 < \phi_{HS} < 0.47$; and $0.24 < \phi_U < 0.40$.

Such a structure makes unrealistic any genetical interpretation of the usual hierarchical analysis of variance.

From approximate covariances between relatives within a single line, the variance-covariance matrix of phenotypes has the structure

$$V = \sum_k d_k \Theta_k,$$

where d_k are matrices of identity coefficients, and Θ_k the variance components to be estimated. Such a model can be related to the general mixed model of analysis of variance, and numerical methods are available, although results about efficiency cannot be directly applied to genetical situations (Henderson, 1953; Hartley and Rao, 1967; Searle, 1970, 1971; Rao, 1971; etc.). Maximum likelihood estimators and quadratic unbiased estimators of variance components have been reviewed for this purpose (Chevalet, 1976a). Designs for which matrices d_k have commutative products with each other are found to be the most interesting ones. They provide a set of sufficient and orthogonal statistics whose

GENOTYPIC VARIANCE COMPONENTS IN SMALL POPULATIONS

expectations are linear combinations of the parameters to be estimated. Though it is often required in theoretical statistics, the property can be only approximately fulfilled in genetics. In practical cases, the aim should be to build a design for which data can be summarized into two independent sets of statistics. The first set represents a "commutative" part, and is made up of orthogonal statistics; the second set is built up from the remaining information and must be treated by general methods.

Such a procedure is applicable with a hierarchical mating design, provided the following additional conditions hold:

(1) Identity coefficients between two offspring depend only on the sires' indices. With the usual notations, where m is the sire index, ℓ the dam index, and k the offspring index within dam family, this condition gives:

$$\text{VAR}(P_{m\ell k}) = v_m \quad \forall \ell, \forall k$$

$$\text{COV}(P_{m\ell k} P_{m\ell k'}) = c_m \quad \forall \ell, \forall k, \forall k' \neq k$$

$$\text{COV}(P_{m\ell k} P_{m\ell' k'}) = b_m \quad \forall \ell, \forall \ell' \neq \ell, \forall k, k'$$

$$\text{COV}(P_{m\ell k} P_{m'\ell' k'}) = e_{mm'} \quad m' \neq m, \forall \ell, \ell', k, k'.$$

(2) The number of offspring per dam depends only on the sire index m.

4.2 An Illustration in Rabbits (Chevalet, 1973, 1976b)

Three growth characteristics were analysed in a line of <u>New Zealand White</u> rabbits, weaning weight at 28 days, weight at slaughter age (77 days) and mean daily weight gain between 28 and 77 days of age. The line was derived from about 15 founders per sex, five generations ahead, and has been reared without selection. Dominance interactions and maternal environmental sources of variation for growth traits in rabbits have been reported by Yao and Eaton (1954) and Venge (1953). The analysis was therefore performed in accordance with a genetical model allowing for those deviations (Gillois, 1964; Willham, 1972). The general expression of phenotypic covariance is made up of 16 components, 12 of which are associated with identity coefficients involving 3 or 4 genes. Approximate expressions for covariances in a single line cannot take into account one component which is associated with a very small identity coefficient (see Section 2). In the line of rabbits, variance and covariance expressions can be restricted to the following ones:

$$\text{VAR}(P_I) \simeq (1 + f_I)V_A + (1 - f_I)V_D$$
$$+ (1 + f_{MI})V_M + (1 + f_{MI} - 2f_I)C_{AM} + V_E$$
$$\text{COV}(P_I P_J) \simeq 2\phi_{I,J} V_A + (\delta_9 + \delta_{12})V_D$$
$$+ 2\phi_{MI,MJ} V_M + 2(\phi_{I,MJ} + \phi_{J,MI})C_{AM}$$

where subscript MI stands for the mother of individual I, V_M is the environmental component of variance arising from the genetical maternal effects on the performance of offspring, C_{AM} is the covariance of these latter effects with direct additive effects, V_E is the variance of one general environmental effect.

According to suggestions formulated in Section 4.1, a hierarchical design was used. Matings were chosen to fulfill the first additional condition stated with respect to kinship coefficients. A few quasi-independent families were available; dams mated to one sire were chosen to be either full sisters, half sisters, or approximately unrelated. Kinship coefficients between mates were small. Identity coefficients were computed for all classes of offspring. Equality conditions required by statistics were quite well fulfilled. However, a great disequilibrium in the collected data was found, females being able to rear from 2 to 58 rabbits up to the weaning age, males having from 27 to 212 offspring. The second condition was not even approximately fulfilled.

Genetical expressions of covariances must not be turned into statistical formulae, except with care. It is seen, indeed, that interaction component C_{AM} cannot be estimated in the planned experiment because of realized values of kinship coefficients, for which $(\phi_{I,MJ} + \phi_{J,MI})$ is found to be almost everywhere a linear function of $\phi_{I,J}$ and $\phi_{MI,MJ}$. Moreover, from comparison with the standard analysis of variance, it is not expected that the design will be efficient with respect to the estimation of the interaction component V_D. Numerical problems cannot be discarded in the discussion of the efficiency of estimation. Dealing in the same way with data disequilibrium, two computer programs were written; one used the maximum likelihood principle, and the other the theory of minimum variance quadratic unbiased estimators (the method of Rao, 1971). The former method leads to tractable computations. It gives, however, no idea about efficiency in disequilibrium cases. The latter method, on the contrary, allows the computing of estimated variances for the estimators of variances components, but requires very heavy and long computations. Comparison of both methods was based on simulation experiments. Pedigrees of rabbits were kept fixed, and segregation of genes was simulated at 72 independent loci. Attaching quantitative values to these genes, one additive character (with heritability of 0.20) and one non-additive character (whose components V_A, F_D, V'_D and C'_{AD} are 0.16, 0.24, 0.19 and -0.40 when divided by the "initial" phenotypic variance $V_E + V_A + V_D$) were simulated and submitted to the estimation procedures.

The results are:

1. In all cases maximum likelihood estimators were computed fastest and a great instability in quadratic methods was observed, due possibly to rounding errors.

2. In the scheme, an estimate of more than 10 per cent for the maternal component is expected to be of significance.

3. Any estimate of the dominance component is questionable.

4. Likelihood estimation of heritability was found to be far better than that provided by the standard analysis of variance, which would ignore relationships between parents. The mean squared deviations between estimated heritability and the true theoretical value of simulated traits were as follows. For the additive trait

($h^2 = .20$), squared deviation was $.0147 = (.121)^2$ by analysis of variance, $.0044 = (.066)^2$ by maximum likelihood estimation. For the "dominant" trait, squared deviations were $.0371 = (.193)^2$ and $.0113 = (.106)^2$, respectively.

Mean effects of parity and litter size at weaning were subtracted before the estimation of variance components was undertaken. Only the distinction between first litters and subsequent ones was recognized, and a linear relationship between mean performance and litter size at weaning was considered. In analysed data these hypotheses seemed to be fair, at least for those results drawn from litters where the number of weaned rabbits was five or more. For weaning weight, heritability was estimated at 20%, a typical value for this trait; 25% of the variance is attributed to environmental genetic maternal effects, including interaction between direct and indirect maternal effects; 40% of the variance is attributed either to dominance interaction, or to some environmental litter effect. For the other two traits no significant indirect maternal effect was found. Assuming additivity of genic effects, as it is usually done for these traits, the known value of 40 to 45% for heritability was obtained, but it was found to be smaller (about 20%) when dominance interactions are assumed; dominance variance would then be of about 12%.

4.3 Further Expected Developments

The foregoing results are strictly dependent upon the mating scheme allowed by the population structure. Nevertheless they prove heritability estimates can at least be improved, when estimation of components involving interactions would have needed more elaborate designs. Appropriate matings would also be needed in order to estimate the covariance between direct and indirect maternal effects, a parameter of great importance in mammals. Finally, several generations under circular mating designs could be appropriate for estimating components C'_{AD} and $V'_{D'}$, introduced when identity coefficients δ_1 and $(\delta_2 + \delta_3 + \delta_4 + \delta_5)$ take rather large values.

At the present time, a more direct use could be made of the ideas and statistical improvements reviewed in this section. In various domestic species, indeed, selection programs tend to be based on closed lines where special structures arise. In a rabbit selection experiment, lines are made up of six sires mated each to about sixteen dams. From random choice of mates among the offspring population the six new sires derive from three male ancestors with probability .23; from four male ancestors with probability .50; from five male ancestors with probability .23. Such population structures are quite similar to that considered in Section 4.2. In cattle breeding such structures are also found, as artificial insemination stations select a small number of related bulls per year.

DISCUSSION

Emphasis has been put on the use of the identity relationship between genes. We feel indeed that involved genetical situations, as found in applied genetics, cannot be handled directly by statistical devices, but need at first a _genetical_ investigation. As far as the usual Fisher model of quantitative inheritance is accepted, the probabilistic approach to correlations based on identity between genes, provides the best mathematical and statistical version of Mendel's laws. However, use of identity coefficients is based on two restrictive assumptions: no selection, and no linkage between loci. In every situation where selection is practiced, expressions based on identity coefficients must be seen as approximate ones, and such approximations have not yet been studied. Linkage can be considered with generalized identity coefficients defined between several loci, but the proportion of recombination must be known, and calculations are very tedious (see Gallais, 1970). Moreover, when going from identity relationships between genes to quantitative correlations, consideration of epistatic deviations is a rather untractable problem (ibid.). Apart from these huge complications, it is worth specifying the significance of genetic parameters that can be estimated from an inbred line or from several inbred lines (Sections 3 and 4).

Genetical expressions for variances and covariances include identity coefficients computed from some "initial" population where individuals have no known relationship. Therefore estimated values of variance components have to be related to this initial population, not to the population in which they were estimated. Those estimated components give information about the character in the base population, and have a predictive value for the average evolution of lines derived from it. Methods of estimation, based either on several lines (Section 3) or one line (Section 4), might provide an improved statistical description of quantitative inheritance. This may be of practical interest, but the genetical significance of those estimates should be verified. Even if such a verification is available--the maintenance of equal estimated components during several generations in a line would make it presumable--no use of them could be made to predict the evolution of the character under selection. At the present time, theories of evolution of selected quantitative characters have been developed, but are based on parameters such as initial frequencies of alleles, degree of dominance and epistasis, proportion of recombination, number of contributing loci. On the other hand, however, values of variance components V_A, V_D, etc. give no information about the preceding useful parameters, but in the special--and unrealistic-- case where all loci contribute in the same way.

Nevertheless, useful improvements are still obtained from a precise analysis of the genetical influences of phenomena like inbreeding and dominance on quantitative variability. Such approaches should result in better heritability estimates, modification of the computation of selection indices and the development of new mating rules in small populations. This work, based on known results in population genetics and statistics, may be humble, but must be done while we wait for the remote day when molecular genetics provides a good understanding of gene interactions and a new model for quantitative inheritance.

ACKNOWLEDGMENT

The authors are particularly grateful to J. R. Sedcole, J. Lewis and E. Pollak for their willingness to correct an early English version.

BIBLIOGRAPHY

(1) Bouffette, J. (1966). Expression de la covariance génotypique chez les tétraploïdes. Thèse 3ème cycle, Fac. Sciences, Lyon.

(2) Chevalet, C. (1971)a. Calcul automatique des coefficients d'identité. Ann. Génét. Sél. anim., 3, 449-462.

(3) Chevalet, C. (1971)b. Calcul a priori, intra- et inter-populations des variances et covariances génotypiques entre apparentés quelconques. Ann. Génét Sélect. anim., 3, 463-477.

(4) Chevalet, C. (1973). Estimation of phenotypic variance components from an inbred flock of rabbits. (Abstract). Genetics, 74, (n° 2, Part 2) s 46.

(5) Chevalet, C. (1974). Nouvelles voies d'approche pour les études génétiques des populations animales. Ist World Congress on Genetics Applied to Livestock Production, Madrid, Garsi ed., Vol. II, 327-343.

(6) Chevalet, C. (1976)a. L'estimation des composantes de la variance phénotypique dans une population consanguine. Ann. Génét. Sél. anim., 8 (2), (in press).

(7) Chevalet, C. (1976)b. Decomposition de la variance phéno-typique de caractères de croissance des lapereaux d'une lignée consanguine. Ann. Génét. Sél. anim., 8 (2), (in press).

(8) Chevalet, C., Gillois, M. (1976). Consanguinité et hétérosis : évolution des moyenes et des variances dans et entre les subdivisions d'une population. Bull. Tech. Dép. Génét. anim. (Inst. Nat. Rech. Agron., Fr.) n° 24, 5-22.

(9) Chevalet, C., Gillois, M. Identity coefficients in finite populations. II: Inbreeding, heterosis and distribution of genetic variance among subdivisions of a population (submitted for publication).

(10) Chevalet, C., Gillois, M., Nassar, R. F. Identity coefficients in finite populations. I: Evolution of identity coefficients in a random mating diploid dioecious population (submitted for publication).

(11) Cockerham, C. C. (1971). Higher order probability functions of identity of alleles by descent. Genetics, 69, 235-246.

(12) Crow, J. F., Kimura, M. (1970). An Introduction to Population Genetics Theory. Harper: New York.

(13) Fisher, R. A. (1918). The correlations between relatives on the supposition of Mendelian inheritance. Trans. Roy. Soc. Edinburgh, 52, 399-433.

(14) Gallais, A. (1970). Covariances entre apparentés quelconques avec linkage et épistasie. I: Expression générale. Ann. Génét. Sél. anim., 2, 281-310.

(15) Gillois, M. (1964). La relation d'identité en génétique. Thèse, Fac. Sciences, Paris, 294 p.

(16) Gillois, M. (1965). Relation d'identité en génétique. Ann. Inst. Henri Poincaré, B, 2, 1-94.

(17) Gillois, M. (1966)a. Le concept d'identité et son importance en génétique. Ann. Génét., 9, 58-65.

(18) Gillois, M. (1966)b. Note sur la variance et la covariance génotypiques entre apparentés. Ann. Inst. Henri Poincaré, B, 2, 349-352.

(19) Gillois, M. (1966)c. La relation de dépendance en Génétique. Ann. Inst. Henri Poincaré, B, 2, 261-278.

(20) Gillois, M. (1966)d. L'homogamie dans une population d'effectif limité. Ann. Inst. Henri Poincaré, B, 2, 299-347.

(21) Gillois, M. (1967)a. La notion de génotype. Ann. Génét., 10, 201-202.

(22) Gillois, M. (1967)b. Les lois conjointes des variables aléatoires génétiques. Ann. Gén., 10, 203-206.

(23) Gillois, M., Bouffette, J., Bouffette, A. R. (1969)a. Etude des population d'effectif limité homogames phénotypiques et panmictiques. Ann. Inst. Henri Poincaré, B, 5, 69-86.

(24) Gillois, M., Bouffette, J., Bouffette, A. R. (1969)b. Covariance génotypique a priori dans les populations homogames. Ann. Inst. Henri Poincaré, B, 5, 87-99.

(25) Harris, D. L. (1964). Genotypic covariances between inbred relatives. Genetics, 50, 1320-1348.

(26) Hartley, H. O., Rao, J. N. K. (1967). Maximum likelihood estimation for the mixed analysis of variance model. Biometrika, 54, 93-108.

(27) Henderson, C. R. (1953). Estimation of variance and covariance components. Biometrics, 9, 226-252.

(28) Malécot, G. (1941). Etude mathématique des populations "mendéliennes." Ann. Univ. Lyon, Sciences, A, 4, 45-60.

(29) Malécot, G. (1946). La consanguinité dans une population limitée. C. R. Acad. Sci. Paris, 222, 841-843.

(30) Malécot, G. (1948). Les Mathématiques de l'Hérédite. Masson et Cie: Paris, 60 p.

(31) Malécot, G. (1969)a. Consanguinité panmictique et consanguinité systématique (coefficients de Wright et de Malécot). Ann. Génét. Sél. anim., 1, 237-242.

(32) Malécot, G. (1969)b. The Mathematics of Heredity. Freeman: San Francisco, 86 p.

(33) Nadot, R. (1971). Les coefficients d'identité. Une méthode de calcul. Génétique et Populations. Hommage à Jean Sutter, P. U. F., Paris, 111-118.

(34) Nadot, R., Vaysseix, G. (1973). Apparentement et identité. Algorithme du calcul des coefficients d'identité. Biometrics, 29, 347-359.

(35) Rao, C. R. (1971)a. Estimation of variance and covariance components. MINQUE theory. J. of Multivariate Analysis, 1, 257-275.

(36) Rao, C. R. (1971)b. Minimum variance quadratic unbiased estimation of variance components. J. of Multivariate Analysis, 1, 445-456.

(37) Searle, S. R. (1970). Large sample variances of maximum likelihood estimations of variance components using unbalanced data. Biometrics, 26, 505-524.

(38) Searle, S. R. (1971). Topics in variance components estimation. Biometrics, 27, 1-76.

(39) Venge, O. (1954). Studies of the maternal influence on the growth of rabbits. Acta Agriculturae Scandinavia, 3, 243-291.

(40) Wright, S. (1922). Coefficients of inbreeding and relationship. Amer., Nat., 56, 330-338.

(41) Wright, S. (1951). The genetical structure of populations. Ann. Eugenics, 15, 323-354.

(42) Wright, S. (1952). The theoretical variance within and among subdivisions of a population that is in a steady state. Genetics, 37, 312-321.

(43) Willham, R. L. (1972). The role of maternal effects in animal breeding. III: Biometrical aspects of maternal effects in animal. J. Anim. Sci., 35, 1288-1293.

(44) Yao, T. S., Eaton, O. N. (1954). Heterosis in the birth weight and the slaughter weight in rabbits. Genetics, 39, 667-676.

The non-linearity of offspring-parent regression

Alan Robertson
INSTITUTE OF ANIMAL GENETICS, EDINBURGH

Asymmetrical responses to artificial selection over several generations are commonplace and plausible explanations are not difficult to find--either in terms of the initial gene frequency of desirable alleles or of scale effects. I deal here with asymmetrical response in the first generation of selection or, rather more generally, with a non-linear regression of offspring on parent. The paper differs somewhat from the talk given at Ames in that some of the difficulties have become clearer to me though I am aware that in its present state it contains loose ends that could be tidied up if there were time, and perhaps even some straightforward errors. The problem has been discussed rather briefly by Kempthorne (1960) and by Curnow (1960) but as far as I know the only recent published work (and this was empirical) is by Nishida and Abe (1974). I encountered this problem in practice in carrying out an offspring mid-parent regression with assortative mating for sternopleural bristles in a highly variable population of Drosophila. The regression was clearly curvilinear and the effective heritability for selection upwards was 0.9 and downwards only 0.4.

I wish to deal with the problem in a character with the same parameters in males and females. I assume either that males and females of the same phenotype are mated (to give an offspring-midparent regression with assortative mating) or that males of a known phenotype are mated to females drawn at random from the population to give an offspring-single parent regression. If in the former we consider only the progeny of extreme parents, this is equivalent to a single generation of selection. If the linear regression coefficient of offspring on parent is not zero, asymmetry of selection response is seen as curvilinearity of regression or, more precisely, as the quadratic term in the regression of offspring on parental phenotype.

We can divide the prediction of progeny values from the phenotypes of the parents into two steps--

i. The regression of gene frequency at relevant loci in the parents on their phenotype.

ii. The functional relationship between progeny mean and gene frequency in parents.

In the absence of genetic interactions, the additive affect of the gene appears in both of these in that with the model

A_2A_2	A_1A_2	A_1A_1
$-a$	d	$+a$
q^2	$2pq$	p^2

the linear regression coefficient of gene frequency on phenotype within population is $(a+d(q-p))pq/\sigma^2$, where σ^2 is the phenotypic variance. The regression within populations must ultimately be non-linear since the dependent variables, as gene frequencies, are bounded between 0 and 1. The expected mean progeny of parents with frequency $p + \delta p$ in both sexes is

$$M + 2(a+d(q-p))\delta p - 2d(\delta p)^2 \qquad (1)$$

where M is the mean with frequency p. If males with frequency $p + \delta p$ are mated to females chosen at random, the expected progeny mean is now linear in δp, being given by

$$M + (a+d(q-p))\delta p$$

Additive Gene Action

Since the functional relationship between gene frequency and progeny mean is here linear, the problem becomes simply a statistical one of the regression of a part on a whole. On the assumption that the heritability is not zero, the symmetry of response depends on the partial regression coefficient of G on P^2. I use g and e for deviations of G and E from their means and assume that they are uncorrelated but not independent i.e. I wish to retain the possibility that the genetic variance is dependent on the environmental level and vice versa so that, although I shall assume $E(ge)$ to be zero, I shall not assume that $E(ge^2)$ and $E(g^2e)$ are zero. The sign of the partial regression of G on P^2 is then the same as that of the covariance of G and P^2, given P. This is equal to $(V_p \operatorname{cov}(GP^2) - \operatorname{cov}(GP)\operatorname{cov}(PP^2))/V_p$. Writing (rs) for the expected value of $g^r e^s$, we have for various expected values

$$V_p = (02) + (20)$$
$$\operatorname{cov}(GP^2) = (30) + 2(21) + (12) + 2m(20)$$
$$\operatorname{cov}(GP) = (20)$$
$$\operatorname{cov}(PP^2) = (30) + 3(21) + 3(12) + (03) + 2m(20) + 2m(02)$$

where m is the overall mean of P. The expected value of the numerator of the expression for the partial covariance is thus

$$(02)(30) - (03)(20) + (21)(2V_e - V_g) + (12)(V_e - 2V_g) . \qquad (2)$$

If g and e are entirely independent, i.e., $(21) = (12) = 0$, then the sign of the partial regression coefficient will be that of

$$\left(\frac{\mu_3}{\mu_2}\right)_g - \left(\frac{\mu_3}{\mu_2}\right)_e .$$

This is a specific case of the statement of Lindley (1947) that if $y = x + w$ with x and w independently distributed, the regression of y on $x + w$ is linear if the cumulants of x are multiples of the cumulants of w. In a genetic context, the result is an algebraic statement of the conclusions of Nishida and Abe. Using a binomial representation of both genetic and environmental variation with the same exponents for each, they found that a linear regression of G on P required skewness of the same sign and magnitude for both g and e.

If we assume that e is normally distributed and that the genetic variance is due to n equivalent loci each with gene frequency p then the ratio of the third to the second moment for the genotypic distribution can be shown to equal $(q-p)(V_g/2np(1-p))^{\frac{1}{2}}$. Thus if p is below 0.5, then the regression coefficient will be positive and it will be greater the fewer the number of loci involved. In common sense terms, the symmetry of the response will depend on the relative distance of the initial population from the two selection limits represented by fixation at all loci.

If the variance of g and e are each dependent on the mean value of the other, then we must consider the other terms in the equation. The curvilinearity now depends on the sign of $E(ge^2)$ and $E(g^2e)$ and on the heritability. Two models of this situation were evaluated empirically. In the first, it was assumed that genotypes with higher values have greater environmental variance, this being assumed to be proportional to $e^{k(G-\bar{G})}$. The partial regression coefficient of G on P^2 was found to be negative when $h^2 > 0.33$ and vice versa, as would be expected from (2) with (21) = 0. In the other model, it was assumed that the phenotype was generated multiplicatively from equivalent underlying normal distributions of genetic and environmental variation. $E(ge^2)$ and $E(g^2e)$ are then equal and positive and it was not surprising to find the quadratic coefficient negative for heritability values greater than 0.5 since (2) is symmetrical in g and e. The probable signs of the two expectations in real life is a matter for speculation. My own feeling is that for the majority of characters they would both be positive. One possible exception would be a character approaching a phenotypic threshold such as egg production in poultry nearing the barrier of one egg a day, when perhaps both might be negative.

Non-additive Gene Action

The regression of progeny mean on parental phenotype will now almost certainly not be linear. I would like to deal briefly with two situations.

i. Dominance. With additive gene action and e normally distributed, the partial regression coefficient of gene frequency on p^2 is zero when the genotypic distribution is symmetrical i.e. $p = 0.5$. With complete dominance ($d = \pm 1$), the coefficient will again be zero with genotypic symmetry i.e. when the recessive gene has a frequency of $1/\sqrt{2}$ so that the two genotypic values are equally frequent. Thus when $d = -1$ (recessivity of the + allele), there will be symmetry at this level at $p = 0.707$ and, as d increases, the critical value of p will decline until at $d = +1$ it is 0.293. Below the critical value of p, the partial regression coefficient is positive.

Added to this we have the quadratic relationship of population mean to p. This implies that in selection or in an offspring mid-parent regression with assortative mating, there is a further contribution to curvilinearity which expression (1) shows has the opposite sign to d. Thus the two factors will augment one another when the recessive allele is at low frequency.

These effects will again depend on the number of such loci contributing to the variance. The larger the number of loci, the more symmetrical the phenotypic distribution and the lower the change of gene frequency on selection with a consequent reduction in curvilinearity. In summary, perhaps the most important possible cause of curvilinearity is a recessive at low or moderate frequencies (the critical value being as high as 0.7) contributing significantly to the genetic variance. Note that in an offspring-single parent regression, the relationship of progeny mean to gene frequency in the parent will now be linear.

   ii. Problems of scale. A general kind of gene interaction may be envisaged in which there is additive gene action but not on the scale of observation. For instance if genes interact in a multiplicative way there might be additive action and a normal distribution due to many loci on an underlying scale, with gene interactions causing an asymmetry of genotypic distribution on the observed scale which would be log-normal. The functional relationship between genotypic mean and the average gene frequency at all loci affecting the character will then be concave upwards. I have carried out some numerical analysis of such models and found that the partial regression of progeny mean on P^2 is invariably positive. One interesting finding appeared. For individuals of a given phenotype I calculated not only their progeny mean but their own mean genotypic value. It then appeared that the asymmetry of the regression of progeny mean on parental phenotype was almost exactly the same as that of the genotypic value of the parents, which depends solely on the relative asymmetries of the genotypic and environmental distributions. In other words, the curvilinearity of the regression of progeny mean on phenotypic value was the same for multiplicative gene action as for additive gene action, given the same skewness of the genotypic distribution.

DISCUSSION

It is clear from the above that asymmetry of response to selection in the first generation can have several causes. I have concentrated solely on the existence of curvilinearity without discussing in any way its magnitude. Further work clearly needs to be done in this direction. I would suggest as a measure of asymmetry in selection the relative response with parents chosen with phenotypic values two standard deviations above and below the mean respectively. This order of selection differential might be achieved in practice with many species.

Although the discussion has been in terms of the asymmetry of the immediate response to selection, the topic is really much wider--the validity of predictions made on the basis of the usual "normal" model of genetic and environmental variation. What other relevant evidence do we have, or could we get, apart from the regression of offspring on parent? Abplanalp's linear heritability estimates (1961) immediately spring to mind--they are in fact regressions of half sib on half sib. But there would appear to be various other sources of evidence. For instance, is the genetic variance between individuals of a given phenotype the same for all values? Will we get the same components of variance within and between families if we rank individuals according to their phenotypic values and carry out a separate analysis for the top and bottom halves of the population?

Finally I would emphasize that I have been considering only situations which were not complicated by differential fitnesses dependent on the value of the quantitative character. There are many examples in Drosophila of selected lines at the limit in which the regression of offspring on parent is highly asymmetrical, being close to zero in the direction of previous selection but being very high indeed for selection in the opposite direction due to the segregation of a lethal gene in the population. This is a totally different phenomenon.

BIBLIOGRAPHY

Abplanalp, H. (1961). Linear heritability estimates. Genet. Res. 21:439-448.

Curnow, R. N. (1960). The regression of true value on estimated value. Biometrika 47:457-460.

Kempthorne, O. (1960). Biometrical relationships between relatives and selection theory, in Biometrical Genetics (Kempthorne, Editor). London. Pergamon Press.

Lindley, D. V. (1947). Regression lines and the linear functional relationship. J. R. Statist. Soc. Sup. 9:218-44.

Nishida, A. and Abe, T. (1974). The distribution of genetic and environmental effects and the linearity of heritability. Can. J. Genet. Cytol. 16:3-10.

V
Theory of Response and Limits to Selection

Artificial selection with a large number of linked loci

Alan Robertson
INSTITUTE OF ANIMAL GENETICS, EDINBURGH

In a recent paper on the effects of linkage in artificial selection (Robertson, 1970), I showed that, for a given initial amount of genetic variance in the selected character and intensity of selection, as the number of loci affecting the character increases the selection process quickly becomes independent of the number of loci in a chromosome of a given length and of their initial frequency. If h^{*2} is the proportion of phenotypic variance due to this chromosome, i is the selection intensity in standard units, N is the effective population size and ℓ is the length of the chromosome in morgans, then I stated that all parameters of the population at the tth generation of selection are functions solely of Nih^*, $N\ell$ and t/N, given that initially there was linkage equilibrium between loci. This conclusion is based partly on theoretical reasoning and partly on a series of computer simulations with a finite number of loci. In the present paper, I shall extend these arguments by presenting results from simulations in which the initial variation is due to an infinite number of loci on the chromosome.

It might be helpful to summarise briefly the results of the earlier paper. It was concerned almost entirely with the advance at the limit in terms of a comparison between the value with free recombination (L_f) and that with no recombination (L_0). The following generalizations appeared (which are in fact almost never true but never far wrong):

(i) If Nih^* is small, $L_f/L_0 = 1 + \frac{2}{3}(Nih^*)^2$;

(ii) As Nih^* increases, L_f/L_0 tends to $2Nih^*/3$;

(iii) If L_ℓ is the limit with intermediate values of recombination, then when $N\ell$ is small,

$$L_\ell/L_0 = 1 + N\ell/3 .$$

Further, we can write approximately

$$L_\ell/L_0 = \tfrac{1}{3} KN\ell/(\tfrac{1}{3}N\ell + K),$$

where $K = L_f/L_0$. Thus the value of $N\ell$ which gives a limit half-way between free and no recombination is roughly $3L_f/L_0$ (or $\ell = 2ih^*$).

(iv) The effect of recombination (as compared to no crossing over) on selection response does not appear in the early generations, for reasons which will become clear later. There is seldom any marked effect on the cumulative response before $2/ih^*$ generations.

What values of these parameters are usual in selection experiments? For typical experiments with <u>Drosophila</u>, say for sternopleural bristles with a heritability of 0.4, we have for the two autosomes, $N = 10$, $i = 1.4$, $h^* = 0.4$ and $\ell = 0.5$, giving $Nih^* = 6$ and $N\ell = 5$. In selection for growth in mice, we might assume that $h^2 = 0.2$. If this variance is equally distributed over the 20 chromosomes, which have an average length of 75 cM then $h^* = 0.1$. For a population with 20 pair matings and a selection differential of 1.2, we have then $Nih^* = 5$ and $N\ell = 30$. We may deduce that existing levels of crossing over would restrict the response in the <u>Drosophila</u> experiment but not much in the case of the mouse selection.

I assume that the genetic variation is contributed equally by an infinite number of loci distributed evenly along the chromosome with no initial correlation between the effects at different loci on the same gamete, i.e. linkage equilibrium In

ARTIFICIAL SELECTION WITH MANY LINKED LOCI   309

the base population, each chromosome then has a value for the selected character, which is normally distributed with variance $V^*_g/2$, where V^*_g is the genetic variance between individuals due to the chromosome. The sampling of individual chromosomes in the zero generation is straightforward.

The production of gametes from selected parents is more complex. Any such gamete, if it has arisen by crossing over, is a combination of sequences from the two parental gametes. Suppose that only one crossover is involved in the production of such a gamete from an individual whose chromosomes had values X_1 and X_2 respectively, so that it contains a proportion p of the first gamete and 1-p of the second. What is the distribution of the values of the two sections of the new gamete? We have to sample a value for a proportion p of the first chromosome, given its total value. We are then, under the assumptions above, dealing with the problems of a sum composed of two independent parts contributing proportions p and 1-p of the total variance. The correlation between the first part and the whole is $\sqrt{p}$, and the regression of the first part on the whole is p, so that the variance of the first part given the whole is $p(1-p)V^*_g/2$. Thus the part can be sampled from a normal distribution with mean pX_1 and variance $p(1-p)V^*_g/2$. Correspondingly, the second part has mean $(1-p)X_2$ and the same variance as the first. By these two processes we have generated a new gamete which is stored in terms of the origin and value and break points of the pieces. Thus the sequence might now read 1 -0.42 0.375 2 +1.24 +0.500 if the chromosome length is 0.5, i.e., the gamete starts with a section of chromosome 1 of length 0.375 and value -0.42 and then has a piece of chromosome 2 of length 0.125 and value +1.24. In the next generation, a further crossover in the first section can be sampled by a similar procedure, except that the variance of the "whole" is now proportional to the length of section concerned. This can be simplified by the maintenance of a "library" in which is stored for each chromosome the values that have been established for its sections by the crossovers involving it which have already occurred. Thus chromosome 1, if it has a total value 1.27, is initially stored in the library as 0 1.27 0.500 and, after the first crossover, above, as 0 -0.42 0.375 1.69 0.500.

The generation of a new gamete from a selected parent is then as follows:

(i) The sampling of the crossover points. If the probability of crossing over is uniform along the chromosome and there is no interference, the distribution of distances between successive crossovers can be shown to be negative exponential. If R is a random number from a uniform distribution between 0 and 1, -log R is a random sample from a negative exponential distribution with mean 1. To start, one of the two parental gametes is chosen at random, the first crossover point is sampled, and then the second, from the first as origin, until the points fall off the end of the chromosome.

(ii) The establishment of the values of the two new sections of chromosome produced by each crossover on both of the parental chromosomes. This involves reference to the library, simulation of a crossover in the relevant section and updating of the library.

(iii) The new gamete is put together from the relevant sections of the parental chromosomes. Its total value is the sum of the values of the sections.

The longer the chromosome the more computer time is taken for each run. In any event, a separate programme has to be written for the case when each of the infinite number of loci segregates independently. In this programme, the average value of the gametes produced by an individual is the mean value of the two gametes which originally formed it and the variance of the issuing gametes now depends on the inbreeding coefficient of the individual and equals to $0.5\ V^*_g(1-F)$. Therefore, whereas for simulation with a finite length of chromosome the homozygosity is actually measured on the two parental gametes which make up an individual, with free recombination the inbreeding coefficient of an individual is necessary to generate gametes from it. The inbreeding coefficient is calculated by maintaining a table of coancestry of all individuals chosen as parents.

It is basic to this approach that the variation is due to a large number of loci spread evenly along the chromosome. The expected change in gene frequency at each locus due to selection is very small and the finite total change is due to the summation of a large number of small effects. The mean heterozygosity over loci does change as a consequence of genetic sampling and of selection. The description of the selection process is then essentially in terms of variances and covariances. The input of each run contains the following:

(i) The number of parents each generation.

(ii) The number of generations of selection.

(iii) The length of the chromosome.

(iv) The intensity of selection.

(v) The heritability of the variation controlled by the chromosome. The genetic variance due to the chromosome is taken as unity and the heritability used to calculate the environmental variation.

(vi) The number of replicate lines required.

The output has varied in different versions but in the main consisted of the values at successive generations of

(i) The average of line means,

(ii) The variance of line means over replicates,

(iii) The average genetic variance within lines,

(iv) The variance over replicates of the within line variance,

(v) The mean homozygosity, relative to that in the initial generation as zero, calculated from an examination of the uniting gametes (except in the case of free recombination),

(vi) The variance of the homozygosity between individuals within lines,

(vii) The covariance within lines between homozygosity and the character value.

The results should be considered in the light of the following
theoretical considerations. With no selection, the genetic variance within lines and the heterozygosity will on average decline by
a fraction 1/2N each generation. The variance between lines is
expected to increase as 2F. The existence of linkage does not
affect these expectations.

Selection changes matters since the genetic variance between
selected individuals and in their progeny is less than that in the
group from which they were selected. Since in this model the mean
gene frequency is altered very little by selection, the reduction
in the progeny variance has a different genetic origin from that
brought about by sampling and is due to linkage disequilibrium
between the loci affecting the character (Bulmer, 1971). Selection
also affects the rate of increase of homozygosis since all selected
individuals will not now have an equal chance of contributing to
future generations--those with the higher genotypic value will contribute more because their progeny will be above average (Robertson,
1961). If H is the measured homozygosity and V_A is the genetic
variance within lines, $V_A/(1-H)$ is a measure of the effect of linkage disequilibrium in reducing genetic variance. Recombination then
leads to a partial recovery of this reduction. Since the response
to selection is dependent on the genetic variance, the effect of
recombination is not seen immediately. The response is proportional
to the genetic variance, which in its turn depends on the effect of
genetic recombination in restoring some of the loss of genetic variation produced by the earlier selection. I commented in the earlier
paper that no effect of the level of recombination on response was
seen before about $2/ih^*$ generations of selection.

We may then examine the effect of linkage--first of all
on the genetic variance within lines, since this is the key
to the selection response. Figures 1, 2 and 3 give the results for
a series of runs with N = 10 and $ih^* = 0.4$, with each value usually
being an average of 50 or more replicates. Figure 1 shows the
decline in genetic variance as selection proceeds, with no crossing

ARTIFICIAL SELECTION WITH MANY LINKED LOCI

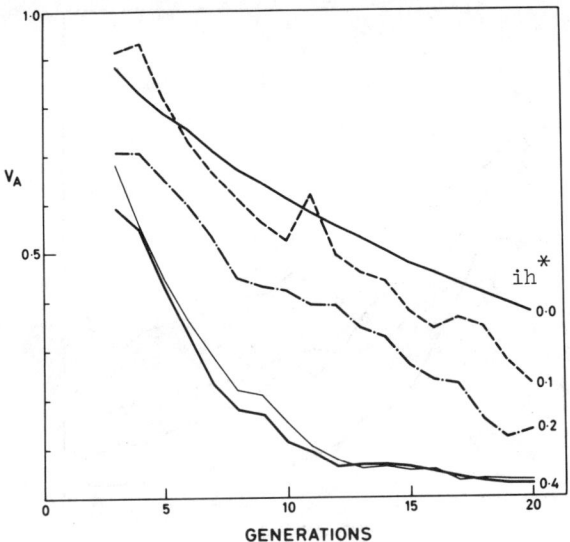

Figure 1. The decline of genetic variance caused by selection, for different values of ih^*, with $N = 10$. The fainter line with $ih^* = 0.4$, shows the values for $N = 20$, $ih^* = 0.2$, with the time scale doubled.

over and for different values of Nih^*. The more intense the selection, the greater the reduction in variance. Of course, there is in this case no recovery by recombination. Figure 2 shows the effect of recombination with $Nih^* = 4$. The more the recombination the higher the genetic variance. The effect of selection is distinguished from that of population size in Figure 3, in which the genetic variance is expressed in terms of the heterozygosity, giving a measure of the linkage disequilibrium produced by selection.

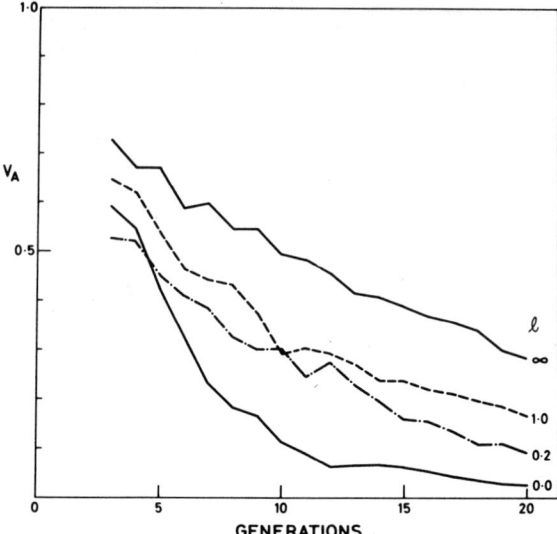

Figure 2. The decline of genetic variance with $N = 10$, $ih^* = 0.4$, with different chromosome lengths.

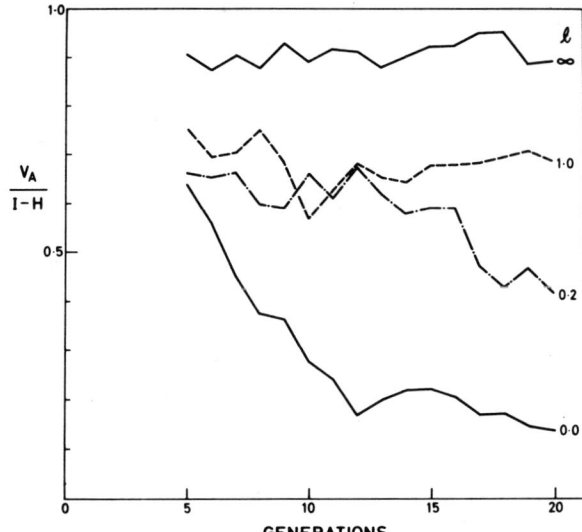

Figure 3. The genetic variance expressed in relation to the heterozygosity (1-H). The deviation of the ratio from unity is due to linkage disequilibrium at loci affecting the character.

With no crossing over, the effect of selection is cumulative and the limiting value of $V_A/(1-H)$ is zero. Bulmer has shown that in infinite populations the joint action of selection and recombination leads to the reduction of the genetic variance to an equilibrium value. This trend to a limiting value of $V_A/(1-H)$ is seen only at the two higher recombination values. It should be noted that even with free recombination the genetic variance is below that expected from the heterozygosity. The disequilibrium produced by selection is then reduced by only one half each generation by recombination. The classical formula of $2Nih\sigma_g$ for the total response, in fact, assumes that recombination completely removes linkage disequilibrium each generation and must therefore be an overestimate.

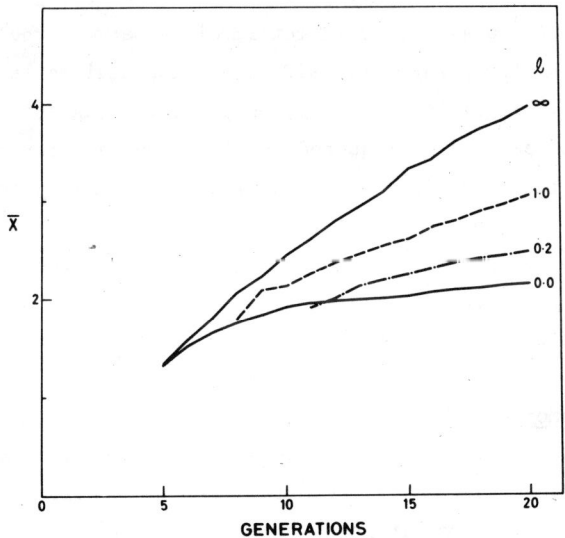

Figure 4. The response to selection, in terms of the initial genetic standard deviation, at $N = 10$, $ih^* = 0.4$ for different chromosome lengths.

Figure 4 shows that there is little effect of recombination on selection response up to the fifth generation. A rather crude calculation, based on Bulmer's formulae, which in finite populations might be expected to underestimate the effect of selection, would suggest that by this time the genetic variance with free recombination should be about 30% higher than with none and that the response to that time should be about 6% larger. The effect of intermediate values of recombination on variance, relative to the value with no crossing over, increases with time, though in absolute terms the difference is greater between the tenth and fifteenth generations.

Figure 4 shows that there is little effect of recombination on selection response at generation 5 (= $2/ih^*$ generations) but that the difference gradually increases with time until, at generation 20, the response with free recombination is about twice that with no recombination and selection with a chromosome length of 1.0 gives a response almost exactly intermediate between the two. The later the time the greater the effect of intermediate levels of recombination. Figure 8 of the 1970 paper would suggest that at the limit the value of $N\ell$ required to give a value intermediate between those with no and free recombination is approximately 6. It is also clear that because of the faster decline of genetic variance with no recombination, the selection limit is then reached much more rapidly--in this case with a half life of about 5 generations, compared to an expectation of 14 (equals 1.4N) with free recombination.

Selection and Homozygosity

With no selection, the expected heterozygosity of individuals is expected to decrease by a fraction $1/2N$ each generation. With selection, homozygosity may increase faster than this because within selected individuals those with a high genotypic value for the selected character will have a higher than average chance of contributing to future generations (Robertson, 1961). Correspondingly, within a generation the individuals with the highest genotypic value tend to have the highest homozygosity. In this simulation, the homozygosity is measured directly and not estimated from pedigrees, except for free recombination. The above effect may be expected to be affected by linkage since, in the absence of recombination, a chromosome with a high value will be continued to be transmitted as a unit and to be selected for over many generations.

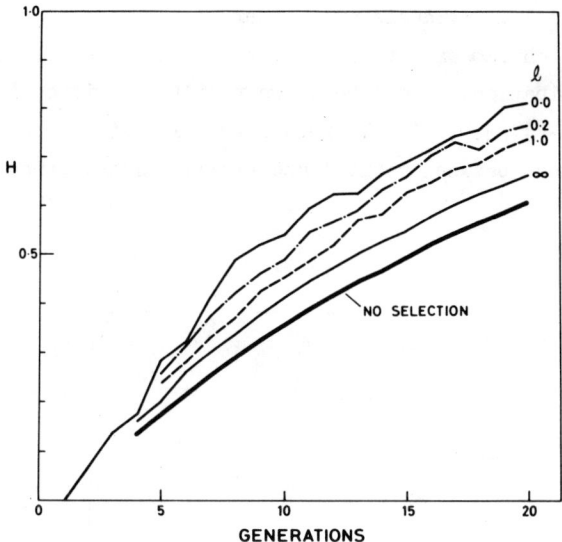

Figure 5. The measured homozygosity of the lines shown in Figure 4.

Figure 5 shows this effect with $Nih^* = 4$, the continuous line giving the expected values with no selection. The homozygosity is zero in the first generation because selfing is prohibited in the simulation. It is clearly dependent on the level of recombination, and the effect of selection on the pedigree inbreeding is small, as represented by the values with free recombination.

The relationship within generations between homozygosity and character value can most usefully be represented as the genetic covariance between the two, as it is then most relevant to the effect on selection response of any reduction in fitness due to homozygosity. Figure 6 shows that the covariance rises rapidly and passes through a maximum after about N/2 generations, being very dependent on the level of recombination. It should be noted that the within-line variance of homozygosity between individuals is very dependent on the level of recombination. If there is no recombination and all individuals are either homozygotes or heterozygotes for this chromosome, then the variance will be F(1-F). With free recombination, it can be shown that the variance is of the order of $(1-F)^2/32N$. I will return to this later in the context of the balance between natural and artificial selection.

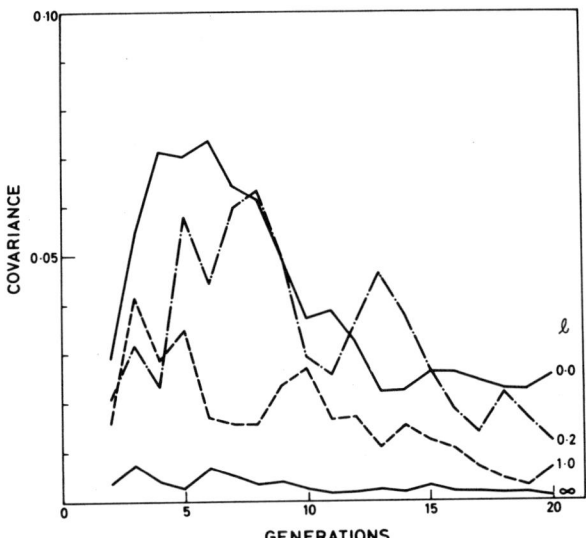

Figure 6. The covariance within generations and lines between homozygosity and value for the selected character for the lines shown in Figure 4.

ARTIFICIAL SELECTION WITH MANY LINKED LOCI

The Variance Between Lines

In the absence of selection, the variance between replicate lines is expected to equal $2FV^*_g$. Selection complicates this relationship in several ways:

(i) Selection increases the decline i heterozygosity.

(ii) The genetic variance within lines varies between lines.

(iii) The average genetic variance within lines declines more rapidly than assumed by the expression because of the establishment of linkage disequilibrium.

(iv) At intermediate generations, there is a positive correlation between homozygosity and the mean of the line.

As usually not more than 100 replicates were run for any combination of variables, it was not possible to examine this problem in any detail. But a cursory examination of the results would suggest that for values of F up to 0.5, the relationship holds fairly well, if we use the value of F calculated from pedigrees. At higher values, however, the deviations from expectation were mostly caused by the more rapid decline of genetic variance in the absence of recombination so that the variance between replicates at low levels of recombination was less than the expression would suggest. This was also found in the 1970 paper in relation to the response at the limit (Table 13) and implies that such selected lines are in linkage disequilibrium in relation to one another.

Artificial and Natural Selection

In the "homeostatic" model of natural selection on quantitative characters, which assumes additive gene action for the quantitative character and overdominance at the same loci for fitness, both the relationship within populations between fitness and the metric character, and that showing the effect of selection on the mean fitness of a line are dependent on the fitness effects at the individual loci (Robertson, 1956). As I am here assuming an infinite number of loci with small effects on the character and, presumably, a finite overall effect of inbreeding on fitness, it follows that within populations the extremes for the metric character are not less fit than average and that there is no direct effect of artificial selection on fitness itself in large populations. But in small populations there is an indirect effect of selection since this increases the level of homozygosity. Thus if the effect of natural selection is small, Table 5 can also be read as the expected decline in fitness with artificial selection. In the early generations the decline differs little from that theoretically expected with no artificial selection, but in the later generations the effect of selection is more noticeable, especially with no recombination, though even then the decline is less than 50% greater. This would hold if the fitness effects were small. But if they are not, natural selection may affect the response to artificial selection. The program was therefore modified to take into account interaction between natural and artificial selection by making the fertility of an individual dependent on its homozygosity--in fact equal to exp(-sxH) where s, the percentage reduction in fitness for each percentage increase in homozygosity, determines the strength of the natural selection. The effect of natural selection on a quantitative character can be shown to be equal to the additive genetic covariance between relative fitness and the character (Robertson, 1966). The expected effect on selection response should then equal the sum of the covariances shown in Figure 6. Various runs have been carried out with s equal to 1, and $Nih^* = 4$. With no recombination the average response at generation 20 was reduced from the value of 2.17 with no opposing natural selection to 1.59 with s = 1, a reduction of about 20%. The sum of the covariance terms from Figure 6 proved to be 0.77, not much greater than the observed effect of natural selection. The small effect of natural selection may be surprising but it must be remembered that in this case we are dealing with whole chromosomes. With intense artificial selection and no opposing natural selection, we might expect to make homozygous the best chromosome in the initial sample. But with such extreme natural selection that homozygous individuals do not survive, we would now expect to have the best two chromosomes segregating and the difference between the best and mean of the best two will not be great. Natural selection did in fact reduce the homozygosity at generation 20 from a value of 0.86 to 0.49.

Natural selection did not reduce the response with free recombination as would be expected since the sum of the covariances in Figure 6 is only 0.06. No effect could be detected even for a chromosome length as low as 20cM. An explanation could perhaps be given in terms of the effect of natural selection in maintaining genetic variation. With no recombination, for instance, the mean homozygosity at generation 20 was 0.49 with s = 1 compared to 0.86 with s = 0. With no recombination, this variation is not usable since genetic variation will be permanently maintained with no response to selection. With a low level of recombination, however, some usable genetic variation is continually being generated by crossing over.

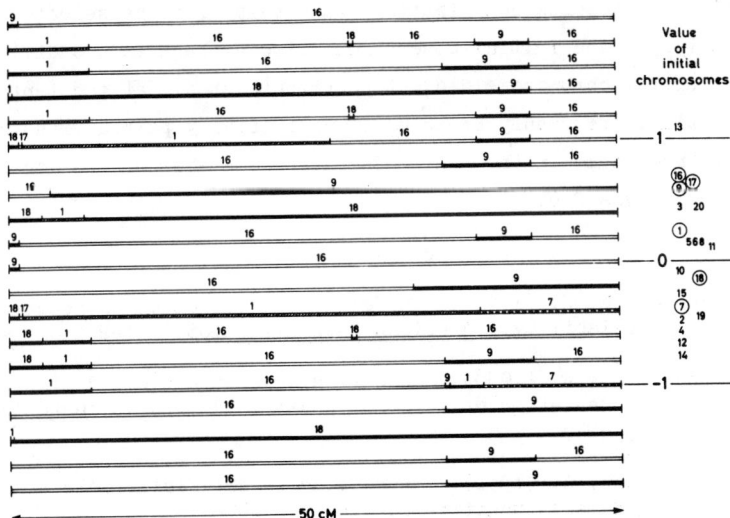

Figure 7. The chromosomes at generation 10 in a typical line selected with $N = 10$, $ih^* = 0.4$, and $\ell = 0.5$. The numbers attached to each section refer to the chromosomes in the initial generation whose values for the character are indicated on the right. Only those circled were represented at generation 10.

Chromosome Structure in a Selection Program

Using this program, we can follow the fate of individual chromosomes in a selection program. Figure 7 illustrates the situation at the tenth generation of selection in a line which had been started with 5 pairs of parents, with a selection intensity of 0.8, a heritability for the chromosome of 0.16 and a chromosome length of 50cM, possible parameters for selection for sternopleural bristle score in Drosophila. The left-hand side of the figure shows the structure of the 20 chromosomes in the population at generation 10 in terms of the parental chromosomes in the initial generation. At the right-hand side are given the values for the character for the 20 chromosomes in the initial generation.

It will be seen that only six of the initial chromosomes contribute any material to the selected line in the tenth generation, when the level of homozygosity was 0.40--in fact, more than half of the total material comes from chromosome 16. Two of the six initial chromosomes contributing had a value for the selected character in the initial generation below average and, oddly, the best initial chromosome did not contribute at all at the tenth generation, presumably because it was initially paired with the worst chromosome in the line. The average number of crossovers per chromosome was 2.6.

BIBLIOGRAPHY

(1) Bulmer, M. G. (1971). The effect of selection on genetic variability. American Naturalist 105: 201-211.

(2) Robertson, A. (1956). Selection against extreme deviants based on deviation or on homozygosity. Journal of Genetics 54: 234-248.

(3) Robertson, A. (1961). Inbreeding in artificial selection programmes. Genetic Research 2: 189-194.

(4) Robertson, A. (1966). A mathematic model of the culling process in dairy cattle. Animal Production 8: 95-108.

(5) Robertson, A. (1970). A theory of limits in artificial selection with many linked loci. In Mathematical Topics in Population Genetics. ed. Kojima. Springer-Verlag, Berlin-New York.

Potential uses of inbreeding to increase selection response

Gordon E. Dickerson
U.S. MEAT ANIMAL RESEARCH CENTER, AGRICULTURAL RESEARCH SERVICE
U.S. DEPARTMENT OF AGRICULTURE, 225 MARVEL BAKER HALL
UNIVERSITY OF NEBRASKA, LINCOLN, NE 68583

N. Bengt H. Lindhé
ASSOCIATION FOR SWEDISH LIVESTOCK BREEDING AND PRODUCTION
S-63184 ESKILSTUNA, HÅLLSTA, SWEDEN

ABSTRACT

Potential usefulness of inbreeding to increase cumulative response to selection for average gene effects on performance depends upon the degree to which variance in average gene effects could be increased between inbred families with minimum offsetting loss of selection intensity per unit of time. For this purpose, cycles of temporary intense inbreeding followed by crossing among inbred families with combination individual plus family selection in each generation seemed likely to be optimum. However, expected increase in cumulative response for growth rate in swine and milk production in dairy cattle from such temporary cycles of inbreeding was discouraging because of (1) the lowered intensity of selection of male or female parents used to produce inbred families in alternate generations and (2) the limited increase in total variance of average gene effects from making inbred matings of highly selected parents. Selection of parents affects distribution of additive genetic variance between and within families and hence relative response expected from family, individual and combined selection. Expected gains from cyclic selection among inbred families are greater when heritability is low and there is little environmental correlation among family members as in egg production of chickens.

INTRODUCTION

The idea that inbreeding can be used somehow to increase effectiveness of selection continues to interest animal breeders, even though both experience and theory are largely negative. In a recent review of evidence concerning inbreeding and heterosis in livestock (Dickerson, 1973), it was suggested that continuous selection within <u>and</u> between families in alternating generations of intensive inbreeding and crossing of selected families should permit appreciable increases in selection response compared with similar selection without inbreeding. The present report considers these ideas more critically, with special reference to application in swine and cattle selection programs.

EFFECTS OF INBREEDING

Variance in gene frequency or additive genotype ($\sigma^2_{q_i}$) under inbreeding without selection is expected to be proportional to (1+F) among individuals of all lines, (1-F) within lines and 2F between lines, where F is Wright's (1921) inbreeding coefficient. Variance of corresponding transmitted breeding values in crosses would be proportional to ($\frac{1-F}{4}$) for individuals within lines and $\frac{F}{2}$ between lines (Dickerson, 1942). However, Robertson (1952) has shown for loci exhibiting complete dominance or slight overdominance and low initial frequency of the recessive allele ($q_0<.1$), that genetic variance within lines ($\sigma^2_{G_w}$) first increases

POTENTIAL USES OF INBREEDING

as F rises to .4 or .5 and then declines toward $\underline{0}$ as F approaches 1. Genetic variance among individuals of all lines ($\sigma^2_{G_t}$) still increases almost linearly with F but that between lines ($\sigma^2_{G_b}$) increases more slowly at first, being $\sigma^2_{G_t} - \sigma^2_{G_w}$. To the extent that there is selective elimination of homozygous recessives, both between- and within-line genetic variances decline after the initial rise with early inbreeding.

The decline in components of reproductive fitness generally associated with inbreeding usually is ascribed to the increased frequency of recessive homozygotes but also may arise from average loss of favorable epistatic effects because of random independent changes in q_i at different loci (i.e., to increased $\sigma^2_{q_i}$ among loci). Both could cause an initial rise in $\sigma^2_{G_w}$ and a loss in reproductive rate. The latter would then reduce the possible intensity of selection as well.

Increased inbreeding (F) then is expected to reduce both intensity (i_w) and effectiveness ($\sigma_{G_w}/\sigma_{P_w}$) of within-line selection and thus reduce rate of response within inbred lines (ΔG_w) except in the elimination of rare unfavorable recessive genes, because

$$\Delta G_w = i_w \cdot r_{G \cdot P} \cdot \sigma_{G_w} = i_w \cdot \sigma^2_{G_w}/\sigma_{P_w}. \tag{1}$$

The larger random sampling changes in gene frequency in small (inbred) populations also are expected to reduce response to within line selection (Robertson, 1960).

The expected increase in genic and genetic variance between inbred lines suggests that cycles of selecting between inbred sublines and crossing of selected lines may be the primary way in which inbreeding might be expected to increase rate of cumulative response to selection for average gene effects. However, compared with continuous selection without inbreeding, cyclic selection among inbred lines of animals is handicapped by (1) limited numbers of lines, (2) time and performance loss required for each cycle of inbreeding and (3) a reduced intensity of selection (especially of males) required to make the inbred matings in each cycle.

CYCLIC SELECTION AMONG INBRED SUBLINES

Response expected from selection between inbred lines (b) relative to that from mass selection without inbreeding (o) is:

$$\frac{\Delta G_b}{\Delta G_o} = \frac{i_b \cdot \sigma_{\bar{P}_n} \cdot h^2_{\bar{P}_n} \cdot Y_o}{i_o \cdot \sigma_{P_o} \cdot h^2_{P_o} \cdot Y_b} \qquad (2)$$

where:

i_b and i_o are standard selection differentials.

Y_b and Y_o are times/cycle of selection.

$\sigma_{\bar{P}_n}$ and σ_{P_o} are line mean and non-inbred individual phenotypic deviations.

$h^2_{\bar{P}_n}$ and h^2_o are heritabilities for line mean and for non-inbred individual phenotypes.

Selection within inbred lines is ignored.

The expected values for $h^2_{\bar{P}_n}$ and $\sigma_{\bar{P}_n}$ in formula (2) are derived from the associations shown in the path coefficient diagram (Figure 1) below.

Thus, expected relative response per unit of time becomes:

$$\frac{\Delta G_b}{\Delta G_o} = \frac{i_b \cdot Y_o}{i_o \cdot Y_b} \cdot \frac{(1+F)[1+(n-1)r_F]}{\sqrt{(1+F \cdot h^2_o)\, n[1+(n-1)t_F]}} \qquad (3)$$

$$= \frac{i_b \cdot Y_o}{i_o \cdot Y_b} \cdot \frac{(1+F)[1+(n-1)r_F]}{\sqrt{n(1+F \cdot h^2_o)+n(n-1)[r_F(1+F)h^2_o+C^2_o]}}$$

When n is moderately large and there is no environmental correlation among line mates ($C^2 = 0$), the ratio $\Delta G_b/\Delta G_o$ approaches $\frac{i_b \cdot Y_o}{i_o \cdot Y_b} \sqrt{(1+F)r_F/h^2_o}$. If full-sib inbreeding and crossing of selected inbred families are practiced in alternate years ($r_F = .6$, $F = .25$ and $Y_o/Y_b = \frac{1}{2}$) and $i_b = i_o$, then

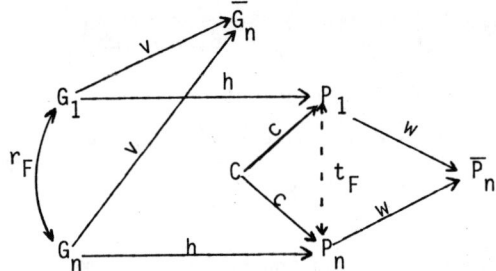

r_F = genic correlation (relationship coefficient, $\frac{2F}{1+F}$) of line members, including effect of common immediate parents.

F = inbreeding coefficient of line members

t_F = phenotypic correlation of line members = $r \cdot h^2 + c^2$

h^2 = heritability of individual phenotype = $(1+F)h_0^2/(1+F \cdot h_0^2)$

c^2 = environmental correlation of line members = $C_0^2/(1+Fh_0^2)$

v = standard partial regression $\bar{G}_n$ on G_i = $1/\sqrt{n[1+(n-1)r_F]}$

w = standard partial regression $\bar{P}_n$ on P_i = $1/\sqrt{n[1+(n-1)t_F]}$

$$\sigma_{\bar{P}_n} = \sigma_{P_0}\sqrt{(1+Fh_0^2)[1+(n-1)t_F]/n}$$

$$h^2_{\bar{P}_n} = b_{\bar{G}_n \bar{P}_n} = \frac{h_0^2(1+F)[1+(n-1)r_F]}{(1+F \cdot h_0^2)[1+(n-1)t_F]}$$

FIG. 1

Path diagram of genetic and phenotypic associations among individuals in an inbred line

$\frac{\Delta G_b}{\Delta G_0} \rightarrow \sqrt{\frac{.1875}{h_0^2}}$. Thus, h_0^2 must be less than .2 for $\Delta G_b/\Delta G_0 \geq 1$. Similarly, for paternal sire x daughter inbred families in alternate generations (r_F = .45), $\Delta G_b/\Delta G_0 \rightarrow \sqrt{.14/h_0^2}$, so that h_0^2 must be <.15 for $\Delta G_b/\Delta G_0 \geq 1$.

When there is important environmental correlation among members of the same inbred family, as for growth rate of litter mates in swine, the ratio of $\Delta G_b/\Delta G_0$ approaches

$\frac{i_b \cdot y_o}{i_o \cdot y_b}$ $(1+F) \cdot r_F \sqrt{(n-1)/n \cdot t_F}$. Again, if $r_F = .6$, $F = .25$, $y_o/y_b = \frac{1}{2}$ and $i_b = i_o$, t_F must be $\leq .12$ for selection among inbred families in <u>alternate</u> years to equal mass selection <u>every</u> year. A $t_F \leq .12$ corresponds to $h_F^2 \leq .2$ if $c_F^2 = 0$ and to smaller h_F^2 values if $c_F^2 > 0$.

These results show that selection among inbred lines is more likely to be helpful when heritability (h^2) and environmental correlation among line members (c^2) are low, as was so clearly demonstrated earlier by Lush (1947). In practice, intensity of selection among inbred lines usually will be less than for mass selection ($i_b < i_o$) and the time per selection cycle more than twice as long ($y_b/y_o > 2$), making the case for use of inbreeding in selection programs even less optimistic. However, there is opportunity to apply selection among individuals within the inbred families and also in the non-inbred generations of each cycle, both of which should make the case for use of inbreeding somewhat <u>more</u> favorable.

INDIVIDUAL AND FAMILY SELECTION, WITH CYCLIC INBREEDING

With combination individual plus family index (I) selection (Lush, 1947), $I = b_1 \cdot P_i + b_2 \cdot \bar{P}_n$. Without inbreeding, expected response per <u>generation</u> of selection is $E(\Delta G)_o$
$= i_{I_o} \cdot R_{G \cdot P_i \bar{P}_n} \cdot \sigma_{G_o}$ where

$$R_{G \cdot P_i \bar{P}_n} = \sqrt{h_o^2 [1 + \frac{(r_o - t_o)^2 (n-1)}{(1-t_o)[1+(n-1)t_o]}]} \quad (4)$$

For combination index selection among one-generation full-sib inbred families, response expected is $E(\Delta G)_F = i_{I_f} \cdot R'_{G \cdot P_i \bar{P}_n} \cdot \sigma_{G_f}$, where.

POTENTIAL USES OF INBREEDING

$$R_{\hat{G} \cdot P_i \bar{P}_n} = \sqrt{h_F^2 [1 + \frac{(r_F - t_F)^2 (n-1)}{(1-t_F)[1+(n-1)t_F]}]}, \quad (5)$$

$$h_F^2 = \frac{h_0^2 (1+F)}{(1+F \cdot h_0^2)}, \quad r_F = .75/1.25 = .6$$

$$t_F = \frac{r_F \cdot h_0^2 (1+F) + c_0^2}{1 + F \cdot h_0^2} \quad \text{and} \quad \sigma_{G_f} = \sigma_{P_0} \cdot h_0 \sqrt{(1+F)}$$

The relative rate of response per unit of time expected from such selection with inbreeding in <u>alternate</u> generations (ΔG_f) to that without inbreeding (ΔG_0) thus would be:

$$\frac{\Delta G_f}{\Delta G_0} = \frac{1}{2} + \frac{(1+F)}{2\sqrt{1+F \cdot h_0^2}} \sqrt{\{1 + \frac{(n-1)(r_F - t_F)^2}{(1-t_F)[1+(n-1)t_F]}\} / \{1 + \frac{(n-1)(r_0 - t_0)^2}{(1-t_0)[1+(n-1)t_0]}\}} \quad (6)$$

With full-sib inbreeding in alternate generations in selection for growth rate in swine ($h_0^2 = .3$, $c_0^2 = .15$, $n = 6$), this ratio ($\Delta G_f / \Delta G_0$) would be about 1.12. However, this expectation ignores (1) the reduced selection intensity for males required to permit full-sib mating of <u>all</u> selected females in alternate generations as well as (2) the smaller increase in additive genetic variance ($\sigma_{G_f}^2 < (1+\bar{F})\sigma_{G_0}^2$) caused by using only selected parents to produce the inbred families. Similar limitations would affect application to sire-daughter inbreeding in alternate generations of selection for milk production in dairy cattle. The expected net effect of such cycles of inbreeding and crossing on cumulative response to selection was explored in more detail for swine and for dairy cattle.

SWINE SELECTION

The model for swine involved combination individual plus family index selection in every generation with or without full-sib inbreeding in alternate generations. Other features were:

1000 first litters/year from 100 sires and 1000 dams.
Discrete generations, age of parents = 1 year.
3 males + 3 females per litter available for breeding, 8 with performance records.

Selection of 120 male and 1200 female replacements/year. Selection for growth efficiency with $h^2 = .3$, $c^2 = .15$, $r = .5$, $t = r \cdot h^2 + c^2 = .3$ in unselected, non-inbred populations.

Selection without Inbreeding

The effect of one generation of selection on parameters controlling selection response is illustrated in Table I.

TABLE I

Effect of Combination Selection <u>without</u> Inbreeding on Parameters

Component	$E(\sigma_i^2)$	Example
Sire, genetic (S_0)	$.25\ h^2(1-R_{G \cdot \overline{PP}_n}^2 \cdot \sigma_{rS}^2)$	$.075[1-.339(.86)] = .053$
Dam, genetic (D_0)	$.25\ h^2(1-R_{G \cdot \overline{PP}_n}^2 \cdot \sigma_{rD}^2)$	$.075[1-.339(.69)] = .058$
Individual, genetic (W_0)	$.5\ h^2$	$.15$
Litter environment (C_0)	c^2	$.15$
Random environment (E_0)	$1-h^2-c^2$	$.55$
Total (T_0)	$T+S_0+D_0-S-D$	$.96$

$h_0^2 = (S_0+D_0+W_0)/T_0 = .271$

$c_0^2 = C_0/T_0 = .156$

$t_0 = \dfrac{(S_0+D_0+C_0)}{T_0} = .271$

$r_0 = \dfrac{S_0+D_0}{S_0+D_0+W_0} = \dfrac{.111}{.261} = .425$

$\sigma_{G_0}^2 = S_0+D_0+W_0 = .261$

$R_{G \cdot \overline{PP}_n} = \sqrt{h_0^2[1+\dfrac{(r_0-t_0)^2(n-1)}{(1-t_0)[1+(n-1)t_0]}]} = .54$ for <u>second</u> generation of selection

where:

$\sigma_{rS}^2 = i_S(i_S-x_S)$ = Reduction in $\sigma_{I_S}^2$ from selection

$\sigma_{rD}^2 = i_D(i_D-x_D)$ = Reduction in $\sigma_{I_D}^2$ from selection

$x_S \cdot x_D$ = + or - deviation from mean at truncation point, in σ units

$i_S = 2.15$ for $p = 120/3000$, $x_S = 1.75$

$i_D = .966$ for $p = 1200/3000$, $x_D = .253$

Note that after the first generation of index selection, r_o changes from .5 to .425 and h_o^2 from .3 to .27 and $R_{G \cdot P\bar{P}_n}$ from .58 to .54. There is a little further expected reduction in these parameters until the population quickly reaches equilibrium under a given intensity and kind of selection, in <u>this</u> case at r_o = .393, h_o^2 = .26 and $R_{G \cdot P\bar{P}_n}$ = .525 in generation 5. In each successive generation, S_o and D_o are calculated as shown in Table I from the h_o^2 and $R_{G \cdot P\bar{P}_n}^2$ of the preceding parental generation, with σ_{rS}^2 and σ_{rD}^2 constant for any specified selection intensity and recognizing that the genic variance among full sibs W_o remains unreduced by selection of parents.

Thus, expected response to index selection for individual plus family performance in this example in the <u>first</u> generation is:

$\Delta G_S = i_S \cdot R_{G \cdot P\bar{P}_n} \cdot \sigma_G$ = 2.15(.582)(1.0) = 1.251 σ_G

$\Delta G_D = i_D \cdot R_{G \cdot P\bar{P}_n} \cdot \sigma_G$ = .966(.582)(1.0) = .562 σ_G

ΔG = (1.251 + .562)/2 = .91 σ_G, where σ_G is the genetic standard deviation in the unselected base population. This rate of response per generation declines quickly toward an expected equilibrium level of about .75 σ_G or about 82% of the initial rate of response.

Selection with Inbreeding in Alternating Generations

In order to make full-sib matings of the best females, it is necessary to save males from more litters than would be necessary without inbreeding and to choose only 1 of 3 males available from each selected litter. The example to be used then involves selecting 1200 females from the best 400/1000 litters and the best male from each of those same 400/1000 litters to make the full-sib matings in alternate generations. Alternatives involving full-sib matings for only the more highly selected females and selection of male replacements from among only these inbred litters in alternate years were nearly as promising and required use of fewer boars for inbreeding. However, expectations for the "all inbred" case are less complicated to present and will illustrate the findings as well.

In Table II, the expected variance structure is shown for a population of 1000 full-sib inbred litters whose non-inbred parents were chosen as described above from the non-inbred population produced by the first generation of individual plus family index selection illustrated in Table I.

Note that inbreeding increases only the between-family genetic variance S_f and D_f, components that otherwise would have been maintained at about the S_o and D_o levels by the selection of immediate parents of the inbred families.

Expected response from selecting the best 120 males and 1200 <u>inbred</u> females then is:

$$\Delta G_S = i_S \cdot R_{G \cdot P\overline{P}_n} \cdot \sigma_{G_F} = 2.15(.6141)(1.026)\sigma_G = 1.3542 \; \sigma_G$$
$$\Delta G_D = i_D \cdot R_{G \cdot P\overline{P}_n} \cdot \sigma_{G_F} = .966(.6141)(1.026)\sigma_G = .6085 \; \sigma_G$$
$$\Delta G_{F_1} = (1.3542 + .6085)/2 = .981 \; \sigma_G$$

When the selected 120 male and 1200 female inbreds are randomly crossed, the expected variance structure for the resulting non-inbred generation is as shown in Table III.

Note in Table III that selection of inbred parents reduces expected sire and dam genetic components (S_o and D_o) in the non-inbred progeny again to the same levels found after one generation of non-inbred index selection (Table I). However, the expected within-family genetic variance (W_o) is reduced in proportion to (1-F) of parents because of reduced heterozygosity of the inbred parents.

Expected response from selection in the non-inbred generation is:

$\Delta G_s = i_b \cdot R_{\overline{G} \cdot \overline{P}_n} \cdot \sigma_{\overline{G}} + i_w \cdot R_{G_i P_i} \cdot \sigma_{G_i} = .966(.5545)(.7049)$
$\qquad + .846(.4121)(.5001) = .5519 \ \sigma_G$

$\Delta G_d = i_b \cdot R_{\overline{G} \cdot \overline{P}_n} \cdot \sigma_{\overline{G}} = .966(.5545)(.7049) = .3776 \ \sigma_G$

$\Delta G_t = (.5519 + .3776)/2 = .4648 \ \sigma_G$

TABLE II

Effect of Full-sib Inbreeding with Selected Parents on Parameters

Component	$E(\sigma_i^2)$	Example
S_f	$.25 \ \sigma_{G_0}^2 (1+2F)[(1-R_{\overline{G} \cdot \overline{P}_n}^2 \cdot \sigma_{rSb}^2)(1+2r_0)/3$ $+ (1-R_{G_i \cdot P_i}^2 \cdot \sigma_{rSw}^2)2(1-r_0)/3]$	$(.25)(.261)(1.5)$ $[.794(.617)$ $+ .8606(.383)]$ $= .0802$
D_f	$.25 \ \sigma_{G_0}^2 (1+2F)[(1-R_{\overline{GP}_n}^2 \cdot \sigma_{rDb}^2)(1+2r_0)/3$ $+ 2(1-r_0)/3]$	$(.25)(.261)(1.5)$ $(.48945 + .3833)$ $= .0854$
W_f	$\sigma_{G_0}^2 (1-r_0)$	$= .150$
G_t		$.3156$
C_f		$= .15$
E_f		$= .55$
T_f		1.0156

$R_{\overline{G} \cdot \overline{P}_n}^2 = h_0^2 \cdot n'[1+(n-1)r_0]^2/n[1+(n-1)t_0][1+(n'-1)r_0], \ n' = 3, \ n = 8$

$R_{G_i P_i}^2 = h_0(\frac{1-r_0}{1-t_0}), \ \sigma_{G_b}^2 = \sigma_{G_0}^2[1+(n'-1)r_0]/n', \ \sigma_{G_w}^2 = (n'-1)(1-r_0)\sigma_{G_0}^2/n'$

$\sigma_{rSb}^2 = \sigma_{rDb}^2 =$ reduction in $\sigma_{G_n}^2$ from selecting best 400/1000 families

$\sigma_{rSw}^2 =$ reduction in $\sigma_{G_w}^2$ from selecting best 1/3 in each litter

For selection among <u>inbred</u> population:

$R_{G \cdot P\overline{P}_n} = \sqrt{.3108[1+\frac{7(.5247-.3108)^2}{.6892[1+7(.3108)]}]} = \sqrt{.3771} = .6141$

$\qquad$ where $h_F^2 = t_F = \frac{.3156}{1.0156} = .3108, \ r_F = .5247$

TABLE III

Effect of Crossing Selected Inbred Parents on Population Parameters

Component	$E(\sigma_i^2)$	Example	
S_0	$.25\ \sigma_{G_f}^2(1-R_{G \cdot \overline{PP}_n}^2 \cdot \sigma_{rS}^2)$	$(.25)(.3156)(.6757)$	$= .0533$
D_0	$.25\ \sigma_{G_f}^2(1-R_{G \cdot \overline{PP}_n} \cdot \sigma_{rD}^2)$	$(.25)(.3156)(.7392)$	$= .0583$
W_0	$.50\ \sigma_G^2(1-F)$	$(.50)(.30)(.75)$	$= .1125$
G_t			$.2241$
C_0			$.15$
E_0			$.55$
T_0			$.9241$

$R_{G \cdot \overline{PP}_n} = \sqrt{.3771} = .6141$ for selection of <u>inbred</u> parents, as shown in Table II

$\sigma_{rS}^2 = .86$ and $\sigma_{rD}^2 = .69$ as in Table I

$r_0 = .1116/.2241 = .4980$, $h_0^2 = .2241/.9241 = .2425$,

$t_0 = (.1116 + .15)/.9241 = .2831$

The lower total genetic variance ($\sigma_{G_t}^2 = .2241$) in the non-inbred generation in turn affects the sire and dam but not the within-family component of genetic variance among the next inbred generation as shown in Table IV.

Expected response from selection **is a little less in the second generation of inbred progeny** than in the first (Table IV vs. II) because of lower total genetic variance (.291 vs. 316):

$\Delta G_s = 2.15(.5733)(.985) = 1.215\ \sigma_G$

$\Delta G_d = .966(.5733)(.985) = .545\ \sigma_G$

$\Delta G_t = (1.215 + .545)/2 = .880\ \sigma_G$

TABLE IV

Variance Structure of Second Inbred Generation

Component	$E(\sigma_i^2)$	Example
S_f	$.25\ \sigma_{G_0}^2(1+2F)[(1-R_{\overline{G}\cdot\overline{P}_n}^2\cdot\sigma_{rSb}^2)(1+2r_0)/3$ $+ (1-R_{G_i\cdot P_i}\cdot\sigma_{rSw}^2)2(1-r_0)/3]$	$.25(.2241)(1.5)$ $[.7882(.6653)$ $+.8892(.3347)]$ $= .0691$
D_f	$.25\ \sigma_{G_0}^2(1+2F)[(1-R_{\overline{GP}_n}^2\cdot\sigma_{rDb}^2)(1+2r_0)/3$ $+ 2(1-r_0)/3]$	$.25(.2241)(1.5)$ $[(.7882)(.6653)$ $+.3347] = .0722$
W_f	$\sigma_{G_0}^2(1-r_0)$	$.15$
G_t		$.2913$
C_f		$.15$
E_f		$.55$
T_f		$.9913$

$R_{\overline{G}\cdot\overline{P}_n}^2 = .2425(3)[1+7(.498)]^2/8[1+7(.2831)][1+2(.498)] = .3075,$

$R_{\overline{G}\cdot\overline{P}_n} = .5545;\qquad R_{G_i\cdot P_i}^2 = .2425(.502)/(.7169) = .1698,$

$R_{G_i\cdot P_i} = .4121;\ \sigma_{\overline{G}_b}^2 = .2241[1+2(.498)]/3 = .1491,\ \sigma_{\overline{G}_b} = .3861;$

$\sigma_{\overline{G}_w}^2 = .2241(2)(.502)/3 = .0750,\ \sigma_{G_w} = .2739$

$h_F^2 = t_F = .2938,\ r_F = .4851,\ R_{G\cdot PP_n} = \sqrt{.2938[1+\frac{7(.4851-.2938)^2}{.7062[1+7(.2938)]}]}$

$=\sqrt{.3287} = .5733$

The mean response per generation over the second cycle (Tables III and IV) is only $.68\ \sigma_G$, lower than the $.75\ \sigma_G$ equilibrium response per generation with no inbreeding (Table I). These results offer little theoretical basis for using alternate generation inbreeding to increase cumulative response to selection for a trait as highly heritable ($h^2 = .3$) and with as much litter environment ($c^2 = .15$) as growth rate in swine. The case would be more favorable for less heritable traits with less environmental correlation among full sibs.

Aside from consideration of inbreeding, the results presented indicate that selection is expected to reduce the genetic correlation among full or half-sibs proportionately more than the corresponding environmental correlation (e.g., r_o from .5 to .4 and t_o from .3 to .26). Thus, the optimum emphasis on family performance is reduced, compared with initial selection from an unselected population (Lush, 1947), and less increase in response over mass selection is expected from including it in index selection. The disappointing experimental results of Wilson (1974) from family and from combination index selection relative to mass selection may be partly explained by such effects of continuous selection on population parameters.

Results from mass selection for egg mass per unit body weight in Japanese quail (MacNeil, et al., 1976) indicate much less cumulative response with full-sib inbreeding in alternate generations than in the non-inbred population, especially during the first 10 generations. Realized heritability was about 20% in both the non-inbred and in the cyclic inbreeding system. Both selection differentials and responses were greater for selection among inbreds than among non-inbreds in each cycle of the alternating system. Cumulative selection applied in the cyclic inbreeding system was only about 62% of that in the non-inbreeding system. Inbreeding depression was large in the inbred generation of each cycle. This Montana experiment illustrates nicely many of the theoretical expectations dealt with in this paper.

DAIRY CATTLE SELECTION

The model used to examine possible usefulness of alternate generation production of sire-daughter inbred sons of elite progeny tested sires was essentially that used by Lindhe (1968) and methods described in Dickerson and Hazel (1944) and Rendel and Robertson (1950) were followed:

POTENTIAL USES OF INBREEDING

Number of milk recorded cows   200,000
Selection of 140 young bulls tested/year:
  From 7/140 tested <u>sires</u> with 100 daughters,
  $i_{BB} = 2.06$, $R_{G_S \bar{P}_S} = .93$
  From 7.5% of 14,000 1st lactation <u>dams</u>,
  $i_{CB} = 1.89$, $R_{G_D \cdot P_D \bar{P}_S} = .62$
Selection of female replacements/year
  From 20/140 tested <u>sires</u> with 100 daughters,
  $i_{BC} = 1.58$, $R_{G_S \bar{P}_S} = .93$
  From 70% of all <u>dams</u> with 1+ records,
  $i_{CC} = .5$, $R_{G_D \cdot P_D} = .50$
Proportion of AI matings/year
  39% to 140 young bulls
  61% to 20 selected tested sires
Ages ($\bar{A}$) of parents when progeny born
  7.00 years for <u>sires</u> of young bulls
  4.75 years for <u>dams</u> of young bulls
  5.15* years for <u>sires</u> of female replacements
  4.20 years for <u>dams</u> of female replacements
 * .39 (2.25) + .61 (7.00)
Heritability of fat corrected milk yield = .25 in <u>unselected</u>
population within herd-year-season.
Genetic coefficient variation = 9% for unselected population

    This model assumes more intense selection for milk production than is generally feasible in practice in the United States (VanVleck, 1977). It therefore assumes more reduction in sire and dam components of genetic variance from selection and less increase in genetic variance from inbreeding than may be typical. However, it will serve to illustrate a method of evaluating possible temporary alternate generation inbreeding in dairy cattle A.I. breeding schemes.

    The expected effect of the selection of parents applied and of inbreeding on the genetic variance among young bulls

chosen to be progeny tested is shown in Table V. Note that
reduction is greater in the sire (74%) than in the dam (32%)

TABLE V

Expected Genetic Variance[a] among Young Bulls
to be Progeny Tested

Component	Non-inbred			Sire x dau. inbred		
Sire (S)	.257[b]	(.25)[c]	= .0642	.257[b]	(.5625)[c]	= .1446
Dam (D)	.678[b]	(.25)	= .1695	.803[b]	(.1875)	= .1505
Within (W)	1.000	(.50)	= .5000	1.000	(.5000)	= .5000
Total σ_G^2	.734	(1.00)	= .734	.636	(1.25)	= .795 (+8.4%)
Total σ_G			.857			.892 (+4.1%)

[a] As fraction of total genetic variance in an unselected, non-inbred population.

[b] Fraction of σ_G^2 remaining among selected sires or dams of young bulls.

[c] Expected fraction of σ_G^2 <u>without selection</u> of parents.

---

component, reducing the total genetic variance to 73% of that
in an unselected population. Among sire-daughter inbred sons
of selected parents, the sire component is expected to more
than double (.1446/.0642) and the dam component to be reduced
little because of less selection among the half-sister females
mated to their sire (.1505/.1695). Total genetic variance is
thus 8.4% greater among the inbred than the non-inbred bull
replacements, but still below that in an unselected population.

The expected phenotypic variance structure of daughters
from inbred, non-inbred, and selected tested sires is also
affected by selection, especially of the sires (Table VI).
This reduces accuracy of selecting among young tested sires
$(r_{G_s \bar{P}_n})$ only slightly because of large numbers of daughters
per sire (n = 100). Accuracy in choosing dams for inbred

matings is based on half-sib means only ($r_{\bar{G}_d \cdot \bar{P}_n}$ = .48).

TABLE VI

Expected Phenotypic Variance for Daughters

Component	Non-inbred bulls	Inbred bulls	Selected tested sires
S	.0625(.734)=.0459	.0625(.795)=.0497	.0625(.292)=.0183
D	.0625(.871)=.0544	.0625(.871)=.0544	.0625(.871)=.0544
W	.125	.1094	.1172
E	.75	.75	.75
T	.975	.963	.940
$r_{\bar{G}_s \bar{P}_n}$ (n=100)	.912	.919	--
$R_{\bar{G}_d \cdot P_d \bar{P}_n}$ (n=100)	.616	.616	--
h^2	.231	.222	.202

<u>Without Inbreeding</u>

For continuous selection without inbreeding, calculation of expected annual response (ΔG_0 = 1.57% of mean) is shown in Table VII. Most response (44%) comes from choosing the best 5% of tested sires to produce sons (SS) but choosing dams (DS) on own plus sisters' records ($r_{\bar{G}_d \cdot P_d \bar{P}_n}$ = .616) also contributes substantially (30%) as does selection of 20/140 best tested bulls (SD) to mate 61% of the cow population (20%). These figures include effect of one prior generation of the selection described. Under continuous selection, genetic variance would decline only a little more.

<u>Sire-daughter Inbred Sons in Alternate Generations</u>

Expected effects of producing sire-daughter inbred sons of best tested sires in alternate generations on mean response per year are shown in Table VIII.

TABLE VII

Expected Response/Year, No Inbreeding

$$\overline{\Delta G}_0 = \frac{\Delta G_{SS} + \Delta G_{DS} + \Delta G_{SD} + \Delta G_{DD}}{A_{SS} + A_{DS} + A_{SD} + A_{DD}}, \text{ where:}$$

For	$\overline{A}$ (yrs.)	Proportion selected (P)	i	× R_{IG}	× σ_G	= ΔG_{ij}
SS	7.00	.05	2.06	× .912	× 7.71[a]	= 14.48[a]
DS	4.75	.075	1.89	× .616	× 8.54	= 9.94
SD	5.15	.143	.61[1.58	× .912	× 7.71]	= 6.78
DD	4.20	.70	.50	× .466	× 8.22	= 1.92

$$\overline{\Delta G}_0 = \frac{(14.48 + 9.94 + 6.78 + 1.92)}{(7.00 + 4.75 + 5.15 + 4.20)} = \frac{33.12}{21.10} = 1.57\%/\text{yr}.$$

[a] As % of mean yield

TABLE VIII

Expected Response/Year, Sire × Dau. Inbred Sons, in Alternate Generations

Generation I. Inbred sons from selected sire × daughters

For	$\overline{A}$ (yrs.)	Proportion selected (P)	i	× R_{IG}	× σ_G	= ΔG_i
SS	7.00	.05	2.06	× .912	× 7.71	= 14.48
DS	4.75	.05	2.06	× .480	× 8.54	= 8.46
SD	5.15	.143	.61[1.58	× .912	× 7.71]	= 6.78
DD	4.20	.70	.50	× .466	× 8.22	= 1.92
	21.10					31.64

Generation II. Non-inbred sons from selected inbred sires

For	$\overline{A}$ (yrs.)	Proportion selected (P)	i	× R_{IG}	× σ_G	= ΔG_i
SS	7.00	.05	2.06	× .918	× 8.03	= 15.19
DS	4.75	.075	1.89	× .616	× 8.54	= 9.94
SD	5.15	.143	.61[1.58	× .918	× 8.03]	= 7.10
DD	4.20	.70	.50	× .466	× 8.22	= 1.92
	21.10					34.15

$$\overline{\Delta G}_I = \frac{31.64 + 34.15}{2(21.10)} = 1.56\% \text{ vs. for } \overline{\Delta G}_0 = 1.57\%$$

At the high intensity of selection assumed in this model, expected rate of response did not increase above that for selection without inbreeding. Loss of response from selection of dams of inbred sons (9.94 to 8.46 in DS) in Generation I was hardly offset by increased genetic variance (8.03 vs. 7.71) and accuracy (.918 vs. .912) in selecting sires of sons (SS) and of daughters (SD) in Generation II.

These results certainly discourage consideration of inbreeding as an aid to selection for milk production in dairy cattle. However, the important reasons for the lack of increase in response were (1) the very intense selection (5%) assumed for sires and dams of bull replacements (SS and DS) and (2) the reduced selection of dams assumed necessary to produce inbred bull replacements (DS). Actual estimates of response in dairy cattle generally have been half or less of those assumed in this selection model (VanVleck, 1977), indicating much less actual selection.

Under such less intense selection, the increase in genetic variance among inbred families of bulls would be more than the 8% assumed in the model. Effects of inbreeding with more realistic intensities of selection may be worth exploring. However, making more complete use of opportunity for selection may be considerably more feasible than implementing a scheme of cyclical inbreeding.

ACKNOWLEDGMENTS

The authors are indebted to Dr. Alan Robertson for helpful suggestions concerning interacting effects of selection and inbreeding on variance structure of populations.

BIBLIOGRAPHY

(1) Dickerson, G. E. (1942). Experimental design for testing inbred lines of swine. J. Anim. Sci. 1, 326-341.

(2) Dickerson, G. E. (1973). Inbreeding and heterosis in animals. Proc. Animal Breeding and Genetics Symposium in Honor of J. L. Lush. Amer. Soc. Anim. Sci., Champaign, Ill. 61820, 54-77.

(3) Dickerson, G. E. & Hazel, L. N. (1944). Effectiveness of selection on progeny performance as a supplement to earlier culling in livestock. J. Agr. Res. 69, 459-476.

(4) Lindhé, B. (1968). Model simulation of A.I. breeding within a dual purpose breed of cattle. Acta. Agr. Scand. XVIII, 1-2, 33-41.

(5) Lush, J. L. (1947). Family merit and individual merit as bases for selection. Amer. Nat. 81, 241 and 362.

(6) MacNeil, M. D., Kress, D. D., Flower, A. E. and Blackwell, R. L. (1976). Effects of mating system on selection response in Japanese quail. Proc. Western Sec. Amer. Soc. Anim. Sci. 27, 9-12.

(7) Rendel, J. M. and Robertson, A. (1950). Estimation of genetic gain in milk yield by selection in a closed herd of dairy cattle. Genetics 50, 1.

(8) Robertson, A. (1952). The effect of inbreeding on variation due to recessive genes. Genetics 37, 189.

(9) Robertson, A. (1960). A theory of limits in artificial selection. Proc. Roy. Soc. B. 1953, 234.

(10) VanVleck, L. D. (1977). Theoretical and actual genetic progress in dairy cattle. Proc. Intern. Conf. Quant. Genetics. Iowa State Univ. Press.

(11) Wilson, S. P. (1974). An experimental comparison of individual, family and combination selection. Genetics 76, 823.

(12) Wright, S. (1921). Systems of mating. Genetics 6, 109.

Variation in response to selection

William G. Hill
INSTITUTE OF ANIMAL GENETICS
EDINBURGH EH9 3JN, SCOTLAND

INTRODUCTION

Variation in response to selection may be utilised by the breeder in selecting among replicate populations to increase response, classically in inbreeding and crossing programmes. Alternatively the variation in response may explain why realised responses from experiments or breeding programmes do not follow prediction and make extrapolations unreliable from small scale experiments. In this paper I shall concentrate on the latter aspects.

Perhaps the major limitation of quantitative genetics theory is that the measurements we can make with any precision on a population, such as the statistical quantities of phenotypic variances, heritabilities and genetic correlations, enable us to predict response formally only for one generation but if parameters do not change too fast, perhaps for rather more. Longer term responses depend on the distribution of effects and frequencies of genes influencing the traits under selection; the information on which is small in Drosophila and almost non-existent in other species. Similarly, whilst it is possible to construct some theory for predicting variation in response between replicate populations over a few generations based on the same statistical quantities, together with effective population size (N), the variance in the long term also depends critically on gene effects and frequencies. Thus a theory based on these observable quantities can not be of a very long term nature, and in this paper discussion is restricted to experiments of not more than about 10 or N/2 generations, whichever is smaller. Beyond these times, there are likely to be large changes in the frequencies of any genes having a substantial effect on the trait. There may also be substantial interactions between artificial and natural selection, and complications due to inbreeding depression, a period of N/2 generations corresponding to an inbreeding coefficient of a little over 20%. Of course, the shorter the time scale, and less intense the selection, the more likely are calculations derived from parameters of the base population to be relevant. A theory of variation in response for such time periods is relevant to several objectives in establishing selection experiments, for example: demonstration of quantitative genetic principles, testing for possible departures from expectation, comparison of alternative selection schemes, estimation of parameters of the base population, and checking on response in commercial populations.

In a series of papers some theory has been developed for predicting the variation in response to selection and, taking account of the correlated error structure of individual generation means, for analysing selection experiments (Hill, 1971, 1972a,b, 1974; Bohren, 1975). An important assumption in Hill's papers was that the variances within populations remained constant over generations, an assumption made because changes in these variances over the time scale were considered to be a second order effect, because they were, in any event, difficult to predict, and for convenience. Regardless of the distribution of gene effects and frequencies there are expected to be some mean changes in within-line variance, due of course to inbreeding, but less obviously to linkage disequilibrium between loci as a result of selection (Bulmer, 1971, 1974). Perhaps more importantly, in the present context, is that genetic drift produces not only a variance between lines in mean performance, but also in variance between lines in within-line variance. The

VARIATION IN RESPONSE TO SELECTION

variance of, say, the additive variance has been previously assumed to be highly dependent on the distribution of gene effects and frequencies, even for additive genes, because it is a term involving fourth powers of effects and frequencies. Recently, however, Bulmer (1976) has argued in a very interesting paper that unless there are very few loci affecting the quantitative trait, most of the variance in additive variance is caused by linkage disequilibrium (including disequilibrium between genes on different chromosomes). Formulae for this variance can be obtained which depend solely on measurable quantities and, although only the conclusions are included here, further details of the arguments are given by Bulmer (1976) and will be extended by Avery and Hill (1977). The basic assumption is that of classical quantitative genetics in that there are many genes affecting the trait, no single one contributing a large part of the variation; for simplicity, most variation is assumed to be due to additive genes.

The theory available to describe variation both in mean and variance rests heavily on that for unselected populations. In the first section of this paper, therefore, populations are considered in which no selection is practised and then these results are extended less formally to populations under selection, reviewing the limitations of the theory of variation in selection response.

VARIATION WITHOUT SELECTION

Consider a set of replicate lines, maintained in the same environment and sampled from a large base population assumed to be in linkage equilibrium, each maintained with discrete populations, random mating within lines, and no mutation, migration or selection. Loci are assumed to be additive, with the additive variance being V_A in the base population and V_{At} in a randomly chosen replicate at generation t, the corresponding phenotypic variances being V_P and V_{Pt}, and heritabilities h^2 and h_t^2. The mean breeding value of the parents of generation t in a randomly chosen replicate is Z_t and the mean phenotype of their M progeny which are recorded is X_t.

Variance in mean

From well known theory (e.g. Falconer, 1960), there is no change in mean performance over replicates, but

$$V(Z_t) = 2[1 - (1-1/2N)^t] V_A = 2F_t V_A, \quad (1)$$

where F_t is the inbreeding coefficient in the next generation. If only generations up to $t \leq N/2$ are considered

$$V(Z_t) = (t/N) V_A \quad (2)$$

approximately (for $t = N/2$ and $N = 20$, for example, $t/2N = 0.25$, $F_t = 0.22$). Also, since sampling at generation $t + 1$ is of genotypes at generation t there is an autocorrelation between means in successive generations. For $k \geq 0$,

$$\text{Cov}(Z_t, Z_{t+k}) = V(Z_t) = 2F_t V_A, \quad (3)$$

$$\text{Corr}(Z_t, Z_{t+k}) = \sqrt{F_t/F_{t+k}} \doteq \sqrt{t/(t+k)}. \quad (4)$$

The variance of the progeny mean X_t equals that of Z_t plus terms accounting for sampling of progeny genotypes from their parents and for environmental variance. The exact value of the latter depends on the distribution of family size, but is not less than $(V_{Pt} - \frac{1}{2}V_{At})/M$, which holds if all families are equally represented, and is unlikely to exceed V_{Pt}/M. Taking the lower figure for simplicity, partly to compensate for the expected decline in V_{Pt} from inbreeding,

$$V(X_t) = tV_A/N + (V_P - \frac{1}{2}V_A)/M \quad (5)$$

approximately. Note that for generation number (t) and mean family size (M/N) greater than unity, most of the variance in (5) is contributed by genetic drift, so the choice of the coefficient of M^{-1} is not important. The covariances between X_t and X_{t+k} are the same as those between Z_t and Z_{t+k}, although increased marginally if the M recorded progeny include the potential parents of the following generation; but the autocorrelations are lower than shown in (4) since $V(X_t) > V(Z_t)$. Formulae such as (5) which show how the variance of means increases with generations and the associated problems of the autocorrelation among them feature in the design of genetic control populations and in analysis of data from them and concurrent selected populations (Hill, 1972a,b,c).

Variance in variance

Turning to the variances, and using the same theory

$$E(V_{At}) = (1-F_t) V_A \doteq (1-t/2N) V_A, \qquad (6)$$

to roughly the same level of approximation. The cumulative variance, V_{At}^*, up to generation t which would, if selection were practised, determine response up to generation t+1 in a replicate, has mean

$$E(V_{At}^*) = E(\sum_{T=0}^{t} V_{AT}) = 2NF_{t+1}V_A \qquad (7)$$

(Robertson, 1960) and, to the same level of approximation,

$$E(V_{At}^*) \doteq (t+1)(1 - \frac{t}{4N}) \qquad (8)$$

Consider now the variance between replicates in V_{At} which, assuming no differences in environmental variance between replicates, will equal the variance in V_{pt}. If at locus i there is an effect a_i (difference between heterozygote and homozygote) on the trait, and the allele contributing high score has frequency q_{it} in some replicate at generation t, then

$$V_{At} = 2 \sum_i a_i^2 q_{it}(1-q_{it}) + 4 \sum_{i<j} \sum a_i a_j D_{ijt} \qquad (9)$$

where D_{ijt} is the linkage disequilibrium at generation t between loci i and j (i.e. the frequency of the pair (i, j) - $q_i q_j$). If there are n loci affecting the trait, there are n terms in $V(V_{At})$ involving single loci, e.g. $V[(q_{it}(1-q_{it})]$, and n(n-1)/2 terms involving pairs of loci, $Cov[q_{it}(1-q_{it}), q_{jt}(1-q_{jt})]$ and $V(D_{ijt})$, together with other crossproduct terms which can be shown to have zero mean in populations starting from equilibrium. Now in the first generation both $V[q_{i1}(1-q_{i1})]$ and $V(D_{ij1})$ are of order 1/2N and increase subsequently. Therefore, providing the number of loci involved is much larger than the population size, it is clear that most of the variation in V_{At} will be contributed by disequilibrium. Formally, we assume there are sufficient loci that the contribution from the variance of frequency at individual loci can be ignored. The above argument is given more rigorously by Bulmer (1976) and Avery and Hill (1977), and the following arguments in the latter paper.

If we make the further strong assumption that there is no association between the distance between loci and their effect on the trait,

$$V(V_{At}) = 2 \overline{r_t^2} [E(V_{At})]^2 \qquad (10)$$

to a good approximation over the time scale ($t \leq N/2$) being considered, where $\overline{r_t^2}$ is the average of the squared correlations among all loci affecting the trait and

$$r_{ijt}^2 = D_{ijt}^2 / [q_{it}(1-q_{it}) \, q_{jt}(1-q_{jt})]. \qquad (11)$$

A simple expression is thus obtained for the coefficient of variation of the additive variance,

$$CV(V_{At}) = \sqrt{2\overline{r_t^2}} \, . \qquad (12)$$

In the first generation of sampling from a population in equilibrium, $r_{ij1}^2 = \overline{r_1^2} = 1/2N$, so

$$CV(V_{A1}) = \sqrt{1/N} \, , \qquad (13)$$

a result which can be deduced merely by considering the estimate of variance from a sample of 2N independent observations (chromosomes) from a normal distribution. A convenient and sufficiently accurate formula for predicting the squared correlation between a pair of loci having crossover frequency between them of c_{ij} is

$$r_{ijt}^2 = \frac{1 - [(1-c_{ij})^2 (1-1/2N)]^t}{1 + 2N \, c_{ij}(2-c_{ij})} \qquad (14)$$

(Sved and Feldman, 1973). For unlinked loci, $c_{ij} = 1/2$, and

VARIATION IN RESPONSE TO SELECTION

$$r^2_{ijt} = \frac{2}{3N}[1-(\frac{1}{4})^t] \qquad (15)$$

approximately, i.e. rapidly reaching 2/3N, whereas for more tightly linked loci the asymptotic value of r^2_{ij} is higher but takes longer to reach. Thus we have that, for $t > 1$,

$$\overline{r^2} > 2/3N \quad \text{and} \quad CV(V_A) > \sqrt{4/3N}. \qquad (16)$$

To evaluate these formulae some assumptions about the distribution of cross-over values has to be made. Since a large number of loci are assumed, a reasonable assumption would seem to be that loci affecting the trait are uniformly distributed along chromosomes, with the number of loci on any chromosome proportional to its map length. If there are m chromosomes of map length ℓ_i, with $\Sigma\ell_i = L$, then a proportion $\Sigma\ell_i^2/L^2$ of randomly chosen pairs of loci are on the same chromosome, with the distribution of distance between such pairs being triangular between 0 and ℓ_i with a higher proportion of pairs close together than apart, and the remaining pairs of loci are on different chromosomes. Thus if the chromosomes are of equal map length a proportion of 1/m pairs are on the same chromosome, and this proportion increases if the chromosomes are of different length. In the following examples the distribution of chromosome length is also assumed to be triangular and is defined such that if, for example, there are 4 chromosomes their lengths are 2L/5, 4L/5, 6L/5 and 8L/5; but numerical results have shown that the results are little affected by varying the distribution of lengths but keeping the mean length constant (Avery and Hill, 1977).

In Table 1 predicted values of $CV(V_{At})$ using this model are given for a range of parameter values, including the case of independent loci shown as $m \to \infty$. There are several striking aspects to the results: firstly, the coefficients of variation are large, greater than $1/\sqrt{N}$, as indicated previously; secondly a large part of the limiting value of $CV(V_{At})$ is reached after only one generation, particularly when there are several chromosomes; thirdly, the value of $CV(V_{At})$ is much less dependent on the number and length of the chromosomes than on the population size. Some qualification of these statements is required, however. The number of chromosomes is more important when the total map length is small than when large, and the asymptotic value of $CV(V_{At})$ is reached more slowly when there are very few chromosomes and the total map length is short.

TABLE 1

Coefficient of variation, $CV(V_{At})\%$, of the additive genetic variance at generation t as a function of population size, N, total map length L, and number of chromosomes, m.

N	L	m	t	1	2	5	10
20	1	1		22.4	28.4	35.6	39.7
		2		22.4	27.4	33.5	37.5
	2.5	1		22.4	26.9	31.0	33.1
		2		22.4	26.5	30.4	32.4
		20		22.4	25.3	27.3	28.6
	10	20		22.4	25.2	26.6	27.2
	>0	→∞		22.4	24.9	25.7	25.7
10	2.5	2		31.6	37.4	42.5	45.0
		20		31.6	35.7	38.3	39.9
50	2.5	2		14.1	16.8	19.3	20.8
		20		14.1	16.1	17.3	18.2

In mammals, in which the haploid chromosome number is typically around 20, it is clear that whatever their total map length turns out to be, $CV(V_{At})$ is not going to greatly exceed $\sqrt{4/3N}$, and will nearly reach that value within four generations. With <u>Drosophila melanogaster</u> there are essentially three chromosomes with a total map length of around $2\frac{1}{2}$, but with no crossing over in males. Thus with an effective population size of 20, $CV(V_{At})$ is likely to reach about 35%, compared with about 26% for mammals. Thus, as Bulmer (1976) noted, the number of chromosomes is not a major consideration.

VARIATION IN RESPONSE TO SELECTION

Checks of this theory with animal populations have not been made, and indeed such a test would be extremely laborious as many replicates and large numbers of individuals in each replicate would be necessary to enable reliable estimates of $CV(V_{At})$. Some checks can be made by Monte Carlo simulation, however. These fall into two categories and each has been used. In the first model a specified number of loci of specified initial frequency and effect were simulated. In the examples shown in Table 2 there are 20 loci of equal effect and initial frequency 0.5,

TABLE 2

Comparison between predicted (P) and observed (O) values for $CV(V_{At})$% using two simulated models of genes on a single chromosome (for details see text)

N	L	c^+		t	1	2	3	4	5	10
			(i) 20 equally spaced loci							
20	→∞		P		22	25	26	26	26	26
		0.5	O (100)‡		22	25	26	28	25	30
20	2.0		P		22	27	30	31	32	35
		0.1	O (100)		21	30	31	35	37	38
20	0.2		P		22	31	36	40	44	54
		0.01	O (100)		21	32	36	38	40	51
			(ii) infinite number of loci							
10	0.2		P		32	43	51	56	61	74
	0.2		O (100)		42	54	70	79	76	89
10	0.5		P		32	41	48	52	55	63
	0.5		O (50)		38	51	55	60	53	81
20	1.0		P		22	28	32	34	36	40
	1.0		O (25)		29	30	30	47	46	51

$^+$recombination fraction between adjacent loci
‡replicates of Monte Carlo simulation

which are assumed to be equally spaced on a single chromosome, although when the recombination fraction between adjacent loci is 0.5 this is equivalent to all loci on a different chromosome.

The other crossover distances used are 0.01 and 0.1 between adjacent loci, equivalent to total map lengths of roughly 0.2 and 2. The second model used was that of Robertson (1976) in which there are conceptually an infinite number of loci, with effects and initial gene frequencies randomly distributed. Formally Robertson's model is the same as that assumed in this theoretical analysis, so differences in results should reflect only minor differences in specification. Checks using data supplied by Alan Robertson for his model are also given in Table 2.

The agreement between the predicted values using (12) and those observed is seen to be good for the 20 locus model, surprisingly less so for the infinite gene model where the observed coefficients of variation of V_{At} exceed those predicted. The likely explanation illustrates a problem of interpretation of the results. Our predicted values and the simulations given in Table 2(i) refer to the variance among the 2N chromosomes, which would be the additive variance in a large population generated in Hardy-Weinberg equilibrium from these chromosomes. Robertson's results are the variance among the N diploids obtained by randomly pairing the N chromosomes. In the first generation, $CV(V_{At})$ among the N diploids is $\sqrt{2/N}$ rather than $\sqrt{2/2N}$ among the 2N chromosomes again using normal distribution theory. Variances in subsequent generations seem to be changed proportionately less, but the exact change has still to be derived. In any estimate of variance from an animal or plant population there is, of course, sampling error. This equals the values given at generation 1 if the sample is of size N, so this value should perhaps be counted as a base point for the subsequent generations.

Autocorrelation of variances

As noted previously, the generation means of replicated finite populations are strongly autocorrelated; the within line variances are also found to be autocorrelated. The detailed analysis will be given elsewhere, but it can be shown that,

VARIATION IN RESPONSE TO SELECTION

assuming a model of many loci and the same assumptions as before,

$$\text{Corr}\ (V_{At}, V_{At+k}) = \overline{(1-c)^k r_t^2} \bigg/ \sqrt{\overline{r_t^2}\ \overline{r_{t+k}^2}}\ ,\quad k \geqslant 0 \qquad (17)$$

where $\overline{(1-c)^k r_t^2}$ is the average taken over all pairs of loci of $(1-c_{ij})^k r_{ijt}^2$, i.e. the correlation of frequencies x the probability of no crossovers between them in k generations. If all loci are unlinked (c = 1/2) then, from (15), $\overline{r_t^2}$ quickly asymptotes and so, after very few generations,

$$\text{Corr}\ (V_{At}, V_{At+k}) = 1/2^k, \qquad (18)$$

i.e. 0.5 for successive generations. This value is thus a lower bound for all realistic models of distributions of chromosome length. Examples are given in Table 3, the case of unlinked loci shown as m→∞.

TABLE 3

Autocorrelation (%) of additive genetic variance between generations t and t+k as a function of population size, N, total map length, L, and number of chromosomes, m.

N	L	m	k t	1 1	5	2 1	5	4 1	5
20	1	1		61.7	79.6	44.2	65.2	27.0	46.7
		2		56.6	75.3	38.3	60.3	22.6	43.4
	2.5	1		54.9	70.0	35.1	51.6	17.8	32.0
		2		53.0	67.9	32.9	49.3	16.3	30.6
		20		46.7	57.0	24.8	35.4	9.1	18.5
	10	20		46.2	54.4	23.9	31.2	7.7	12.9
	>0	→∞		44.8	50.0	21.9	25.0	5.4	6.2
10	2.5	2		53.2	67.8	33.2	49.2	16.5	30.4
		20		46.8	56.9	24.9	35.1	9.1	18.2
50	2.5	2		52.9	68.0	32.8	49.4	16.2	30.7
		20		46.6	57.1	24.7	35.5	9.0	18.7

The results of Table 3 show that the autocorrelation between successive generations is almost independent of population size, but rises as the loci become more tightly linked, the effects, relative to unlinked loci, being greater as, k, the number of generations apart, increases. For <u>Drosophila melanogaster</u> the autocorrelation of variances four generations apart is likely to be nearly 40%. Thus estimates of genetic variance in a base population made from successive generations of, say, a control population derived from it, convey less information than the standard errors of the individual generation estimates might suggest. As with $CV(V_{At})$ the autocorrelations are affected by the method used to determine the variance, whether in terms of diploids or chromosomes, the latter giving higher values.

Variance of cumulative variance

The formulae used to obtain the autocorrelation of variances are required to compute the variance, $V(V^*_{At})$ of the cumulative variance to generation t (eq. 8), since

$$V(V^*_{At}) = V(\sum_{T=0}^{t} V_{AT}) = \sum_T V(V_{AT}) + 2\sum\sum_{T<T'} Cov(V_{AT}, V_{AT'}) \qquad (19)$$

Typical results are shown in Table 4. Here the results have been standardised to give

$$CV(V^*_{At}) = \sqrt{V(V^*_{At})}/E(V^*_{At})$$

and with our model are not dependent on the initial variance, the initial gene frequencies or distribution of effects.

With unlinked loci $CV(V^*_{At})$ reaches its maximum in very few generations and then declines, the maximum being near 15% with $N = 20$. With tighter linkage the maximum is reached later and is rather higher, but the increase over free recombination is not very large. The general picture is thus of a value of around one-half of that obtained for $CV(V_{At})$ with free recombination, which is reached very quickly and then retained, the value being approximately $\sqrt{1/3N}$.

The convention in this analysis has been to take the mean additive variance at generation 0 as V_A, but with a variance of V_A equal to zero, so $CV(V^*_{A0}) = 0$ and $CV(V^*_{A1}) = \sqrt{[V(V_{A1})]}/[V_{A0}+E(V_{A1})]$.

VARIATION IN RESPONSE TO SELECTION

An alternative would have been to take the sampled first generation as the base point, giving $CV(V_{A1}^*) = \sqrt{V(V_{A1})}/E(V_{A1})$ since no conceptual infinite population at generation 0 could be sampled, and the values of $CV(V_{A1}^*)$ would then have been considerably higher.

TABLE 4

Coefficient of variation, $CV(V_{At}^*)\%$, of the cumulative additive genetic variance to generation t as a function of population size, N, total map length, L, and number of chromosomes, m.

N	L	m	t	1	2	5	10
20	1	1		11.0	15.0	20.1	22.5
		2		11.0	14.5	18.6	20.7
	2.5	1		11.0	14.3	17.2	17.8
		2		11.0	14.1	16.7	17.2
		20		11.0	13.4	14.5	13.8
	10	20		11.0	13.4	14.1	12.8
	>0	→∞		11.0	13.2	13.5	11.5
10	2.5	2		15.4	19.6	23.1	23.4
		20		15.4	18.7	20.0	18.9
50	2.5	2		7.0	9.0	10.7	11.1
		20		7.0	8.6	9.3	8.9

VARIATION WITH SELECTION

The formulae used above for the case of no selection were obtained by analyses of moments of the gene frequency distribution. With selection, however, expectations of each order of moments depend on moments of higher order, for example $E(V_{At})$ depends on third moments at generation t-1. It is therefore not possible to obtain exact general formulae for means and variances when selection is practised, even in the special limiting case of many loci of very small individual effect, and resort has to be made to imprecise verbal arguments before taking over formulae from the case of no selection. Assuming many loci and 10 or less generations with inbreeding coefficients of 20% or less, the effects of selection on the variances may not be too drastic. Some of these problems will be clarified by Robertson (1976) and it is hoped to expand on these preliminary results subsequently, using checks by Monte Carlo simulation. Several possible effects of selection on the variance between replicates will be considered in turn.

In addition to changing the population mean, truncation selection is also expected to change variance in infinite populations (Felsenstein, 1965; Bulmer, 1971, 1974). The reduction in additive variance is due to linkage disequilibrium; in the first generation it equals $\frac{1}{2}Kh^4$ where K is the change in phenotypic variance among selected individuals relative to the whole population, and h^2 is the heritability; in subsequent generations there is a further loss of $\frac{1}{2}Kh_t^4$, but some of the variance lost previously is regenerated by recombination. Therefore the proportional loss in variance increases with increase in heritability, selection intensity and linkage. Let us take as a base point the case of free recombination when, in infinite population,

$$V_{A,t+1} - V_A = \frac{1}{2}(V_{At} - V_A) + \frac{1}{2}Kh_t^4 \qquad (20)$$

(Bulmer, 1971). For example, with $V_P = 1$ and $h^2 = 0.5$ and a typical value of $K = -0.8V_{Pt}$, then $V_A = 0.5$ and (20) gives $V_{A1} = 0.4$, $V_{A2} = 0.38$, the asymptotic value of 0.37 being approached in two generations. With tighter linkage the asymptote is reached later, and the assumption of constancy of gene frequencies becomes less tenable. It is clear that this effect of selection on variance will still be present in finite populations, thus it contributes to a change in the mean rate of response and to a reduction in the variance of mean performance, Z_t and X_t between replicates, this variance being proportional to that within lines. There is also presumably a reduction in $V(V_{At})$ from this source, but the coefficient of variation $CV(V_{At})$ may be less affected since both its numerator and denominator are changed in the same direction.

In previous discussions of the variance in response between replicates, the formulae derived have shown that the variance among selected lines was less than among randomly chosen ones, since the selected parents varied less amongst themselves in phenotype and thus in breeding value (Hill, 1971, 1974). In the extreme case where all are of the same phenotype, $V(Z_1) = (1-h^2)V_A/N$, but with truncation selection a term has to be added to take account of the variation between replicates in the mean phenotype of the selected individuals. This additional term is roughly $(0.2+p)V_A/N$, where p is the proportion selected (Hill, 1974). Thus we replace, in the first generation, $V(Z_1) = V_A/N$ from (2) for the case of no selection by

$$V(Z_1) = [(1-h^2) + (p+0.2)h^2] V_A/N \qquad (21)$$

for truncation selection. The reduction in variance is due to linkage disequilibrium and is the same phenomenon as described by Bulmer (1971). Here it is applied solely to the between replicate variance, but it is clear that there is also a reduction in the within replicate variance which further reduces the between replicate variance subsequently. It is important to note, however, that while (21) shows a reduction in variance relative to random replacement this assumes that the populations are of the same effective size. By increasing the variance of family size, since more individuals from "good" families are selected, the effective population size is likely to be reduced by selection (Robertson, 1961).

In the previous section it was shown that a considerable variation between individual replicate lines in their additive genetic variance was produced by genetic drift. Whilst the theory was developed for unselected populations it is likely that it holds at least approximately in selected populations, certainly when selection is weak and heritability low. The variation in variance therefore contributes to the variation in selection response and replicate means. From this source alone with mass selection,

$$V(X_t) = V(Z_t) = V(\sum_{T=0}^{t-1} \bar{i} \ V_{AT}/\sqrt{V_{PT}}) , \qquad (22)$$

where $\bar{i}$ is the standardised selection differential, which for simplicity is assumed to remain constant. As a first approximation (holding strictly only when coefficients of variation are small)

$$V(V_{AT}/\sqrt{V_{PT}}) = \frac{1}{V_{PT}}(1-\tfrac{1}{2}h^2_T)^2 V(V_{AT})$$

$$= \frac{1}{V_P}(1-\tfrac{1}{2}h^2)^2 V(V_{AT}) \quad (23)$$

approximately, providing the time scale is not too long. Hence (22) reduces to, using the definition (8),

$$V(Z_t) \doteq \frac{\bar{i}^2}{V_p}(1-\tfrac{1}{2}h^2)^2 V(\sum_{T=0}^{t-1} V_{AT}) = \frac{\bar{i}^2}{V_p}(1-\tfrac{1}{2}h^2)^2 V(V^*_{At-1}). \quad (24)$$

Now the mean response is, again assuming changes in V_{PT} are small,

$$E(Z_t) \doteq \frac{\bar{i}}{V_p} E(V^*_{At-1}) \quad (25)$$

so, as a result of variation in variance alone,

$$CV(Z_t) \doteq (1-\tfrac{1}{2}h^2) CV(V^*_{At-1}) \quad (26)$$

Therefore, after reduction by $1-\tfrac{1}{2}h^2$, results for the coefficient of variation of response due to variation between lines in additive variance are given in Table 4, albeit with many approximations, among these the use of $CV(V^*_{At})$ values obtained with no selection.

Making further simplifying assumptions, let us put together the two sources of variation in Z, that arising from drift in mean directly, taken for simplicity as $(t/N)V_A$ from (2) rather than (21) and that indirectly from variation in within-line variance. These two sources are assumed to be uncorrelated, and again for simplicity we take $E(V^*_{At}) = tV_A$, by reducing (9). Then

$$V(Z_t) \doteq \frac{t}{N}V_A + \frac{\bar{i}^2(1-\tfrac{1}{2}h^2)^2 V(V^*_{At-1})}{V_p}$$

$$\doteq \frac{t}{N}V_A[1+Nt\bar{i}^2h^2(1-\tfrac{1}{2}h^2)^2 CV^2(V^*_{At-1})], \quad (27)$$

with the variation in observed mean inflated by sampling of progeny.

VARIATION IN RESPONSE TO SELECTION

As shown by Table 4, $CV(V_{At}^*)$ rises quickly and tends to asymptote if there is tight linkage or drop slowly with free recombination. It becomes a proportionately larger source of variation as t increases. Let us take as a typical example $N = 20$, $\bar{i}^2 = 2$, $h^2 = \frac{1}{4}$, giving

$$V(Z_t) = [1+10t(7CV(V_{At-1}^*)/8)^2]tV_A/20 = (1+\alpha)tV_A/20 , \qquad (28)$$

say. From Table 4, with free recombination ($m \to \infty$) and $t = 3$, 6 and 11, $CV(V_{At}^*) = 0.132$, 0.135 and 0.115 giving $\alpha = 0.40$, 0.84 and 1.11 respectively; while for $L = 2.5$ and $m = 2$, $CV(V_{At}^*) = 0.141$, 0.167 and 0.172 giving $\alpha = 0.45$, 1.28 and 2.49 respectively.

The contribution to the variation in mean from variation in variance is therefore not trivial in initial generations and soon becomes important. Furthermore, since the variances in successive generations are positively autocorrelated, the pattern of responses are likely to be affected; we should expect to see some replicate lines responding more than others consistently for several generations, a phenomenon often observed (e.g. Falconer, 1973).

What then is a satisfactory formula to adopt for the variation in response to selection? Relative to random replacement, selection has been shown to reduce variance due to reduction in the variance among selected individuals, but to increase it due to variation in genetic variance, tighter linkage making both corrections larger. It seems clear that the one proposed by me previously (Hill, 1974) is likely to be an underestimate, since variation in variance is ignored. Perhaps the simplest and most robust formula is just $V(Z_t) = 2F_t V_A$ or tV_A/N as with no selection, assuming the other corrections cancel each other. Robertson (1976) finds this formula gives a reasonable fit to his "infinite locus" selection results.

DISCUSSION

In this paper it has been demonstrated that considerable variation between replicated lines in their genetic variance can be

expected and the theory is thus a natural extension of that for
variance between replicates in mean performance. However,
following Bulmer (1976), it has been argued that most of this
variation in within-line variance, unlike that in mean in unselected populations, is due to linkage disequilibrium. In species
with many chromosomes the major contribution to the variance of
additive genetic variance is from disequilibrium between genes on
different chromosomes. Thus we should not ignore linkage disequilibrium effects on the behaviour of quantitative traits even
in mammals. Furthermore, it has been demonstrated that variances
at different generations in individual replicates are highly
correlated, not being less than one-half in successive generations.
The assumption that many loci affect the trait does not seem
unrealistic, but the assumption that all are additive is much less
so. Problems of variation in genetic variance have not been
tackled for dominant or epistatic genes, and generality is unlikely
to be achieved (even if the mathematical difficulties of handling
eighth or higher moments could be overcome).

Whilst a precise theory, within the limitations of the model,
can be developed for populations in which parents are chosen at
random, the extension to the more interesting case of selection
is less satisfactory. What we have done is simply to consider
the influence of each of the effects of selection on the variance
of response, and since it is clear that some of these oppose each
other, namely the reduction due to mean negative linkage disequilibrium versus the increase due to variance of disequilibrium,
have tentatively suggested that in short term selection experiments
the variance among selected replicates is not greatly different
from that among unselected controls. In most practical applications,
however, precise estimates of variation in response are not required.
If an experiment is sufficiently replicated, the variance of
response can be estimated directly from the experiment so it is
only in the design of selection experiments or analysis of those

with one or few replicates that the theory needed. Certainly in design, in which several unknown parameters such as heritability need to be specified a priori when calculating the predicted response, a general impression of the size of the error variance rather than any exact value is required. As emphasised in the introduction any such predictions based solely on parameters of the base population can be used only for experiments of a few generations duration; the underlying genetic structure must subsequently be critical.

The check on the utility of any theory is, of course, its experimental verification. No tests have as yet been made on the variance in within line variance, and the sampling errors of such estimates are likely to make any check very laborious or quite inconclusive. There have been, however, many selection experiments with laboratory animals and of these a few have been replicated sufficiently to provide estimates of variance between replicate means at successive generations, albeit with few degrees of freedom. For those published or with data available in 1973 an analysis was undertaken in this Institute by Yuksel (1974). Space does not permit a full description of his results so I shall merely summarise his conclusions.

The observed variances, particularly in the high body weight selected lines, in Falconer's (1973) replicated mouse experiment were higher than predicted whereas the data of Eisen et al (1970) with mice fitted well to prediction. In Drosophila experiments with selection for bristle number (Clayton et al., 1957; Frankham et al., 1968; Lopez-Fanjul and Hill, 1973; Madelena and Robertson, 1974) the predicted variances were almost always below observation when selection was for high bristle number, whereas with selection for low bristle number the predicted variances were above those observed, certainly after the first two or three generations. Yuksel's calculations were based on formulae given by Hill (1971), and so no terms were included either for variation in selection differential (the term $0.2h^2 V_A/N$ in (21)) or for mean and variance of changes in variance, and effective population sizes may have been over-estimated. If the calculations were redone, rather higher variances would be predicted. Baker et al. (1975) also analysed their selection experiment with rats to obtain estimates of drift variance using formulae of Hill (1971). The implication from their paper (p.187) would be that more variation between replicates was found than could be explained by this drift variance.

The main difficulty found in Yuksel's analysis, particularly in the Drosophila experiments, was that there was a rapid directional change in within-line variance in the direction of selection, presumably from scale effects. No terms for change in scale have been included in the predictions, though this could be done. It would seem, however, that when designing an experiment the criterion of number and size of individual replicates is likely to be the predicted mean response, relative to its predicted standard error. With a simple logarithmic scale, at least, the ratio of these predictions is unlikely to be affected by the scale transformation.

A theory for predicting variation in response is most likely to be of use in design of selection experiments. Many aspects of this problem have been considered by Bohren (1975), who also included discussions of inbreeding depression and limits, for example, ignored in my previous analysis (Hill, 1971). On some topics discussed both by Bohren (1975) and Hill (1971) qualitatively different answers are obtained. Whilst it is clear that if parameters change rapidly as a result of selection, experiments of only one generation's duration give unbiassed estimates of parameters of the base population, while those of longer duration do not. If, however, the parameters can be assumed not to change greatly, Hill (1971) shows that longer experiments give lower sampling errors, for example of a realised heritability or response per generation. The genetic drift variance increases in proportion to t (eq. (5) and (21)), so the drift variance of response per

VARIATION IN RESPONSE TO SELECTION

generation, $V(X_t/t)$, declines in proportion to $1/t$. Bohren (1975), however, (his eq. 10) includes an additional drift term which equals t^2 times that of the first generation, and I have been unable to follow the logic of this; Bohren therefore concludes that single generation experiments are statistically more efficient. In this paper it has been shown (28) that the variation in response due to variation in variance increases with increasing length of the experiment, and its contribution to variance of response per generation would depend on $t^2 CV^2(V_{At}^*)$ i.e. soon become proportional to t^2. This term can not be equated, however, to the drift variance in mean occurring in the first generation. The contribution of both drift between replicate means and of measurement error (the variance of X_t about Z_t) do not increase in proportion to t^2, so the experiments of more than one duration are likely to be more efficient from a variance viewpoint, but may give biassed estimates if parameters change. Further work on these problems of design is clearly still required.

SUMMARY

Using a model of many additive genes, formulae are given for the variance and autocorrelation of the additive genetic variance in replicated unselected finite populations effective of size N. The coefficient of variation of the additive variance exceeds $\sqrt{4/3N}$ and the autocorrelation between successive generations exceeds 0.5 after a few generations. The implications of this variance and autocorrelation on the variation between replicates in response to selection are discussed.

ACKNOWLEDGEMENTS

I am grateful to Dr P.J. Avery for making many unpublished results available, to Prof. Alan Robertson for unpublished data and to both for helpful discussion, and to Dr M.G. Bulmer for a copy of his 1976 paper prior to publication.

BIBLIOGRAPHY

[1] Avery P.J. & Hill, W.G. (1977). Variation in genetic parameters in small populations. (In preparation).

[2] Baker, R.L., Chapman, A.B. and Wardell, R.T. (1975). Direct response to selection for post weaning gain in the rat. Genetics 80, 171-189.

[3] Bohren, B.B. (1975). Designing artificial selection experiments for specific objectives. Genetics 80, 205-220.

[4] Bulmer, M.G. (1971). The effect of selection on genetic variability. Amer. Natur. 105, 201-211.

[5] Bulmer, M.G. (1974). Linkage disequilibrium and genetic variability. Genet. Res. 23, 281-289.

[6] Bulmer, M.G. (1976). The effect of selection on genetic variability: A simulation study. Genet. Res. (in press).

[7] Clayton, G.A., Morris, J.A. & Robertson, A. (1957). An experimental check on quantitative genetical theory. I. Short term responses to selection. J. Genet. 55, 131-151.

[8] Eisen, E.J., Legates, J.E. & Robison, O.W. (1970). Selection for 12-day litter weight in mice. Genetics 64, 511-532.

[9] Falconer, D.S. (1960). Introduction to Quantitative Genetics. Oliver and Boyd, Edinburgh.

[10] Falconer, D.S. (1973). Replicated selection for body weight in mice. Genet. Res. 22, 291-321.

[11] Felsenstein, J. (1965). The effect of linkage on directional selection. Genetics 52, 349-363.

[12] Frankham, R., Jones, L.P. & Barker, J.S.F. (1968). The effects of population size and selection intensity in selection for a quantitative character in Drosophila. I. Short term responses to selection. Genet. Res. 12, 237-248.

[13] Hill, W.G. (1971). Design and efficiency of selection experiments for estimating genetic parameters. Biometrics 27, 293-311.

[14] Hill, W.G. (1972a). Estimation of realised heritabilities from selection experiments. I. Divergent selection. Biometrics 28, 741-765.

[15] Hill, W.G. (1972b). Estimation of realised heritabilities from selection experiments. II. Selection in one direction. Biometrics 28, 767-780.

[16] Hill, W.G. (1972c). Estimation of genetic change. I. General theory and design of control populations. Anim. Breed. Abstr. 40, 1-15.

[17] Hill, W.G. (1974). Variability of response to selection in genetic experiments. Biometrics 30, 363-366.

[18] Lopez-Fanjul, C. & Hill, W.G. (1973). Genetic differences between populations of Drosophila melanogaster for a quantitative trait. I. Laboratory populations. Genet. Res. 22, 51-68.

[19] Madelena, F.E. & Robertson, A. (1974. Population structure in artificial selection: studies with artificial selection. Genet. Res. 24, 113-126.

[20] Robertson, A. (1960). A theory of limits in artificial selection. Proc. Roy Soc. Lond. 153, 234-249.

[21] Robertson, A. (1961). Inbreeding in artificial selection programmes. Genet. Res. 2, 189-194.

[22] Robertson, A. (1976). Effects of linkage and population size on selection response and limits. (These proceedings).

[23] Sved, J.A. & Feldman, M.W. (1973). Correlation and probability methods for one and two loci. Theor. Pop. Biol. 4, 129-132.

[24] Yuksel, E. (1974). Check on a quantitative genetic theory. Unpublished M.Sc. thesis, University of Edinburgh.

Selection with overlapping generations

William G. Hill
INSTITUTE OF ANIMAL GENETICS
EDINBURGH EH9 3JN, SCOTLAND

INTRODUCTION

For simplicity, population and quantitative geneticists often use models in which generations are discrete; but with the main exception of the annual plants, generations usually overlap. Even when we use an overlapping generation theory, it is frequently assumed that the population has reached a stable age distribution and that the rate of genetic change is constant. In cattle and sheep populations the low rate of female reproduction requires keeping breeding females for many years, even in an efficient artificial selection programme; thus after several years in which selection has been practised, some females may be from the foundation population and have been subjected to no selection at all, while some young animals may be from the third generation. How good are models which ignore this structure?

In this paper I shall restrict discussion solely to the problem of artificial selection in animal populations. I shall also make several simplifying assumptions in common with most of the animal breeding literature on overlapping generation theory which are not unreasonable for the type of programme to be discussed, for example one planned for 10-20 years in cattle, representing less than 5 generations. Heritabilities and genetic variances will be assumed to remain unchanged, so that phenotypic and genetic selection differentials are constant. Studies have been made of long term evolutionary problems in which frequencies at single genes are considered and require nonlinear analytic methods, but these will not be discussed here.

This paper is in three separate sections. The first deals with recent developments in what can be termed "classical" or "asymptotic" theory, namely where the same programme has been continued for a sufficient period of time that the population is improving at a constant rate each generation. In the second section discussion centres on methods of predicting response before such an asymptotic rate is reached. This would apply, for example, to new schemes or programmes where selection objectives or population structures have been altered. Using this theory it is possible to check the utility of asymptotic predictions. Finally, the computation of rates of inbreeding will be discussed because this is relevant to optimum choice of population age structure, can be handled by similar methods and the same questions of short term or asymptotic rates arise. In the earlier sections the population is assumed to be large enough that drift can be ignored.

ASYMPTOTIC THEORY

The original formulae for predicting rates of response to artificial selection were obtained by Dickerson and Hazel (1944) and generalised by Rendel and Robertson (1950). Essentially they showed that the rate of response in a continuing breeding programme equalled the ratio of the mean genetic selection differential of the parents to the mean age of the parents when their progeny were born (generation interval), each quantity being computed over the four pathways of genes from male and female parents to male and female progeny. These formulae have been used in subsequent analyses of potential

SELECTION WITH OVERLAPPING GENERATIONS

breeding programmes in order to find the one giving the highest predicted rate of response as, for example, in decisions on whether to adopt performance or progeny testing. Most such analyses have been of an _ad_ _hoc_ nature; but recently Ollivier (1974) has formalised calculations of optimum replacement rates both for traits recorded once on each animal, such as juvenile growth rate, and for traits which have repeated measurements through the animal's lifetime. Not surprisingly, the calculations show that the more prolific the species, the more rapidly should generations be turned over, and this turnover should be faster in males.

One of the difficulties in selecting animals from populations in which there is an overlap of generations is in comparing progeny from parents of different ages. If the population is improving, animals from younger parents are, on average, superior to those from older parents, the difference depending on the rate of genetic progress. Genetic differences between such progeny groups may, however, be confounded with differences in maternal environment (Bichard, Pease, Swales and Ozkutuk, 1973). Bichard et al. show how the selection intensities among progeny of different age groups should be adjusted to take more from the younger parents, giving an increased rate of response over that with selection of equal numbers from each parental age group. Such a procedure is clearly less efficient, however, than a scheme in which there is sequential culling of parents according to their predicted breeding value. In principle, the optimum scheme is simply that where the predicted breeding values of individuals of all ages are compared at regular intervals, and the poorest culled, regardless of their age. In practice, however, such a scheme may be difficult to operate. For dairy bulls in which breeding values can be fairly accurately assessed by progeny testing, individual bulls can be compared directly using Henderson's (e.g., 1973) BLUP procedure or other method, and some of the problems of comparison among age groups do not arise. These comparisons also do not rely on there being a steady rate of genetic progress.

The issues of optimum selection intensity relative to generation interval apply at the nucleus level in a breeding programme, but in species such as pigs there may be one or more generations of multiplication of breeding stock to commercial level, with an overlap of generations and perhaps a delay of many years between selection of young animals at nucleus level and the final culling of old animals at multiplier level. The difference in genetic merit between the nucleus and commercial stock can be measured as a "lag", the number of years of selection this difference represents. Bichard (1971) has discussed how this lag is affected by alternative replacement policies and use of A. I. This analysis is in terms of the asymptotic lag, and assumes that the same nucleus breeding and multiplication programme continues indefinitely, whereas in practice the multiplication process expands and contracts with changing sales patterns.

SHORT-TERM THEORY

A number of authors have computed predictions of response from selection before the asymptotic response has been reached, I believe the first being Searle (1961). Such analyses have been stimulated recently by the adoption of discounting techniques in monetary evaluation of breeding programmes, where the value of response obtained is assumed to decrease each year into the future. With the heavy discounting associated with high interest rates it could be critical to obtain fairly precise predictions of response, say, two years hence, without the need to assume this is twice the asymptotic rate. The

SELECTION WITH OVERLAPPING GENERATIONS

methods are of two basic kinds. That most commonly adopted, to which I shall return, is to consider the mean breeding value of all individuals of each age and sex group in the population, perhaps subdivided further into potential breeders, multiplication and commercial stock. The alternative method, due to McClintock and Cunningham (1974) is to consider the number, genetic contribution (i.e., one-half for progeny, one-quarter for grand-progeny) and year of birth of descendants derived from one insemination by a bull or by one individual. Thus they add up the total number of "standard" expressions of an individual's genotype up to a specified time, and these can be discounted if required. In this way selection decisions at the level of single animals can be evaluated which is particularly useful when considering bulls for potential use in A. I. where such animals have a large individual influence.

Calculations for the alternative method of keeping track of the mean breeding value of each age group in the population can be done in various ways. A simple formal method using matrix algebra has been developed independently by Hill (1974) and Elsen and Mocquot (1974). Hill's paper considers matrix algebra in more detail while Elsen and Mocquot's goes into greater depth on construction of the matrices and on applications. With non-overlapping generations one scalar (single number) is sufficient to specify the population mean and another to specify the mean breeding value of selected parents. With an overlap, a vector is required to describe the genetic merit of each age group and a matrix to define the contribution of genes between age groups. As an example, discussed less formally by Hill (1971), assume a beef herd is maintained such that all progeny are got by 2 year old bulls and one-third from each of 2, 3 and 4 year old cows.

Young animals are selected on a performance test of growth rate. Let $x_{(t)}$ denote a column vector, whose elements are the mean breeding values at time t of males of age 1, 2, females of age 1, 2, 3 and 4 in that order. The passage of genes among the age and sex groups is specified by the matrix P where

$$P = \begin{array}{c} \\ \text{Age} \\ \\ \\ \\ \\ \\ \\ \end{array} \begin{array}{cc} \multicolumn{2}{c}{\text{Males}} \\ 1 & 2 \end{array} \begin{array}{cccc} \multicolumn{4}{c}{\text{Females}} \\ 1 & 2 & 3 & 4 \end{array}$$

$$P = \left(\begin{array}{cc|cccc} 0 & \frac{1}{2} & 0 & \frac{1}{6} & \frac{1}{6} & \frac{1}{6} \\ 1 & 0 & 0 & 0 & 0 & 0 \\ \hline 0 & \frac{1}{2} & 0 & \frac{1}{6} & \frac{1}{6} & \frac{1}{6} \\ 0 & 0 & 1 & 0 & 0 & 0 \\ 0 & 0 & 0 & 1 & 0 & 0 \\ 0 & 0 & 0 & 0 & 1 & 0 \end{array} \right)$$

and the element p_{ij} denotes the proportion of genes contributed by age-sex class j at year t to age-sex class i at year t+1. The first row of each block of P denotes the transfer of genes by reproduction, the remainder by ageing. Thus P is a modified "Leslie" matrix (Leslie, 1945) adapted to include two sexes and with all row totals equal to unity. It is clear that, if no selection is practised,

$$x_{(t)} = Px_{(t-1)} = P^t x_{(0)}.$$

If, by selection, the breeding values of individuals in the initial year are incremented by the vector s of genetic selection differentials, the increment among their descendants at year t will thus be $P^t s$. To separate the effects of selection from ageing, it is convenient to define a matrix Q which specifies the passage of genes due to ageing alone. Q is thus the

SELECTION WITH OVERLAPPING GENERATIONS

same as P but with the first row of each block set to zero. The response at year t from the one selection is thus the vector

$$r_{(t)} = (P^t - Q^t)s.$$

With continued selection of the same intensity, the cumulative response is

$$R_{(t)} = \sum_{T=1}^{t} (P^T - Q^T)s.$$

It is possible to show that, as t increases, $r_{(t)} = R_{(t)} - R_{(t-1)}$ approaches the asymptotic values given by Rendel and Robertson (1950). Further details of the analysis are given by Hill (1974).

As an example (a), let us assume (following Hill, 1971) that selection is for live weight at 400 days, a trait with a phenotypic standard deviation of 40 kg and a heritability of 0.4. Assume 7% of bulls are selected on performance, giving a standardised selection differential $\bar{i}$ of 1.9, and all females are required for replacement. The generation interval is 2.5 years, so the asymptotic rate of response is $\frac{1}{2}$ x 1.9 x 40 x 0.4/2.5 = 6.08 kg/year. The actual response using the matrix, and the asymptotic predictions are compared in Figure 1. As a further illustration (b), consider a second programme in which males are retained for two breeding seasons, and females for five. Again assuming equal calving rates and no cow mortality, the first row of each block of P is now

	Males			Females					
Age	1	2	3	1	2	3	4	5	6
	(0	0.25	0.25	0	0.1	0.1	0.1	0.1	0.1)

Assuming $3\frac{1}{2}$% of males selected and 60% of females, $\bar{i}$ = 2.2 and 0.6 respectively, and the generation interval is now 3.25 years,

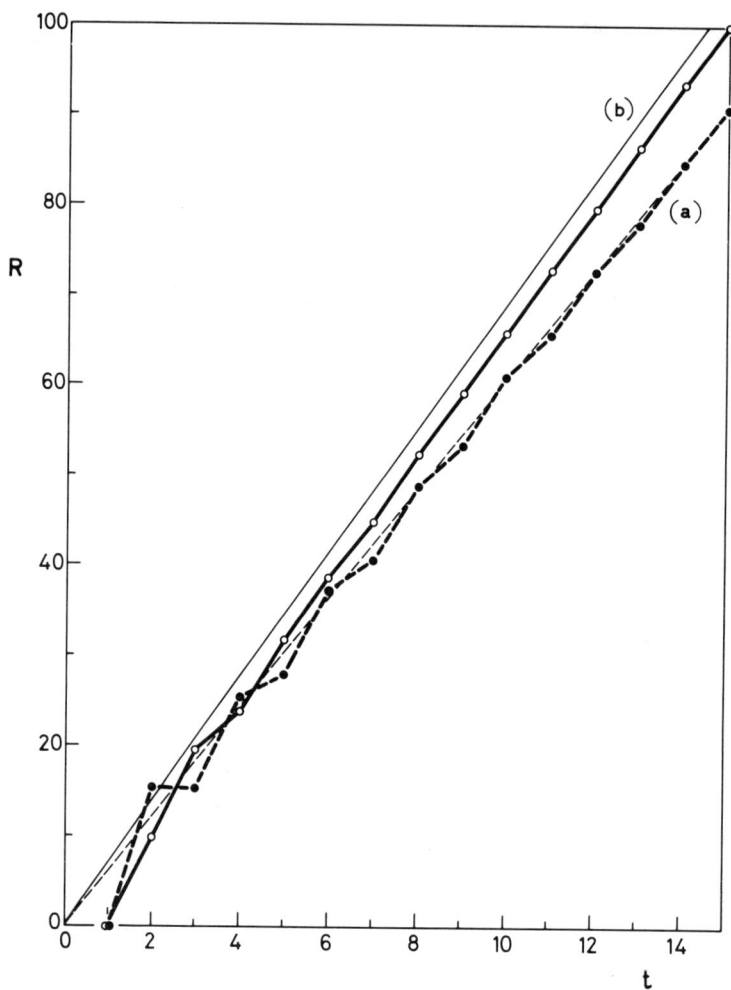

FIG. 1.

Predicted response (R) in successive years (t) to continued selection for 400 day weight (kg) in beef cattle, using both exact (thick lines) and asymptotic (fine lines) formulae.

Scheme	Age at Breeding	
	males	females
(a)	2	2,3,4
(b)	2,3	2,3,4,5,6

giving an asymptotic rate of response of 6.89 kg/year, deliberately chosen to be similar to that in the other scheme. Predicted responses are also given for this scheme in Figure 1.

In both examples the asymptotic predictions are seen to be within a few percent of the exact values after some six years, much of the difference in model (b) between the two predictions being of a constant value. Further, comparisons between schemes (a) and (b) are soon made adequately on the basis of the asymptotic values. The matrix P in each case has a single eigenvalue of unity, which is associated with the asymptotic rate of response to selection. The approach to this asymptote therefore depends on the larger non-unit eigenvalues of P. In case (a) the largest non-unit root is real and equals -0.839, in case (b) it is complex, the real part being -0.441, the imaginary part ± 0.552 and the modulus 0.707. The root in (a) is larger presumably because the males are used just once, and there is a slow flow of genes between individuals born one year apart. In both cases, the contribution from eigenvectors associated with the non-unit eigenvalues are clearly going to be small after 5 to 10 years.

As noted previously, much of the interest in the actual pattern of response with overlapping generations has been in monetary evaluation of breeding programmes. Hill (1971) and Brascamp (1975) have shown that quite good predictions of total discounted returns, even for evaluation periods equivalent to only two or three generations, can be made using asymptotic rather than actual response predictions.

INBREEDING AND GENETIC DRIFT

An overlapping of generations influences other population parameters in addition to selection response, for example rates of inbreeding and the associated variance among conceptual replicate populations due to genetic drift. Again most published analyses have been for populations assumed to have been maintained sufficiently long that the level of inbreeding is increasing at an asymptotic rate. It turns out that the asymptotic rate of inbreeding per generation with overlapping generations is the same as that if generations are discrete and with the same number of individuals entering the population per generation and the same distribution of lifetime family size (Hill, 1972). This last proviso is an important one. It is customary to assume as a base point in computing rates of inbreeding that there are no real differences in viability and fertility between individuals so that the distribution of family size is approximately Poisson and Wright's classic formula for the rate as $(8 \times males)^{-1} + (8 \times females)^{-1}$ applies. But with overlapping generations, even if there are no fertility differences among survivors and there is random death of breeding individuals, the inbreeding rate will be higher than given by this formula since the distribution of lifetime family size is not Poisson. With an exponential distribution of deaths the rate can be nearly three times as high as in the simple formula (Felsenstein, 1971).

In a newly established population, it will be many years before there is any inbreeding among the older individuals, and because of the breeding structure, the probability of identity by descent of randomly chosen genes from individuals of different age groups will not be the same as for genes from individuals of the same age group. Felsenstein (1971) recognised this, but computed only asymptotic rates. Recently, D. L. Johnson (personal communication) has obtained formulae for computing levels of inbreeding at intermediate generations for general models, the procedure using the same matrix P as defined above. Let $F_{(t)}$ be a square matrix of the same dimension as P with elements equal to the probabilities of identity of randomly chosen genes from the specified age group at time t. It can be shown that

$$F_{(t)} = P F_{(t-1)} P' + D$$

where P' is the transpose of P and D is a diagonal matrix whose elements depend on the number of individuals in the age groups. The asymptotic rate of inbreeding is approached about as fast as is the asymptotic rate of response to selection.

CONCLUSIONS AND SUMMARY

In animal breeding applications of quantitative genetics it is usual for generations to overlap. Standard formulae for rates of genetic progress and of inbreeding give asymptotic rates, assuming the population to be maintained in the same way for many years. It is found, however, that when progress from selection or the inbreeding coefficient is predicted using exact formulae taking account of the pattern of change each year, the predictions are never far from those of asymptotic theory.

BIBLIOGRAPHY

(1) Bichard, M. (1971). Dissemination of genetic improvement through a livestock industry. Anim. Prod. 13:401-411.

(2) Bichard, M., Pease, A. H. R., Swales, P. H. and Ozkutuk, K. (1973). Selection in a population with overlapping generations. Anim. Prod. 17:215-227.

(3) Brascamp, E. W. (1975). Model calculations concerning economic optimalisation of AI-breeding with cattle. Agric. Res. Rep. 846. Centre Agric. Publishing and Documentation, Wageningen, Netherlands.

(4) Dickerson, G. E. and Hazel, L. N. (1944). Effectiveness of selection on progeny performance as a supplement to earlier culling of livestock. J. Agr. Res. 69: 459-476.

(5) Elsen, J. M. and Mocquot, J. C. (1974). Recherches pour une rationalisation technique et économique des schémes de sélection des bovins et ovins. Bull. Tech. Dep. Genet. Anim. (Inst. Natn. Rech. Agron., Fr.) No. 17.

(6) Felsenstein, J. (1971). Inbreeding and variance effective numbers in populations with overlapping generations. Genetics 68:581-597.

(7) Henderson, C. R. (1973). Sire evaluation and genetic trends. In Proc. Anim. Breed Genet. Symp., Blacksburg, Virginia, pp. 10-41. Amer. Soc. Anim. Sci., Champaign, Illinois.

(8) Hill, W. G. (1971). Investment appraisal for national breeding programmes. Anim. Prod. 13:37-50.

(9) Hill, W. G. (1972). Effective size of populations with overlapping generations. Theor. Pop. Biol. 3: 278-289.

(10) Hill, W. G. (1974). Prediction and evaluation of response to selection with overlapping generations. Anim. Prod. 18:117-139.

(11) Leslie, P. H. (1945). On the use of matrices in certain population mathematics. Biometrika 33:183-212.

(12) McClintock, A. E. and Cunningham, E. P. (1974). Selection in dual purpose dairy cattle populations: defining the breeding objective. Anim. Prod. 18:237-247.

(13) Ollivier, L. (1974). Optimum replacement rates in animal breeding. Anim. Prod. 19:257-271.

(14) Rendel, J. M. and Robertson, A. (1950). Estimation of genetic gain in milk yield by selection in a closed herd of dairy cattle. J. Genet. 50:1-8.

(15) Searle, S. R. (1961). Estimating herd improvement from selection programmes. J. Dairy Sci. 44:1103-1112.

Selective advance in populations with overlapping generations

Edward Pollak
STATISTICAL LABORATORY
IOWA STATE UNIVERSITY, AMES, IA 50011

INTRODUCTION

The selection of individuals upon the basis of their own phenotypes is widely used in plant and animal breeding. In most of the theory that has been developed for this situation, it is assumed that the population under consideration is infinite, that generations do not overlap, and that there is no epistasis or linkage disequilibrium. Then if there is, in addition, random mating of parents, followed by the selection of some of the offspring, the expected response from selection is

$$r = \frac{\sigma_A^2}{\sigma_P^2} I = h^2 I . \qquad (1)$$

In expression (1)

r = the selective advance (or response)
  = the mean of the offspring of saved individuals
  − the mean of the population of offspring that would result without selection,
I = the selection differential
  = the mean of the selected parents
  − the mean of the population from which the parents came,

Journal Paper No. J-8679 of the Iowa Agriculture and Home Economics Experiment Station, Ames, Iowa. Project 1669. Partial support by National Institute of Health, Grant GM 13827.

σ_A^2 = the additive genetic variance in the quantitative character for which there is selection,

σ_P^2 = the phenotypic variance in the quantitative character,

h^2 = Lush's heritability in narrow sense.

One way in which to obtain expression (1) is to derive a linear predictor for the phenotypic value of the offspring, given that there are known phenotypic values for the parents. Then if males are chosen independently of the females to which they are mated, the regression equation is

$$\hat{O} = \overline{O} + \frac{\text{Cov}(O, S)}{\text{Var}(S)} (S - \overline{S}) + \frac{\text{Cov}(O, D)}{\text{Var}(D)} (D - \overline{D}), \qquad (2)$$

where

$\overline{O}$ = the mean of all offspring in the random mated populations,

$\hat{O}$ = the predicted phenotypic value of an offspring,

S = the phenotypic value of the sire of the offspring,

$\overline{S}$ = the mean of all sires,

D = the phenotypic value of the dam of the offspring,

$\overline{D}$ = the mean of all dams.

It is well known (see, for example, Kempthorne (1957, ch. 15) that, if there is no epistasis and no linkage disequilibrium, the genetic covariance between parents and offspring, when they are both measured in the same attribute, is equal to $\sigma_A^2/2$. Thus, if father, mother, and offspring are measured with respect to the same quantitative character and the environments in which they live are chosen independently of each other and of genotypes,

$$\text{Cov}(O, S) = \text{Cov}(O, D) = \tfrac{1}{2}\sigma_A^2,$$
$$\text{Var}(S) = \text{Var}(D) = \sigma_P^2,$$
$$\overline{S} = \overline{D} = \overline{O} = \overline{P} = \text{the phenotypic mean.}$$

Therefore

$$\hat{O} = \overline{P} + \tfrac{1}{2}\frac{\sigma_A^2}{\sigma_P^2}(S - \overline{P}) + \tfrac{1}{2}\frac{\sigma_A^2}{\sigma_P^2}(D - \overline{P})$$

$$= \overline{P} + \frac{\sigma_A^2}{\sigma_P^2}\left\{\frac{S+D}{2} - \overline{P}\right\}, \qquad (3)$$

which is one version of (1).

Now if phenotypic variability is differently distributed among males than it is among females, we can no longer expect (2) to hold. In some instances it may be possible to salvage this simple expression. For example, it may be the case that the expression of a character in a male is b times the corresponding measurement in a female with the same genotype. Then the measurements in one sex may be multiplicatively adjusted to be equivalent to what one would have expected to observe in an individual of the same genotype in the other sex. This has been done, for example, by Falconer (1960, p. 177) in his discussion of data on abdominal bristle numbers in Drosophila melanogaster. I doubt, however, that such a simple procedure would work in general.

The only reason that I have gone into this detailed discussion of a theory that is so well known is that it will be seen to illustrate, in a simple form, some difficulties that arise in more elaborate theories that have been developed for populations with overlapping generations. These theories now will be discussed.

SOME THEORIES FOR POPULATIONS WITH OVERLAPPING GENERATIONS

The first paper in which there is a discussion of the rate of genetic improvement to be expected in a population with overlapping generations seems to be that of Dickerson and Hazel (1944). These authors stated that

$$\Delta G = \text{the average annual gain} = \frac{\Delta P}{T}, \qquad (4)$$

where ΔP is the average genetic gain in T years and T is the average age of parents when their offspring are born. Rendel and Robertson (1950) gave the more general expression

$$\Delta G = \frac{I_{11} + I_{21} + I_{12} + I_{22}}{L_{11} + L_{21} + L_{12} + L_{22}}, \qquad (5)$$

where ΔG has the same meaning as in (4),

I_{uv} = the genetic superiority of individuals of sex u, chosen to breed sex v, compared with the average of individuals of sex u,

L_{uv} = the mean age of sex u individuals when their offspring of sex v are born,

u, v = 1, 2, where 1 represents male and 2 female.

Both (4) and (5) are approximate in the sense that they are derived under the assumption that there is a linear relationship between genetic gain and the amount of time in which the population has been subjected to selection. Hill (1974) has recently given a more detailed analysis of the problem, taking account of the fact that the genes from a group of selected individuals may take several years to be represented in all age groups of the population. One would then not expect the rate of improvement in the mean performance of the population to be constant, at least in the short run, although in the long run the rate of progress from continued selection may approach a fixed value.

Hill showed, among other things, that r_t, the response in the first age group at time t from selection at time 0, approaches (5) as $t \to \infty$ if it is assumed, as by Rendel and Robertson, that there is only selection among individuals of the first age group. He also showed that if there is repeated selection of individuals of the first age group, then R_t, the response at time t among individuals of the first age group, has the property that

$$\frac{R_t}{tr_t} \to 1, \quad t \to \infty, \tag{6}$$

as had been conjectured earlier by Hinks (1971) and Hill (1971). Hence the rate of response to repeated selection, given by R_t/t, approaches the classical prediction equation (5) of Rendel and Robertson as t becomes very large.

Hill's detailed analysis provides a more logically complete and rigorous development than what had been available before for the theory of selection of a quantitative character in a population with overlapping generations. There nevertheless remain some difficulties, even in his formulation. These will be discussed in the following three sections, as a new derivation of some of his results is given. The emphasis will be on making explicit what seem to be underlying implicit assumptions in Hill's development, whereas there will be only a brief mention of those parts of the argument that already have been explicitly worked out by him.

THE CONSTRUCTION OF A LINEAR PREDICTION EQUATION FOR RESPONSE FROM SELECTION

Expressions (4) and (5) and the more general ones derived by Hill are linear prediction equations. Such equations must contain covariances between relatives. I shall take the point of view that if an ancestor and a descendant are observed at different ages, the character in the descendant is different from the character in the ancestor. Thus, in general, we will be interested in covariances between ancestors, measured in one character, and descendants, measured in another. The results can then be specialized in those instances in which it is possible to make some sort of age or sex correction.

We shall assume that there is initially an infinite diploid random mating population in which there is no linkage disequilibrium. In general, let us consider individuals V and Z, which are a random pair of members of the population with a certain pattern of relationship, e.g., grandmother and granddaughter. Suppose that V is measured in attribute X and Z in attribute Y. Then, if the environments in which V and Z live are chosen independently of each other and of the genotypes, the covariance between V, measured in X, and Z, measured in Y, is

$$\text{Cov}(V_X, Z_Y) = 2r_{VZ}\sigma_{A;XY} + u_{VZ}\sigma_{D;XY} \qquad (7)$$
$$+ 4r_{VZ}^2 \sigma_{AA;XY} + 2r_{VZ} u_{VZ} \sigma_{AD;XY} + u_{VZ}^2 \sigma_{DD;XY} + \cdots$$

where

$\sigma_{A;XY}$ = the additive genetic covariance in X and Y,
$\sigma_{D;XY}$ = the dominance covariance in X and Y,
$\sigma_{AA;XY}$ = the additive x additive covariance in X and Y,
$\sigma_{AD;XY}$ = the additive x dominance covariance in X and Y,
$\sigma_{DD;XY}$ = the dominance x dominance covariance in X and Y,
r_{VZ} = the probability that a randomly chosen gene from V is identical by descent to a randomly chosen gene,

at the same locus, from Z,

u_{VZ} = the probability that one gene of V is identical by descent to one gene of Z and the second gene of V, at the same locus, is identical by descent to the second gene of Z.

If q_i is the probability of allele A_i and α_i and β_i are, respectively, the effects of A_i with reference to X and Y, we have in the one-locus case that

$$\sigma_{A;XY} = 2\Sigma_i q_i \alpha_i \beta_i \quad . \tag{8}$$

Also, if $d_{i\ell}$ and $\delta_{i\ell}$ are, respectively, dominance deviations associated with the genotype $A_i A_\ell$ at a locus with reference to attributes X and Y,

$$\sigma_{D;XY} = \Sigma_i \Sigma_\ell q_i q_\ell d_{i\ell} \delta_{i\ell} \; , \tag{9}$$

If many loci affect the character, $\sigma_{A;XY}$ and $\sigma_{D;XY}$ are equal to what is obtained by summing expressions like (8) and (9), respectively, over all loci. The epistatic covariance components, have, like $\sigma_{A;XY}$ and $\sigma_{D;XY}$, a structure that is entirely analogous to the corresponding variance components. The only difference between them and the variance components is that cross products replace squares.

Perhaps (7) is known to several people attending this conference. But, since I have not seen a written derivation of it anywhere, even by Mode and Robinson (1959), one is given in the appendix. The expression is implicit in the work of Mode and Robinson, who do give special cases.

Of particular interest to us will be the special case in which V is an n-th generation descendant of Z. Then $u_{VZ} = 0$ and $r_{VZ} = 1/2^{n+1}$. Hence, in this case,

$$\text{Cov}(V_X, Z_Y) = \frac{1}{2^n} \sigma_{A;XY} + \frac{1}{2^{2n}} \sigma_{AA;XY} + \cdots \quad . \tag{10}$$

Following Hill's notation, we classify individuals into $h + k$ age-sex classes, where the symbols $1, \ldots, h$ denote males of age groups 1 through h and h + 1, ..., h + k denote females of age groups 1 through k. The age of a parent is measured at the time the child is born. In what follows, we shall assume that there is weak selection, applied only at time 0. Then expression (10), which was derived for a random mating population without selection, is approximately valid.

We consider the regression of descendants of age-sex class i' (type i' for short) at time t on all possible types of ancestors living at time 0. One of the ways in which a gene at time 0 can pass to an individual of age-sex class i' at time t is that it be passed in n generations to the descendant from an ancestor of age-sex class j that is used to breed sex v progeny. Because we are assuming an infinite random mating population, such genes are passed on independently of genes with different histories, so that we can consider the regression coefficient associated with these genes separately from other regression coefficients.

The numerator of this regression coefficient is, by (10), equal to

$$\frac{1}{2^n} \sigma_{A;i'j} + \frac{1}{2^{2n}} \sigma_{AA;i'j} + \cdots$$

where $\sigma_{A;i'j}$ and $\sigma_{AA;i'j}, \ldots$ are, respectively, equal to the additive genetic covariance, the additive x additive genetic covariance, etc., of relatives of types i' and j. In the denominator there is the phenotypic variance of the mean of the ancestors of age-sex class j that were used to breed sex v progeny. In the infinite random mating population, there are 2^n n-th generation ancestors. Hence

E[number of n-th generation ancestors, of a type i' individual at time t, that were of type j and used to breed sex v progeny]

$= Q(i', j; t, 0; v \mid n) 2^n$,

where

$Q(i', j; t, 0; v | n)$

= P (a gene in a descendant of type i' at time t comes from a type j ancestor that was used to breed progeny of sex v | n generations separate the ancestor and descendant).

Therefore the regression coefficient is

$$[\frac{1}{2^n} \sigma_{A;i'j} + \frac{1}{2^{2n}} \sigma_{AA;i'j} + \ldots] / [\frac{1}{Q(i', j; t, 0; v | n) 2^n} \sigma_{P_j}^2]$$

$$= Q(i'j; t, 0; v | n)[\sigma_{A;i'j} + \sigma_{AA;i'j}/2^n + \ldots]/\sigma_{P_j}^2$$

where $\sigma_{P_j}^2$ is the phenotypic variance of individuals of type j.

Actually, the preceding argument is not completely in conformity with linear regression theory, because I have used an expected number of ancestors rather than a fixed number. This seems reasonable, however, because the associated probability does end up in the numerator, where it should if we are predicting an expected advance from selection. In addition, if t is large, n will be also, and then the ratio of the number of ancestors to the expected number is unlikely to deviate much from unity. It should also be noted that, as n increases, the terms involving the epistatic covariance components in the numerator of the regression coefficient become more and more negligible in size.

Coming back to the argument, we note that all ancestors of type j that were used to breed sex v progeny have the same selection differential associated with them, regardless of how many generations separate them from their descendant of type i' at time t. Let this selection differential be $I_{j,v}$. This is the difference between the mean of individuals of age-sex class j that were saved to breed sex v progeny and the mean of all individuals of the same age-sex class.

Then, if there is no epistasis, and

$\pi_{n,t}$ = P (n generations separate the ancestor and the descendant at time t),

we find that the contribution of ancestors of type j to the linear predictor is

$$\sum_{n=0}^{\infty} \pi_{n,t} Q(i', j; t, 0; v \mid n) \frac{\sigma_{A;i'j}}{\sigma^2_{P_j}} I_{j,v}$$

$$= Q(i'j; t, 0; v) \frac{\sigma_{A;i'j}}{\sigma^2_{P_j}} I_{j,v} ,$$

where

$Q(i', j; t, 0; v)$

= P(a gene in a descendant of type i' at time t comes from a type j ancestor that was used to breed progeny of sex v).

The selective advance, or response, of individuals of age-sex class i' at time t is equal to the predicted mean of such individuals, minus the mean of individuals of this age-sex class in the original population. In the absence of epistasis, it is equal to

$$r_{i'}(t) = \sum_{j=1}^{h+k} \{Q(i', j; t, 0; 1) I_{j,1} + Q(i', j; t, 0; 2) I_{j,2}\} \frac{\sigma_{A;i'j}}{\sigma^2_{P_j}} \quad (11)$$

SELECTIVE RESPONSES IN THE LONG RUN

In this section there will be a heuristic derivation of $r_{i'}$, the limiting response in age-sex class i' as t tends to infinity. A more rigorous matrix algebra derivation is given by Hill (1974), for a special case. He also is able with his method to obtain explicit expressions for $r_{i'}(t)$ for every value of t. The argument that will be used here will not reveal what $r_{i'}(t)$ is when t is finite, but will, I hope, be intuitively appealing.

It is clear from the form of (11) that the only terms on the right side that depend upon t are the probabilities $Q(i', j; t, 0; v)$. Thus $\lim_{t \to \infty} r_{i'}(t)$ can be obtained if we obtain the limiting values of these probabilities. This now will be done.

Clearly, a gene that is in a descendant of age-sex class i' at time t is in a descendant of age-sex class i' + 1 at time t+1.

Hence,
$$Q(i', j; t, 0; v) = Q(i'+1, j; t+1, 0; v)$$
and thus, as $t \to \infty$, we would not expect the limiting probabilities to depend upon i'. Our problem therefore reduces to the calculation of the quantities

$$Q(j;v) = \lim_{t \to \infty} Q(i', j; t, 0; v)$$

= P (a gene in an individual in the distant future is a copy of a gene in an individual of age-sex class j at time 0 that was used to breed progeny of sex v).

Now, following Hill (1974), we let

p_{1j} = P (a gene in a male of age 1 at time t comes from a parent of age-sex class j at time t-1)

and

$p_{h+1,j}$ = P (a gene in a female of age 1 at time t comes from a parent of age-sex class j at time t-1),

where $j = 1, \ldots, h, h+1, \ldots, h+k$, and subscripts 1 through h are associated with ages of males and subscripts h+1 through h+k with ages of females.

Until now, we have considered the contributions to the distant future of individuals of age-sex class j at time 0 that were used to breed progeny of sex v. We may also assess their contributions with reference to time 1. Let such founding members breed progeny of sex v throughout their reproductive lives. They then may influence the future as individuals of age-sex class j + 1, which occurs with probability $Q(j+1;v)$, unless j = h or j = h+k. In the latter cases, the probabilities are equal to 0. Secondly, they may have influence through their offspring, events that occur with probabilities $p_{1j}Q_1$ and $p_{h+1,j}Q_2$, respectively, if v = 1 or 2, where

Q_v = P (a gene in an individual of the distant future is a copy of a gene in an individual of age 1 and sex v).

Therefore

$$Q(j,1) = Q(j+1;1) + p_{1j}Q_1, \quad j = 1,\ldots, h-1, h+1,\ldots, h+k-1$$

$$Q(h;1) = p_{1h}Q_1$$

$$Q(h+k;1) = p_{1,h+k}Q_1 \tag{12}$$

$$Q(j;2) = Q(j+1;2) + p_{h+1,j}Q_2, \quad j = 1,\ldots, h-1, h+1,\ldots, h+k-1$$

$$Q(h;2) = p_{h+1,h}Q_2$$

$$Q(h+k;2) = p_{h+1,h+k}Q_2$$

We also expect, inasmuch as the proportions of parents in the various age-sex classes do not change with time, that

$$Q(1;1) + Q(1;2) = Q_1$$
$$Q(h+1;1) + Q(h+1;2) = Q_2 \tag{13}$$

The solution of the system of equations (12) is given by

$$Q(j;1) = Q_1 \sum_{w=j}^{h} p_{1w}, \quad j = 1,\ldots, h,$$

$$= Q_1 \sum_{w=j}^{h+k} p_{1w}, \quad j = h+1,\ldots, h+k,$$

$$Q(j;2) = Q_2 \sum_{w=j}^{h} p_{h+1,w}, \quad j = 1,\ldots, h, \tag{14}$$

$$Q_2 \sum_{w=j}^{h+k} p_{h+1,w}, \quad j = h+1,\ldots, h+k.$$

Now half the genes in individuals of age group 1 originate from males and half from females. Hence

$$\sum_{w=1}^{h} p_{1w} = \sum_{w=h+1}^{h+k} p_{1w} = \sum_{w=1}^{h} p_{h+1,w} = \sum_{w=h+1}^{h+k} p_{h+1,w} = \tfrac{1}{2}, \tag{15}$$

so that we have from (14) that

$$Q(1;1) = Q(h+1;1) = \tfrac{1}{2}Q_1,$$
$$Q(1;2) = Q(h+1;2) = \tfrac{1}{2}Q_2.$$

Therefore, by (13), $Q_1 = Q_2 = Q$, and because the expressions given by (14) are a set of probabilities that add to 1,

$$1 = Q\{\sum_{j=1}^{h}\sum_{w=j}^{h} p_{1w} + \sum_{j=h+1}^{h+k}\sum_{w=j}^{h+k} p_{1w} + \sum_{j=1}^{h}\sum_{w=j}^{h} p_{h+1,w}$$

$$+ \sum_{j=h+1}^{h+k}\sum_{w=j}^{h+k} p_{h+1,w}\}$$

$$= Q\{\sum_{w=1}^{h} w p_{1w} + \sum_{w=h+1}^{h+k}(w-h)p_{1w} + \sum_{w=1}^{h} w p_{h+1,w}$$

$$+ \sum_{w=h+1}^{h+k}(w-h)p_{h+1,w}\}$$

$$= Q\{\sum_{w=1}^{h} w p_{1w} + \sum_{w'=1}^{k} w' p_{1,w'+h} + \sum_{w=1}^{h} w p_{h+1,w}$$

$$+ \sum_{w'=1}^{k} w' p_{h+1,h+w'}\}.$$

By using (15), we see that

$2\sum_{w=1}^{h} w p_{1w} = L_{11}$ = the mean age of males when their male progeny are born,

$2\sum_{w=1}^{k} w p_{1,h+w} = L_{21}$ = the mean age of females when their male progeny are born,

$2\sum_{w=1}^{h} w p_{h+1,w} = L_{12}$ = the mean age of males when their female progeny are born,

$2\sum_{w=1}^{k} w p_{h+1,h+w} = L_{22}$ = the mean age of females when their female progeny are born.

Therefore

$$Q = \{2 \tfrac{1}{4}(L_{11} + L_{21} + L_{12} + L_{22})\}^{-1} = (2L)^{-1} \qquad (16)$$

where

L = the mean age of parents of newborn progeny
  = the generation interval.

It now follows from the definition of $Q(j;v)$, (11), (14) and (16) that

$$r_{i'} = \lim_{t \to \infty} r_{i'}(t) = \sum_{j=1}^{h+k} [Q(j;1) I_{j,1} + Q(j;2) I_{j,2}] \frac{\sigma_{A;i'j}}{\sigma_{P_j}^2} \qquad (17)$$

$$= \frac{1}{2L} \{\sum_{j=1}^{h+k} [v_{mj} I_{j,1} + v_{fj} I_{j,2}] \frac{\sigma_{A;i'j}}{\sigma_{P_j}^2} \}, \quad i' = 1, \ldots, h+k,$$

where, consistent with Hill's notation, we have

$v_{mj} = \sum_{w=j}^{h} p_{1w}$, $j = 1, \ldots, h$,

$\phantom{v_{mj}} = \sum_{w=j}^{h+k} p_{1w}$, $j = h+1, \ldots, h+k$,

$v_{fj} = \sum_{w=j}^{h} p_{h+1,w}$, $j = 1, \ldots, h$,

$\phantom{v_{fj}} = \sum_{w=j}^{h+k} p_{h+1,w}$, $j = h+1, \ldots, h+k$.

The definition of $I_{j,v}$ implies, of course, that the quantities $I_{j,1}$ and $I_{j,2}$ are, respectively, selection differentials associated with breeders of males and females.

It is easy to show that (17) is consistent with (2). For, with discrete generations, $L = 1$, and there are two age-sex classes, with indices 1 and 2 standing for males and females, respectively.

Hence, by using (7), we see that (2) may be put in the form

$$r_{i'} = \tfrac{1}{2} \frac{\sigma_{A;i'1}}{\sigma^2_{P_1}} I_1 + \tfrac{1}{2} \frac{\sigma_{A;i'2}}{\sigma^2_{P_2}} I_2, \quad i' = 1, 2,$$

because we have not assumed that there are different selection differentials for breeders of males and females. So, of course, I_v is the selection differential for parents of sex v and, in addition,

$\tfrac{1}{2}$ = P(a gene in the offspring is a copy of a gene in a parent of sex v), v = 1, 2.

THE COMPARISON OF (17) WITH HILL'S RESULTS

Expressions (17) differ from Hill's results in that he finds that $r_{i'}$ is the same for all values if i', whereas, by (17), this is not necessarily so. The contrast lies in his having the genetic selection differential of ancestors of age-sex class j, used to breed offspring of sex v, in place of $I_{j,v} \sigma_{A;i'j}/\sigma^2_{P_j}$. Rendel and Robertson (1950) define the genetic selection differential to be equal to the heritability times the corresponding phenotypic selection differential, whereas Hill (1974) defines it to be the superiority in breeding value of a group of selected individuals of a particular age-sex class, relative to the mean of that age-sex class. Judging by the numerical example in Hill's paper, these definitions are consistent and involve setting $\sigma_{A;i'j} = \sigma^2_A$ and $\sigma^2_{P_j} = \sigma^2_P$ for i' and j. Thus, it seems that these authors assume either that the same attribute is being measured in all age-sex classes, or that measurements in one age-sex class can be adjusted to be equivalent to measurements in any other class.

Presumably, such an adjustment would be most feasible if the only attribute of interest is a single measurement taken on an animal at "maturity." This state of affairs may be approximated if animals are raised for meat and are all marketed for slaughter at approximately the same age. The only adjustment that would need to be made in this case would be for sex.

With dairy cattle, for example, the situation is much more complex, even if we neglect the obvious difficulty that milk production cannot be observed in bulls and that males can only be evaluated by taking measurements on their female relatives. For, even if we only considered cows, we would be interested in milk yields for several lactations, not just the first. There is indeed an extensive literature on the age-yield relation in dairy cattle. As Freeman (1973) points out in his review paper on this subject, this massive effort has not yet resulted in a completely satisfactory solution, although progress is being made.

Another approach that could be taken would be to set up experiments to estimate all the variances and covariances or all the ratios $\sigma_{A;ij}/\sigma^2_{P_j}$ in (17). This procedure may, however, require large-scale and very expensive experiments if there are several age-sex classes. There may also be difficulties of a statistical nature. If, for example, the quantities $\sigma_{A;i'j'}/\sigma^2_{P_j}$ were to be estimated by the computation of linear regressions of offspring on parents, the performance records of the offspring would have to be adjusted for any improvements in management techniques that had occurred since the parents were measured.

DISCUSSION

It has been assumed throughout the development given in this paper that the covariances between relatives of any two particular age-sex classes, and the phenotypic variances of the age-sex classes, do not change with time. This assumption, which seems also to underlie Hill's argument, seems to me to be essential to keep the mathematics manageable. In biological terms, it means that we are assuming that selection is slow enough and completed in a short enough time so that covariances and variances remain rather close to what they were in the original unselected population. One consequence of this is that it is reasonable to assert as Hill (1974) does in his discussion of response to repeated selection, that the response at time t from selection at time t_0 is equal to that at time $t - t_0$ from selection at time 0.

It is then of interest to know if $r_i'(t)$ approaches r_i' fairly rapidly. Hill (1976) has shown for some examples, with rates of turnover that could easily arise in animal breeding practice, that the approach of $r_i'(t)$ to its limiting value r_i' is indeed rapid. Thus the issue raised in the preceding paragraph may not be of serious concern.

It should be noted, however, that expression (11) only holds either if there is no epistasis, or if t is large enough so that it is highly probable that there are a large number of generations separating ancestors and their descendants in age-sex class i'. Thus, if there is considerable epistasis, the approach of $r_{ji'}(t)$ to the expression for $r_{ji'}$ given by (17) may be slowed down. Nevertheless, if there is weak selection, so that the phenotypic variances and genetic covariances do not change much with time, the right side of (17) may actually be approached, whereas, in the presence of epistasis, the discrete generation analog of (17) would not hold.

We note also that, at autosomal loci in a population with discrete generations, exactly one half of the genes in an individual come, for example, from the father. On the other hand, the terms $Q(j;v)$ in (17) are only expected proportions of genes originating from various age-sex-breeding classes. Therefore there is an extra element of sampling variability in the selective advances in populations with overlapping generations, when they are compared with selective advances in populations with discrete generations.

Another assumption in the theory that has been developed here is that, once an individual has been chosen to breed progeny of sex v, it will be used in only that way throughout the rest of its reproductive life. In my development, this assumption is used in setting up the first and fourth of equations (12). It appears in Hill's argument when he sets the genetic selection differential of an individual of age-sex class j, used to breed sex v offspring, equal to the genetic selection differential of an individual of age-sex class j + 1, used to breed sex v offspring, one time unit later.

This second assumption seems unrealistic because, for example, a very superior cow may be chosen to be the mother of bulls, but a sensible breeder presumably would also save her female calves. Thus (17) would have to be generalized in some way if it is desired to include the possible use of an animal to breed offspring of both sexes throughout its life.

In one sense (17) is already quite general, because there is nothing in its form that requires offspring to be measured in the same biological attribute as its parents. So, for example, an offspring might be measured in body weight, while its mother was measured for milk production. In this case $\sigma_{A;i'j}$ could be taken to be the covariance between the weight of the descendant of type i' and the milk production of the female ancestor of type j.

SUMMARY

We consider an infinite diploid population, measured at times $0, 1, 2, \ldots$, in which there are h age groups among males and k age groups among females. In addition, there is a possible division of each of the h + k age-sex classes into two types: individuals used to breed male offspring and individuals used to breed female offspring. It is assumed that, once an individual is chosen to breed a particular sex of offspring, it is used in that way throughout the remainder of its reproductive life.

We assume that, initially, there is random mating, with no linkage disequilibrium, and that there is then weak selection applied only once, at time 0. The difference between the mean of individuals of age-sex class j that were saved to breed sex v progeny and the mean of all individuals of the same age-sex class is called the selection differential associated with this age-sex class. It is denoted by $I_{j,v}$.

The selective advance, or response, of individuals of age-sex class i' at time t is equal to the predicted mean of such individuals minus the mean of individuals of this age-sex class in the original population. This is denoted by $r_{i'}(t)$.

The view is taken that, if a descendant and an ancestor are observed at different ages, the character in the descendant is different from the character in the ancestor. A general expression is given for the covariance between a random pair of relatives with a certain pattern of relationship, if one is measured in attribute X and the other in attribute Y. This expression is used in constructing linear prediction equations for the $r_{ji'}(t)$ in terms of given values of the $I_{j,v}$. Expressions for the limiting responses $r_{i'} = \lim_{t \to \infty} r_{ji}(t)$ are then derived, and these turn out not necessarily to be the same for all values of i'. There is a discussion of the relationship of these results to others in the literature.

ACKNOWLEDGMENT

I am indebted to Professor Oscar Kempthorne for suggesting the proof of (7) that is given in the appendix. It is superior to my original proof, particularly with regard to its easier incorporation of epistatic effects. He also made helpful comments that prompted me to write down underlying assumptions in more detail and with more clarity.

BIBLIOGRAPHY

Dickerson, G. E. and L. N. Hazel (1944). Effectiveness of selection on progeny performance as a supplement to earlier culling in livestock. J. Agric. Res., 69:459-476.

Falconer, D. S. (1960). Introduction to Quantitative Genetics, 1970 reprinting. The Ronald Press Co., New York.

Freeman, A. E. (1973). Age adjustment of production records; history and basic problems. J. Dairy Sci., 56:941-946.

Hill, W. G. (1971). Investment appraisal for national breeding programmes. Anim. Prod., 13:37-50.

Hill, W. G. (1974). Prediction and evaluation of response to selection with overlapping generations. Anim. Prod., 18:117-139.

Hill, W. G. (1976). Selection with overlapping generations. Proceedings of the International Conference on Quantitative Genetics.

Hinks, C. J. M. (1971). The genetic and financial consequences of selection amongst dairy bulls in artificial insemination. Anim. Prod., 13:209-218.

Kempthorne, O. (1957). An Introduction to Genetic Statistics. John Wiley and Sons, Inc., New York.

Mode, C. J. and H. F. Robinson (1959). Pleiotropism and genetic variance and covariance. Biometrics 15:518-537.

Rendel, J. M. and A. Robertson (1950). Estimation of genetic gain in milk yield by selection in a closed herd of dairy cattle. J. Genet., 50:1-8.

APPENDIX

We assume that the population is in linkage equilibrium and that the environments in which individuals V and Z live are chosen independently of each other and of the genotype. It may then be shown (cf. Kempthorne (1957, ch. 19)) that

$$\text{Cov}(V_{X+Y}, Z_{X+Y}) = 2r_{VZ}\sigma_{A;X+Y,X+Y} + u_{VZ}\sigma_{D;X+Y,X+Y}$$
$$+ (2r_{VZ})^2 \sigma_{AA;X+Y,X+Y} + 2r_{VZ}u_{VZ}\sigma_{AD;X+Y,X+Y} + \ldots,$$

$$\text{Cov}(V_X, Z_X) = 2r_{VZ}\sigma_{A;XX} + u_{VZ}\sigma_{D;XX}$$
$$+ (2r_{VZ})^2 \sigma_{AA;XX} + 2r_{VZ}u_{VZ}\sigma_{AD;XX} + u^2_{VZ}\sigma_{DD;XX} + \ldots,$$

$$\text{Cov}(V_Y, Z_Y) = 2r_{VZ}\sigma_{A;YY} + u_{VZ}\sigma_{D;YY}$$
$$+ (2r_{VZ})^2 \sigma_{AA;YY} + 2r_{VZ}u_{VZ}\sigma_{AD;YY} + u^2_{VZ}\sigma_{DD;YY} + \ldots,$$

where, in our notation, $\sigma_{A;XX}$ is, for example, equal to the additive genetic variance in attribute X. We also know, from elementary properties of variances and covariances, that

$$\text{Cov}(V_{X+Y}, Z_{X+Y}) = \text{Cov}(V_X, Z_X) + \text{Cov}(V_X, Z_Y)$$
$$+ \text{Cov}(V_Y, Z_X) + \text{Cov}(V_Y, Z_Y),$$

and that, for example,

$$\sigma_{A;X+Y,X+Y} = \sigma_{A;XX} + 2\sigma_{A;XY} + \sigma_{A;YY},$$
$$\sigma_{D;X+Y,X+Y} = \sigma_{D;XX} + 2\sigma_{D;XY} + \sigma_{D;YY},$$
$$\sigma_{AA;X+Y,X+Y} = \sigma_{AA;XX} + 2\sigma_{AA;XY} + \sigma_{AA;YY}.$$

Therefore, to complete the proof of (7), we must show that

$$\text{Cov}(V_X, Z_Y) = \text{Cov}(V_Y, Z_X).$$

RESPONSE WITH OVERLAPPING GENERATIONS

To do this, we note that if

$p(g, h) = P(\text{genotype of V is } g, \text{ genotype of Z is } h)$,

then, if there is no selection,

$p(g, h) = p(h, g)$.

Hence, if we let g_X and h_Y stand, respectively, for the coded genotypic values of g in X and h in Y

$$\text{Cov}(V_X, Z_Y) = \sum_{\{g, h\}} p(g, h) g_X h_Y$$

and, similarly,

$$\text{Cov}(V_Y, Z_X) = \sum_{\{g, h\}} p(g, h) g_Y h_X$$

$$= \sum_{\{g, h\}} p(h, g) h_X g_Y .$$

Because these two sums are taken over all genotypes, they are clearly the same, and

$$\text{Cov}(V_X Z_Y) + \text{Cov}(V_Y, Z_X) = 2\text{Cov}(V_X, Z_Y)$$

$$= 2 \{ 2r_{VZ}\sigma_{A;XY} + u_{VZ}\sigma_{D;XY} + (2r_{VZ})^2 \sigma_{AA;XY}$$

$$+ 2r_{VZ} u_{VZ} \sigma_{AD;XY} + u^2_{VZ} \sigma_{DD;XY} + \ldots \} .$$

This completes the proof of (7).

Selection limits in self-fertilizing populations following the cross of homozygous lines

Theodore B. Bailey, Jr.
DEPARTMENT OF STATISTICS
IOWA STATE UNIVERSITY, AMES, IA 50011

1. INTRODUCTION

The genetic improvement of plant and animal populations depends on the presence and availability of favorable alleles in the species. The process of collecting the superior alleles present in a species into a single population or individual typically involves the formation of a foundation population, followed by some form of recurrent selection. Foundation populations may be formed in a variety of ways. For example, two or more source populations may be combined to form a segregating population. Alternatively, if one of the source populations were clearly superior to the other(s), the construction of a foundation population might include one or more backcrosses to that source population.

The selection scheme to be investigated in this research consists of within-line selection during the formation of single seed descent lines. In evaluating the effectiveness of selection procedures in population improvement programs it is natural to examine criteria such as duration, rates and limits of response. In the case of intra-line selection considered here, the duration of response, often expressed in terms of the half-life of the

---

Journal Paper No. J-8740, Iowa Agriculture and Home Economics Experiment Station, Ames, Iowa, Project 0101.

expected change in gene frequency, is not expected to be greater than two generations because of the approximately 50 percent reduction in genetic variation each generation. As a result of the short duration of response, rates of response are not of special interest, except as they relate to the limit of response. Limits to selection are typically studied in terms of the probability of fixation of favorable alleles.

The effectiveness of the total breeding program, including both a) formation of the foundation population and b) effectiveness of selection, can be expressed in terms of the level of performance of the final selected population relative to the best source population. For example, if one could predict that a particular breeding program had a very low probability of producing a population at the selection limit that was better than the best source population, he would obviously abandon that procedure and search for a more promising program.

Results presented in this paper are specifically concerned with the development of inbred lines following the cross of two homozygous lines. The foundation population from which the lines are initiated may be the F_2 obtained by selfing the F_1, or it may consist of the individuals formed by one or more backcrosses to the superior line.

2. DEVELOPMENT OF INBRED LINES FROM AN F_1

The genetic state of the lines with successive generations of selfing and within line selection can be described by using matrices. Let $f_j^{(n)}$ be the frequency of the three genotypes ++, +-, and -- where $j = 2$, 1 or 0 specifies the number of favorable alleles in the genotype. For example, $f_2^{(n)}$ is the frequency of individuals in generation n homozygous for the favorable allele. Then $f^{(n)}$ will denote the vector of expected frequencies of the three genotypes in generation n. Further,

$$f^{(n+1)} = Tf^{(n)} = T^2 f^{(n-1)} = T^n f^{(1)}, \qquad (1)$$

SELECTION LIMITS IN SELFING POPULATIONS

where T is the probability transition matrix, $\{p_{ij}\}$, and p_{ij} is the probability of genotype i in generation n+1 given an individual of genotype j in generation n. The values of six of these probabilities can be written down immediately for the homozygous genotypes: $p_{22} = p_{00} = 1.0$ and $p_{12} = p_{02} = p_{20} = p_{10} = 0.0$. A heterozygous parent in generation n can produce ++, +- and -- genotypes in generation n+1 with frequencies 1/4, 1/2 and 1/4, respectively. If the relative fitnesses of these genotypes are assigned values of $1+s/2$, 1/2 and $1-s/2$, then the values of p_{21}, p_{11} and p_{01} corresponding to the three genotypes are $(1 + s/2)/4$, 1/2 and $(1 - s/2)/4$. In these expressions $\underline{s}$ is the difference in relative fitness between the two homozygous genotypes.

Given the preceding quantities, and that the line is initiated from a single heterozygous F_1 individual, the expected frequencies of the three genotypes in generation n are given by $f_2^{(n)}$, $f_1^{(n)}$ and $f_0^{(n)}$. In particular, it can be shown that as n becomes large the probability that a homozygous line contains the favorable allele approaches as a limit

$$P = (1 + s/2)/2 \ . \qquad (2)$$

This quantity also corresponds to the probability of fixation of the favorable allele.

If selection is completely random, the value of $\underline{s}$ in expression (2) is equal to zero, and the probability that the favorable allele will be present in the selected homozygous line is 0.5. To determine the effect of selection on the probability of fixation of the favorable allele, it is necessary to specify reasonable values of $\underline{s}$. Bailey and Comstock (1976) presented a model that provides estimates of $\underline{s}$ that depend on the heritability (in the F_2), the number of loci affecting the trait and intensity of selection. Table I gives some values of $\underline{s}$ and the associated probabilities of fixation as given by (2).

When the selection criterion is based on phenotype of single individuals, a heritability of 0.25 is perhaps as large as could be expected for within line selection, given the multifactorial mode of inheritance being considered. The heritability could always be increased through the use of progeny tests, although probably not much above a value of 0.5.

TABLE I

Values of s and P When the Selection Intensity is Ten Percent.

Number of loci ℓ	h^2	s	P
20	.25	.566	.641
20	.50	.830	.707
60	.25	.316	.579
60	.50	.459	.615
-	.00	.00	.500

The values of P in Table I, which do not exceed .71 even with a relatively small number of loci and a fairly high heritability, give the probability of fixation for a single locus. Consequences of the model described would apply to more than one locus in the line if one assumes independent assortment of loci. If in addition P is the same for all loci, the binomial frequency distribution can be used to determine the probabilities of fixation for any number of loci.

Consider the distribution of number of loci containing the favorable allele in a line developed without selection from an F_1 with twenty heterozygous loci. The expected number of loci fixed for the best allele is ten, so that the genotypic value of such an individual under these conditions would be exactly the same as the F_1. The chance of an individual with a large (or small) number of loci with the best allele is small. Less than two individuals in ten thousand would be expected to contain as many as eighteen (or as few as two) loci homozygous for the favorable allele. A corresponding distribution for a trait influenced by sixty loci is similar in that it is centered at the value of the F_1. However, with sixty segregating loci there is a very small chance for an individual to be homozygous at any large proportion (>70 percent) of the loci (Table II).

It is of interest to consider what effect a selection scheme, such as the one described, would have on the distribution of loci homozygous for the favorable allele in the derived lines. The expected frequency of selected lines containing various numbers of loci homozygous for the favorable allele for the case of a quantitative trait influenced by sixty loci is presented in Table II. Little discussion is needed except to note that intra-line selection can be expected to result in lines with greater numbers of loci containing the favorable allele (compared with no selection). However, the probability of a line containing more than seventy percent of the loci fixed favorable is small. So for instance, if we have a F_1 population of soybeans with a mean yield of 30 bushels per acre, and with a maximum in a homozygous line being 60, the expected frequency of lines yielding greater than 42 is only about 2 percent if heritability is 25 percent; even if heritability is 50 percent, the expected frequency is less than 7 percent.

TABLE II

Probabilities of Homozygous Lines with More than $\underline{k}$ Loci Homozygous for the Favorable Allele, given sixty heterzygous loci in the F_1.

k	No Selection	Ten Percent Selection Intensity	
		$h^2 = .25$	$h^2 = .50$
30	.4487	.8660	.9534
33	.1831	.6293	.8157
36	.0462	.3248	.5443
39	.0067	.1057	.2455
42	.0005	.0196	.0660
45	- *	.0019	.0094
48	-	-	.0006

* - indicates probability less than .0001

It is also informative to evaluate the performance of the lines produced by a breeding program relative to the source populations (in this case the two homozygous parents). Two parents of equal value would each have thirty loci containing the favorable allele (and thirty loci with the unfavorable allele) if the F_1 were heterozygous for sixty loci. With equal-valued parents, Table II indicates that one could be confident of recovering a line superior to the parent lines.

In contrast, consider the case where one parent contains 42 loci with the favorable allele, while the other parent possesses eighteen different loci with the favorable allele. If the heritability for the trait under selection were 0.5, an average line produced by a selection program as described above would be expected to have only 36.90 loci (compared with 42 in the best parent) fixed for the favorable allele. Also, less than seven lines in one hundred would be expected to be as good as or better than the best parent. Of course if heritability is less than 0.5, or if selection is not practiced, the probability of obtaining superior genotypes is reduced (Table II).

The probabilities given in Table II show that, whatever the values of the two parent lines, there is vitually no chance of recovering the perfect genotype (i.e., all 60 loci containing the favorable allele). Indeed, under the conditions specified, it is very unlikely that a line containing more than 80 percent of the loci with the favorable allele would be recovered.

3. DEVELOPMENT OF LINES AFTER BACKCROSSING

The results of the previous section strongly suggest that if one of the two parent lines (source populations) used in forming the F_1 is clearly superior to the other parent, then it is very unlikely that the selection program under consideration will yield a line better than the best parent. Under these conditions, it is of interest to consider the effect of selection on single-seed descent lines initiated from populations formed by one or more backcrosses to the superior parent.

For purposes of discussion, the superior and inferior parents will be referred to as recurrent and donor parents, respectively. We will focus our attention only on those loci in which the recurrent and donor parents differ in their allelic contents. It follows that the loci in F_1 plants produced by one or more backcrosses are of two types: a) loci in which the recurrent parent contained the favorable allele and b) loci in which the donor parent contained the favorable allele.

Let $\underline{m}$ be the number of loci containing the favorable allele in the recurrent parent and the unfavorable allele in the donor parent. Then, assuming all loci assort independently, the probability that $\underline{r}$ loci ($r \leq m$) are fixed for the favorable allele in a line produced by a selection program as described is given by expression (3)

$$P_r = \frac{m!}{r!(m-r)!} P^r (1-P)^{m-r} . \qquad (3)$$

Here, P is the probability that a locus contains the favorable allele in the homozygous line, where

$$P = f_1 P_1 + f_2 P_2 , \text{ and}$$

P_1 is the probability of fixation of the favorable allele in the homozygous line given that the locus was heterozygous in the F_1 plant,

P_2 is the probability of fixation of the favorable allele in the homozygous line given that the locus was homozygous favorable in the F_1 plant (=1.0), and

f_1, f_2 are the probabilities that the F_1 plants are heterozygous or homozygous favorable.

In an analogous fashion, let $\underline{n}$ be the number of loci fixed for the unfavorable allele in the recurrent parent but the favorable allele in the donor parent. If Q is the probability that a locus is fixed for the favorable allele in a line developed by the program under consideration, then the probability of s loci ($s \leq n$) being homozygous for the favorable allele is given by expression (4)

$$Q_s = \frac{n!}{s!(n-s)!} Q^s (1-Q)^{n-s} \qquad (4)$$

where

$Q = f_0 Q_0 + f_1 Q_1$, and

Q_0 is the probability of fixation of the favorable allele in the homozygous line given that it was homozygous unfavorable in the F_1 plant (=0.0),

Q_1 is the probability of fixation of the favorable allele in the homozygous line given that the locus was heterozygous in the F_1 plant, and

f_0, f_1 are the probabilities that the F_1 plant is homozygous unfavorable or heterozygous.

Values for f_0, f_1 and f_2 for populations resulting from various numbers of backcrosses are presented in Table III.

TABLE III
Values of f_0, f_1 and f_2 for Backcross Populations Produced by t Backcrosses.

Backcross Population	Number of Backcrosses t	f_0	f_1	f_2
B1	1	1/2	1/2	1/2
B2	2	3/4	1/4	3/4
B3	3	7/8	1/8	7/8
⋮	⋮	⋮	⋮	⋮
Bt	t	$1-(1/2)^t$	$(1/2)^t$	$1-(1/2)^t$

The two events of r out of m loci being fixed favorable and s out of n loci being fixed favorable are independent under the model described. It follows that the probability of

SELECTION LIMITS IN SELFING POPULATIONS

obtaining a line with r+s loci homozygous for the favorable allele is

$$P_r Q_s = \frac{m!n!}{r!s!(m-r)!(n-s)!} P^r(1-P)^{m-r} Q^s(1-Q)^{n-s}, \qquad (5)$$

and the probability that a homozygous line contains more loci with the favorable allele than the best parent is

$$\sum_{r,s} P_r Q_s, \qquad \begin{array}{l} r + s > m, \\ r + s \leq \ell = m + n. \end{array} \qquad (6)$$

Use of expressions (3), (4), (5) and (6) requires an estimate of P_1. Values of P_1 can be obtained by using expression (2) when appropriate values of $\underline{s}$, the selection coefficient, corresponding to the various backcross generations are provided. The magnitude of $\underline{s}$ depends on the quantity u/σ, where $\underline{u}$ is the effect of a single (+) allele so that the genetic means for ++, +- and -- individuals are $2u$, u and 0, respectively. Here σ is the phenotypic standard deviation of the selection criterion. It should be noted that the actual variance required involves all loci <u>except</u> the one under consideration. The difference in the two variances for the work presented in this paper is of trivial importance. The phenotypic standard deviation in F_2 populations derived from the various backcrosses will be less than σ in the original F_2 (i.e., the population derived from the F_1 formed by crossing the two parent lines) since each successive backcross decreases the number of segregating loci by half. This change in phenotypic variance can be described in terms of the heritability of the original F_2 population. In the model being considered, the additive genetic variance, σ_g^2, is $\ell u^2/2$, or $\ell/2$ where ℓ is the number of segregating loci and $\underline{u}$ is assigned a value of 1.0. The phenotypic variance is then $\sigma^2 = \sigma_g^2 + \sigma_e^2$ where σ_e^2 represents all nongenetic variance including genotype-environment interaction. If h^2 is the heritability in the original F_2, then the corresponding phenotypic variance can be written as $\sigma^2[h^2 + (1 - h^2)]$. The phenotypic variance in an F_2 population after t backcrosses is then $\sigma^2[(1/2)^t h^2 + (1 - h^2)]$. The total phenotypic variance for various F_2 populations is presented in Table IV.

TABLE IV

Phenotypic Variance of F_2 Populations Derived from B1 to B5 Backcrosses, Expressed in Terms of h^2, the Heritability in the F1-F2 Population.

F_2 Population	Phenotypic Variance σ^2	$1/\sigma^2$
F1-F2	1	1
B1-F2	$1-h^2/2$	$2/(2-h^2)$
B2-F2	$1-3h^2/4$	$4/(4-3h^2)$
B3-F2	$1-7h^2/8$	$8/(8-7h^2)$
B4-F2	$1-15h^2/16$	$16/(16-15h^2)$
B5-F2	$1-31h^2/32$	$32/(32-31h^2)$

The preceding results can be used to examine the effects of intra-line selection on single seed descent lines initiated from various backcross populations. Consider the case where $\ell = 20$, $m = 16$ and $n = 4$. Table V gives the values of P_1, P and Q when $h^2 = 0.5$ and a selection intensity of ten percent is assumed. By using these quantities, the probabilities of lines containing more than 0.8ℓ loci (i.e., 16 loci) homozygous favorable were calculated with expression (6) and are presented in Table VI. Similar computations corresponding to $\ell = 60$, $m = 48$ and $n = 12$ were made and also are presented.

TABLE V

Values of P_1 (and Q_1), P and Q for Various Backcross Populations when $\ell = 20$ or $\ell = 60$. The Best Individual out of Ten is Selected to Continue the Line and $h^2 = 0.5$.

Number of loci, ℓ		Backcross Population				
		B1	B2	B3	B4	B5
20	$P_1 = Q_1$	.731	.753	.768	.776	.781
	P	.865	.938	.971	.986	.993
	Q	.365	.188	.096	.049	.024
60	$P_1 = Q_1$	.632	.644	.652	.657	.659
	P	.816	.911	.957	.979	.989
	Q	.316	.161	.082	.041	.021

TABLE VI

Probabilities of Lines Homozygous for the Favorable Allele at more than $0.8\ \ell$ Loci for Lines Initiated from various Populations.*
A Selection Intensity of Ten Percent was Used.

Population	$\ell = 20$; m = 16, n = 4		$\ell = 60$; m = 48, n = 12	
	No Selection	$h^2 = 0.5$	No Selection	$h^2 = 0.5$
F1	.001	.121	-	.001
B1	.028	.241	.001	.034
B2	.072	.271	.020	.108
B3	.089	.222	.058	.163
B4	.074	.146	.088	.173
B5	.048	.085	.086	.140

*The probability of fixation of the favorable allele for the six populations was obtained from Table I (F1) and Table V (B1-B5).

For the four cases considered in Table VI, some amount of backcrossing resulted in an increase in the chance of obtaining a line superior to the recurrent parent, relative to results expected with no backcrossing. The optimum number of backcrosses varied between one and four. As the number of loci increased from twenty to sixty, the optimum number of backcrosses increased whether selection was practiced or not. Given m = 48, n = 12 and $h^2 = 0.5$, Table VI shows that seventeen percent of the lines are expected to be superior to the best parent given four backcrosses. Stated in other words, eighty-three lines out of one-hundred produced under these conditions would not be expected to be better than the recurrent parent. Furthermore, the probability that a $B4$ derived line would contain more than fifty loci homozygous favorable is only 0.0044. This represents a fairly low chance of recovering a line that has as few as two more loci fixed favorable than the best parent. Moreover, there is essentially no chance of obtaining a line with more than fifty-four loci homozygous favorable (i.e., exceeding ninety percent of the potential). The situation is not much better when $m = 16$ and $n = 4$, even with $h^2 = 0.5$; less than two percent of the lines would be expected to contain more than eighteen loci homozygous favorable.

It is also interesting to note that when $m = 48$, $n = 12$ and $h^2 = 0.5$, the expected frequency of lines with more than fifty loci homozygous favorable is < 0.0001, 0.0052, 0.0149 and 0.0119 when the lines are derived from the F2, B1, B2 and B3, respectively. This suggests that the optimum number of backcrosses required to obtain a line that exceeds the recurrent parent by a specified amount may be less than the optimum number of backcrosses required to recover a line that simply exceeds the best parent. This same phenomenon was observed when $\ell = 20$.

Consider the situation where one of the parents contains most of the favorable alleles, i.e., ninety percent. Given $\ell = 20$ and $h^2 = 0.5$, use of expression (6) indicates that the optimum number of backcrosses for recovery of a line exceeding the best parent is two. Given two backcrosses in this situation, the probabilities of lines containing more than eighteen and nineteen loci favorable are 0.122 and 0.011. Thus, there is about an eleven percent chance that a single favorable allele would be obtained from the donor, but only a one percent chance of recovering both the favorable alleles of the donor. In the corresponding

SELECTION LIMITS IN SELFING POPULATIONS 411

situation when m = 54 and n = 6, the optimum number of backcrosses is four. Then, the probability of recovering a single favorable allele from the donor (i.e., fifty-five loci fixed for the favorable allele) is 0.070, but the probability of getting two or more is only 0.007. These results are consistent with the results of Reddy and Comstock (1976) in their analysis of the backcross breeding method. They concluded that depending on the type of selection program utilized, one, but not more than five, favorable alleles could be transferred from the donor to the recipient.

It should be noted that the results obtained by using expression (6) refer to the distribution of lines after they are completely inbred. The problem of actually selecting the superior line out of all the lines produced by the program must also be overcome.

4. CONCLUSION

The effect of intra-line selection in the development of single-seed descent lines following the cross of two homozygous parents was investigated. The lines were initiated from foundation populations whose formation may have involved one or more backcrosses. The choice of foundation population was shown to influence the chance that a breeding program would yield superior genotypes. Given that one parent was better than the other, the use of backcrossing increased the probability of obtaining a line better than the best parent.

The effect of selection was shown to increase the expected frequency of loci fixed for the favorable allele in derived lines, compared with lines developed without selection. However, if the parents are of approximately equal value, it is not likely that the selection described would result in a line that contained a large proportion (eighty percent) of the total genetic potential offered by the two parent lines. Given that one parent is clearly superior to the other, it is concluded that the programs considered can at best produce a line that has only a few (two or three) more loci fixed favorable than the best parent.

BIBLIOGRAPHY

Bailey, T. B. and R. E. Comstock (1976). Linkage and the synthesis of better genotypes in self-fertilizing species. Crop Science, 16:363-370.

Reddy, B. V. S. and R. E. Comstock (1976). Simulation of the backcross breeding method. I. The effects of heritability and gene number on fixation of desired alleles. Crop Science, 16:825-830.

Selection for populations of interacting genotypes

Bruce Griffing
DEPARTMENT OF GENETICS
THE OHIO STATE UNIVERSITY, COLUMBUS, OH 43210

ABSTRACT

A theoretical basis for accommodating genotypic interaction in artificial selection theory is presented. The classical truncation selection model for quantitative genetic variables is extended to include genotypic interaction. This interaction model is used to evaluate various strategies that involve individual, group and index selection procedures applied to random and non-random group structures. The criteria for evaluation are: (i) ensuring that the change in population mean due to one cycle of selection is non-negative, and (ii) ensuring that the selection procedure is efficient. Important factors to consider in choosing each selection procedure are given.

I. INTRODUCTION

Plant crops represent attempts by man to obtain maximum production from a given environmental space. In order to utilize the space optimally, crowding is essential in both annual and perennial crops. With crowding, interactions occur among individuals within the population (for reviews see among others, Sakai, 1955; Simmons, 1962; Allard and Adams, 1969a; and Jacquard, 1975).

In single species populations, the interactions are primarily of a competitive nature, and under these conditions unexpected results from selection may occur. For example, Wiebe et al. (1963), who studied the effects of interplant competition on selection for yield in barley, concluded that:

> "Significant reversals in relative yield were found to exist in comparison between the same genotype, VV or vv, when grown in pure stand and in an advanced generation, thus, indicating that the poorest plants should be saved from an advanced hybrid population rather than the good ones when yield is the criterion for selection. If this phenomenon has a degree of universality, then it may explain why breeding for increased yield has progressed so slowly."

This barley experiment indicates that the aggressive competitor has an advantaged in mixed stands but yields poorly in a pure stand. In terminology to be defined later, this result implies a negative relationship between the two dimensions of competitive gene action, 'direct' and 'associate'. Such a negative relationship, in turn, implies that positive individual selection may result in a negative change in the mean. Continued selection could result in fixation of the least desirable genotype.

The theoretical solution to this plant breeding dilemma is to construct a biological model that accommodates interaction and then to identify selection procedures that give the desired results when applied to this more realistic model. However, few theoretical plant breeding studies have dealt with genotypic interaction.

Sakai (1955) considered various possible configurations of plants and the competition patterns resulting from such configurations. He developed a genetic model that described population change when the plants exhibited complementary competition. In the present terminology Sakai's model was restricted to closed groups of size two generated by a base population with an arbitrary number of autogenous elements (elements reproducing only like kind). Because the model is complementary, it is strictly additive, i.e., direct x associate interaction effects are zero. Consequences of selection with such a model are not complicated: selection invariably results in fixing a single element of the base population.

Schutz et al. (1968) extended the Sakai model to include direct x associate interactions. This accounted for a greater variety of long-term consequences of selection, e.g., equilibria of various kinds, as well as fixation. These authors used their model to investigate, by computer simulation, the long-term consequences of natural selection when different kinds of direct x associate interactions were involved. Allard and Adams (1969a and 1969b), also using computer simulation, studied the long-term consequences of natural selection operating on autogenous closed groups of size two. Their model was essentially that of Schutz et al. (1968). The problem of equilibria in autogenous closed groups of size two, was solved algebraically by Cockerham and Burrows (1971).

The studies reviewed so far have been entirely concerned with autogenous base populations. The remaining studies involve random-mating base populations. Schutz and Usanis (1969) extended the Sakai-type model to include closed groups of size two which are formed from a random mating base population. The base population, in turn was generated by two alleles at a single locus. These authors used computer simulation to study the long-term consequences of selection when the model assumed different sets of genetic parameters. Cockerham et al. (1972) made an exhaustive algebraic study of the long-term consequences of selection on a generalization of this model.

In the model developed by Griffing (1967), the notion of generating all possible configurations (groups) within which genotypic interaction may occur was adopted from the early work of Sakai. However, the genetic model employed was that currently used in quantitative genetics. The objective was to construct an interaction model useful for quantitative genetics, particularly as it pertains to truncation selection theory. Classical (non-interaction) theory then becomes a special case of group (interaction) theory when the associate main and interaction effects are assumed to be zero.

Once a modeling system accommodating the problem under consideration was constructed, the next step was to search for optimal selection procedures for such a system. The first obvious condition of an optimal selection procedure is that <u>positive</u> selection cannot result in a <u>negative</u> change in the population mean. It was demonstrated that with an interaction model, such a result was possible (Griffing, 1967). However, it was also shown that a desirable solution could be obtained by selecting on a group rather than on an individual basis. The second condition of an optimal selection procedure is efficiency. Unfortunately, although group selection ensures that $\Delta\mu \geq 0$, it can be inefficient. Therefore, it was necessary to search for other selection methods which were more efficient. The remaining studies of the series by Griffing (1968a, 1968b, 1969, 1976a, and 1976b), and also a study by Gallais (1976), were concerned with various aspects of this problem.

II. SELECTION APPLIED TO POPULATION OF INTERACTION GENOTYPES.

A. Biological Models which Accommodate Interaction.

Plants are fixed in space and, therefore, each plant interacts with a small number of adjacent plants. Hence for an interaction model to be meaningful, it must incorporate the notion of small groups within which interactions between group members can occur. Most crop plants are planted according to a hill, row or drill planting arrangement. These planting patterns provide practical examples of two classes of groups, closed and open.

With closed groups it is assumed that genotypic interactions are confined to members within groups and that the position in the group does not influence the magnitude of interaction. The hill planting arrangement provides an approximate example of the closed group model. The more complicated open group model permits position effects within the group and interaction between plants inside and outside a given group. Row and drill planting patterns provide practical examples of the open group model.

The basic results of selection are similar in open and closed groups. However, the results using open groups are more complicated because more than one class of associate effects must be considered.

Hence to simplify the following presentation, only the closed group model will be considered.

B. Selection Procedures.

This study is concerned with recurrent selection procedures involving the individual, group, or index as the basic unit of evaluation. The selection procedures operate with regard to either random or non-random groups. The non-random classification includes full-sib, clonal, isogenic and isogenic-clonal group structures. Since the plant breeding objective is to produce an elite random group population, non-random groups are used merely to increase the efficiency of the selection procedure. (For a more detailed discussion of how these non-random groups are obtained and used in an overall recurrent selection program see Griffing 1976a and 1976b).

In the following presentation the plan is to discuss first individual selection on random groups to demonstrate that the interaction model can produce the anomalous situation in which positive selection produces a negative change in the population mean. It will be shown that this situation can be corrected by transferring the basis of selection to the group. However, because group selection can be inefficient, a search is made for other strategies that not only ensure $\Delta\mu \geq 0$ but are also efficient. These strategies ultimately involve selection indices in certain non-random group structures. Because the selection index theoretically ensures that $\Delta\mu \geq 0$ and maximizes the genetic change in the population mean, it is an optimum solution to the complex genotypic interaction problem.

A single-locus gene model is used in the preliminary analyses. In the final analysis, however, the consequences of selection are given in terms of covariances among relatives. These covariances are directly estimable and can be interpreted in terms of gene models of any degree of complexity.

C. Comparisons of Selection Methods in Terms of a Single-Locus Model.

In this section a single-locus model is used to explore the consequences of individual and group selection operating on random and non-random groups.

1. <u>Individual and Group Selection with Random Groups of Size n.</u>
Let

$$\sum p_i p_j (A_i A_j) = \text{genotypic array at a single locus for individuals in the base population.}$$

Then the array of random groups of size n is obtained as

$$[\sum p_i p_j (A_i A_j)]^n = \sum (p_{i_1} p_{j_1}) \cdots (p_{i_n} p_{j_n})(A_{i_1} A_{j_1}, A_{i_2} A_{j_2}, \cdots, A_{i_n} A_{j_n}).$$

Although only a single-locus model in the base population is considered, a n-locus model is necessary to characterize completely the genotypic interactions among the n group members. Hence the gene model for $A_{i_1} A_{j_1}$ in the group $(A_{i_1} A_{j_1}, \cdots, A_{i_n} A_{j_n})$ is

$$_{i_1 j_1} x_{i_2 j_2}, \cdots, _{i_n j_n} = \mu + \overbrace{d_{i_1} + d_{j_1} + (dd)_{i_1 j_1}}^{\text{Direct Effects}} + \overbrace{\sum_{t=2}^{n} [a_{i_t} + a_{j_t} + (aa)_{i_t j_t}]}^{\text{Associate Effects}}$$

+ (other interaction effects).

The direct and associate variances and covariance used in prediction are:

$$\sigma_{dd}^2 = (2)\sum p_{i_t}(d_{i_t})^2, \quad \sigma_{da} = (2)\sum p_{i_t}(d_{i_t})(a_{i_t}), \text{ and } \sigma_{aa}^2 = (2)\sum p_{i_t}(a_{i_t})^2.$$

a. <u>Consequences of Individual Selection.</u>

Since an individual is evaluated without reference to its group, the selective value of $A_i A_j$ is

$$w_{ij} = 1 + (\bar{t}/\sigma)_{\text{ind.}} (_{ij} x .., \underline{\quad}, .. -\mu),$$

where $\bar{t}$ = standard selection differential, σ = phenotypic standard deviation and the subscript 'ind.' indicates that $\bar{t}$ and σ refer to individual values.

The change in gene frequency is a function only of the direct gene effect, i.e.

$$\Delta p_i = (\bar{t}/\sigma)_{\text{ind.}} (p_i)(d_i),$$

and the change in the group population mean is

$$\Delta \mu \simeq (\bar{t}/\sigma)_{\text{ind.}} [\sigma_{dd}^2 + (n-1)\sigma_{da}]. \quad\text{———— (1)}$$

In (1), σ_{da} can be negative and of greater magnitude than σ^2_{dd}. In this case, the incongruous situation occurs in which <u>positive</u> selection results in a <u>negative</u> change in the population mean. Thus the interaction model can accommodate the phenomenon reported in the barley competition studies (Wiebe, et al., 1963).

   b. Consequences of Group Selection.

When the entire group is accepted or rejected on the basis of its mean value, the change in gene frequency becomes a function of both the direct and associate dimensions of gene activity, i.e.

$$\Delta p_i = (\pm/\sigma)_{gr.} (p_i)(1/n)[d_i+(n-1)a_i].$$

The change in the group population mean is

$$\Delta \mu \simeq (\pm/\sigma)_{gr.} (1/n)[\sigma^2_{dd}+2(n-1)\sigma_{da}+(n-1)^2\sigma^2_{aa}] \geq 0.$$

Thus for any group size, transferring the basis of selection from the individual to the group ensures that $\Delta \mu \geq 0$. However, group selection can be inefficient. The next section briefly considers the strategy of using non-random groups to increase the efficiency of selection.

   2. <u>Individual and Group Selection with Non-Random Groups of Size n.</u>

In some plant species, clonal propogation and/or extraction of doubled haploids is possible. These procedures can be used to generate groups exhibiting maximum homogeneity and/or homozygosity. This section gives the results of the use of these extreme forms of non-random groups as well as the results for the standard full-sib group structure. The purpose is to show how use of non-random groups contributes toward (i) ensuring that $\Delta \mu \geq 0$, (ii) increasing the efficiency of the selection procedures.

Table I, which gives the changes in gene frequency, indicates that the problem of ensuring that $\Delta \mu \geq 0$ can be obtained in either of two ways. First, all forms of group selection result in $\Delta \mu \geq 0$. Second, individual selection can ensure that $\Delta \mu \geq 0$ when it operates on certain non-random groups. {Note that $\Delta \mu \geq 0$ when Δp_i is a function of $[d_i+(n-1)a_i]$}.

TABLE I

Gene Frequency Changes for Individual and Group Selection Involving Random and Non-Random Group Structures

Group Structure	Individual Selection	Group Selection
Random	$_{RI}\Delta p_i = {_IR}(\frac{i}{\sigma})_{ind.}(p_i)(d_i)$	$_{RG}\Delta p_i = {_IR}(\frac{i}{\sigma})_{gr.}(p_i)(\frac{1}{n})[d_i+(n-1)a_i]$

NON-RANDOM GROUP STRUCTURE

Isogenic	$_{II}\Delta p_i = {_I}(\frac{i}{\sigma})_{ind.}(p_i)(2)(d_i)$	$_{IG}\Delta p_i = {_I}(\frac{i}{\sigma})_{gr.}(p_i)(\frac{2}{n})[d_i+(n-1)a_i]$
Full-Sib	$_{FSI}\Delta p_i = {_{FS}}(\frac{i}{\sigma})_{ind.}(p_i)[d_i+(\frac{n-1}{2})a_i]$	$_{FSG}\Delta p_i = {_{FS}}(\frac{i}{\sigma})_{gr.}(p_i)(\frac{n+1}{2n})[d_i+(n-1)a_i]$
Clonal	$_{CI}\Delta p_i = {_C}(\frac{i}{\sigma})_{ind.}(p_i)[d_i+(n-1)a_i]$	$_{CG}\Delta p_i = {_C}(\frac{i}{\sigma})_{gr.}(p_i)[d_i+(n-1)a_i]$
Isogenic-Clonal	$_{ICI}\Delta p_i = {_{IC}}(\frac{i}{\sigma})_{ind.}(p_i)(2)[d_i+(n-1)a_i]$	$_{ICG}\Delta p_i = {_{IC}}(\frac{i}{\sigma})_{gr.}(p_i)(2)[d_i+(n-1)a_i]$

Table I also provides the basic information for comparing relative efficiencies of selection involving different kinds of groups. Space does not permit a detailed examination of the efficiency problem, but efficiency ratios and computer simulation studies indicate that selection involving non-random groups can be much more efficient than selection involving random groups (Griffing, 1976b).

D. <u>Comparisons of Selection Methods in Terms of a Generalized Genetic Model</u>.

In previous studies which were briefly reviewed above, the consequences of individual and group selection were explored for a variety of random and non-random group structures using a single-locus gene model. These results were useful in that they demonstrated the nature of the responses to selection in terms of direct and associate gene effects. These analyses, in turn, facilitated the search for efficient and reliable selection procedures.

However, two further extensions of the theory are necessary before the total response pattern can be determined and before the theory can be used practically. The first extension has to do with generalizing the gene model to include more than a single locus in order to evaluate the total response to selection. The other extension has to do with the practical problem of stating the consequences of selection in terms of parameters than can be estimated directly. Both of these problems can be solved by transferring the model-building unit from that of the gene to that of the individual genotype. This permits the consequences of selection to be formulated in terms of covariances among relatives. The covariances are directly estimable and can be interpreted in terms of gene models of any degree of complexity.

In the following presentation parent and progeny populations are characterized in terms of individual and group arrays and the necessary parent, progeny and joint parent-progeny parameters are defined. The consequences of selection are then derived in terms of selection indices. This procedure provides a method whereby the results of individual, group and the combination of individual and group selection, through an index specifically designed to provide maximum genetic gain, can be given and compared in one unified theory.

Finally, the index selection results based on the interaction model are compared with the classical index selection results based on the non-interaction model. The classical results are those derived by Lush (1947) and later discussed in various textbooks (Lerner (1958), Falconer (1960) and Turner and Young (1969), among others).

1. Formulation of the Parent and Progeny Population in Terms of the Individual as the Genetic Unit.

a. Parent Population.

Characterization of the parent population of groups must reflect the kind of group structure on which selection operates. Thus the specification of parental parameters will differ for each kind of group. Since space does not permit an elaboration of all types of parental groups, only random groups will be considered here. However, the results for all group structures will be given later. Let,

$\sum f_m H_m$ = genotypic array of the parent base population,

$\sum f_{m_1} f_{m_2} \cdots f_{m_n} (H_{m_1}, H_{m_2}, \cdots, H_{m_n})$ = random group array for groups of size n generated by the parental base array,

and $\sum f_{m_1} f_{m_2} \cdots f_{m_n} (_{m_1} h_{m_2}, \ldots, m_n)$ = array of parental genotypic values.

Various parental means are of interest. These are:

$_m h._{\_\_\_.}$ = direct mean for parental genotype H_m,

$._{m}h._{\_\_.} = ._{\cdot}h._{m.\_\_.}$ = etc. associate mean for H_m, and

$.h._{\_\_.} = \mu_p$ = parental population mean.

b. Progeny Population.

The progeny base population is derived in terms of a full-sib structure. To do this let

$\sum q_{rst}(H_{rst})$ = full-sib array from the mating, $H_r \times H_s$, and

$\sum f_r f_s q_{rst}(H_{rst})$ = progeny base population formulated as the array of full-sib families.

POPULATIONS OF INTERACTING GENOTYPES

Then the random progeny groups of size n are obtained as

$$\left(\sum f_r f_s q_{rst}(H_{rst})\right)^n = \sum \left(\prod_{u=1}^{n}(f_{r_u} f_{s_u} q_{r_u s_u t_u})\right)(H_{r_1 s_1 t_1}, \ldots, H_{r_n s_n t_n}).$$

The array of progeny genotypic values is:

$$\sum \left(\prod_{u=1}^{n}(f_{r_u} f_{s_u} q_{r_u s_u t_u})\right)(_{r_1 s_1 t_1} h_{r_2 s_2 t_2}, \ldots, _{r_n s_n t_n}).$$

Various progeny means of interest are:

$_{r..}h_{...,\_,...}$ = direct mean for half-sib progeny of parent H_r,

$_{...}h_{r..,\_,...}$ = $_{...}h_{...,r..,\_,...}$ = etc. = associate mean for half-sib progeny of parent H_r,

$_{...}h_{...,\_,...}$ = μ = progeny population mean, and

$G_r = (2)[(_{r..}h_{...,\_,...}-\mu) + (n-1)(_{...}h_{r..,\_,...}-\mu)]$

= breeding value of H_r measured as both sire and dam.

c. <u>Joint Parent - Progeny Parameters</u>.

Four different kinds of covariances between parents and offspring need to be distinguished for prediction purposes. These are:

$\text{Cov}(P_d O_d) = E(_r h._{\_}.-\mu_p)(_{r..}h_{...,\_,...}-\mu)$

$\text{Cov}(P_d O_a) = E(_r h._{\_}.-\mu_p)(_{...}h_{r..,\_,...}-\mu)$

$\text{Cov}(P_a O_d) = E(.h_{r.\_}.-\mu_p)(_{r..}h_{...,\_,...}-\mu)$, and

$\text{Cov}(P_a O_a) = E(.h_{r.\_}.-\mu_p)(_{...}h_{r..,\_,...}-\mu)$.

Two functions of covariances are useful:

$X = \text{Cov}(P_d O_d) + (n-1)\text{Cov}(P_d O_a)$, and

$Y = \text{Cov}(P_a O_d) + (n-1)\text{Cov}(P_a O_a)$.

2. <u>Construction of a Selection Index Based on Group Structure</u>.

The genotypic value for the ℓth genotype in the kth parental group can be represented by the model

$\phi_{k\ell} = \mu_p + \tau_k + \gamma_{k\ell}$, where

$\mu_p = \phi_{..}$ = parental population mean,

$\tau_k = \phi_{k.} - \phi_{..}$ = deviation of the kth group mean, and

$\gamma_{k\ell} = \phi_{k\ell} - \phi_{k.}$ = deviation of the $k\ell$th genotypic value from the kth group mean.

Then a selection index for the $k\ell$th genotype can be defined as

$$I_{k\ell} = \beta_1(\tau_k) + \beta_2(\gamma_{k\ell}). \qquad (2)$$

The genotype, H_m, occurs in different groups and thereby generates a subpopulation of indices the mean of which is designated as I_m.

3. Consequences of Selection with an Index Having Arbitrary Weights.

Assume that the selection value of H_m is

$$w_m = 1 + (\bar{i}/\sigma)_I (I_m).$$

Change in the progeny group mean due to a single cycle of truncation selection is

$$\Delta\mu = \sum \left\{ \prod_{u=1}^{n} [(f_{r_u} w_{r_u})(f_{s_u} w_{s_u})(q_{r_u s_u t_u})] \right\} (_{r_1 s_1 t_1} h_{r_2 s_2 t_2}, \underline{\quad}, r_n s_n t_n) - \mu$$

$$\simeq (\bar{i}/\sigma)_I \sum f_r(I_r) \left\{ (2)[(_{r..}h..., \underline{\quad}, ...-\mu) + (n-1)(...h_{r..}, \underline{\quad}, ...-\mu)] \right\}$$

$$= (\bar{i}/\sigma)_I \text{Cov}(GI)$$

$$= (\bar{i}/\sigma)_I \left[\beta_1 \text{Cov}(G\tau) + \beta_2 \text{Cov}(G\gamma) \right] \qquad (3)$$

where $\text{Cov}(G\tau) = (2)\left[(1/n)(X-Y)+Y\right]$, and

$\text{Cov}(G\gamma) = (2)\left[(n-1)/n\right](X-Y).$

Equation (3) may be put into matrix notation

$$\Delta\mu \simeq (\bar{i})_I \left(\frac{B'G}{(B'PB)^{1/2}} \right), \qquad (4)$$

where $B = \begin{pmatrix} \beta_1 \\ \beta_2 \end{pmatrix}$ = vector of index weights,

$G = \begin{pmatrix} \text{Cov}(G\tau) \\ \text{Cov}(G\gamma) \end{pmatrix}$ = vector of covariances, and

$P = \begin{pmatrix} \sigma_\tau^2 & 0 \\ 0 & \sigma_\gamma^2 \end{pmatrix}$ = variance - covariance matrix.

The result as represented in (3) or (4) is valid for any type of group structure. The differences in response due to different parental group structures are reflected in the compositions of $Cov(G\tau)$, $Cov(G\gamma)$, σ_τ^2 and σ_γ^2.

The development of the optimum index theory is interrupted at this stage (arbitrary β's) in order to derive the results for individual and group selection and to make comparisons among the responses to these selection procedures elicited by different group structures.

a. <u>Use of Index Theory to Derive a General Formulation for Individual Selection</u>.

To characterize individual selection put $\beta_1 = \beta_2 = 1$ in (2) and (3). Then the selection index reduces to

$$I_{k\ell} = \phi_{k\ell} - \phi_{..},$$

and the change in population mean becomes

$$\Delta\mu_{ind.} \simeq (\overline{I}/\sigma)_{ind.} [Cov(G\tau) + Cov(G\gamma)], \quad\quad (5)$$

or in terms of parent-offspring covariances,

$$\Delta\mu_{ind.} \simeq (\overline{I}/\sigma)_{ind.} (2)[Cov(P_d O_d) + (n-1)Cov(P_d O_a)]. \quad (6)$$

The genetic interpretation of the parent-offspring covariances depends on the particular parental group structures involved.

With no genotypic interaction (6) reduces to

$$\Delta\mu_{ind.} = (\overline{I}/\sigma)_{ind.} (2)[Cov(PO)].$$

b. <u>Use of Index Theory to Derive a General Formulation for Group Selection</u>.

To characterize group selection put $\beta_1 = 1$ and $\beta_2 = 0$ in (2) and (3). The selection index for the kth group then reduces to

$$I_k = \phi_{k.} - \phi_{..},$$

and the change in population mean is

$$\Delta\mu_{gr.} \simeq (\overline{I}/\sigma)_{gr.} [Cov(G\tau)]. \quad\quad (7)$$

Equation (7) may be recast in terms of parent-offspring covariances as

$$\Delta\mu \simeq (\overline{I}/\sigma)_{gr.} (2/n)\{Cov(P_d O_d)+(n-1)[Cov(P_d O_a)+Cov(P_a O_d)]$$
$$+(n-1)^2 Cov(P_a O_a)\}. \quad\quad (8)$$

As in the case of individual selection, the genetic interpretation of the parent-offspring covariances depends on the parental group structure involved.

With no genotypic interaction (8) reduces to

$$\Delta\mu_{gr.} \simeq (\bar{1}/\sigma)_{gr.} (2/n)[\text{Cov}(PO)].$$

c. **Results of Individual and Group Selection for Different Group Structures in Terms of Cov(Gτ) and Cov(Gγ).**

The different group structures considered in this study are characterized by different degrees of homozygosity and homogeneity as expressed among group members. These aspects of non-randomness can be expressed more exactly in terms of standard coefficients of relationship. To accomplish this let

group = $\{I = (a,b), J = (c,d), \text{etc.}\}$

and denote the standard coefficients of genetic relationship among group members as

$f_I = P(a\equiv b)$ = inbreeding coefficient,

$f_{IJ} = (1/4)[P(a\equiv c) + P(a\equiv d) + P(b\equiv c) + P(b\equiv d)]$ = coefficient of consanguinity, and

$r_{IJ} = [(2)f_{IJ}]/[(1+f_I)(1+f_J)]^{1/2}$ = coefficient of relationship.

Then Cov(Gτ) and Cov(Gγ) can be determined for group structures involving any degree of homozygosity (f) and homogeneity (r) by use of the following formulae (interpreted in terms of a single-locus model):

$\text{Cov}(G\tau) = \{[1+(n-1)r](1+f)\}(1/n)(A)$

and $\text{Cov}(G\gamma) = \{(1-r)(1+f)\}(1/n)A'),$

where $A = [\sigma_{dd}^2 + 2(n-1)\sigma_{da} + (n-1)^2\sigma_{aa}^2],$

and $A' = [(n-1)\sigma_{dd}^2 + (n-1)(n-2)\sigma_{da} - (n-1)^2\sigma_{aa}^2].$

Table II presents the compositions of Cov(Gτ) and Cov(Gγ) for each of the different kinds of random and non-random groups generated by combinations of (f = 0,1) and r = (0, 1/2, 1). These covariances can be used to obtain the changes in population mean resulting from individual selection (equation 5) and group selection (equation 7) for each kind of group.

TABLE II

Composition of $Cov(G_T)$ and $Cov(G\gamma)$ for Various Kinds of Group Structures

		DEGREE OF HOMOGENEITY		
	COVARIANCES	RANDOM ($r=0$)	FULL-SIB ($r=1/2$)	CLONOL ($r=1$)
Non-Isogenic ($f=0$)	$Cov(G_T)$	$[(1/n)]A$	$[(n+1)/(2n)]A$	A
	$Cov(G\gamma)$	$[(1/n)]A'$	$[1/(2n)]A'$	0
	Total	$[(1/n)](A+A')$	$[1/(2n)][(n+1)A+A']$	A
Isogenic ($f=1$)	$Cov(G_T)$	$(2/n)A$	$[(n+1)/n]A$	$2A$
	$Cov(G\gamma)$	$(2/n)A'$	$[1/n]A'$	0
	Total	$(2/n)(A+A')$	$(1/n)[(n+1)A+A']$	$2A$

DEGREE OF HOMOZYGOSITY

4. **Consequences of Selection with an Index that Combines Within and Between Group Information to Maximize the Change in Population Mean.**

The result of index selection as represented in equation (4) is valid for an index with arbitrary weights. Clearly it is desirable to consider an index for which the β's are chosen in order to maximize $\Delta\mu$. To accomplish this objective, it is necessary to determine the vector B which satisfies the equation,

$$\frac{\partial L}{\partial B} = 0,$$

where $L = \ln(\bar{\tau}) + \ln(B'G) - (1/2)\ln(B'PB)$.

The normal equation $\quad\quad PB = G,$ \hfill (9)

resulting from this maximization procedure yields the optimum weights

$$\hat{\beta}_1 = [\text{Cov}(G\tau)/\sigma_\tau^2] \quad \text{and} \quad \hat{\beta}_2 = [\text{Cov}(G\gamma)/\sigma_\gamma^2].$$

When (9) is substituted into (4), the maximum for $\Delta\mu$ is obtained:

$$\Delta\mu_{\text{max.}} = (\bar{\tau})_I [\sigma_I], \hfill (10)$$

where $\quad\quad \sigma_I = \{\beta_1^2 \sigma_\tau^2 + \beta_2^2 \sigma_\gamma^2\}^{1/2}.$

Equation (10) may be recast as

$$\Delta\mu_{\text{max.}} \simeq (\bar{\tau})_I [\beta_1 \text{Cov}(G\tau) + \beta_2 \text{Cov}(G\gamma)]^{1/2}.$$

It is clear that $\Delta\mu_{\text{max.}}$ is composed of two statistically independent components each of which is non-negative. The first component is a function of the deviations of group means from the population mean, and the second component is a function of the deviations of individual from their group mean.

The theoretical advantage of the optimum index is that it resolves the problems inherent in individual and group selection. In the case of individual selection, the contribution of $\text{Cov}(G\gamma)$ can be negative and of sufficient magnitude to cause the change in the population mean to be negative. Use of the optimum index corrects such a situation because if $\text{Cov}(G\gamma)$ is negative, β_2 is also negative, and the contribution from this source of variation is non-negative, i.e.

$$\beta_2 [\text{Cov}(G\gamma)] = \beta_2^2 \sigma_\gamma^2 \geq 0.$$

POPULATIONS OF INTERACTING GENOTYPES

On the other hand, if the associate effects are negligible relative to the direct effects, thus making group selection inefficient, then $Cov(G\tau)$ is relatively small. This results in a small value for β_1. Consequently the index shifts the emphasis toward use of the direct effects through the contribution of $\beta_2 [Cov(G\gamma)]$.

Therefore, theoretically, the selection index with optimum weights provides the ideal solution for any group structure in that it ensures $\Delta\mu \geq 0$ and at the same time yields a maximum value for $\Delta\mu$.

5. Adapting the Optimum Index to the Experimental Situation.

In an experimental situation it is assumed that N groups each with n members are chosen at random from the conceptual parent group population. Truncation selection is performed with the index

$$I^*_{ij} = b_1(x_{i.}-x_{..}) + b_2(x_{ij}-x_{i.})$$
$$= b_1(\tau^*) + b_2(\gamma^*),$$

where $(x_{i.}-x_{..}) = \tau^* =$ deviation of family mean from overall mean, and

$(x_{ij}-x_{i.}) = \gamma^* =$ deviation of individual value from family mean.

For any group structure it can be shown that

$$\sigma^2_{\tau^*} = [(N-1)/N][(1/N)(\sigma^2_P - {}_P\sigma_P) + {}_P\sigma_P],$$
$$\sigma^2_{\gamma^*} = [(n-1)/n](\sigma^2_P - {}_P\sigma_P),$$
$$Cov(G\tau^*) = 2[(N-1)/N][(1/n)(X-Y)+Y], \text{ and}$$
$$Cov(G\gamma^*) = 2[(n-1)/n](X-Y),$$

where $\sigma^2_P = \sigma^2_G + \sigma^2_E =$ phenotypic variance,

${}_P\sigma_P = {}_G\sigma_G + {}_E\sigma_E =$ phenotypic covariance among group members,

$X = Cov(P_d O_d) + (n-1)Cov(P_d O_a)$, and

$Y = Cov(P_a O_d) + (n-1)Cov(P_a O_a)$.

The expected values for the statistics associated with the index yielding maximum genetic gain are:

$$E(b_1) = \frac{Cov(G\tau^*)}{\sigma^2_{\tau^*}} = \frac{(2)[(1/n)(X-Y)+Y]}{[(1/n)(\sigma^2_P - {}_P\sigma_P) + {}_P\sigma_P]},$$

$$E(b_2) = \frac{Cov(G\tau^*)}{\sigma_{\gamma^*}^2} = \frac{2(X-Y)}{(\sigma_P^2 - {}_P\sigma_P)},$$

and

$$E(\Delta\mu_{max.}) = (\overline{I})_I(2)\left\{\frac{[(N-1)/N][(1/n)(X-Y)+Y]^2}{[(1/n)(\sigma_P^2 - {}_P\sigma_P) + {}_P\sigma_P]} + \frac{[(n-1)/n](X-Y)^2}{(\sigma_P^2 - {}_P\sigma_P)}\right\}^{1/2}.$$

The above formulation is applicable to all non-clonal group structures. However, evaluation of index selection must be made separately for each kind of group since the genetic expectations of variances and covariances depend on the group structure.

6. Comparison of the Optimum Selection Index Statistics for the Non-Interaction and Interaction Models.

The Lusherian theory (Lush, 1947) of combined between and within family selection based on a non-interaction model leads to an index which is formulated by Falconer (1960) as

$$I = h_f^2 P_f + h_w^2 P_w,$$

where h_f^2 = heritability of family means,

h_w^2 = heritability of within-family deviations,

P_f = deviation of the family mean from the population mean,

and P_w = deviation of the individual from the family mean.

Table III gives an interpretation of the family and within-family heritabilities at two levels (i.e. in terms of the parent-offspring covariances and in terms of a single-locus gene model) for each of the classical (non-interaction) and group (interaction) models.

It is clear that the values for the regression coefficients could be considerably different for the two models, if in fact genotypic interaction exists.

III. CONCLUSION

The objective of this study is to provide alternative theoretical solutions to the plant breeding dilemma caused by genotypic interactions. These strategies involve individual, group and index selection

TABLE III

Comparison of Combined Selection Index Statistics for Classical (Non-Interaction)
and Group (Interaction) Models

Classical (Non-Interaction)	Group (Interaction)
$E(b_1) = h_f^2 \quad E(b_1) = \dfrac{Cov(GT^*)}{\sigma^2_{T^*}}$	
$= \dfrac{(2/n)Cov(PO)}{(1/n)(\sigma^2_p - \sigma_{p_p p_p}) + \sigma_{p_p p_p}}$	$= \dfrac{(2/n)[Cov(P_d O_d) + (n-1)[Cov(P_d O_a) + Cov(P_a O_d)] + (n-1)^2 Cov(P_a O_a)]}{(1/n)(\sigma^2_p - \sigma_{p_p p_p}) + \sigma_{p_p p_p}}$
$= \dfrac{[1+(n-1)r](1+f)(1/n)\sigma^2_A}{[1+(n-1)t](1/n)\sigma^2_p}$	$= \dfrac{[1+(n-1)r](1+f)(1/n)\{\sigma^2_{dd} + 2(n-1)\sigma_{da} + (n-1)^2 \sigma^2_{aa}\}}{[1+(n-1)t](1/n)\sigma^2_p}$
$E(b_2) = h_w^2 \quad E(b_2) = \dfrac{Cov(G\gamma^*)}{\sigma^2_{\gamma^*}}$	
$= \dfrac{(2)Cov(PO)}{\sigma^2_p - \sigma_{p_p p_p}}$	$= (2) \dfrac{\left\{ Cov(P_d O_d) + (n-1)[Cov(P_d O_a)] - Cov(P_a O_d) - (n-1)Cov(P_a O_a) \right\}}{\sigma^2_p - \sigma_{p_p p_p}}$
$= \dfrac{(1-r)(1+f)\sigma^2_A}{(1-t)\sigma^2_p}$	$= \dfrac{(1-r)(1+f)[\sigma^2_{dd} + (n-2)\sigma_{da} - (n-1)\sigma^2_{aa}]}{(1-t)\sigma^2_p}$

Where: $t = \dfrac{\sigma_{p_p p_p}}{\sigma^2_p}$

procedures operating on random and non-random group structures. Factors to consider in choosing and utilizing each of the selection methods can be summarized as follows.

Individual Selection: Individual selection in a randomly planted bulk population composed of different genotypes is not advisable unless it is known that interplant competition does not exist. If, however, the decision is made to continue individual selection under conditions of interplant competition, then use of non-random groups should be considered. In this regard, choice of the dimension of non-randomness (f or r) to be used is important and can be determined by expressing the change in population mean due to one cycle of individual selection as follows:

$$\Delta\mu_{ind.} \simeq (\bar{I}/\sigma)_{ind.}[(1+f)/n]\{[1+(n-1)r]A+(1-r)A^{\prime}\} \quad\text{———} \quad (11)$$

where $A \geq 0$, but $A^{\prime}$ can be negative. Equation (11) clearly shows r to be the parameter of genetic relationship that reduces the relative influence of $A^{\prime}$. Thus individual selection should be used only with group structures having sizable values for r regardless of the f value. Full-sib groups are an obvious choice. Individual matings among selected plants in the previous cycle can be used to provide the basis for full-sib groups. This procedure would not lengthen the selection cycle.

Group Selection: The advantage of group selection over individual selection is that it is perfectly safe when applied to any group structure. The disadvantage of group selection is that it may be inefficient. This can be corrected to a large extent by use of non-random groups. The advantage of group selection over index selection is that the practical difficulty of obtaining reliable estimates of index weights is avoided.

Optimum Index Selection: For any non-clonal group structure, use of the index that optimally combines within and between group information theoretically yields the best selection results. The practical difficulty in use of the index is that reliable estimates of the four parent-offspring covariances, $[Cov(P_d O_d)$, $Cov(P_d O_a)$, $Cov(P_a O_d)$ and

$Cov(P_a O_a)$], have to be obtained in order to calculate the index weights. Considerable time and effort may be required for this estimation procedure.

Finally, in this era of considerable scientific interest in cell biology in general, and tissue culture in particular, the tremendous advantage of using clonal groups should be mentioned. For populations of any f value, use of clonal groups yields all the advantages of index selection without the necessity of estimating parent-offspring covariances.

BIBLIOGRAPHY

[1] Allard, R. W. & Adams, J. (1969a). The role of intergenotypic interactions in plant breeding. *XII Int. Congr. Genet., Proc. 3*, 349-70.

[2] Allard, R. W. & Adams, J. (1969b). Population studies in predominantly self-pollinating species. XIII Intergenotypic competition and population structure in barley and wheat. *Amer. Nat. 103*, 621-45.

[3] Cockerham, C. C. & Burrows, P. M. (1971). Populations of interacting autogenous components. *Amer. Nat. 105*, 13-29.

[4] Cockerham, C. C., Burrows, P. M., Young, S. S. & Prout, T. (1972). Frequency-dependent selection in randomly mating populations. *Amer. Nat. 106*, 493-515.

[5] Falconer, D. S. (1960). *Introduction to Quantitative Genetics.* Edinburgh: Oliver and Boyd.

[6] Gallais, A. (1976). Effects of competition on means, variances and covariances in quantitative genetics with an application to general combining ability selection. *Theor. Appl. Genet. 47*, 189-95.

[7] Griffing, B. (1967). Selection in reference to biological groups. I. Individual and group selection applied to populations of unordered groups. *Aust. J. Biol. Sci. 20*, 127-39.

[8] Griffing, B. (1968a). Selection in reference to biological groups. II. Consequences of selection in groups of one size when evaluated in groups of a different size. *Aust. J. Biol. Sci. 21*, 1163-70.

[9] Griffing, B. (1968b). Selection in reference to biological groups. III. Generalized results of individual and group selection in terms of parent-offspring covariances. *Aust. J. Biol. Sci. 21*, 1171-78.

[10] Griffing, B. (1969). Selection in reference to biological groups. IV. Application of selection index theory. *Aust. J. Biol. Sci. 22*, 131-42.

[11] Griffing, B. (1976a). Selection in reference to biological groups. V. Analysis of full-sib groups. *Genetics 82*, 703-22.

[12] Griffing, B. (1976b). Selection in reference to biological groups. VI. Use of extreme forms of non-random groups to increase selection efficiency. *Genetics 82*, 723-31.

[13] Jacquard, P. (1975). Intraspecific competition and potentialities of yield. *Ann. Amelior Plantes 25*, 3-24.

[14] Lerner, I. M. (1958). *The Genetic Basis of Selection*. New York: John Wiley and Sons.

[15] Lush, J. L. (1947). Family merit and individual merit as bases for selection, Parts I and II. *Amer. Nat. 81*, 241-61, 362-79.

[16] Sakai, K. (1955). Competition in plants and its relation to selection. *Cold Spring Harbor Symp. Quan. Biol. 20*, 137-57.

[17] Schutz, W. M. & Usanis, S. A. (1969). Inter-genotypic competition in plant populations. II. Maintenance of allelic polymorphisms with frequency-dependent selection and mixed selfing and random mating. *Genetics 61*, 875-91.

[18] Schutz, W. M., Brim, C. A. & Usanis, S. A. (1968). Intergenotypic competition in plant populations. I. Feedback systems with stable equilibria in populations of autogamous homozygous lines. *Crop Sci. 8*, 61-66.

[19] Simmonds, N. W. (1962). Variability in crop plants, its use and conservation. *Biol. Rev. 37*, 422-65.

[20] Turner, H. N. & Young, S. S. Y. (1969). *Quantitative Genetics in Sheep Breeding*. Melbourne: Macmillan and Co.

[21] Wiebe, G. A., Petr, F. C. & Stevens, H. (1963). Interplant competition between barley genotypes. *Statistical Genetics and Plant Breeding* (W.D. Hanson & H. F. Robinson, Eds.)(N.A.S./N.R.C. Pub. No. 982).

Opposition to artificial selection caused by natural selection at linked loci

J. A. Sved
SCHOOL OF BIOLOGICAL SCIENCES
UNIVERSITY OF SYDNEY, AUSTRALIA

ABSTRACT

This paper considers models in which selectively neutral genes affecting a quantitative character are linked to genes affecting fitness. A two locus model is used to show that directional selection, in the presence of linkage disequilibrium, will cause a negative correlation to be built up between genotypes at the two loci. As a result of this negative correlation, selection tends to act in opposition at the two loci. However multiple locus computer simulation has shown that, if only finite-size linkage disequilibrium is assumed, the retardation of artificial selection response due to the opposition of this type of natural selection will probably be small. It appears that only quite closely linked genes contribute substantially to this retardation. Stabilising natural selection acts to prevent the loss of favorable genes by drift as well as to retard their fixation, which partly accounts for its failure to retard the overall selection response more strongly.

INTRODUCTION

Progress made under artificial selection for a quantitative character may be opposed by natural selection for, broadly speaking, two reasons:

(1) <u>Direct or pleiotropic effects</u>. Genes affecting the character in question may themselves be of selective importance, either through their effect on the character or through pleiotropic effects. Natural selection would presumably directly oppose any change in frequency of such genes.

(2) <u>Linkage effects</u>. The genes affecting the character may not themselves be subject to natural selection. However the frequencies of other genes throughout the genome may be altered during the artificial selection process, and it is these changes which may be opposed by natural selection.

This paper will be restricted to a consideration of linkage effects. Being indirect effects, it may seem that these should be secondary to the direct effects of class (1). However the fact that any genes of importance to fitness may contribute to class (2) makes it difficult to rule out potential cumulative effects of this class. Also in the absence of any knowledge about the correlated effects of artificial selection for any particular character, effects of class (2) constitute a base level of natural selection which must be expected in all cases. Therefore before conclusions can be drawn about direct effects, it is desirable to try to estimate the potential for linkage effects.

Linkage effects are not expected merely because of linkage between genes affecting the quantitative character and genes affecting fitness. It is necessary in addition that there be linkage disequilibrium (Lewontin & Kojima, 1960), or non-zero correlation of frequencies, between genes at the two sets of loci. Linkage effects have been defined above in a way which specifically precludes the existence of any selective interaction. Therefore linkage disequilibrium is not generally to be expected in an infinite population (Lewontin & Kojima, 1960). However finite-size linkage disequilibrium (Hill & Robertson, 1968; Sved, 1968;

Ohta & Kimura, 1969), which is generally expected to be of a limited magnitude, may become an important factor owing to the small size of most selected populations. Linkage disequilibrium where it occurs throughout this paper is usually assumed to be attributable to this cause.

TWO LOCUS THEORY

We consider a two locus model in which directional (artifcial) selection occurs at one locus and stabilising (natural) selection occurs at a second closely linked locus. The model is treated deterministically, but at the same time it is assumed that linkage disequilibrium occurs because of random fluctuation of frequencies. The model is intended to be neither rigorous nor realistic. It is useful simply as a device to illustrate the manner in which the opposition of natural selection to artificial selection may build up.

The simplest model is one with two loci, A and B, each containing two alleles. The frequencies of the four possible gametes, AB, Ab, aB and ab are defined as x_1, x_2, x_3 and x_4 respectively. The individual gene frequencies may then be defined in terms of the gamete frequencies, e.g. $p_A = x_1 + x_2$, $p_b = x_2 + x_4$ etc. The usual measure of linkage disequilibrium D, is equal to $x_1 x_4 - x_2 x_3$. A more useful measure for disequilibrium which will be used later is the correlation of frequencies, r (Hill & Robertson, 1968) which is defined in terms of D as $r^2 = D^2/p_A p_a p_B p_b$. The $\underline{x}$ values can be written in terms of $\underline{p}$ and D values, e.g. $x_1 = p_A p_B + D$ etc. Artificial selection is assumed to act on the A locus, the A gene being the favored gene, and the heterozygote Aa is assumed to be intermediate in selective value between the two homozygotes AA and aa. Stabilising selection occurs at the B locus, and for a start this will be assumed to take the form of symmetrical heterozygote advantage, i.e. an equal advantage of Bb over both BB and bb.

As mentioned previously, selection at the B locus will have an effect on the A locus only if there is some correlation between the two loci. The present model allows this correlation to be defined more specifically. It is clear that selective consequences depend on a correlation between the occurrence of the favorable gene at the A locus and heterozygosity at the B locus. A negative correlation implies that natural selection will tend to oppose artificial selection.

We will calculate the covariance, which is closely related to the correlation, between the value at the A locus, α_A, i.e. 1, ½ and 0 for AA, Aa and aa respectively, and the value at the B locus, α_B, i.e. 0, 1 and 0 for BB, Bb and bb respectively. The required covariance is equal to $\Sigma f. \alpha_A \alpha_B - (\Sigma f.\alpha_A).(\Sigma f.\alpha_B)$, which from Table 1 is $2x_1 x_2 + x_1 x_4 + x_2 x_3 - p_A \cdot 2 p_b p_B$. Substituting for the x values in terms of p and D, this simplifies to $D.(p_b - p_B)$.

TABLE I

Genotype Frequencies and Values for Random Mating Two Locus Model with Directional Selection at A Locus and Heterozygote Advantage at B Locus.

GENOTYPE	FREQ. (f)	A VALUE (α_A)	B VALUE (α_B)	$f.\alpha_A$	$f.\alpha_B$	$f.\alpha_A \alpha_B$
AB/AB	x_1^2	1	0	x_1^2	.	.
AB/Ab	$2x_1 x_2$	1	1	$2x_1 x_2$	$2x_1 x_2$	$2x_1 x_2$
AB/aB	$2x_1 x_3$	½	0	$x_1 x_3$	.	.
AB/ab	$2x_1 x_4$	½	1	$x_1 x_4$	$2x_1 x_4$	$x_1 x_4$
Ab/Ab	x_2^2	1	0	x_2^2	.	.
Ab/aB	$2x_2 x_3$	½	1	$x_2 x_3$	$2x_2 x_3$	$x_2 x_3$
Ab/ab	$2x_2 x_4$	½	0	$x_2 x_4$	.	.
aB/aB	x_3^2	0	0	.	.	.
aB/ab	$2x_3 x_4$	0	1	.	$2x_3 x_4$	.
ab/ab	x_4^2	0	0	.	.	.
				p_A	$2 p_b p_B$	$2x_1 x_2 + x_1 x_4 + x_2 x_3$

OPPOSITION OF NATURAL TO ARTIFICIAL SELECTION

An algebraic argument may be used to show that directional selection tends to make $D \cdot (p_b - p_B)$ negative. This is given in the Appendix. We consider here in addition a non-mathematical argument which brings out more clearly the reasons for the negative covariance. Both components of the covariance, D and $(p_b - p_B)$, may be either negative, positive or zero. We must assume first that due to the finite size of the population, D is in general non-zero (Hill & Robertson, 1968; Sved, 1968). Given the symmetrical nature of selection at the B locus, D can then with equal probability be either positive or negative. The term $p_b - p_B$ may also be zero, and in fact the action of natural selection at the B locus will be to reduce the value of this term towards zero. However there is a systematic pressure attributable to selection at the A locus which acts to move the genes at the B locus away from their equilibrium, and thus to make the value of $p_b - p_B$ non-zero. Furthermore the direction of this pressure depends on the sign of D. If D is positive, so that for example the frequency of the AB gamete is in excess, an increase in the frequency of A will result in a net increase in the frequency of B. Thus the sign of $p_b - p_B$ will tend to be the opposite to that of D. The same argument holds for a negative value of D. Thus the overall value of $D \cdot (p_b - p_B)$ will tend to be negative.

The argument given so far depends critically on the assumption of symmetrical heterozygote advantage at the B locus. However an analogous argument can be given for the asymmetrical case, although it is no longer possible to argue simply in terms of heterozygosity at the B locus. It is now necessary instead to study the correlation of selective values at the A and B loci. If the selective values at the A locus are taken as $1 : 1 - \frac{s}{2} : 1 - s$ respectively, and at the B locus as $1 - t_1 : 1 : 1 - t_2$, then proceeding as previously it can be shown that the covariance of selective values at the two loci is $sD \cdot (t_2 p_b - t_1 p_B)$. The term $t_2 p_b - t_1 p_B$ of this expression is exactly analogous to the $p_b - p_B$ term of the previous expression; natural selection at the B locus tends to reduce it to zero, and by the same arguments as previously it can be seen that its sign tends to be the opposite of that of D. Thus the overall covariance will be negative, so that selection at the B locus will tend to be in opposition to that at the A locus.

Although the two locus model may be useful for visualising the nature of the opposition of natural selection, it is not clear that two locus arguments can easily be extended to derive quantitative results of any value. The model considered here cannot really be compared with long-term deterministic models, e.g. those analysed by Karlin & Carmelli (1975), since the effects being examined here are ultimately attributable to random fluctuations. Considered as a purely deterministic model, the equilibria are trivial, and it is only the rates of approach which could be of interest. Furthermore it is far from clear that a two locus model can adequately represent the potential complexity given by two interspersed sets of loci, which is the situation of real interest. In order to obtain any understanding of such models, as well as to test the validity of some of the arguments given above, it seems necessary to turn to computer simulation.

MULTIPLE LOCUS MODELS

The initial objective of the computer simulation study was to model a single chromosome containing a number of loci affecting a quantitative character (referred to hereafter for convenience as quantitative loci), and a somewhat larger number of loci subject to natural selection (fitness loci). The numbers eventually chosen were 12 quantitative loci and 96 fitness loci, a choice partly dictated by the number of binary bits per word of the computer used. Decisions also had to be made on large numbers of other parameters of the model as follows:

Recombination Values

A chromosome of 50 map units was assumed in most of the populations. This is the smallest possible chromosome length given an obligatory chiasma per chromosome pair. Two models of gene distribution were used (Fig. 1): (1) An equal spacing of loci, with each gap between successive loci being 50/107 map units. (2) A random distribution of loci over a chromosome of the same map length,

where c is the recombination frequency. This is a convenient intermediate value, since it represents to a reasonable degree of accuracy the expected limiting value of r^2 (Hill, 1976). It is referred to in connection with the results as the random disequilibrium level, although its use in this context as the initial disequilibrium level involves the assumption that the base population is of the same size as the population drawn for selection.

Initial Gene Frequencies

All runs were made with the fitness genes starting at their expected equilibrium of 50%. Initial frequencies are also important for quantitative loci (Robertson, 1970), and a range of values was tried for these (0.1, 0.25 and 0.5). In all cases however, the same gene frequency was assigned to all quantitative loci in a particular run.

Length of Runs

All runs were made to last for 60 generations, with the following pattern of artificial selection:
(1) An initial period of 25 generations of selection.
(2) A period of 15 generations during which artificial selection was relaxed, but natural selection continued.
(3) A further period of 20 generations of selection.

RESULTS

Single chromosome models

The high variability of selection response is indicated in Fig. 2, in which single runs are plotted for populations with and without natural selection. In this example the quantitative loci are started at 25% frequency and, as in all examples, their average frequency is used directly as the measure of selection response. There appears to be some effect of natural selection, but this is largely obscured by the high degree of overlap between the two sets of results. It seems clear from this example that the effects of natural selection are sufficiently small that they can only be accurately defined by averaging over a number of replicate populations.

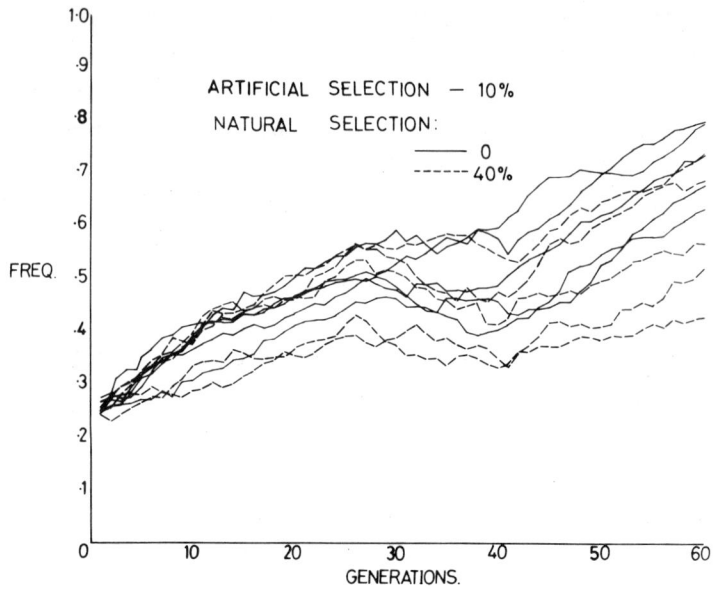

FIG. 2

Selection response for single populations with and without the opposition of natural selection. Populations were started with random linkage disequilibrium. Equally spaced loci were assumed, with artificial selection following natural selection.

The results of averaging the response over 50 replicate populations are given in Fig. 3. Natural selection, especially at the higher intensity, causes a retardation, although obviously not sufficient to lead to any sort of plateau. A numerical estimate of the retardation is given by calculating the realised heritability, as estimated by the regression of selection response on accumulated selection differential (Falconer, 1961, p202). This calculation may be made for both Period 1 (gen. 1-25) and Period 3 (gen. 41-60) where artificial selection is being practised. For Period 1, the realised heritability with no natural selection is 1.00, which is of course the expectation given the complete heritability assigned in the model. For 10% and 40% natural

OPPOSITION OF NATURAL TO ARTIFICIAL SELECTION 445

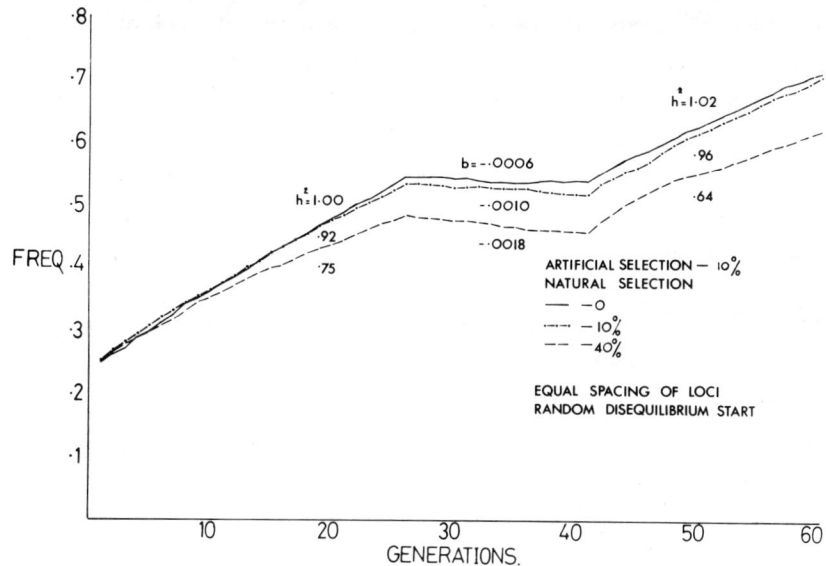

FIG. 3

Selection response averaged over 50 populations, conditions as in Fig. 2.

selection, the realised heritabilities are 0.92 and 0.75 respectively. A similar pattern is observed in Period 3.

A slightly different treatment is necessary in Period 2 where artificial selection is not being practised and calculation of heritability is therefore not relevant. Instead a calculation can be made of the regression of gene frequency on generation. As seen from Fig. 3, there is a tendency for the mean value to regress under the influence of natural selection. However the effect is a small one, and scarcely exceeds the regression when there is no natural selection. The return towards the equilibrium in the latter case is expected from the arguments of Griffing (1960). Truncation selection generates a high additive x additive variance component, and this leads to an immediate response which is subsequently lost through recombination.

Another possible area of effect of natural selection concerns the variance of selection responses. Two measures of variability were studied: (1) the variance in realised heritability values between runs, and (2) the variance about the regression line within runs. Some tendency was found for natural selection to increase both of these parameters, but the increase was small compared to the variability between replicate runs and apparently of little consequence.

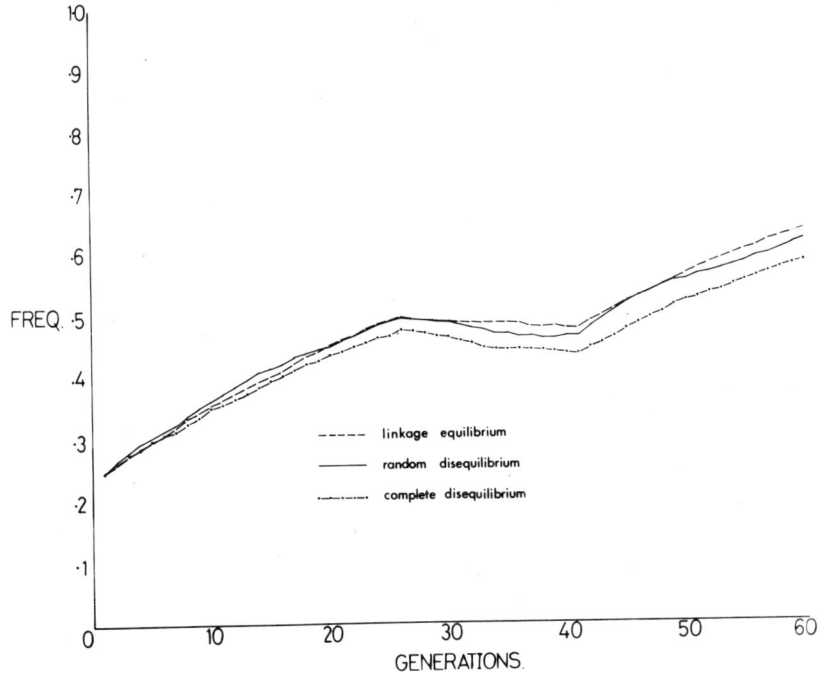

FIG. 4

Effects of various initial disequilibrium values. The three curves are each based on 25 replicate runs, with 40% truncation for natural selection in all cases.

As mentioned previously, it was anticipated that the amount of starting linkage disequilibrium would be of importance. The results of Fig. 4 show that starting disequilibrium does have some effect, although not a crucial one. As expected, the retardation is highest when the disequilibrium is highest. There is little difference between the starting conditions of linkage equilibrium and random linkage disequilibrium.

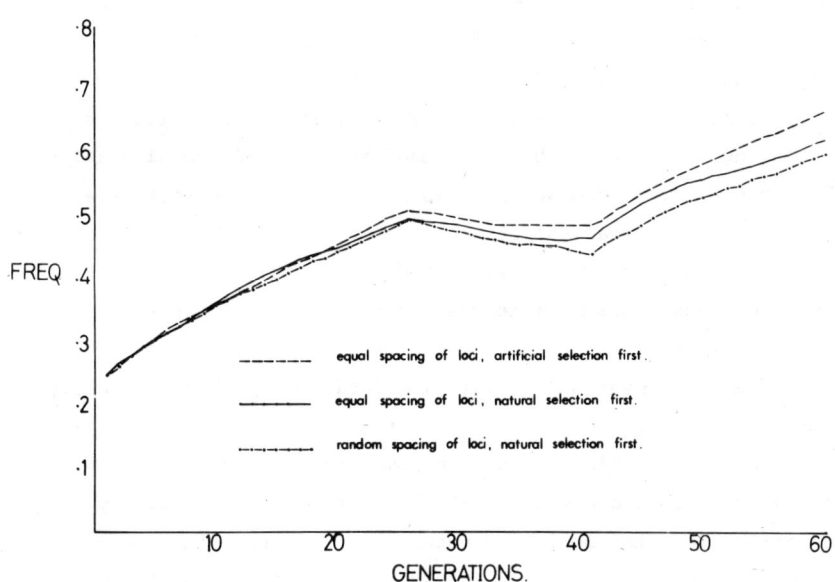

FIG. 5

Effect of changing the order of natural selection and artificial selection, and of random versus equal spacing of loci.

Two possible variations of the model are studied in Fig. 5. The first shows the effect of varying the order of natural selection and artificial selection. The results indicate that natural selection has less effect when it follows artificial selection in each generation, although it is not easy to see why this should be so. This finding may in fact be an artifact attributable to different selective intensities for artificial selection in the two cases. Given that a whole number of individuals is culled, it is not possible to equalise exactly the selective intensities in the two cases. The actual truncation values in the two cases were 5/55 = 9.1% and 9.92 = 9.8% respectively, and this difference may

be sufficient to account for the higher response in the latter
case.

The second aspect of Fig. 5 concerns the comparison between
equal and random spacing of loci along the chromosome (Fig. 1).
A greater retardation is achieved by a random spacing model, which
is readily understood in terms of occasional very close linkage
of fitness and quantitative loci. The random model is probably
the more realistic of the two, so that the results given for the
equal spacing model, which has been used for most of the simulations
of the present paper, probably represent a slight under-estimate
of the effect of natural selection.

Several parameters were calculated during the simulation runs
to try to get some insight into the underlying reasons for the
observed responses. One of these was the correlation between the
number of favored genes at quantitative loci and the heterozygosity
at fitness loci, which is analogous to the two locus parameter
discussed earlier. This and other parameters were calculated at
three stages in each generation, at the zygote stage before any
selection, between the action of natural and artificial selection,
and after both kinds of selection. The results given in Fig. 6
show the correlation at the zygote stage for the same runs as
presented in Fig. 3. Five generation averages are plotted in order
to reduce the magnitude of fluctuations. In the runs without natural
selection it is seen that during the initial period of directional
selection, a negative correlation of around -0.1 is reached, and that
during later generations there is little systematic change from this
level. In the runs with 40% natural selection the build-up of
negative correlation is slower, and the correlation is reduced almost
to zero during the period of relaxation. The two locus arguments
would predict such a reduction, based on the return of gene
frequencies at fitness loci towards their equilibrium during this
period. The absence of a negative correlation at the start of the
second period of artificial selection is seen to be reflected by the
shape of the response curve of Fig. 3. These results explain to
some extent why natural selection has a comparatively low retarding

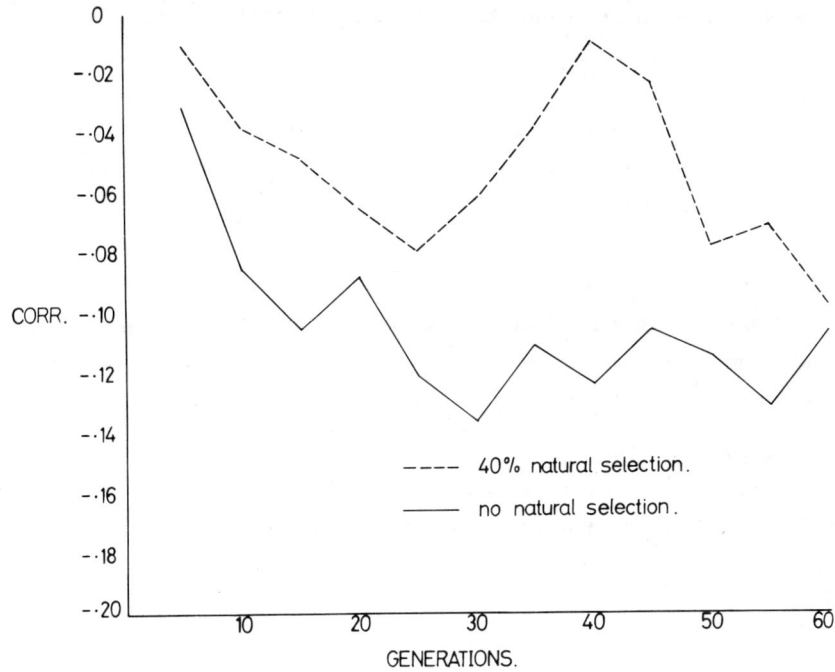

FIG. 6

Correlation between the number of favored genes and heterozygosity for 50 populations with and without natural selection.

effect on artificial selection response, since higher levels of natural selection result in a lower negative correlation on which the retardation depends.

Another interesting aspect of the results concerned loci at which the favored gene was lost through genetic drift. The most extreme example of this came from runs in which the favored gene started at 50% frequency. In a total of 150 such populations with no natural selection and with various degrees of initial disequilibrium, there were altogether 57 cases, 3.2% of all loci, in which the favored gene was lost. The comparable numbers of favored genes lost from 150 populations with 10% and 40% natural selection were 2 and 0 respectively. This prevention of loss of the favored gene of course opposes the retardation of artificial selection by natural selection.

The variance in frequency between quantitative loci was also studied as a more accurate measure of the between locus variability due to drift. A similar pattern was found with this statistic, the variance for 40% natural selection being reduced by about 50% from the variance for no natural selection. These tendencies are readily understood in terms of the arguments given by Sved (1968) and Ohta & Kimura (1970) concerning the stabilising effects of loci with heterozygote advantage on linked neutral loci. As mentioned earlier, the reduction in between locus variability does not lead to any reduction in the variance of overall selection response.

Multiple Chromosome Models

The results of Fig. 7 compare equivalent runs on single and multiple chromosome models with and without natural selection.

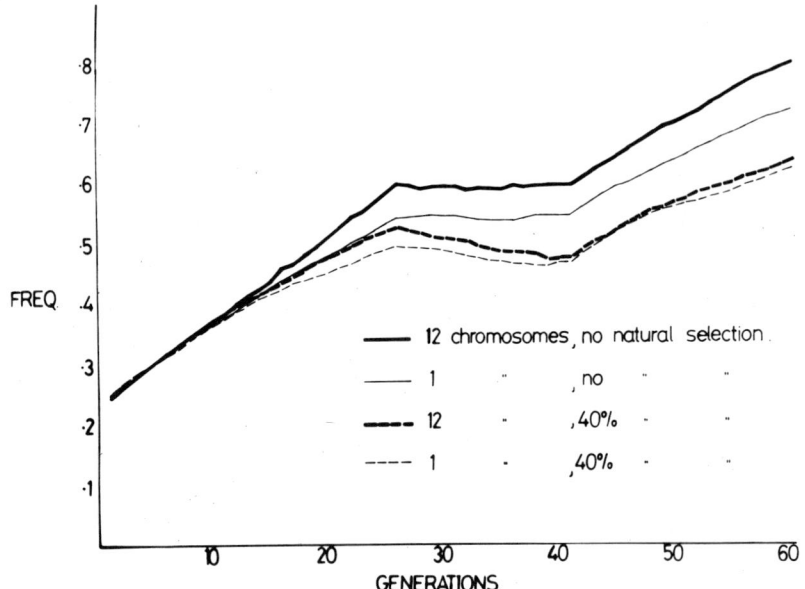

FIG. 7

Comparison of the effect of natural selection in single and multiple chromosome models.

OPPOSITION OF NATURAL TO ARTIFICIAL SELECTION

The comparison of the effects of natural selection is complicated by the differences in the absence of natural selection (c.f. Robertson, 1970). However the retardation achieved by natural selection is if anything less for a single chromosome than for multiple chromosomes. This would seem to indicate that loci other than those closely linked to the quantitative loci are having little effect on the retardation process.

Some multiple chromosome runs were also made in which each quantitative locus was linked to only one fitness locus rather than eight fitness loci. This comparison is complicated by the problem of deciding what are comparable map distances in the two cases.

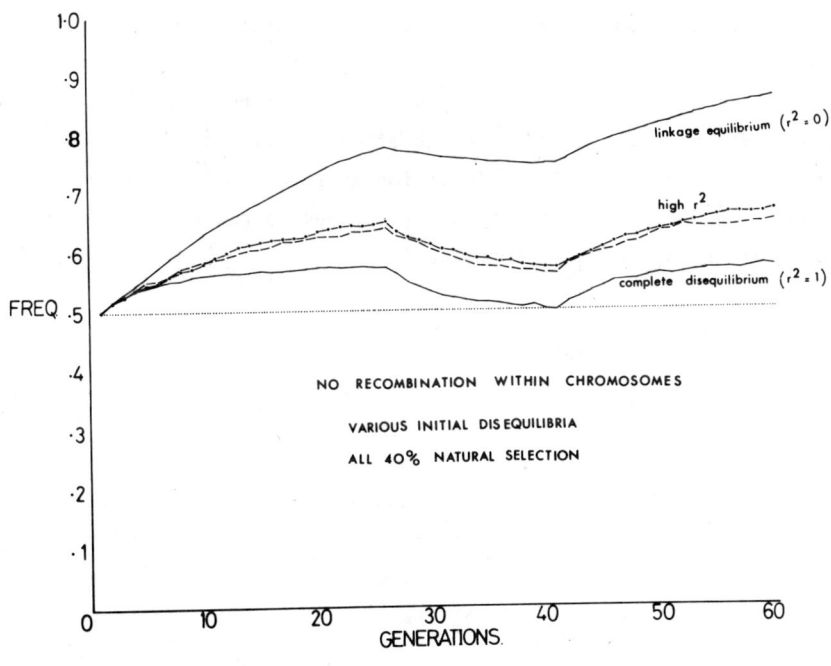

FIG. 8

Extreme multiple chromosome runs, with no recombination within chromosomes.

However if the same adjacent locus spacing was maintained in the two models, it was found that a similar level of retardation was achieved. Thus there is overall very little evidence that the added complexity of multiple locus models results in any marked changes to the pattern of retardation given by two locus models.

The final results to be presented (Fig. 8) concern an extreme multiple chromosome model in which there is no recombination between the quantitative and fitness loci. If in addition there is complete disequilibrium in the starting population ($r^2 = 1$) it is seen that a plateau is rapidly reached, and regression towards the starting value occurs when artificial selection is relaxed. In this extreme form, the model becomes indistinguishable from models in which natural selection has direct effects on the quantitative loci. The introduction of a few recombinant chromosomes into the starting population (high r^2) is sufficient to ensure that the population does not return to its starting value after the relaxation of artificial selection, while a population starting in linkage equilibrium ($r^2 = 0$) shows little tendency to reach a plateau even in the absence of any further recombination.

DISCUSSION

Probably the most speculative feature of the simulation models is the manner in which natural selection has been introduced. There is very little direct evidence for the widespread occurrence of overdominance, let alone symmetrical overdominance (Lewontin, 1974). Nevertheless there is evidence that natural selection, possibly of a similar kind, does occur. The evidence which has motivated the models of the present study comes from experiments with Drosophila. It has shown (see e.g. Sved, 1975) that flies which are homozygous for any wild type chromosome suffer an enormous impairment of fitness in population cages, perhaps as much as 80 - 90% on average. Such inbreeding depression is of the kind to be expected if there is overdominance. It could however be equally well caused by dominance (deleterious recessives), and the main purpose of this discussion is to try to determine how such differences in interpretation might affect the arguments of this paper.

Based on single locus models, the consequences of dominance and overdominance for fitness would seem to be very different. However it has been shown (Ohta, 1971; Sved, 1972) that linkage between deleterious genes does in fact produce some stability of the genome. This can be understood at a slightly different level if it is accepted that the effect of directional selection is to increase homozygosity, particularly in the latter stages of the selection process. The inbreeding experiments then suggest that such tendencies will be opposed by natural selection no matter what its underlying basis. In general it can be argued that any form of stablising selection or homeostasis would yield a similar conclusion.

The advantage of the symmetrical overdominance model for computing is the simplicity with which a relatively stable polymorphic population model can be produced. Intuitively it would seem that the symmetrical overdominance model, since it achieves almost complete stability at the fitness loci, ought to give an upper limit to the amount of retardation than can be achieved. This would be a convenient argument for the present paper, since it means that if the symmetrical overdominance model is able to produce only limited retardation at linked loci, the same should be true a forteriori for any other form of stabilising selection.

In conclusion the simulation results presented in this paper obviously cover only a small fraction of the possible range of parameter values. However the lack of any substantial effects in these or any of the other runs made, save in the extreme disequilibrium runs of Fig. 8, seems to indicate that linkage effects in general cannot be too large. In particular the reports sometimes made of plateaus in selection experiments followed by substantial returns towards the original values on relaxation of selection are not readily explicable in these terms. Direct effects of the quantitative genes on fitness need to be added to the model in some way to produce such effects.

BIBLIOGRAPHY

(1) Falconer, D.S. (1961). Introduction to Quantitative Genetics. Edinburgh: Oliver & Boyd.

(2) Griffing, B. (1960). Theoretical consequences of truncation selection based on the individual phenotype. Aust. J. Biol. Sci. 13, 307-343.

(3) Hill, W.G. (1976). Non-random associations of neutral linked genes in finite populations. Population Genetics and Ecology (S. Karlin & E. Nevo, Eds.). New York: Academic Press, 339-376.

(4) Hill, W.G. & Robertson, A. (1968). Linkage disequilibrium in finite populations. Theor. App. Genet. 38, 226-231.

(5) Karlin, S. & Carmelli, D. (1975). Some population genetic models combining artificial and natural selection pressures: II. Two-locus theory. Theor. Pop. Biol. 7, 123-148.

(6) King, J.L. (1967). Continuously distributed factors affecting fitness. Genetics 55, 483-492.

(7) Lewontin, R.C. (1974). The Genetic Basis of Evolutionary Change. New York: Columbia U.P.

(8) Lewontin, R.C. & Kojima, K. (1960). The evolutionary dynamics of complex polymorphisms. Evolution 14, 116-129.

(9) Ohta, T. (1971). Associative overdominance caused by linked detrimental mutations. Genet. Res. 18, 277-286.

(10) Ohta, T. & Kimura, M. (1969). Linkage disequilibrium due to random genetic drift. Genet. Res. 13, 47-55.

(11) Ohta, T. & Kimura, M. (1970). Development of associative over-dominance through linkage disequilibrium in finite populations. Genet. Res. 16, 165-177.

(12) Robertson, A. (1970). A theory of limits in artificial selection with many linked loci. Mathematical Topics in Population Genetics (K. Kojima, Ed.). Berlin: Springer-Verlag, 246-288.

(13) Sved, J.A. (1968). The stability of linked systems of loci with a small population size. Genetics 59, 543-563.

(14) Sved, J.A. (1972). Heterosis at the level of the chromosome and at the level of the gene. Theor. Pop. Biol. 3, 491-506.

(15) Sved, J.A. (1975). Fitness of third chromosome homozygotes in Drosophila melanogaster. Genet. Res. 25, 197-200.

APPENDIX

We wish to consider the effect of directional selection at the A locus on the parameter combination $D \cdot (P_b - P_B)$. We begin by calculating the effect of directional selection on the four gamete frequencies. If we assume that the frequency of the favored gene, A, is increased by a small amount δ, then since the relative frequencies of the gametes AB and Ab are unaltered by such selection, it can be shown (c.f. Sved, 1968) that the frequencies of these two gametes after selection are respectively $x_1 + \delta \cdot \frac{x_1}{x_1 + x_2}$ and $x_2 + \delta \cdot \frac{x_2}{x_1 + x_2}$. Similarly the frequencies of the gametes aB and ab become $x_3 - \delta \cdot \frac{x_3}{x_3 + x_4}$ and $x_4 - \delta \cdot \frac{x_4}{x_3 + x_4}$ respectively.

We now study the expected change of D and $p_b - p_B$. The value of D after directional selection becomes

$$(x_1 + \delta \cdot \frac{x_1}{x_1 + x_2})(x_4 - \delta \cdot \frac{x_4}{x_3 + x_4}) - (x_2 + \delta \cdot \frac{x_2}{x_1 + x_2})(x_3 - \delta \cdot \frac{x_4}{x_3 + x_4}).$$

Ignoring a term in δ^2, this simplifies to

$$D + \delta \cdot D \cdot \frac{(p_a - p_A)}{p_a p_A}$$

Similarly the new value of $p_b - p_B$ becomes

$$(x_2 + \delta \cdot \frac{x_2}{x_1 + x_2}) + (x_4 - \delta \cdot \frac{x_4}{x_3 + x_4}) - (x_1 + \delta \cdot \frac{x_1}{x_1 + x_2}) - (x_3 - \delta \cdot \frac{x_3}{x_3 + x_4}).$$

This simplifies to

$$(p_b - p_B) - \frac{2\delta \cdot D}{p_a p_A}$$

The quantity $D \cdot (p_b - p_B)$ after selection is thus equal to

$$D \cdot (p_b - p_B) - \frac{2\delta \cdot D^2}{p_a p_A} + \frac{\delta \cdot D \cdot (p_a - p_A)(p_b - p_B)}{p_a p_A},$$

again ignoring a term in δ^2.

The negative contribution to the covariance comes from the D^2 term. The contribution of the last term may be either positive or negative, and under some circumstances it may outweigh the contribution of the D^2 term. However it can readily be shown that if the B locus starts off at or sufficiently close to its equilibrium, such that the contribution of the D term is initially lower than that of the D^2 term, then the covariance can never become positive during the process of substituting the A gene.

An analogous argument can once again be given for the case of asymmetrical selection at the B locus. The change in the appropriate covariance term, $sD.(t_2p_b - t_1p_B)$, in this case comes to

$$- \frac{\delta.sD^2.(t_1 + t_2)}{p_a p_A} + \frac{\delta.sD.(t_2p_b - t_1p_B)(p_a - p_A)}{p_a p_A} ,$$

and the consequences are again similar to those of the symmetrical model.

VI
Results and Theory with Plants

76 generations of selection for oil and protein percentage in maize

J. W. Dudley
AGRONOMY DEPARTMENT
UNIVERSITY OF ILLINOIS, URBANA, IL 61801

INTRODUCTION

Seventy-six generations of selection for high oil, low oil, high protein, and low protein in corn (Zea mays L.) have been completed. A recent paper (Dudley, Lambert, and Alexander, 1974) summarized a detailed analysis of the data from 70 generations of selection. The purpose of this paper is to present the results of 76 generations of selection and their implications to the questions of:

    1) limits of selection,
    2) population gene frequencies,
    3) number of loci,
    4) effects of linkage on selection progress,
    5) types of gene action, and
    6) genetic diversity.

MATERIALS AND METHODS

Details of the selection procedure, chemical and statistical analyses have been presented (Dudley, Lambert, and Alexander, 1974) and only pertinent parts are included here. Selection was initiated in 1896 by analyzing 163 open-pollinated ears of the variety 'Burr's White' for percent oil and percent protein. The 24 ears highest in protein, the 12 ears lowest in protein, the 24 ears highest in oil, and the 12 ears lowest in oil were selected to initiate the Illinois high protein (IHP), low protein (ILP), high oil (IHO), and low oil (ILO) strains respectively. The forward selection phase of the experiment was divided into 4 segments as follows:

Segment 1. Generations 0-9, mass selection based on chemical composition. Number of ears analyzed and selected varied but approximately 20% of the ears analyzed were selected. Each strain was grown in a separate isolated field.

Segment 2. Generations 10-25. 120 ears per strain were analyzed and 24 were saved. Seed from each ear was planted ear-to-row. Alternate rows were detasseled and 20 ears were analyzed from each of the six highest yielding rows. Four ears were saved per row.

Segment 3. Generations 26-52 in IHP and ILP; generations 26-58 in IHO and ILO. Twelve selected ears were arbitrarily divided into two lots (A and B) of six ears. Seed within each lot was bulked and planted in the nursery. Silks in lot A were pollinated by a bulk sample of pollen from 15-20 plants in B while silks in B were pollinated with pollen from A. Thirty ears from each lot were analyzed and the 12 most extreme of the 60 ears analyzed were saved.

Segment 4. Generations 53-76 in IHP and ILP; 59-76 in IHO and ILO. Procedure was the same as in segment 3 but 80 lbs of N per acre were added to the soil.

In generation 48, reverse selection was initiated in each strain (Leng, 1962b) to form four new strains: reverse high protein (RHP), reverse low protein (RLP), reverse high oil (RHO),

and reverse low oil (RLO). The procedure was the same as in the
forward selection experiment except that selection was for low
protein in the high protein strain, high oil in the low oil strain,
etc. After seven generations of selection in RHO, a new strain,
designated switchback high oil (SHO) was initiated by selecting
the 12 ears in RHO that were highest in oil. The reverse phase of
the experiment was divided into two segments based on the date of
application of N as follows: (1) generations 0-4 in RHP and RLP
and generations 0-10 in RHO and RLO; (2) generations 5-28 in RHP
and RLP and generations 11-28 in RHO and RLO.

STATISTICAL ANALYSIS

Each generation a bulk sample of IHO and ILO was analyzed for
percent protein. Because protein was affected by year-to-year
weather fluctuations and the correlated response between protein
and oil was negligible, the protein means for each generation
were adjusted by subtracting the mean protein percentage of IHO
and ILO. The protein data are presented as adjusted percent protein. The oil data were not adjusted because of the small year-to-year variation in percent oil.

Realized heritabilities for each strain were calculated from
regression of cumulative selection differential on generation
means for each segment using the procedure of dummy variables in
multiple regression described by Draper and Smith (1966, p. 140,
ex. 2).

Additive genetic variance (σ_A^2) for each segment was estimated
from the realized heritabilities and the pooled estimates of
phenotypic variance (σ_P^2) from the generations included in the segment. Estimates of σ_A^2 for the original population were calculated
as the average of the estimates from segment 1 of the high and low
strains. Because percent oil in a kernel is determined by the
genotype of the kernel, selection among ears is equivalent to
selection among half-sib family means. Thus, $\sigma_A^2 = 8\sigma_P^2 h^2$ for segments 1, 3 and 4 for percent oil instead of $2\sigma_P^2 h^2$ as given by
Dudley, Lambert, and Alexander, 1974. Because percent protein in

the kernel is determined by the genotype of the mother plant, selection among ears is equivalent to mass selection and $\sigma_A^2 = 2\sigma_P^2 h^2$ as previously reported.

RESULTS

Limits to Selection

The limit to selection has not been reached in any of the 9 strains as evidenced by:

1) effectiveness of reverse selection and switchback selection (Figs. 1, 2, 3, 4);
2) significant estimates of realized heritability in the last segment of all 9 strains (Tables I and II);
3) significant estimates of genetic variation among half-sib family means in IHP, ILP, IHO, and ILO after 65 generations of selection (Dudley and Lambert, 1969).

FIG. 1

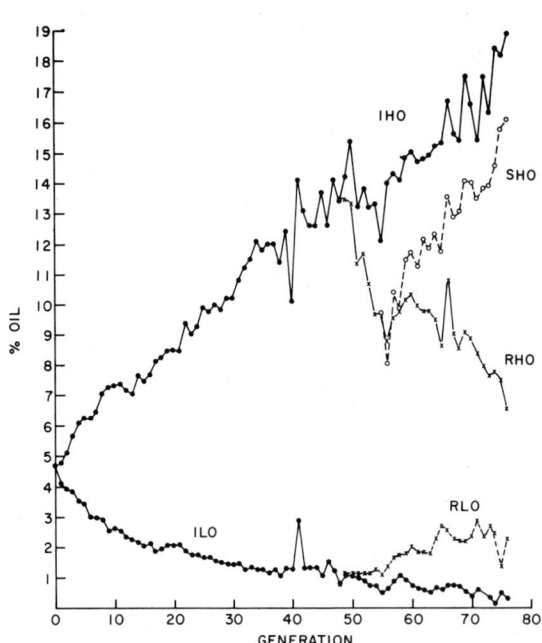

Mean percent oil for IHO, ILO, RHO, RLO, and SHO plotted against generations.

FIG. 2

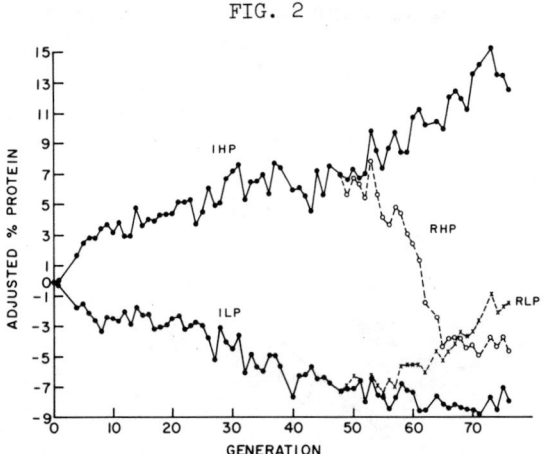

Mean adjusted percent protein for IHP, ILP, RHP, and RLP plotted against generations.

FIG. 3

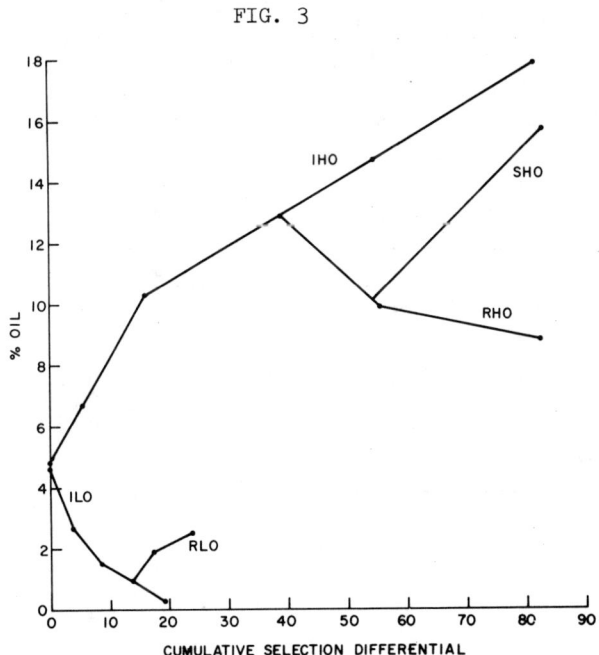

Plot of realized heritabilities for the oil strains.

FIG. 4

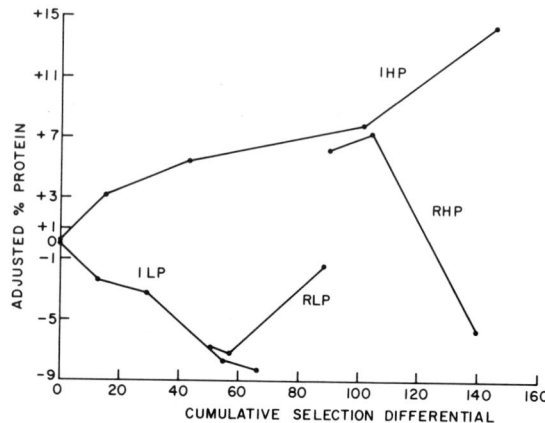

Plot of realized heritabilities for the protein strains.

TABLE I

Realized Heritabilities for the Oil Strain

Segment	Strain				
	IHO	RHO	SHO	ILO	RLO
1	.32±.10			.50±.05	
2	.34±.04			.23±.03	
3	.11±.01	.18±.04		.10±.01	.37±.08
4	.12±.02	.07±.02	.19±.01	.15±.04	.09±.03

TABLE II

Realized Heritabilities for the Protein Strains

Segment	Strain			
	IHP	RHP	ILP	RLP
1	.20±.05		.18±.04	
2	.08±.02		.05±.02	
3	.04±.01		.17±.01	
4	.15±.01	.37±.02	.07±.02	.17±.01

Although there is still genetic variation for percent oil in ILO, it is apparent that, with a mean of .3% oil, progress is nearly exhausted since the lower limit cannot be less than zero.

At intermediate gene frequencies, realized heritability for reverse selection should be similar to realized heritability for forward selection in the segment preceding initiation of reverse selection assuming environmental variance remains constant. This was true for RLP. However, for RHP and RLO realized heritability for reverse selection was significantly higher than for the preceding segment of forward selection. For RHO realized heritability was significantly lower than in the preceding segment of IHO. The reasons for this are not clear. In the case of IHP and RHP, progress in segment 3 of IHP was low, presumably because of limiting soil N levels, since realized heritability in segment 4 was significantly higher than in segment 3.

Gains at the end of 76 generations were 20, 6, 20 and 12 x σ_A for IHO, ILO, IHP, and ILP respectively. These gains, expressed as percent of the original mean are 279, 92, 133, and 78 percent. Although gains of this magnitude at first seem surprising, calculation of expected ultimate limits to selection with free recombination using the equation:

$$G/\sigma_A = [2n(1-q)/q]^{1/2} \quad \text{(Robertson, 1970)}$$

where G is the ultimate gain, n is the number of loci, and q is the frequency of the favorable allele, for varying values of n and q (Table III) shows that to gain $20\sigma_A$ requires 200 loci if q = .5 or between 50 and 100 if q is .25. Thus, the gains achieved are explainable when reasonable gene frequencies and numbers of loci are assumed.

TABLE III

Ultimate Limits to Selection in Terms of σ_A with Varying Values of n (Number of Loci) and q (Frequency of the Favorable Allele).

n	.1	.25	q .5	.75	.9
10	13	8	4	3	2
50	30	17	10	6	3
100	42	24	14	8	5
200	60	35	20	12	7

Estimates of Gene Frequency

No good method of estimating gene frequency for quantitative traits is available. However, if divergent selection experiments are carried to the limit, it should be possible to obtain an estimate (of unknown reliability) of average gene frequency using the equation for ultimate gain with no linkage given by Robertson (1970) as follows:

Let ultimate gain in the upward direction be G_H and in the downward direction be G_L. Since $G_H = [2n(1-q)/q]^{\frac{1}{2}}\sigma_A$ and $G_L = [2nq/(1-q)]^{\frac{1}{2}}\sigma_A$ then $G_H/G_L = [2n(1-q)/q]^{\frac{1}{2}}/[2nq/(1-q)]^{\frac{1}{2}} = (1-q)/q$ and $q = 1/[(G_H/G_L)+1]$.

Thus, q can be estimated knowing only the gain at the limit for selection in both directions. Assuming that the 9 strains in this experiment had reached the limit, estimates of q were obtained for the three possible comparisons among the protein strains and the four possible comparisons among the oil strains (Table IV).

TABLE IV

Estimates of q at Different Stages in the Selection Experiment Assuming the Ultimate Limit was Reached.

% Oil		% Protein	
Source	q	Source	q
IHO/RHO	.51	IHP/RHP	.62
SHO/RHO	.32	IHP/ILP	.37
IHO/ILO	.25	RLP/ILP	.19
RLO/ILO	.28		

Estimates of q = .25 for % oil and q = .37 for % protein in the original population were obtained. After 48 generations of selection in IHP (see IHP/RHP comparison) q = .62 and in ILP q = .19. The results for percent oil were inconsistent in that the estimated q after 48 generations of downward selection (RLO/ILO) was similar to the estimate for the original population. These estimates are tentative since the limits to selection have not been reached. However, the results do suggest that initial average gene frequencies were less than .5 for both traits and for percent oil considerably less than .5 since the present estimate is .25, and appreciable progress in the upward direction can likely still be obtained while little additional progress downward is possible. Thus, the ratio of G_H/G_L may increase which will reduce the estimate of q.

Number of Effective Factors

Student (1934) estimated that 33 loci controlled percent oil based on data through 28 generations. However, his estimate of σ_A^2 was crude. Assuming that the high selected strains are homozygous for the (+) alleles for the selected traits, estimates of the number of effective factors controlling percent oil and percent protein were obtained using the expression

$$K_1 = [\tfrac{1}{2}(\overline{P}_1 - \overline{P}_2)]^2 / [1/2q(1-q)](\sigma_A^2)$$

which is a modified form of the equation given by Mather and Jinks

(1971) where K_1 is the number of effective factors, $\bar{P}_1$ is the mean of the high strain, $\bar{P}_2$ is the mean of the low strain, σ_A^2 is estimated as outlined in Materials and Methods, and values of q are those given in Table IV. This procedure assumes no dominance but removes the restriction that $q = .5$.

Estimates obtained suggest that the number of effective factors controlling percent protein (122) is larger than the number controlling percent oil (54) (Table V).

TABLE V

Estimates of Number of Effective Factors Controlling Percent Oil and Percent Protein.

% Oil		% Protein	
Source	K_1	Source	K_1
IHO vs ILO	54	IHP vs ILP	122
IHO vs RHO	9	IHP vs RHP	47
SHO vs RHO	3	ILP vs RLP	18
RLO vs ILO	6		

The estimates of numbers of factors differentiating the reverse strains from the strain from which they were selected were made using estimates of σ_A^2 from the segment of the experiment following initiation of reverse selection. The sum of the number of effective factors differentiating IHP from RHP and ILP from RLP is less than the total estimated from IHP vs. ILP even though RHP has lower percent protein than RLP and all the genetic difference between IHP and ILP is accounted for by the comparisons IHP vs. RHP plus ILP vs. RLP.

Additional estimates of the number of factors controlling percent oil were obtained from a Design III experiment (Comstock and Robinson, 1952) using estimates of σ_A^2 and the dominance variance σ_D^2 (Moreno-Gonzalez, Dudley, and Lambert, 1975) from the random-mated F_6 generation of the cross IHO x ILO. From the F_6

data, estimates of σ_A^2 and σ_D^2 were .812 and .189 respectively. Estimates of K_1, K_2, and k using the procedure described by Mather and Jinks (1971) (p. 323) were 35, 14, and 71 respectively, where K_1 is as previously described, $K_2 = (\overline{F}_1 - MP)^2/4\sigma_D^2$ and $k = 3K_1K_2/(4K_2 - K_1)$. The K_1 estimates from the Design III and from the IHO/ILO comparison are in reasonable agreement when one considers that linkage disequilibrium in the IHO x ILO cross may not be completely dissipated by the F_6.

Because of the simplifying assumptions necessary (see Mather and Jinks, 1971) these are minimum estimates of the number of loci involved. However, the data suggest that the number of effective factors controlling percent protein may be higher than the number controlling percent oil.

Effects of Linkage

Although there is no apparent way of measuring the effect linkage has had on progress in this experiment, a few observations can be made. Following Hanson (1959) the average unbroken segment length of chromosome after 76 generations was calculated as .013 crossover units. Thus, there has been ample opportunity for essentially free recombination to occur. Using Robertson's (1970) result that linkage has little effect before 2/ih* generations where i = approximately 1.4 for this experiment and $h^* = h/H^{\frac{1}{2}}$, where H is the haploid number of the species, the number of generations before linkage should show an effect was calculated as 7 for IHO and ILO and 10 for IHP and ILP. Since the breeding system was changed at generation 9, it is not clear whether the reduced realized heritability in segment 2 of both IHP and ILP (Table III) resulted from the change in breeding system or effects of linkage. For IHO it is apparent that if linkage had an effect it was much later than 7 generations since realized heritability remained the same through 25 generations.

Evidence from the design III study (Moreno-Gonzalez, Dudley, and Lambert, 1975) clearly demonstrated the presence of coupling phase linkages in the cross of IHO x ILO since the additive genetic variance in F_2 was nearly twice the F_6 estimate. Thus, linkage for genes controlling percent oil exists but its effect on progress is unknown.

Types of Gene Action

Analysis of a diallel cross (including parents) among all 9 strains in the experiment using design II of Gardner and Eberhart (1966) showed 5% and 7% of the entry sum of squares were accounted for by heterosis for percent oil and percent protein, respectively (Dudley, de la Roche, and Lambert, 1976). F_1-MP values for percent oil indicated that dominance for high oil was present in crosses of RHO with IHO and SHO while dominance for low oil was present in crosses of ILO and RLO with IHO and SHO (Table VI).

TABLE VI

Estimates of Heterosis (F_1-MP) for Percent Oil from Crosses among Oil Strains.

	RHO	SHO	ILO	RLO
IHO	1.9*	-.2	-3.3*	-2.0*
RHO		1.6*	-.8*	.1
SHO			-2.4*	-2.2*
ILO				.2

* = Significant at the .05 probability level.

Although the low percentage of the entry sum of squares accounted for by heterosis in the diallel analysis suggests that dominance is relatively unimportant for percent oil, evaluation of the estimates of σ_A^2 and σ_D^2 from the Design III suggests a relatively high proportion of the genetic variance results from loci showing dominance. In the F_6 of the cross IHO x ILO, gene frequency is approximately .5. Only loci showing dominance contribute to σ_D^2 (ignoring epistasis). For loci with complete dominance, $\sigma_D^2 = 1/3$ of the total genetic variance and $\sigma_A^2 = 2/3$ when q = .5. Thus the additive genetic variance contributed by loci showing dominance equals twice the dominance variance. In the F_6 of IHO x ILO σ_D^2 = .189 and σ_A^2 = .812. Therefore, the additive genetic variance contributed by loci showing dominance is .378 or 46.5% of the total additive genetic variance. If all loci contribute equally to the genetic variance, then nearly half the loci by which IHO and ILO differ show dominance. If some loci show only partial dominance, then the proportion of loci showing dominance is greater than indicated.

Genetic Diversity

Much has been written about the lack of genetic diversity in major cultivated plant species as a barrier to continuing genetic improvement. The implication is that future progress is being hampered by a lack of genetic variability in existing germplasm which could be remedied by collection and use of exotic germplasm or other means. What do the results of this experiment imply for the future improvement of economic species? If the mean of an open-pollinated corn variety could be increased to a level of 3-4 times its original mean as was done for oil in IHO and protein in IHP, selected populations could, conservatively, yield in the 300 bushel per acre range with possible additional improvement from use of appropriate hybrids between populations.

There appears to be little need for concern about exhausting genetic variability for grain yield since with gene frequencies between .25 and .5 only from 50-200 loci are required to allow a gain of $20\sigma_A$. One would expect that there are more loci segregating for yield in most random-mating populations than for percent oil or protein. Thus, so far as quantitative traits are concerned, the most pressing problems may not be those of obtaining more genetic variability, but those of learning how to most efficiently concentrate the favorable alleles we now have and then proceeding to do it.

The preceding argument, of course, does not deny the importance of having available diverse genetic materials as potential sources of major genes for traits such as resistance to disease and insect pests. In some cases, resistance to such pests may not be present in adapted germplasm and availability of divergent germplasm may be extremely valuable.

SUMMARY AND CONCLUSIONS

1. 76 generations of high and low selection for percent oil and protein in corn has not exhausted genetic variation for either trait in either direction even though progress of 20 σ_A was made in the high direction for both oil and protein. Evaluation of theoretical limits to selection suggest that these results are well within the range of response expected with reasonable gene frequencies and numbers of loci.

2. A method of estimating gene frequency from divergent selection experiments which have reached their upper limits is presented. Application of this method to the data from this experiment indicate gene frequencies for favorable alleles in the original population were below .37 for percent protein and below .25 for percent oil.

3. Estimates of number of effective factors suggest a minimum of 54 loci differentiating IHO and ILO and 122 loci differentiating IHP and ILP.

4. Evidence suggests dominance for both high and low percent oil and that nearly half the loci differentiating IHO and ILO show some degree of dominance.

5. The results of this experiment suggest that the potential for improvement of random mating corn populations may be well beyond what is commonly considered possible.

BIBLIOGRAPHY

(1) Comstock, R.E. & Robinson, H.F. (1948). Estimates of average dominance of genes. p. 494-516. In J. W. Gowen (ed.) Heterosis. Iowa State College Press, Ames.

(2) Draper, N.R. & Smith, H. (1966). Applied Regression Analysis. John Wiley & Sons, Inc., New York.

(3) Dudley, J.W. & Lambert, R.J. (1969). Genetic variability after 65 generations of selection in Illinois high oil, low oil, high protein, and low protein strains of Zea mays L. Crop Sci. 9, 179-181.

(4) Dudley, J.W., Lambert, R.J. & Alexander, D.E. (1974). Seventy generations of selection for oil and protein concentration in the maize kernel. p. 181-212. In J. W. Dudley (ed.) Seventy Generations of Selection for Oil and Protein in Maize. Crop Sci. Soc. of Am., Madison, Wis.

(5) Dudley, J.W., Lambert, R.J. & de la Roche, I. (197-). Genetic analysis of crosses among corn strains divergently selected for percent oil and protein. (Accepted for publication by Crop Sci.)

(6) Gardner, C.O. & Eberhart, S.A. (1966). Analysis and interpretation of the variety cross diallel and related populations. Biometrics 22, 439-452.

(7) Hanson, W.D. (1959). Theoretical distribution of the initial linkage block lengths intact in the gametes of a population intermated for n generations. Genetics 44, 839-846.

(8) Mather, K. & Jinks, J.L. (1971). Biometrical Genetics. Cornell University Press, Ithaca, New York. 382 p.

(9) Moreno-Gonzalez, J., Dudley, J.W. & Lambert, R.J. (1975). A design III study of linkage disequilibrium for percent oil in maize. Crop Sci. 15, 840-843.

(10) Robertson, A. (1970). A theory of limits in artificial selection with many linked loci. In Ken-ichi Kojima (ed.) Mathematical Topics in Population Genetics. Biomathematics 1, 246-288. Springer-Verlag, New York-Heidelberg-Berlin.

(11) 'Student'. (1934). A calculation of the minimum number of genes in Winter's selection experiment. Ann. Eugenics 6, 77-82.

Quantitative genetic studies and population improvement in maize and sorghum

C. O. Gardner
DEPARTMENT OF AGRONOMY
UNIVERSITY OF NEBRASKA, LINCOLN, NE 68583

ABSTRACT

Development of quantitative genetic theory and empirical studies have been useful to plant breeders in choosing base populations for breeding and in choosing breeding procedures most likely to succeed. Recurrent selection programs in maize have successfully increased yields, and realized responses have been in excellent agreement with that expected base on extensive genetic variance studies. Discrepancies that occur are likely due to inadequate sampling in estimating variances and expected gains or in conducting the selection experiment itself. Small populations lead to inbreeding and fixation of undesirable alleles.

Recent results in selection studies in the Hays Golden variety of maize have been disappointing, but are believed to be explained by reduction in genetic variability due to selection, inability to identify superior genotypes in new more variable isolations, and genotype-environment interaction in the recent hot drouthy years.

Published as Paper No. 5262, Journal Series, Nebraska Agricultural Experiment Station.

Quantitative genetic studies in sorghum indicate existence
of considerable non-additive genetic variance, that S_1 family
selection is likely to be the most effective breeding system,
that frequencies of favorable alleles tend to be below 0.5, and
that exotic germ plasm can be useful in breeding programs.

INTRODUCTION

The ultimate goal of the plant breeder is to utilize the
germ plasm at his disposal to synthesize and select new cultivars
which will maximize yields of food, feed and fiber. In the past
quarter century, quantitative geneticists have played a major
role in providing basic information and in developing procedures
for use by plant breeders in making major decisions. Techniques
have been developed for estimating population parameters, which
have been used for predicting the outcome of different recurrent
selection systems. This has made it possible for breeders to
choose from among the many such alternative schemes available for
population improvement and for the eventual development of useful
cultivars.

The purpose of this paper is to briefly comment on applications
of quantitative genetic theory used to solve some of the problems
in maize breeding and to indicate that similar techniques are now
being used in grain sorghum at Nebraska.

GENETIC VARIANCE STUDIES IN MAIZE

Information on the kinds and magnitudes of the different types
of genetic variation in populations as well as knowledge about the
means is essential in order to draw conclusions concerning the choice
of breeding material and the breeding system to be used to improve
populations. Structured mating designs have been suggested to provide the individuals and/or families needed to conduct such studies

by Comstock and Robinson (1948, 1952), Cockerham (1963) and others. Analyses of variance and covariance of data collected provide the population parameter estimates needed for further analysis and genetic interpretation. In addition to the problems inherent in quantitative genetic theory, many biological problems also arise in making the required matings, in designing and conducting the experiments, and in interpreting the results. All too often, biases are introduced which make genetic interpretations difficult if not impossible. Often the vagaries of weather factors which cannot be controlled have drastic and unpredictable results, even when experiments are grown under irrigation.

Many genetic variance studies have been conducted on many different plant populations, but estimates often have large standard errors because the sample size used was too small. At Nebraska genetic variances for grain yield have been estimated from four different samples of families from the corn variety Hays Golden (Table I). The nested mating design (Design I) was used in each case. All four were large experiments (256 to 400 families) replicated in two environments, yet the conclusions reached differ substantially. Additive genetic variance ($4\sigma_m^2$) from sample 1 is estimated to be 5.4 times greater than that of sample 2. Variances do not appear to be directly correlated with yield level. The information from the four experiments has been pooled to provide the best possible estimates of genetic variances given in Table I. Additive genetic variance, σ_A^2, is substantially larger than non-additive, σ_D^2. This is generally true for all crops and all traits studied. Non-additive genetic variance tends to be relatively higher for grain yield than most other traits and tends to be relatively higher for cross-fertilizing species than for self-fertilizing ones.

Heritability of grain yield in Hays Golden is estimated to be 15%, and expected gain from mass selection for high grain yield is 3.08% (Table I). Without extensive sampling and testing, such precise estimates would be impossible.

TABLE I

Genetic and environmental components of variance estimated for the maize variety Hays Golden.

Sample	Mean Yield g/plant	Components of variance estimated*				
		σ^2_m	$\sigma^2_{f(m)}$	σ^2_{me}	$\sigma^2_{f(m)e}$	σ^2
1	297	402	360	41	10	432
2	325	74	129	164	99	568
3	350	249	286	148	121	541
4	277	120	265	79	102	628
Pooled	308	216	263	113	88	542
		σ^2_A	σ^2_D	σ^2_P	h^2	Δg
		864	188	5798	0.15	3.08%

* σ^2_m = male component, $\sigma^2_{f(m)}$ = females within males component, σ^2_{me} = male x environment interaction component, $\sigma^2_{f(m)e}$ = females within males x environment interaction component, σ^2 = error component, σ^2_A = additive genetic variance, σ^2_D = non-additive genetic variance, σ^2_P = total phenotypic variance, h^2 = heritability, and Δg = gain expected from selection.

COMPARISON OF PREDICTED AND REALIZED GAINS FROM SELECTION FOR YIELD IN MAIZE

Many plant selection experiments reported fail to give predicted gains to compare with realized gains. From the standpoint of evaluating selection theory, such information is needed. Where predicted gains have been reported, the observed results have generally been in good agreement with expected results.

In our mass selection studies in Nebraska, a realized gain of 3.00% per generation predicted as a linear response over the first 15 generations of selection (Gardner (1969a, 1973)) is in excellent agreement with the expected 3.08% response calculated from pooled data in Table I. Results are shown in Fig. 1.

FIG. 1

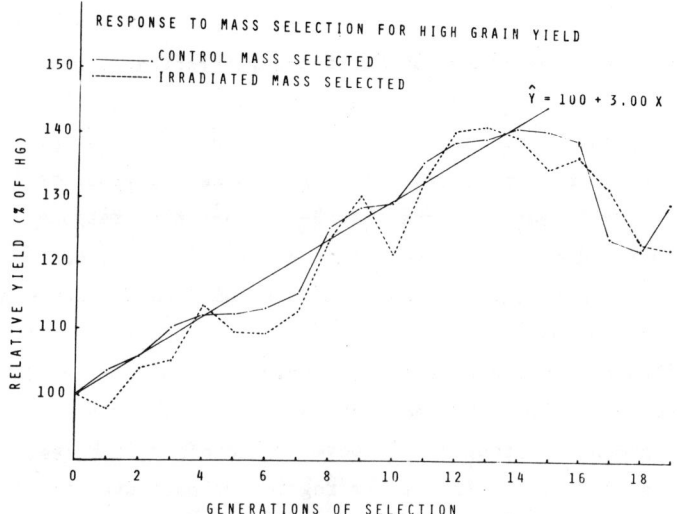

Moll and Stuber (1971) practiced full-sib family selection for six generations. Observed responses in the Jarvis and the Indian Chief varieties of 3.75% and 2.65% were in excellent agreement with expected responses of 4.74% and 2.77%, respectively. However, in their composite formed by crossing these two varieties and random mating the offspring for four generations, they observed a gain of only 3.91% compared to the expected 6.26% based on genetic variance studies.

Moll and Stuber (1971) and Eberhart, Debela, and Hallauer (1973) reported observed responses to reciprocal recurrent selection for high grain yield to be 3.19% and 4.47% compared to expected responses of 2.77% and 7.2%, respectively. The response obtained by Eberhart et al. is somewhat below that expected.

Webel and Lonnquist (1967) reported observed and expected results of 9.44 and 8.39% per cycle, respectively, from four generations of modified ear-to-row selection in Hays Golden, which suggests excellent agreement between observed and expected results. Compton and Bahadur (1977) continued the same selection study through 10 generations and reported an observed response (linear regression estimate) of 5.26% per generation compared to an expected response of 4.87%. Again this is excellent agreement, but their expected response is little more than half that predicted by Webel and Lonnquist. Also their observed response is definitely curvilinear. I have also been evaluating response from the modified ear-to-row selection program in relation to the mass selection program. When the same combined data used to predict gain from mass selection are also used to predict gain from modified ear-to-row selection, a gain of 4.53% per cycle is expected. When all of the data collected over the years including my own with that cited above are combined, the results are as shown in Fig. 2. The linear regression estimate of 4.62% gain per cycle is in excellent agreement with the predicted 4.53%; however, the response does not appear to be linear. An excellent fit to a quadratic response curve is indicated. Gains in early generations appear to be much higher than expected. A check on procedure followed reveals that the first generation was actually full-sib family selection, which should give twice the gain expected from half-sib family selection. A change in selection criterion at generation 8 to a selection index which included lodged plants and dropped ears (Compton and Bahadur (1977)) may account for lack of continued yield increases. Increasing frequencies of favorable alleles and the resulting reduction in additive genetic variance could cause the population to plateau but not decrease in yield. Genotype-environment interaction discussed later may be a factor in the response observed.

FIG. 2

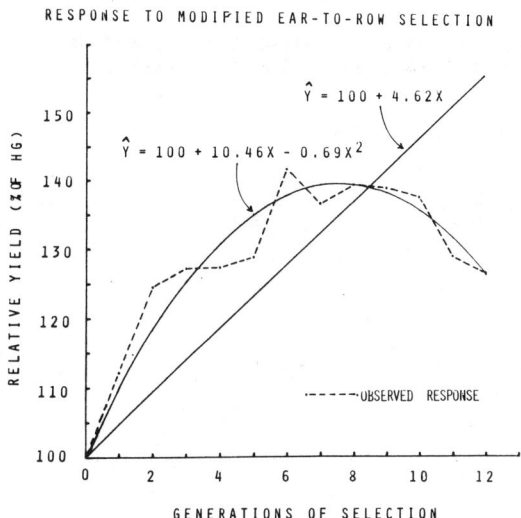

RESPONSE TO MODIFIED EAR-TO-ROW SELECTION

$\hat{Y} = 100 + 4.62X$

$\hat{Y} = 100 + 10.46X - 0.69X^2$

------ OBSERVED RESPONSE

RELATIVE YIELD (% HG)

GENERATIONS OF SELECTION

RECENT RESPONSE TO MASS SELECTION

Fig. 1 indicates that both mass selected populations tended to plateau at about generation 13 and that yields started decreasing at generation 17. Fluctuations noted throughout the first 16 generations do not represent significant departures from a linear trend, but the decrease in yield noted in recent generations merits consideration. Several hypotheses can be suggested. One is that the base population maintained by growing a large sample in isolation every four years suddenly improved perhaps due to unintentional selection or contamination. This can be discounted by an examination of the annual yield performance of Hays Golden shown in Fig. 3. There is no evidence that the population has changed. The drastic fluctuations in yield of Hays Golden from year to year is the reason for expressing yields of selected generations relative to Hays Golden. It would be very difficult to interpret absolute yields.

FIG. 3

Annual mean yields of the Hays Golden variety and Nebr. 501D hybrid check in mass selection evaluation trials.

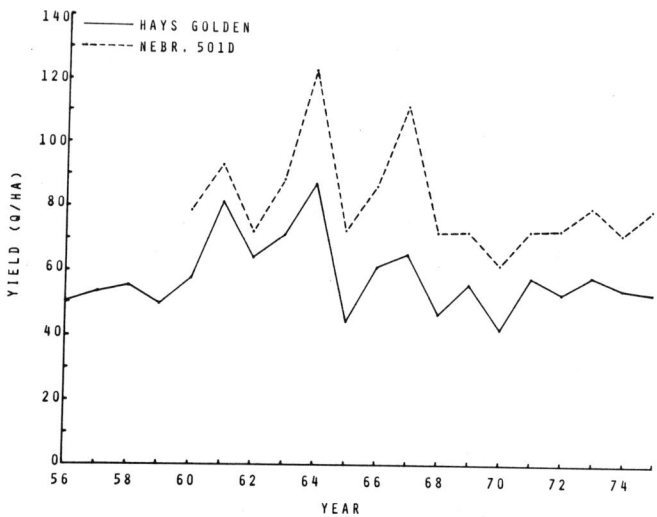

Another hypothesis is that campus expansion forced a shift from the isolated nurseries which were uniform in soil type, had minimum environmental variation, and were easy to water uniformly when necessary. New isolations were extremely variable in soil and other environmental factors and were difficult to water at all, let alone uniformly. Severe heat and drouth in recent years have accentuated the variability. Under such conditions, superior genotypes cannot easily be identified. If selection had been for favorable linked epistatic combinations of genes for which selection was no longer effective because of extreme environmental variations, breakdown of such gene combinations could lead to reduced yields.

Still another hypothesis is that the yields of the selected generations in the annual irrigated yield evaluation trials were more adversely affected by heat and lack of rainfall than were the yields of Hays Golden; i.e., the selected generations interacted more with environments. Over the years, a number of experiments have been conducted involving material from selected generations and from Hays Golden. When grown under some form of environmental stress, selected generations have not performed nearly as well in relation to Hays Golden as they have in favorable corn-growing years. While Hays Golden seems to lack the ability to respond to more favorable growing conditions, it still has the capacity to perform well under heat and drouth situations for which it is well known.

CHANGES IN GENETIC VARIANCE

With many loci controlling grain yield and with initial frequencies of many favorable alleles at rather low levels, additive genetic variance could be maintained at a fairly high relatively constant level, and a linear response such as that observed in the mass selection program would be expected. Genetic variance studies involving material from selected generations and Hays Golden have been reported by Lonnquist, Cota and Gardner (1966), Gardner (1969b) and Harris, Gardner and Compton (1972). Relative values from those and more recent studies are summarized in Table II. At generation 6, additive genetic variance was higher in selected generations than in Hays Golden. At generation 10, the results from 1967, a favorable year, indicated no apparent differences; however, results from 1968, a less favorable year, and from the combined analysis indicated a decline in additive genetic variance with selection. At generation 15, no further change could be detected. S_1 lines per se, their testcross hybrids, and S_6 line hybrids derived from selected generations 9 and Hays Golden all indicated reduced genetic variation in selected generations. Hence, less progress is expected from selection in more recent generations, but it should still be approximately half that of the early generations.

TABLE II

Relative genetic variation estimated from families derived from Hays Golden (HG) and from Control (C) and Irradiated (I) selected populations of maize.

Popn.	Additive Genetic Variance			Gen. 9 Random Lines			$S_6 \times S_6$ Random Hybrids
	Gen.6	Gen.10	Gen.15	S_1 Per Se	S_1 Testcross Hybrids (1)	(2)	
HG	100	100	100	100	100	100	100
C	136	30	60	40	22	30	77
I	263	89	49	67	47	14	57

EVIDENCE FOR EPISTASIS

Random sets of S_6 inbred lines from generation 9 of the two selected populations and from Hays Golden were divided into subsets of four lines within each population. All possible single, three-way, and double-cross hybrids were produced for each subset and were tested in 1974 and 1975. The general genetic model of Eberhart and Gardner (1966) was used to analyze hybrid means. Total variation attributed to each of the different kinds of gene effects was computed. Mean squares for additive x additive epistasis and for deviations from the model due to dominance types of epistasis and/or linkage were highly significant for yield in all three populations. Relative measures of variation are given in Table III. It is interesting to note that there are no great differences in distribution of relative variances among the three populations; however, irradiation may have induced small differences. These data support the hypothesis that epistasis is important in heterosis in corn yields, but they do not support the hypothesis that selection has been for favorable linked epistatic combinations of genes, unless such selection occurred in generations 10 through 16.

TABLE III

Variation among random-line hybrids attributable to different genetic effects according to the general genetic model of Eberhart and Gardner (1966). Lincoln, Nebraska 1974-1975. (Martin and Gardner, Unpublished).

	% of total variation among hybrid means		
Genetic Effect	HG	C-9	I-9
General (Additive)	41	41	43
Specific (Dominance)	8	8	13
Additive x Additive Epistasis	21	19	20
Deviations from Model due to Dominance Types of Epistasis and/or Linkage	30	32	24
	100	100	100
Relative Total Variation	100	89	73

ADDITIONAL RESEARCH NEEDS

In spite of a vast amount of research done on this one random mating population, Hays Golden, there are still many unanswered questions as to the kinds of genetic changes that have occurred to give the continuous linear response observed over 15 generations and the apparent recent loss in yield. We are developing random lines from generation 15 of selected populations. Also we now have five generations of relaxed selection (no selection pressure) beginning at generation 15, which will be evaluated in the next two years. Our previous experience where generations of relaxed selection were compared to comparable selected generations from which they were derived showed no apparent yield reduction.

In addition to further quantitative genetic studies, isozyme studies planned may provide information on gene frequency changes, linkage relationships, and epistasis.

QUANTITATIVE GENETIC STUDIES IN SORGHUM

Quantitative genetic studies in sorghum, which is largely self-fertilized, have been done in random-mating sorghum populations created by use of genes which cause male sterility (Nordquist, Webster, Gardner, and Ross (1973)). An extensive study made of the NP3R population by Jan-orn, Gardner and Ross (1976) involved sets of full-sib, half-sib and S_1 families, which are easy to produce but which have some obvious differences with respect to their male fertility-sterility relationships. If we assume no effects of genes controlling male sterility on grain yield and assume no epistasis and linkage equilibrium, the expectations of the estimates obtained are as follows using Falconer's symbols:

Family type	Component	Estimate	Expectation
Half-sib	σ^2_{HS}	12.7 ± 2.6	$\frac{1}{4}\sigma^2_A$
Full-sib	σ^2_{FS}	43.0 ± 5.9	$\frac{1}{2}\sigma^2_A + \frac{1}{4}\sigma^2_D$
S_1	$\sigma^2_{S_1}$	39.9 ± 6.0	$\sigma^2_{A*} + \frac{1}{4}\sigma^2_D$

The difference between σ^2_A and σ^2_{A*} is as follows:

$$\sigma^2_A = \sum_{i=1}^n 2p_i q_i [a_i + (q_i - p_i) d_i]^2$$

$$\sigma^2_{A*} = \sum_{i=1}^n 2p_i q_i [a_i + \tfrac{1}{2}(q_i - p_i) d_i]^2$$

Estimates of σ^2_A, σ^2_D and σ^2_{A*} are 50.9 ± 10.3, 70.1 ± 30.3, and 22.4 ± 9.8, respectively. The estimate of dominance variance is surprisingly large relative to additive genetic variance, and σ^2_{A*} is surprisingly low relative to σ^2_A. If there is no dominance,

QUANTITATIVE GENETIC STUDIES IN MAIZE AND SORGHUM 487

$\sigma^2_{A*}=\sigma^2_A$. If dominance exists, $\sigma^2_A>\sigma^2_{A*}$ when p<q, and $\sigma^2_A<\sigma^2_{A*}$ when p>q. Results suggest that the frequencies of many favorable genes are less than 0.5. Predicted gains from family selection for high grain yield indicate that S_1 family selection is likely to be the most effective system, but empirical data are needed for verification.

Incorporation of exotic germ plasm into the NP3R population more than doubled additive genetic variability in grain yield with no reduction in mean performance (Eckebil, Ross, Gardner and Maranville (1977), Ross, Eckebil, Kofoid, and Gardner (1977)). S_1 family selection is being practiced in both the original population and the broader based one.

Several years will be required to obtain empirical data necessary to compare mass selection and family selection systems in sorghum, but such studies are now under way directed by Dr. W. M. Ross.

BIBLIOGRAPHY

(1) Cockerham, C. C. (1963). Estimation of genetic variances. Statistical Genetics and Plant Breeding (W. D. Hanson & H. F. Robinson, Eds.). Washington, D.C.: National Academy of Science-National Research Council Publ. 982, 53-94.

(2) Compton, W. A. & Bahadur, K. (1977). 10 cycles of progress from modified ear-to-row selection in corn (Zea mays L.). Crop Science (Submitted).

(3) Comstock, R. E. and Robinson, H. F. (1948). The components of genetic variance in populations of biparental progenies and their use in estimating the average degree of dominance. Biometrics 4:254-266.

(4) Comstock, R. E. and Robinson, H. F. (1952). Estimation of average dominance of genes. Heterosis (J. W. Gowen, Ed.). Ames, IA: Iowa State College Press, 494-516.

(5) Eberhart, S. A., Debela, S., & Hallauer, A. R. (1973). Reciprocal recurrent selection in the BSSS and BSCB1 maize populations and half-sib selection in BSSS. Crop Sci. 13:451-456.

(6) Eberhart, S. A., & Gardner, C. O. (1966). A general model for genetic effects. Biometrics 22:864-881.

(7) Eckebil, J. P., Ross, W. M., Gardner, C. O., & Maranville, J. W. (1977). Heritability estimates, genetic correlations, and predicted gains from S_1 progeny tests in three grain sorghum random-mating populations. Crop Science (Submitted).

(8) Gardner, C. O. (1969a). The role of mass selection and mutagenic treatment in modern corn breeding. Proc. 24th Ann. Corn and Sorghum Research Conference. American Seed Trade Assn., 15-21.

(9) Gardner, C. O. (1969b). Genetic variation in irradiated and control populations of corn after ten cycles of mass selection for high grain yield. Induced Mutations in Plants. Vienna: International Atomic Energy Agency, 469-477.

(10) Gardner, C. O. (1973). Evaluation of mass selection and of seed irradiation with mass selection for population improvement in maize. Genetics 74:s88-s89.

(11) Harris, R. E., Gardner, C. O. & Compton, W. A. (1972). Effect of mass selection and irradiation in corn measured by random S_1 lines and their testcrosses. Crop Sci. 12:594-598.

(12) Jan-orn, J., Gardner, C. O., and Ross, W. M. (1976). Quantitative genetic studies of the NP3R random-mating grain sorghum population. Crop Sci. 16:489-496.

(13) Lonnquist, J. H., Cota, O., & Gardner, C. O. (1966). Effect of mass selection and thermal neutron irradiation on genetic variances in a variety of corn (Zea mays L.). Crop Sci. 6:330-332.

(14) Moll, R. H. & Stuber, C. W. (1971). Comparison of response to alternative selection procedures initiated with two populations of maize (Zea mays L.). Crop Sci. 11:706-711.

(15) Nordquist, P. T., Webster, O. J., Gardner, C. O., Ross, W. M. (1973). Registration of three sorghum germplasm random-mating populations. Crop Sci. 13:132.

(16) Ross, W. M., Eckebil, J. P., Kofoid, K. D. & Gardner, C. O. (1977). Some quantitative characteristics of five sorghum random-mating populations. Maydica (Submitted).

(17) Webel, O. D. & Lonnquist, J. H. (1967). An evaluation of modified ear-to-row selection in a population of corn (Zea mays L.). Crop Sci. 7:651-655.

Quantitative genetics and practical corn breeding

Steve A. Eberhart
FUNK SEEDS INTERNATIONAL
BLOOMINGTON, IL 61701

Dramatic improvements in the performance of corn varieties have been obtained by the "Art" of corn breeding through phenotypic selection since corn was first domesticated. The development of statistical genetic theory has provided the foundation for a new level of corn breeding, but unfortunately, the application of this theory to improve corn breeding methodology has been minimal to date. Excellent progress has been obtained, however, in the few instances where breeding programs have been designed to maximize gain from selection with information based on statistical genetic theory.

In practical corn breeding programs, improvement must be obtained for yield, standability, disease resistance and insect resistance. Simultaneous selection for several characters invariably reduces the rate of gain for each character unless there are large positive genetic correlations among all desired traits. The first selection programs were designed for single trait selection in open-pollinated varieties and in F_2's of single cross hybrids in order to varify that gains could be made and that observed gains were similar to expected gains. When the effective population size was reasonably large, observed gains have been similar to expected gains (Moll and Stuber, 1971; Gardner, 1968; Webel and Lonnquist, 1967). In other cases, gain was somewhat less than expected (Eberhart, et al. 1973; Russell, et al. 1973; Darrah, et al., In Press; Burton, et al. 1971; Horner, et al. 1973).

In corn, the commercial product is an F_1 hybrid. Unfortunately, very little direct information is available on hybrid improvement through the application of statistical genetic theory. Considerable indirect information, however, can be assembled to suggest that greater progress could be expected if breeding programs were planned according to information now available.

At one time the hypothesis of over-dominance in corn was widely acclaimed and considerable effort has been devoted to empirically identifying inbred lines that will "nick" to give maximum yields. Most statistical genetic studies, however, suggest that complete dominance may be much more important than over-dominance. The Design III is a powerful design for measuring the average level of dominance. Although estimates of parameters from F_2 populations suggest over-dominance ($\bar{a}>1$), estimates of $\bar{a}$ from the same populations following random-mating were much lower, indicating complete or partial dominance (Gardner, 1963; Moll, et al. 1964; Moll and Robinson, 1967).

Empirical population improvement studies have failed to support the hypothesis of over-dominance. Dr. G. F. Sprague initiated two studies in Iowa in which two breeding populations were improved by recurrent selection with the same inbred line as a tester. If gene action were primarily over-dominance, each population by line cross would show rapid improvement, but the population by population cross would show little, if any, improvement. However, Russell, et al. (1973) reported that five cycles of selection had improved the population by population cross at a much greater rate than for either population crossed to the inbred tester.

A reciprocal recurrent selection study at Iowa provides additional empirical information. The population cross of BSCB1(R)C5 with BSSS(HT)C6 and even BSSS(R)C5 x BSSS(HT)C6 yielded as much as the population cross, BSCB1(R)C5 x BSSS(R)C5 (Table I). BSSS(R)C5 was improved by reciprocal recurrent selection whereas BSSS(HT)C6 was improved with the double-cross Iowa 13 as a tester (Russell and Eberhart, 1975). If over-dominance had been the most important type of gene action, the BSCB1(R)C5 x BSSS(R)C5 population cross should have been much higher yielding than the BSCB1(R)C5 x BSSS(HT)C6 cross.

TABLE I

Performance of Population Crosses in Iowa*

Hybrid	Yield (Q/Ha)	Moisture %	Lodging %	
			Roots	Stalks
BSSS(R)C5 x BSCB1(R)C5	65.5	23.6	4	16
BSSS(HT)C6 x BSCB1(R)C5	66.5	22.5	4	12
BSSS(HT)C6 x BSSS(R)C5	66.9	25.2	16	9
L.S.D. (P=.05)	5.7	0.8	6	6

*From Russell and Eberhart (1975)

Recent results from an evaluation of the products of full-sib reciprocal recurrent selection by Hoegemeyer and Hallauer (1976) support the relative importance of partial to complete dominance versus over-dominance. The lines developed by pair-wise selection for five generations yielded as well when crossed to other lines as they did in the selected cross. If over-dominance were the predominate type of gene action, the specific combining ability should have been much greater than the general combining ability.

Studies with near isogenic lines by Russell and Eberhart (1970), and Russell (1971) suggest over-dominance, but the possibility of pseudo-over-dominance from repulsion-phase linkages cannot be ruled out. If over-dominance or over-dominant epistatic types of gene action do not occur, the ultimate goal will be a

homozygous pure-line with all the favorable alleles rather than
a hybrid. When the large number of loci involved with yield,
standability and resistance to diseases and insects are considered,
however, it is easy to show that the development of an ideal genotype is virtually impossible without some type of recurrent selection to gradually increase the frequency of desirable genotypes
(Lonnquist, 1951). Even with recurrent selection, many cycles
will be required. Meanwhile, hybrids can be used to take advantage of the favorable alleles in both parental lines. On the
other hand, if over-dominance and over-dominant types of epistasis
occur at a few loci, hybrids must be used to obtain maximum yields.
Because reciprocal recurrent selection as suggested by Comstock,
Robinson & Harvey (1949) will produce superior hybrids whether
gene action is primarily complete dominance or over-dominance,
this breeding method is probably the best one available for
developing improved corn hybrids. Moll and Stuber (1971) reported
greater yield improvement of the variety cross from reciprocal
recurrent selection than from full-sib selection in Jarvis and
Indian Chief.

If gene frequencies are between 0.2 and 0.8 and adequate
effective population sizes are maintained, genetic variances are
expected to change very slowly for traits controlled by many loci,
each with a small effect. Under this situation, the rate of improvement of F_1 hybrids should be similar to the improvement of
the population cross (Sprague and Eberhart, In Press). Eberhart,
et al. (1973) have reported 2.73 q/ha/cycle (4.6% per cycle) improvement in the BSSS x BSCB1 population cross with five cycles
of reciprocal recurrent selection in Iowa. Darrah, et al. (In
Press) obtained 4.2 q/ha/cycle (7% per cycle) improvement in the
Kitale II x Ec573 cross in Kenya. Moll and Stuber (1971) reported
3.5% per cycle improvement in the Jarvis x Indian Chief cross.
These studies conclusively demonstrate that single trait selection
by reciprocal recurrent selection is effective in increasing the
yield of the population cross.

Moll, Bari and Stuber (In Preparation) compared single crosses of lines derived from the sixth cycle of reciprocal recurrent selection with single crosses from the original Jarvis and Indian Chief varieties. Their results showed higher yields for the single crosses from the improved populations. Harris, Gardner & Compton (1972) reported that topcrosses of unselected S_1 lines from Hays Golden populations improved by nine cycles of mass selection yielded 17% more than comparable topcrosses involving S_1 lines from the original Hays Golden variety. The topcrosses of the lines from the improved populations were 15 cm taller than those from the original popualtion, however.

Other recurrent selection studies have demonstrated excellent progress from recurrent selection for stalk rot resistance (Jinahyan and Russell, 1969); European corn borer resistance (Penny, Scott, and Guthrie, 1967); and leaf blight resistance (Jenkins, Robert, and Findley, 1954). Gain from selection has been asymptotic in three to five cycles as the limit to progress is approached.

Although inbred lines have been developed from populations improved by recurrent selection, only a few of them have produced agronomically acceptable hybrids. However, a survey conducted by the American Seed Trade Association has shown that these elite lines have been widely used by commercial companies (Table II). B14 and B37 were derived by USDA and Iowa State University breeders from two of the S_1 lines recombined to form the improved Stiff Stalk Synthetics BSSS(HT)C1. B14A, A632 and A634 are backcross derivatives of B14. N28 was developed by Dr. J. Lonnquist at the University of Nebraska from an S_1 line used to form the second cycle for a recurrent selection in Stiff Stalk Synthetic, NSSS II. B73 was derived from the fourth cycle of BSSS(HT), and usage in 1976 and 1977 will be much greater than in 1975.

TABLE II

Results from an American Seed Trade Survey
on Usage in Hybrid Seed Production of Some of
the Public Lines Derived from Stiff Stalk Synthetic

Inbred Line	Source	Usage in Hybrid Seed Production (lbs.)	Percentage 1975 Seed Requirement
B14	BSSS CO	997,000	0.1
B14A		15,286,000	1.6
A632		147,058,000	15.2
A634		5,795,000	0.6
B37	BSSS CO	65,816,000	6.8
N28	NSSS I	29,916,000	3.1
B73	BSSS(HT)C4	30,048,000	3.1

Although yields of the improved variety crosses are comparable to commercial hybrids, standability is not (Table III). When hybrids are developed from the improved breeding populations, considerable improvement can be made in yield as demonstrated by Russell and Eberhart (1975) and Darrah, et al. (In Press). The EAH6302 or an equivalent three-way cross hybrid developed by the Kenya Seed Company (Table IV) is expected to be used in commercial production, but standability is not satisfactory even though some improvement has been made.

Successful development of commercial corn hybrids demands simultaneous improvement of standability and insect and disease resistance in addition to yield. When the various agronomic traits are manifest at different growth stages, Young (1964) points out that multi-stage selection can be an efficient procedure. Various types of index selection have been proposed, but few reports of practical application are available. A selection index to maximize improvement of the economic worth of an individual was proposed by Smith (1936), and this index has been used in animal improvement. Hazel and Lush (1942) demonstrated that such an index is more

TABLE III

Performance of Original and Improved Population
Cross Hybrids from Six Trials in Iowa, 1972-73*

Hybrid	Yield (Q/ha)	Stalk Lodging %
BSSS CO x BS12 CO	57	26
BSSS CO x BSK CO	62	19
BSSS(R) C6 x BS12(HI)C5	85	21
BSSS(HT)C7 x BSK(S)C5	75	27
BSSS(HT)C7 x BSTL	80	13
BS12(HI)C5 x BSTL	75	19
B37 x OH43	79	5
N28 x MO17	92	4
L.S.D.(P=.05)	8	11

*From Hallauer and Malithano (1976)

TABLE IV

Performance of an Improved Synthetic and Hybrids in Kenya*

	Yield (Q/Ha)	Lodging % Rots	Lodging % Stalks	Ears/ 100 Plants	Days to Tassel	Ear Height
Kitale II	45.8	7	38	91	116	201
H611	59.1	10	39	93	115	221
H611(R)C3	71.6	5	36	110	115	220
H611(R)C3	87.3	9	27	111	105	214
EAH6302	96.7	6	23	132	107	220
L.S.D.(P=.05)	9.9	7	8	11	2	9

*From Darrah and Penny (In Press)   EAH6302 is (5 x 30)50

efficient than tandem selection. Lack of reliable estimates of phenotypic and genotypic variances and covariances can reduce the usefulness of index selection. When progeny tests with an adequate number of plants per progenies and enough replications and locations are used in corn, heritabilities will be high. Falconer (1960) has pointed out that an index based solely on economic weights (a base index) is just as satisfactory as the conventional index when heritabilities are high.

Realistic choices of economic weights, however, is very difficult in corn breeding. Pesek and Baker (1969) proposed a modified index in which desired gains are defined for each character, and an index is computed as a function of genetic variances and covariances and the desired gain. Suwantaradon et al. (1975) reported that correlation coefficients of rather small values effected the expected gain from selection. The expected gains for root size were opposite in sign to the economic weight for the conventional index and the base index in their study. Their results indicated that expected gains were similar for the conventional and the base index. Although the modified index forced the expected gains to reach the desired goal for all characters simultaneously, expected gain for yield was even less than for the other two indexes. Subandi, Compton, and Empig (1973) suggested a multiplicative type of index for selection of erect plants, increased yield, and a reduction in dropped ears. This index has been used in a modified ear-to-row selection experiment. Although lodging has been reduced in the latest cycles, the rate of yield improvement appears to be less than in the early cycles (Compton and Bahadur, In Press). Horner, et al. (1976) report good progress for simultaneous selection for reduced lodging, lower ear height and increased yield in the population and in topcrosses from seven cycles of recurrent selection with a single-cross tester.

QUANTITATIVE GENETICS AND CORN BREEDING 499

SUMMARY

A breeding system has been developed through the application of quantitative genetic theory to corn improvement. This reciprocal recurrent selection system has been applied by corn breeders working in public institutions in Iowa, North Carolina, Nebraska, and Kenya, and higher yielding corn hybrids have been developed. However, insufficient selection pressures for standability, disease resistance, and insect resistance have restricted the commercial usefulness of these improved hybrids. The use of selection indexes and multi-stage selection for multiple traits can be expected to improve the usefulness of the products of future breeding programs.

BIBLIOGRAPHY

[1] Burton, J. W., Penny, L. H., Hallauer, A. R., & Eberhart, S. A. (1971). Evaluation of synthetic populations developed from a maize population (BSK) by two methods of recurrent selection. Crop Sci. 11, 361-367.

[2] Compton, W. A., & Bahadur, (In Press). Ten cycles of progress from modified ear-to-row selection in corn (Zea mays L.). Crop Sci.

[3] Comstock, R. E., Robinson, H. F., & Harvey, P. H. (1949). A breeding procedure designed to make maximum use of both general and specific combining ability. Agron. J. 41, 360-367.

[4] Darrah, L. L., Eberhart, S. A., & Penny, L. H. (In Press). Six years of maize selection in Kitale II, Ec573, and Kitale Composite A using methods of the comprehensive breeding system. Theoretical and Applied Genetics.

[5] Darrah, L. L., & Penny, L. H. (In Press). Inbred line extraction from improved breeding populations. E. A. Agr. For. J.

[6] Eberhart, S. A., Debela, S., & Hallauer, A. R. (1973). Reciprocal recurrent selection in the BSSS and BSCB1 maize population and half-sib selection in BSSS. Crop Sci. 13, 451-456.

[7] Falconer, D. S. (1960). Introduction to quantitative genetics. New York: Ronald Press Co., 365 pp.

[8] Gardner, C. O. (1963). Estimates of genetic parameters in cross fertilizing plants and their importance in plant breeding. pp. 225-252. In W. D. Hanson and H. F. Robinson (eds.) Statistical Genetics and Plant Breeding. Nat. Acad. Sci.-Nat. Res. Counc. Publ. 982. Washington, D.C.

[9] Gardner, C. O. (1968). Mutation studies involving quantitative traits. In the Present State of Mutation Breeding, Gamma Field Symposium No. 7, 57-77.

[10] Hallauer, A. R., & Malithano, D. (1976). Evaluation of maize varieties and their potential as breeding populations. Euphytica 25, 117-127.

[11] Harris, R. E., Gardner, C. O., & Compton, W. A. (1972). Effects of mass selection and irradiation in corn measured by random S_1 lines and their testcrosses. Crop Sci. 12, 594-598.

[12] Hazel, L. N., & Lush, J. H. (1942). The efficiency of three methods of selection. J. Hered. 33, 393-399.

[13] Hoegemeyer, T. C., & Hallauer, A. R. (1976). Selection among and within full-sib families to develop single crosses of maize. Crop Sci. 16, 76-81.

[14] Horner, E. S., Lundy, H. W., Lutrick, M. C., & Chapman, W. H. (1973). Comparison of three methods of recurrent selection in maize. Crop Sci. 13, 485-489.

[15] Horner, E. S., Lutrick, M. C., Chapman, W. H., & Martin, F. G. (1976). Effect of recurrent selection for combining ability with a single cross tester in maize. Crop Sci. 16, 5-8.

[16] Jenkins, M. T., Robert, A. L., & Findley, W. R., Jr. (1954). Recurrent selection as a method for concentrating genes for resistance to Helminthosporium turicum leaf blight in corn. Agron. J. 46, 89-94.

[17] Jinahyon, S., & Russell, W. A. (1969). Evaluation of recurrent selection for stalk-rot resistance in an open-pollinated variety of maize. Iowa State J. Sci. 43, 229-237.

[18] Lonnquist, J. H. (1951). Recurrent selection as a means of modifying combining ability in corn. Agron. J. 43, 311-315.

[19] Moll, R. H., Bari, A., & Stuber, C. W. (In Preparation). Frequency distributions of yield of hybrids of Zea mays L. between inbreds from two varieties before and after reciprocal recurrent selection. (To be submitted to Crop Sci.).

[20] Moll, R. H., Lindsey, M. F., & Robinson, H. F. (1964). Estimates of genetic variances and level of dominance in maize. Genetics 49, 411-423.

[21] Moll, R. H., & Robinson, H. F. (1967). Quantitative genetic investigations of yield in maize. Der Züchter 37, 192-199.

[22] Moll, R. H., & Stuber, C. W. (1971). Comparisons of response to alternative selection procedures initiated with two populations of maize (Zea mays L.). Crop Sci. 11, 706-711.

[23] Penny, L. H., Scott, G. E., & Guthrie, W. D. (1967). Recurrent selection for European corn borer resistance in maize. Crop Sci. 7, 407-409.

[24] Pesek, J., & Baker, R. J. (1969). Desired improvement in relation to selection indexes. Can. J. Plant Sci. 49, 803-804.

[25] Russell, W. A. (1971). Types of gene action at three gene loci in sub-lines of a maize inbred line. Can. J. Genet. Cytol. 13, 322-334.

[26] Russell, W. A., & Eberhart, S. A. (1970). Effects of three gene loci in the inheritance of quantitative characters in maize. Crop Sci. 10, 165-169.

[27] Russell, W. A., & Eberhart, S. A. (1975). Hybrid performance of selected maize lines from reciprocal recurrent selection and testcross selection programs. Crop Sci. 15, 1-4.

[28] Russell, W. A., Eberhart, S. A., & Vega, U. A. (1973). Recurrent selection for specific combining ability for yield in two maize populations. Crop Sci. 13, 257-261.

[29] Smith, L. H. (1936). A discriminant function for plant selection. Ann. Eug. 7, 240-250.

[30] Sprague, G. F., & Eberhart, S. A. (In Press). Corn breeding in G. F. Sprague (ed.) Corn and Corn Improvement. Amer. Soc. Agron.

[31] Subandi, W., Compton, W. A., & Empig, L. T. (1973). Comparison of the efficiencies of selection indices for three traits in two variety crosses of corn. Crop Sci. 13, 184-186.

[32] Suwantaradon, K., Eberhart, S. A., Mock, J. J., Owens, J. C., & Guthrie, W. D. (1975). Index selection for several agronomic traits in the BSSS2 maize population. Crop Sci. 15, 827-833.

[33] Webel, O. D., & Lonnquist, J. H. (1967). An evaluation of modified ear-to-row selection in a population of corn (Zea mays L.). Crop Sci. 7, 651-655.

[34] Young, S. S. Y. (1964). Multi-stage selection for genetic gain. Heredity 19, 131-145.

Single character and index mass selection with random mating in a naturally self-fertilizing species

D. F. Matzinger, C. Clark Cockerham, and E. A. Wernsman
NORTH CAROLINA STATE UNIVERSITY, RALEIGH, NC 27607

INTRODUCTION

Quantitative genetic studies in naturally self-fertilizing species have made little use of recurrent selection procedures which involve random mating. It is rare to find estimates of genetic variances from random mating populations for these species; usually estimates are derived from F_2 and advanced generations arising from hybridization of two homozygous lines. In selection studies, progenies are usually advanced from one generation to the next by self-fertilization rather than by random mating of selected plants or families.

A number of different designs have been utilized to estimate genetic variances in populations of <u>Nicotiana tabacum</u> L., a naturally self-fertilizing species. Initial studies yielding estimates of additive genetic variance (σ_g^2), dominance variance (σ_d^2), and additive x additive epistatic variance (σ_{gg}^2) from the F_2 generation of crosses of pure lines indicated a preponderance of σ_g^2 with little evidence of σ_d^2 or σ_{gg}^2 (e.g. Matzinger, 1968). Further evidence of the importance of σ_g^2 was obtained from diallel crosses of eight tobacco varieties evaluated in the F_1 and F_2 generations with the largest portion of the genetic variance occurring as general combining ability (Matzinger, Mann, and Cockerham, 1962). Estimates of interactions of general and specific combining ability effects with years and locations were small in the diallel study and genotype x environment interactions have been found to be minimal in tobacco variety evaluation studies (Jones, Matzinger, and Collins, 1960).

Earlier studies of recurrent family selection in populations formed by crossing two homozygous varieties yielded selection responses in good agreement with predictions for the direct character under selection and for the correlated characters (e.g. Matzinger, Wernsman, and Cockerham, 1972).

A broad-base tobacco population, Black Shank Synthetic, was created as a source population for genetic studies by intermating the eight varieties evaluated in the diallel. The varieties were intercrossed to form an 8-line hybrid and this was followed for additional generations of random mating by pair crossing 200 plants each generation. Genetic variances were estimated following two generations of random mating and predictions suggested that mass selection should be an efficient selection procedure in this population (Legg, Matzinger, and Mann, 1965). Four cycles of mass recurrent selection for increased yield gave a linear increase of 4.29% per cycle, slightly in excess of the 3.52% gain predicted (Matzinger and Wernsman, 1968). Correlated responses were in reasonable agreement with predictions.

In a number of these studies, many antagonistic genetic correlations occurred, often in the range of -.5 to -.7. Such correlations become troublesome in providing useful populations following single trait selection. The present study was designed to test the utility of a selection index to select in opposition to a genetic correlation.

PROCEDURE

The Black Shank Synthetic population was advanced by six generations of random mating. Full-sib families were evaluated to provide estimates of genetic and environmental variances. Two correlated characters chosen for the selection study were plant height and number of leaves, r_g = .67. Three independent selection studies were initiated for 1) decreased plant height, 2) increased number of leaves, and 3) an index for increased leaves on shorter plants. The purpose of the index was to select in opposition to the genetic correlation.

For each selection criteria, 400 plants were grown in 20 rows of 20 plants per row. Prior data from uniformity studies suggested that the best environmental control would be achieved by practicing within-row selection. Three plants were selected from each row providing a selection intensity of 3/20. The 60 selected plants were pair crossed at random to provide seed for the next cycle the following year.

Index Selection. The procedure for establishing the selection index follows Hazel (1943). The two characters in the index were plant height (X_1) and number of leaves (X_2). The aggregate genotype (H) of a plant is defined as

$$H = a_1 G_1 + a_2 G_2$$

in which a_1 and a_2 are relative economic weights of characters X_1 and X_2 and G_1 and G_2 represent the additive genetic contribution of X_1 and X_2. An estimated index for each plant is of the form

$$I = b_1 X_1 + b_2 X_2.$$

The index weights, b_i, were estimated by

$$b = V_p^{-1} V_g \underline{a}$$

in which V_p is the phenotypic variance-covariance matrix, V_g is the genetic variance-covariance matrix and $\underline{a}$ is the vector of economic weights. Estimates of V_p and V_g were obtained from a study of 160 full-sib families. A number of different methods have been proposed in the literature to obtain estimates of economic weights. In the present case economic weights were obtained by allowing a 1% increase in the mean number of leaves in the population equal to a 1% decrease in plant height. This reduces to $a_1 = -1$ and $a_2 = \bar{X}_1/\bar{X}_2$.

The estimates of V_g obtained from the full-sib families were utilized for all cycles of selection. Estimates of V_p were obtained from each cycle as the experiment progressed. Since selection would be expected to change the ratio of plant height/number of leaves, new estimates of the economic weights were obtained each cycle. The economic weights, a_i, and index weights, b_i, are presented in Table I.

TABLE I

Relative Economic Weights and Index Coefficients In Index Selection

Cycle	a_1	a_2	b_1	b_2
C_0	-1	7.62	-.2346	2.1560
C_1	-1	7.62	-.3015	2.7655
C_2	-1	5.98	-.7280	3.0960
C_3	-1	7.24	-.4770	2.6192
C_4	-1	6.89	-.4029	1.5557

Predicted Response. The expected progress from selection in the single character experiments was

$$\Delta = \frac{k\,\sigma_g^2}{\sigma_p}$$

in which k was the standardized selection differential, σ_g^2 was the additive genetic variance, and σ_p was the phenotypic standard deviation among individuals.

The expected response from index selection was

$$\Delta_H = a_1\Delta_1 + a_2\Delta_2$$

in which Δ_1 and Δ_2 are the expected changes in the two component characters.

$$\Delta_1 = \frac{k(b_1\sigma_{g_1}^2 + b_2\sigma_{g_{12}})}{\sigma_I}$$

$$\Delta_2 = \frac{k(b_1 \sigma_{g_{12}} + b_2 \sigma_{g_2}^2)}{\sigma_I}$$

The standardized selection differential, k, was 1.47 for all experiments. In addition to the full-sib family experiment which provided estimates of V_g in constructing the indexes, two other experiments involving full-sib and self families were available to provide additional estimates of genetic variances and covariances and phenotypic variances in the base population. Therefore, for purposes of predicting gain from selection in all three experiments, pooled estimates of V_g and V_p were utilized.

For index selection, Δ_H, Δ_1, and Δ_2 were predicted for each cycle of selection, utilizing the a_i and b_i actually used in that particular cycle. The b_i were determined utilizing estimates of additive genetic variance from the one full-sib family study only.

Selection Evaluation. Remnant seed of the 30 pair-crosses from each cycle of the three selection studies provided plants to prepare fresh seed for evaluation. All cycles were advanced at random by 30 pair-crosses. Seed from cycles 1 and 2, C_1 and C_2, had poor germination and were discarded. Within each selection study cycles C_0, C_3, C_4, and C_5 were evaluated in 4 replications at each of 2 locations and 2 years. Three plots of C_0 were grown within a replication, each assigned at random to one of the three selection studies. Measurements were obtained on growth characters and chemical constituents in addition to the two index characters.

Selection response was determined as the regression of cycle means on cycle number. The actual selection differential remained constant over cycles for plant height selection and index selection, but increased in later cycles for leaf number selection. As expected, this increase in the selection differential over cycles was accompanied by an increase in the phenotypic standard deviation so that the actual standardized selection differential remained fairly constant.

RESULTS

Combined analyses of variance over the four environments provided evidence of significant differences among cycle means for plant height, number of leaves, the aggregate genotype, yield, days to flower, leaf width, and internode length. No significant differences were obtained for total alkaloids, sugars, and leaf length. Significant estimates of genotype x environment interactions were obtained for the two primary characters, plant height and leaf number; however, the variance components for interactions were only about 10% of the components for cycle means.

Plant Height Selection. Selection for reduced plant height provided a linear decrease of 6.52 cm/cycle, somewhat less than the predicted decrease of 10.19 cm (Table II). Number of leaves decreased .46 leaves/plant/cycle and this was about one-half of the -.95 predicted. Significant correlated response occurred for all of the other characters for which there was a significant predicted correlated response.

TABLE II

Predicted and Observed Response To
Selection For Decreased Plant Height

Characters	Predicted	Observed
Height, cm	$-10.19^{\dagger}$	-6.52^{**}
Leaves, no.	$-.95^{\dagger}$	$-.46^{*}$
Yield, g	$-4.26^{\dagger}$	-4.16^{*}
Flower, days	$-2.16^{\dagger}$	$-.70^{*}$
Alkaloids, %	$.169^{\dagger}$	$.086^{**}$
Sugars, %	$-.660$	$-.044$
L. Length, cm	$-.023^{\dagger}$	$-.012$
L. Width, cm	$-.693^{\dagger}$	$-.993^{*}$
Internode, cm	$-.290^{\dagger}$	$-.213^{*}$

† Variance or covariance used in predicting response was twice its standard error.

*,** Significantly different from zero at 5% and 1% levels, respectively.

Leaf Number Selection. A linear increase of 1.28 leaves/plant/
cycle was in close agreement with the prediction of 1.35 leaves
(Table III). The correlated response of 4.74 cm increased height
was slightly less than the 6.63 cm predicted. A delay in time
of flowering and a decrease in average internode length were in
agreement with predictions. An increase in yield and a narrowing
of the leaves was less than that predicted and was not determined
to be a significant response.

TABLE III

Predicted and Observed Response To
Selection For Increased Number Of Leaves

Characters	Predicted	Observed
Leaves, no.	1.35†	1.28**
Height, cm	6.63†	4.74**
Yield, g	3.23†	1.36
Flower, days	1.89†	2.27**
Alkaloids, %	-.094	.007
Sugars, %	.271	-.121
L. Length, cm	.126	-.045
L. Width, cm	-1.205†	-.841
Internode, cm	-.253†	-1.74**

† Variance or covariance used in predicting response was twice
its standard error.

** Significantly different from zero at 1% level.

Index Selection. Predicted response in H and in the two characters in the index are presented in Table IV. The average
prediction over five cycles of selection was a decrease in plant
height of 4.26 cm and an increase of .212 leaves per plant. The
observed response per cycle was a decrease of 2.00 cm in height
and an increase of .541 leaves per plant. The observed changes
in the two index characters were weighted by the average economic weights of $a_1 = -1$, and $a_2 = 7.07$ to obtain an observed

selection response in H of 5.83 compared with the average prediction from the five cycles of 5.94.

TABLE IV

Average Predicted and Observed Response
Of H and Component Characters In Index

	Δ_H	Δ_1	Δ_2
Predicted	5.94	-4.26	.212
Observed	5.83**	-2.00*	.541*

*, ** Significantly different from zero at 5% and 1% levels, respectively.

Few significant correlated changes in other characters were obtained following index selection (Table V). The leaves became narrower and the average internode length was decreased.

TABLE V

Predicted and Observed Response Of
Correlated Characters Following Index Selection

Characters	Predicted	Observed
Yield, g	-1.41	-2.00
Flower, days	-.500	.080
Alkaloids, %	.084	.036
Sugars, %	-.406	.248
L. Length, cm	.082	-.150
L. Width, cm	-1.66	-1.85*
Internode, cm	-.486	-.285**

*, ** Significantly different from zero at 5% and 1% levels, respectively.

Comparisons of the three selection studies can be visualized from figures showing actual selection response. In Figure 1, selection for decreased plant height, increased

SELECTION WITH RANDOM MATING IN A SELFING SPECIES 511

number of leaves, and index selection all produced an increase in H with index selection providing the greatest change. Figures 2 and 3, for plant height and number of leaves, respectively, allow comparisons of direct selection and index selection of the two characters in the index. Selection for shorter plants produced plants with fewer leaves whereas selection for increased leaf number also increased plant height. Index selection led to increased leaf number on shorter plants. The percentage change in these characters is presented in Table VI. Direct selection for shorter plants reduced height 4.9%/cycle and selection for leaf number increased leaves 7.0%/cycle. Correlated response of each of these characters was about one-half that obtained from direct selection. Index selection reduced plant height 31% of the reduction obtained from selecting only for reduced height and increased leaf number 42% of that obtained from selecting solely for increased number of leaves.

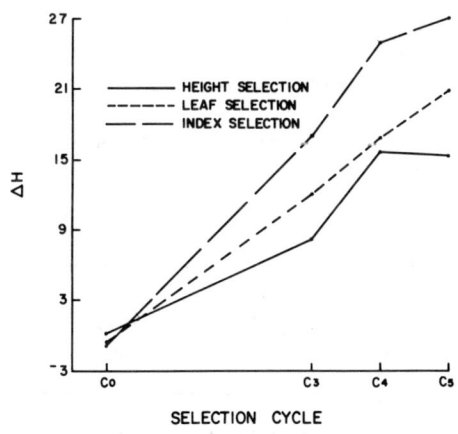

FIG. 1

Cycle Means for Aggregate Genotype (H) following 3 Types of Selection.

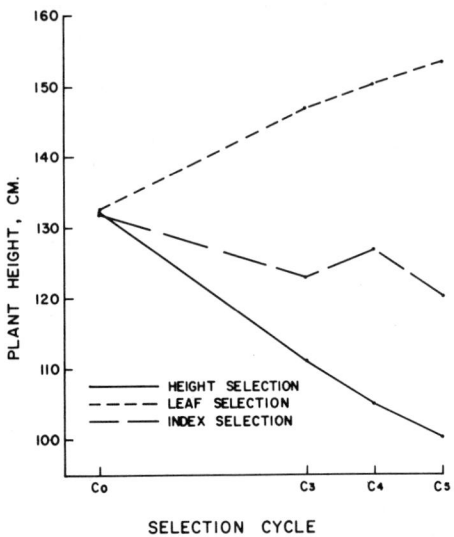

FIG. 2

Cycle Means for Plant Height following 3 Types of Selection.

TABLE VI

Comparative Selection Response In Plant Height
and Number of Leaves Following Direct and Index Selection

| Selection | Height | | Leaves | |
Criteria	cm/cycle	% of C_0	no./cycle	% of C_0
Short Plants	-6.52**	-4.9	-.46*	-2.5
Increased Leaves	4.74**	3.6	1.28**	7.0
Index	-2.00*	-1.5	.54*	2.9

*, ** Significantly different from zero at 5% and 1% levels, respectively.

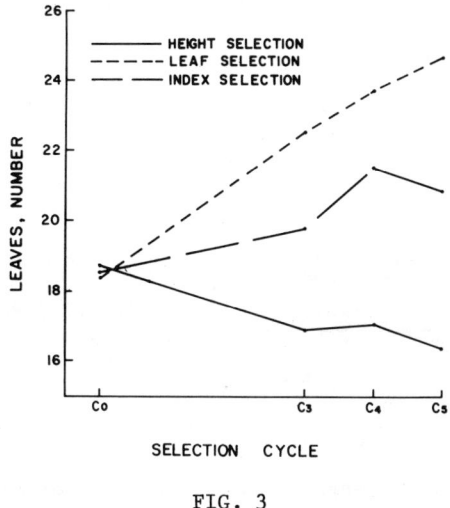

FIG. 3

Cycle Means for Number of Leaves following 3 Types of Selection.

Internode length was reduced in all three selection procedures (Figure 4). The greatest decrease was obtained from index selection since increased leaves on shorter plants would be a form of selection expected to decrease average internode length which was estimated by dividing the plant height by the number of leaves. This is the same value utilized as a_2 for the economic weight.

Comparisons of individual plant phenotypic variances for each cycle and each selection study are presented in Table VII. There appears to be no general change in the variances and covariance of plant height and number of leaves in the index population. There was a correlation between the means and phenotypic variance in the selection line for shorter plants and the line for increased leaf number.

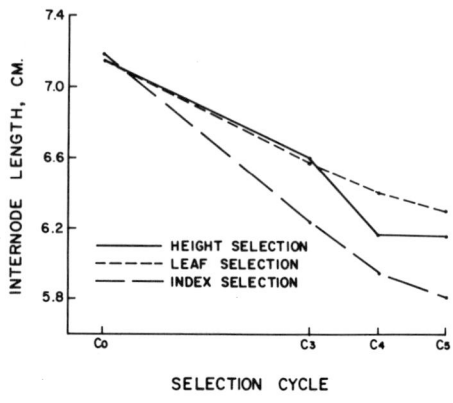

FIG. 4

Cycle Means for Internode Length following 3 Types of Selection.

TABLE VII

Individual Plant Phenotypic Variances In Each Cycle

Cycle	Index Selection			Height Selection	Leaves Selection
	Height	Leaves	Cov.	Height	Leaves
C_0	337.04	8.043	34.151	377.04	8.043
C_1	284.79	4.891	8.274	231.56	8.001
C_2	388.34	13.979	55.944	184.35	19.601
C_3	256.41	6.839	18.540	186.42	28.851
C_4	279.29	9.466	18.536	189.92	28.819

DISCUSSION

Response to direct selection for shorter plants in one study and for increased number of leaves in the other single character study were linear over five cycles of selection. In the short height study, the heritability estimated from variances in the base population was 34.2% compared with a realized heritability of 32.2%. Observed gain was less than that predicted because the actual selection differential was less than predicted using the base population phenotypic variance. The agreement between predicted and observed response for selection of increased number of leaves was excellent but the realized heritability of 19.2% was considerably less than 31.7% estimated from the base population. In this case the actual selection differential increased in later cycles because of increasing phenotypic variance. The realized genetic correlation between plant height and number of leaves estimated from the two single trait selection studies was .51 compared with the estimate of .67 in the base population.

With index selection, estimates of predicted Δ_H remained relatively stable as predicted from each cycle of selection and agreed well with the observed response. The realized heritability of the aggregate breeding value was 17.6%. As stated earlier, the economic weights were allowed to change each cycle since a_2 = (plant height/number of leaves) and the index was constructed to change this ratio. If one used the average of a_i for the five cycles for purposes of predicting change in H, a_1 = -1 and a_2 = 7.07, the predicted Δ_H was 5.76 compared with 5.94 using a_i from each cycle. The observed change of 5.83 was intermediate and in close agreement with both predictions. The relative predicted response of the two characters in the index changed from C_0 to C_4. Predictions in earlier cycles were in closer agreement with observed response. The use of equating percentage of the mean changes of two index characters to provide economic weights is only one of the several methods of weighting. Pesek and Baker (1969) developed the procedure of using desired genetic gains for each trait in the index. Moll, Stuber, and Hanson (1975) applied this concept for selecting against correlated traits of yield and ear height. Lack of close agreement between predicted and observed response was attributed to a nonlinear genetic association between the two traits.

Recurrent selection procedures with random mating appear to offer a tremendous potential for the improvement of populations of self-fertilizing species. The realized per cycle response of 4.9% for plant height and 7.0% for number of leaves would usually be considered respectable progress for any selection system. Index selection with mass selection was also successful in the joint improvement of traits in opposition to their genetic correlation. In the present studies, these gains were realized with an experiment that required an extremely small investment. The experiment was purposely kept small and simple to test the ability to make progress with minimum effort. Only 400 plants were evaluated in each study. Characters were chosen which could be evaluated during the growing season in time to make crosses on selected plants. Plants were selected within blocks (rows) to minimize environmental variability, similar to the procedure of Gardner (1961). Since both male and female parents were selected for intercrossing the total additive genetic variance was utilized.

Genotype x environment interaction throughout the five cycles of selection must not have been of great importance. Single plant selection provides the least sampling of environments possible in a given year and each cycle was in a new environment the following year. The final comparison of all cycles in four environments yielded only slight evidence of interaction and selection responses were linear over cycles.

Estimates of genetic variances have not been obtained from the selected populations. Estimates of phenotypic variances have not decreased except in the population selected for short plants. The continued linear response to selection suggests that genetic variability continues to remain in all three selection populations.

In many naturally self-fertilizing species there is great difficulty in using recurrent selection procedures which involve intermating because of large labor requirements to get hybrid seed. Data from tobacco selection experiments would suggest that some additional effort at this stage of a selection program may in fact be more economical over a number of generations. There appears to be a considerable release of genetic variability and recurrent selection does not rapidly reduce the variability for a number of cycles, leaving useful variation for further selection.

ACKNOWLEDGMENTS

Paper No. 5105 of the Journal Series of the North Carolina Agricultural Experiment Station, Raleigh, North Carolina. This investigation was supported in part by NIH Research Grant No. GM 11546 from the National Institute of General Medical Sciences.

BIBLIOGRAPHY

[1] Gardner, C. O. (1961). An evaluation of effects of mass selection and seed irradiation with thermal neutrons on yield of corn. Crop Sci. 1, 241-45.

[2] Hazel, L. N. (1943). The genetic basis for constructing selection indexes. Genetics 28, 476-90.

[3] Jones, G. L., Matzinger, D. F. & Collins, W. K. (1960). A comparison of flue-cured tobacco varieties repeated over locations and years with implications on optimum plot allocation. Agron. J. 52, 195-99.

[4] Legg, P. D., Matzinger, D. F. & Mann, T. J. (1965). Genetic variation and covariation in a Nicotiana tabacum L. synthetic two generations after synthesis. Crop Sci. 5, 30-3.

[5] Matzinger, D. F. (1968). Genetic variability in flue-cured varieties of Nicotiana tabacum L. III. SC58 X Dixie Bright 244. Crop Sci. 8, 732-5.

[6] Matzinger, D. F., Mann, T. J. & Cockerham, C. C. (1962). Diallel crosses in Nicotiana tabacum. Crop Sci. 2, 383-6.

[7] Matzinger, D. F. & Wernsman, E. A. (1968). Four cycles of mass selection in a synthetic variety of an autogamous species Nicotiana tabacum L. Crop Sci. 8, 239-43.

[8] Matzinger, D. F., Wernsman, E. A. & Cockerham, C. C. (1972). Recurrent family selection and correlated response in Nicotiana tabacum L. I. 'Dixie Bright 244' X 'Coker 139'. Crop Sci. 12, 40-3.

[9] Moll, R. H., Stuber, C. W. & Hanson, W. D. (1975). Correlated responses and responses to index selection involving yield and ear height of maize. Crop Sci. 15, 243-8.

[10] Pesek, J. & Baker, R. J. (1969). Desired improvement in relation to selection indices. Can. J. Plant Sci. 49, 803-4.

An experimental check of quantitative genetics on an autotetraploid plant, *Medicago sativa*, L., with special reference to the identity by descent relationship

A. Gallais
STATION D'AMÉLIORATION DES PLANTES FOURRAGÈRES
INSTITUT NATIONAL DE LA RECHERCHE AGRONOMIQUE
86600 LUSIGNAN, FRANCE

INTRODUCTION

In the theory of quantitative genetics and in that of breeding methods, the effects of systems of mating on means and variances (or covariances) are important to consider. Since the work of Fisher (1918) the theory on the effects of systems of mating has been considerably developed. However there are few experimental tests of this theory on a large scale. In Drosophila we can cite the experiments of Taneja (1963) and Kidwell (1966) devoted to verifying the usefulness of inbreeding before crossing in increasing the covariances between half and full sibs. In cross-fertilized plants, for which to develop hybrid varieties it is classical to use inbreeding before crossing, we are not aware of large experiments, to verify that inbreeding always does increase both the variance among the inbred families and the variance among crosses. Clearly the theoretical conclusions rest upon certain genetic assumptions not always satisfied. However they are sufficiently important to the strategy of plant or animal breeding to justify large experiments.

The lack of comprehensive experiments may be due to the fact that the development of a general theory for inbred diploid populations is still relatively recent, dating from the work of Gillois (1964) and independently that of Harris (1964). For autotetraploids, a generalization was made by Bouffette (1966), and Gallais (1967, 1968) has given some theoretical results on the effects of inbreeding and crossing on the means and the variances among different types of families.

To test the theory for autotetraploids we have experimentally studied the system of inbreeding followed by crossing as reflected in means, variances and covariances. In the interpretation of the results we will show the usefulness of several identity coefficients of rank 1 (for one zygote) and for rank 2 (for two zygotes). We give here only the preliminary results.

BRIEF REVIEW OF THE THEORY

General Formulation

In appendix 1 we give the states of identity between the four genes of a zygote, and all possible states of identity between 2 zygotes for 2 to 8 homologous genes. Following the general approach reviewed in Gallais (1976 b) we derive the expressions of the mean and covariance.

The complete one locus model leads to four parameters for the expression of the mean and to 61 parameters for the expression of the covariance. For the latter, restricting the interactions between alleles to the 1st degree there are only 5 parameters, the same as for diploids, and 18 parameters with interactions restricted to the second degree, i.e., without interactions between four alleles. With biallelism the complete one locus model leads to 32 parameters and with gene frequencies of one half there are only five. With the complete model, the covariances between zygotes from crosses of inbred unrelated plants involved only 26 parameters.

EXPERIMENTAL CHECK OF THEORY IN AN AUTOTETRAPLOID

The expression of the mean shows that a deviation from linearity with the coefficient of inbreeding will mean that there is interaction between more than two alleles and/or that there is epistasis. However a linear relationship cannot be interpreted as indicating the absence of such effects.

Application to the System of Inbreeding Followed by Crossing

From the general approach Gallais (1968) has given some theoretical results in terms of genetic effects for means and variances under self-fertilization and crossing after selfing. We summarize here some original results of interest for our experimental test (see also Gallais, 1970, 1975):

The variance among independent inbred families, measured by the covariance (cov n-1; n, n) - according to the notation of Horner (1956) - can increase first and then decrease, with advanced generations of self-fertilization (n) if the variances of interactions between non-identical alleles are greater than the variances of interactions between identical alleles.

The more inbred the parents, the weaker will be the single crosses.

The variance among the crosses of two unrelated plants can increase or decrease with the level of inbreeding of the parents. It will decrease when the variance associated with more than 2 alleles is important relative to additive and digenic variances.

THE EXPERIMENT

Principles

From a population of a natural cross-fertilized autotetraploid plant, Medicago sativa, L., we have studied in two steps the effect on means and covariances between relatives of inbreeding and of crossing after different numbers of generations of self-fertilization.

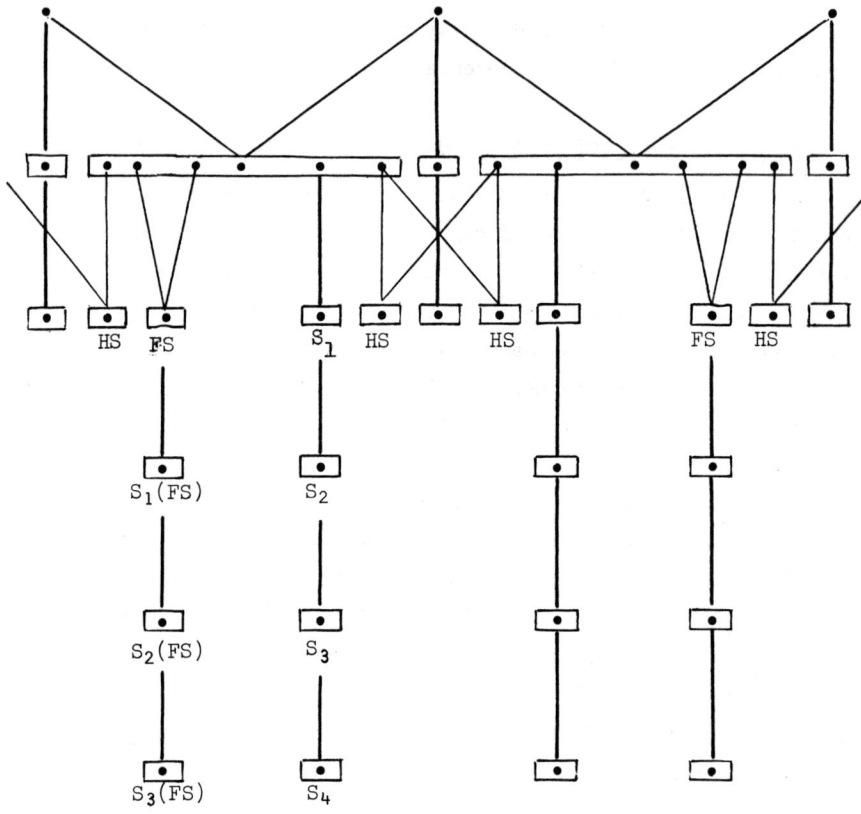

FIG. 1

The mating design used in producing 10 levels of inbreeding and 130 degrees of kinship. The families are represented by a rectangle. HS and FS are respectively progenies from half and full-sib mating. S_1, S_2... are generations of selfing from S_0 plants. S_i(FS) means generations of selfing from a FS plant.

For the inbreeding phase we developed without conscious selection, 10 levels of inbreeding (13 with the crosses) and 130 degrees of kindship according to the genealogies diagrammed in Figure 1. This plan applied to 4 groups of 22 plants crossed in a cyclic manner allows a great number of repetitions for each kinship state.

For the hybridization phase different crossing designs were used. These were in the notation of Cockerham (1963):
- $S_0 \times S_0$: A/B and (A/B) (C/D), which gives 6 covariances
- $S_1 \times S_1$, $S_2 \times S_2$, $S_3 \times S_3$: A/B, which gives 14 covariances (2 for each level and 8 between levels, due to the connection by selfing between the A and the B of the different levels).

The experiment was laid out in the spring in spaced plants (without competition) with 3 or 4 repetitions of 10 plants for each family. We give here results for the total green matter yield on 4 or 5 cuts on two years (the seedling year and the next) and for the height of the plants measured in the seedling year for one or two regrowths.

Statistical Interpretation of the Results

1. We fitted several linear models, $Y = b X$, in which the dependent variate Y was the mean or the covariance and the independent variates X the coefficients of identity of rank 1 or rank 2. The parameters b are genetic components of the mean or of the variance. Regression models $Y = a + b X$ were also fitted. For the covariances a step-wise regression was used to choose among all coefficients a subset which would explain the observed results.

The unweighted least square procedure was employed in our first approach. The assumption of homogeneity of variances of all entries was perhaps not satisfied due to the simultaneous use of mean squares and of cross products. As mentioned by Chi et al (1969) weighting of the mean squares by $\sqrt{.5}$ would make the variances more nearly equal. The examination of residues tends to show that the assumption of homogeneity of variance could be accepted. However the least squares estimates were not minimum variance since some covariances between mean products involved the same branch of a family.

Another important difficulty of the least squares analysis in our experiment is that some coefficients of identity are highly correlated. This leads to high variances of some regression coefficients and to some instabilities of the estimates. This problem was discussed by Marquardt (1970) who proposed ridge regression. The ridge estimator has a smaller variance than the least squares estimator, but it is biased.

2. As many of the coefficients are highly correlated we have tried to reduce the set of coefficients by a principal component analysis, using the first four principal components in the regression models (orthogonal regression, Tomassone, 1968). The meaning of the principal components is related to the metric used in the space of the kinships.

We also computed the analysis of correspondence (Benzécri, 1973, Hill, 1973, 1974) to see if for the kinships of the experiment, all the coefficients of identity of rank 2, between the 8 homologous genes, are necessary. The method established some correspondence between the relationships and the states of rank 2.

PRELIMINARY RESULTS

Means

There is a strong effect of inbreeding on yield and a lesser one on height (Table I). Both characters tend to show a curvilinear relationship with the coefficient of inbreeding F. The regression model with only F explains 96.8% of the variation among the 13 means, but deviations from the model are significant.

For yield, an extrapolation for values of F towards 1 shows a necessary curvilinear relationship which cannot be explained with the model M2 restricting interactions between alleles to the first order. The complete one-locus model explained 99.8% of the variation and reduces the deviations significantly, and these deviations are then just at the limit of significance (Table II).

The estimation of the values of the five states of identity show a high advantage in yield of the tri- and tetragenic states in comparison to the others. This was less pronounced for height. Due to strong correlations between the coefficients of identity of the nulliplex, simplex and duplex states, the precision of the associated estimates was low (Table III), and the estimates very susceptible to a small change in the entries. However, the estimates obtained have a clear biological meaning.

For the crosses it is clearly verified that the more inbred the parents, the weaker is the progeny (Table IV). This is only possible for an autopolyploid species, and was already verified by Demarly (1964), Gallais and Guy (1971), Busbice (1972), Hill (1975). The deviations between observed values and values calculated with the F model show the probable role of interactions between more than two alleles.

Variances

The variances among inbred families (cov n-1; n,n) increase first and then tend to decrease with the level of inbreeding (Table I). This is more pronounced for yield than for height, and is suppressed by a logarithmic transformation. However the statistical basis of this transformation may be questioned.

The general combining ability variances (covariances between half sibs) do not increase regularly with the level of inbreeding of the parents of the crosses, but increase from S_0 to S_1 and then tend to decrease or remain stable. The same trend is observed for the covariances between full sibs.

TABLE I

Effects of Different Levels of Inbreeding on the Means
and on the Variances among Family Means

F and ϕ are respectively the coefficient of inbreeding and the classical coefficient of kinship.

Kinship	F	ϕ	Yield		Height	
			Means	Variances	Means	Variances
$S_0 \times S_0$	0.	0.125	100	412	100	305
HS	0.042	0.156	92.2	486	94.3	240
FS	0.083	0.187	87.5	559	92.2	330
S_1	0.167	0.250	68.8	548	81.2	457
$S_1(FS)$	0.236	0.312	56.5	673	72	674
S_2	0.306	0.375	40.7	504	59.4	600
$S_2(FS)$	0.363	0.427	36.8	450	56.1	692
S_3	0.422	0.479	24.9	413	48.3	490
$S_3(FS)$	0.469	0.522	21*	420	48.3	250
S_4	0.518	0.566	18*	319	47.8	320

*adjusted value for selection effect of a decrease in the number of families due to self sterility.

TABLE II

Analysis of Variance of Deviations for the Complete One-Locus Model (M 4) and for the Model with Interactions restricted to the First Order (M 2). Yield Data.

Source of variation	F observed	F (0.01)
Deviations from M 2	33.1	2.64
Deviations from M 4	3.3	3.01
M 4 - M 2	8.2	3.78

EXPERIMENTAL CHECK OF THEORY IN AN AUTOTETRAPLOID

TABLE III

Estimates of the Values of the five States of Identity for a Zygote

The precision was calculated taking the deviation from the model as residual.

State of identity	Yield	Height
i j k l	100 ± 4	100 ± 1.6
i i j k	77 ± 13.4	84 ± 5.2
i i j j	N.S.	56 ± 19.6
i i i j	N.S.	20 ± 16.8
i i i i	N.S.	35 ± 18.4

N.S. Nonsignificant estimates

TABLE IV

Effects of Inbreeding of the Parents on the Performance of Crosses and on the Estimates of cov HS and cov FS

Crosses	F	Yield			Height		
		Mean	cov HS	cov FS	Mean	cov HS	cov FS
$S_0 \times S_0$	0.	100	138	206	100	116	300
$S_1 \times S_1$	0.055	93	173	266	94.4	243	347
$S_2 \times S_2$	0.102	85	113	381	87.5	60	210
$S_3 \times S_3$	0.141	73	137	178	83.6	125	215

Such observed results correspond to those theoretically predicted for some genetic situations. They show that the classical coefficient of kinship will not be sufficient to explain the change in the second degree statistics.

Interpretation of the Covariances for the Inbreeding Phase

When we used all coefficients of identity of rank 2 from

order 2 to order 8, it appears that several coefficients have a higher "correlation" with the observed results than the classical coefficient of kinship (Table V). For yield and height this coefficient explains about 40% of the variation and the coefficient of the variance of interactions between 3 alleles explained 81 and 60% of the variation, for yield and height respectively. The difference is highly significant.

TABLE V

Correlations between Covariances and some Coefficients of Identity for three Characters for the Inbreeding Phase

States of identity	Yield	Height	Log-Yield
i|i	0.64	0.64	0.83
ij|ij	0.74	0.70	0.85
ii|iii	0.35	0.28	0.78
iij|iij	0.55	0.53	0.90
ijk|ijk	0.90	0.77	0.62
iii|iii	0.32	0.20	0.73
ijkl|ijkl	0.66	0.40	0.12

It may be mentioned that the logarithm of the yield is well explained by the classical coefficient of kinship. Thus, as expected if yield is a multiplicative character, the logarithmic transformation increases the role of additivity. It mainly increases the weight of coefficients of identity between two consanguineous zygotes highly correlated with the coefficient of kinship.

The model with additivity and interactions between alleles restricted to the first order fits our data very well. $R^2 = 0.86$, 0.76, and 0.67 respectively for yield, Log-yield and height. Least squares estimates of the components of variance are given in Table VI. They confirm the effect of the Log transformation. The lack of precision is mainly due to very high

TABLE VI

Estimates of the Components of Variance with only Additivity and Digenic Interactions for the Inbreeding Phase

Components	Yield	Height	Log-Yield
16 $E(\alpha_i^2)$	57 ± 23	40 ± 15	N.S.
48 $E(\alpha_i \beta_{ii})$	-309 ± 63	N.S.	N.S.
36 $E(\beta_{ii}^2)$	223 ± 63	N.S.	182 ± 102
36 $E(\beta_{ij}^2)$	146 ± 32	36 ± 24	92 ± 27
36 $[E\beta_{ii}]^2$	16 ± 7	N.S.	134 ± 95

N.S. Nonsignificant estimates

"correlations" between the coefficients involved (Table VII). A small change in the entries can then lead to very different estimates of the components, in spite of the high coefficient of determination of the model. This is the field of application of ridge regression. Another reason for the bad fit can be an inappropriate model. We know from the study of the means that the model M2 with only digenic interactions is not the best. We have therefore fitted a model M3 including interactions between three alleles, which has 18 parameters. It explains 90, 89, 78% of the variation for yield, Log-yield and height respectively. More coherent estimates for the different characters were obtained but with a poor precision for some parameters. This is not only due to the numbers of parameters, but also to their interdependence. The reduction of deviation from the model M2 when M3 is used is highly significant (Table VIII). Due to the restricted size of the central memory of the computer used, the complete one locus model was not fitted. The possible effect of epistasis has not been explored.

With the M2 or M3 models, the predicted values for the variances among inbred families confirm the tendency of change of

TABLE VII

Correlations between some Coefficients of Identity for the Inbreeding Phase

	i|i	i|ii	ii|ii	ij|ij	ijk|ijk
i|ii	0.93				
ii|ii	0.86	0.97			
ij|ij	0.97	0.88	0.81		
ijk|ijk	0.71	0.50	0.41	0.82	
ijkl|ijkl	0.22	0.02	-0.10	0.31	0.69

TABLE VIII

Analysis of Variance of Regression for the M_2 and M_3 Models

Cause of variation	Degrees of freedom	F for Yield	F for Height	F 0.01
M 3	16*	62	24	2.2
M 2	5	184	64	3.2
$M_3 - M_2$	11	6.4	6.3	2.4
Residual M 3	109			

*16 and not 18 due to a quasi-singularity of the matrix of coefficients

observed values with level of inbreeding: first an increase and then a decrease for yield and height, and a regular increase for Log-yield.

The use of coefficients of rank 1 to explain the change of variances among families due to the level of inbreeding tends to show that heterozygous states are more variable than homozygous states. This resembles results predicted in the case of pure overdominance.

EXPERIMENTAL CHECK OF THEORY IN AN AUTOTETRAPLOID 531

Interpretation of Covariances for the Crossing Phase

The comparison of the covariances between half sibs and the covariances between full sibs shows that the specific combining ability variance is not significant whatever the level of inbreeding. There is "additivity" of the gametic effects.

The general regression analysis shows the classical coefficient of kinship to be more important than for the inbreeding phase (R^2 = 0.64 for yield and height, and 0.81 for Log-yield). This means a greater additivity at the level of crosses than at the level of inbred families. Again as with inbreeding the Log transformation increases the additivity and the role of coefficient of identity between two consanguineous zygotes (Table IX). However with non-transformed data such coefficients are less important than for the inbreeding phase, and it appears that the coefficients of identity associated with variances of interactions between 2, 3, and 4 alleles explained the results as do the classical coefficient of kinship (coefficient of the additive variance). This is due to the fact that in our experiment the "correlations" between the classical coefficient of kinship and the others cited are higher under crossing than under inbreeding (Table X). Such inherent "correlations" will reduce the sensitivity of the model in detecting the effect of different kinds of interactions between alleles, i.e., to choose between several models. The same situation will occur with epistasis.

For the covariances between crosses, another difficulty is the restricted number of entries (20). However it was possible to fit the M2 model (Additivity + Digenic interactions). The coefficient of determination (R^2) was 0.81, 0.81, and 0.88 respectively for yield, height, and Log-yield. A step-wise regression for Log-yield isolated a combination of three coefficients associated with states i|ii, i|ijj, ijkl|ijkl which accounted for 93.5% of the variation. Estimates of components

TABLE IX

Correlations between Covariances and some Coefficients of Identity for three Characters for the Crossing Phase

States of Identity	Height	Log-Yield
i|i	0.78	0.90
i|ii	0.49	0.91
ij|ij	0.67	0.86
ii|ijj	0.48	0.90
ijk|ijk	0.68	0.85
ijkk|ijkk	0.42	0.82
ijkl|ijkl	0.72	0.84

TABLE X

Correlations between some Coefficients of Identity in the Crossing Phase

	i|i	i|ii	ii|ii	ij|ij	ijk|ijk
i|ii	0.89				
ii|ii	0.84	0.98			
ij|ij	0.91	0.85	0.81		
ijk|ijk	0.89	0.81	0.76	0.99	
ijkl|ijkl	0.87	0.75	0.68	0.96	0.99

of variance for the M2 model are given in the Table XI. As for the estimates of the inbreeding phase, there is a poor precision, and this is always mainly due to the high correlations among coefficients used in the model. Furthermore, we know from the studies of means and covariances for the inbreeding phase that the M2 model is not the best to describe our data.

EXPERIMENTAL CHECK OF THEORY IN AN AUTOTETRAPLOID 533

TABLE XI

Estimates of the Components of Variance with the M2 model for the Crossing Phase

Component	Height	Log-Yield
$16\ E(\alpha_i^2)$	216 ± 100	198 ± 188
$48\ E(\alpha_i \beta_{ii})$	N.S.	N.S.
$36\ E(\beta_{ii}^2)$	N.S.	N.S.
$36\ E(\beta_{ij}^2)$	N.S.	N.S.
$36\ \{E(\beta_{ii})\}^2$	N.S.	N.S.

The calculated values for the covariances between half sibs and full sibs are in close agreement with observed values, which suggests that the observed trend in their change with parent inbreeding is not due to a great experimental error. Such a trend was not observed by Rotili (1976) who, however, practiced selection during the inbreeding phase. However, his results show no great significant increase in the variances among crosses from $S_2 \times S_2$ to $S_4 \times S_4$, in comparison to the increase from $S_0 \times S_0$ to $S_2 \times S_2$.

Approaches to Reduce the Number of Parameters

The application of principal component analysis shows that for the two sets of data, inbreeding and crossing, only 4 components can explain 95% of the "variance" of the set of coefficients, and that the kinships are near the subspace defined by these components. Due to the limitations of the computer that was used, the entire set of 61 coefficients has not been studied. We have studied only some subsets, and in particular the subset with 18 parameters corresponding to the model M3 (including trigenic interactions). For this set with inbreeding data the three first principal components explain respectively 81%, 10.1% and 5.9% of the total "variance" (or inertia). Furthermore it clearly appears that the second component is a second degree function of the first, which can be considered as a general measure of the kinship well expressed by the classical coefficient of kinship. The variance of additive x additive epistasis will be then confounded with the parameters of the one-locus model.

The first principal components used as new variates lead to a good degree of determination of the observed results (from 0.50 to 0.80). In Table XII we give the results obtained with inbreeding data for a set of 29 coefficients including the 18 parameters of the M3 model, and 11 coefficients associated with tetragenic interactions. The same type of result has also been obtained with the coefficients of the crossing phase. The first two principal components had the same meanings in both sets of data. In our opinion, even if this approach is considered as not very rigorous it shows clearly that only some coefficients are necessary to explain the observed results, and then to predict them.

TABLE XII

Correlations between the first Principal Components of a set of Coefficients of Identity of rank 2 and observed Covariances between Relatives

Component	C_1	C_2	C_3	C_4	R^2
% "variance"	58	24	10	3.4	95.4
Yield	0.604	-0.464	0.422	0.275	0.834
Log-Yield	0.813	-0.156	0.156	0.057	0.719
Height	0.630	-0.226	0.447	-0.012	0.665

EXPERIMENTAL CHECK OF THEORY IN AN AUTOTETRAPLOID 535

The application of correspondence analysis to the entire set of the 66 coefficients of identity of rank 2, order 8, corresponds to the field of applicability of this method, as for a given degree of kinship the sum of the coefficients will be the unity. Such coefficients can be considered as frequencies.

The preliminary results for the inbreeding data show that to have a good explanation (80%) of the distances between the kinships or between the states of identity it is necessary to consider the first six factors. They explain respectively, 23.5, 19.5, 15.2, 9.7, 7.2, 6.2% of the "variation" (inertia). Due to the closeness of eigenvalues, it is difficult to interpret each factor. All states of identity are near to the subspace determined by these factors. Measuring this proximity by the squared cosine ($\cos^2$) of the angle between the vector of a state and the factorial hyperplane, it appears that 45 states have a $\cos^2$ greater than 0.70, and only 5 have a $\cos^2$ less than 0.60, with only 2 inferior to 0.50 (but greater than 0.40). In the same way only 3 kinships among 126 have a $\cos^2$ inferior to 0.50, but superior to 0.40. Again, as with the principal component analysis, it appears that the second factor is a second degree function of the first.

The simultaneous representation on the same graph, the first factorial plane, of the kinships and of the states of identity leads to some obvious results. The less "inbred" kinships can be characterized by states of identity between zygotes with non-identical genes, and the more "inbred" by states with identical genes in each zygote. The intermediate kinships can be characterized by states with two or more identical genes in one zygote and non-identical genes in the other, or identical but independent of those of the first zygote.

The correspondence analysis leads to a greater number of factors than does the principal component analysis. This does

not mean that they are all useful to explain the change in the
covariances between relatives. It seems to us that the interest
of this method is to show that it is possible to characterize
types of relationships by a reduced number of coefficients. This
idea must be developed.

CONCLUSION

This experiment and its analysis show good agreement with
developed theory and justify the interest in theoretical development. The usefulness of the coefficients of identity is shown,
and it appears that a restricted number of coefficients can
explain the observed results.

From a breeding point of view it is confirmed in agreement with theory, that, for autopolyploids, inbreeding does not
necessarily increase the genotypic variance among families or
the variance among single crosses. However problems of scaling
can exist.

From a biological, or genetic, point of view, it may be
mentioned that there coexists a strong inbreeding depression
and a high degree of additivity at the level of crosses. The
model used tends to lead to the conclusion that this would be
due to a kind of pure overdominance. However, this may also
be due to linkage between favorable dominant and unfavorable
recessive genes, i.e., the dominance hypothesis of heterosis
(Gallais, 1975). Such an observation can also be made for
several diploid allogamous species (corn, forage grasses...).

From a statistical and quantitative genetics point of view,
it appears that it is difficult to distinguish different models,
i.e., different genetic explanations of results, and to have a
good precision on the estimates. Furthermore, the estimates are
different according to the genealogies used. These difficulties
are due to the great number of parameters and to their high interdependence. This will also be the case for all experiments

devoted to estimating the variances for different kinds of interactions between genes, with the possible complication of different dependence relationships among genes, e.g., inbreeding panmictic and linkage disequilibrium. Statistical and experimental methods to reduce the number of parameters and their interdependence must be developed. The consequence of these difficulties is that the true nature of the phenomenon will be difficult to get at. The interest for a breeder of a comprehensive analysis may be questioned. What interests him is prediction. From this point of view, which is mainly statistical rather than truly genetic, only some parameters will be useful.

ACKNOWLEDGMENTS

We are greatly indebted to Dr. Millier, Station de Biométrie, Nancy, France, and to his colleagues for the statistical analysis of the experiment. I also thank Mr. H. S. Easton, on leave from Grasslands Division, D.S.I.R., New Zealand for his correction of the English text.

BIBLIOGRAPHY

Benzecri, J.P. & al. (1973). *L'analyse des Données*. II. *L'analyse des Correspondances*. Paris : Dunod.

Bouffette, J. (1966). Expression de la covariance génotypique chez les tétraploïdes. Thèse doctorat 3ème cycle. Fac. Sci., Lyon.

Bouffette, J. (1966). Correlation génétique entre deux zygotes tétraploïdes quelconques. *C.R. Acad. Sc. Paris*, 263, 220-222.

Busbice, T.H., Hill, R.R. Jr & Carnahan, H.L. (1972). Genetics and breeding procedures. *In* Hanson, C.H. (Ed).*Alfalfa Science and Technology*. Am. Soc. of Agron., Madison, Wis., 283-318.

Chi, R.K., Eberhart, S.A. & Penny, L.H. (1969). Covariances among relatives in a maize variety. *Genetics*, 63, 511-520.

Cockerham, C.C. (1963). Estimation of genetic variances. *In*. Hanson, W.D. & Robinson, H.F. (Ed). *Statistical Genetics and Plant Breeding*. Pub. 982. N. A. S., N.R.C., Washington, D.C., 53-94.

Demarly, Y. (1963). Génétique des tétraploïdes et amélioration des plantes. Thèse Fac. Sci. Paris. *Ann. Amélior. Plantes*, 13, 307-400.

Fisher, R. (1918). The correlation between relatives on the supposition of Mendelian inheritance. *Trans. Roy. Soc.*, Edinburgh, 52, 399-433.

Gallais, A. (1967). Moyenne des populations tétraploïdes. *Ann. Amélior. Plantes*, 17, 215-228.

Gallais, A. (1968). Interactions between alleles and their variability in autotetraploid cross-fertilized plant. Consequences for selection. *Proc. 5th Eucarpia Congress*. Milan 312-323. *Genetica Agraria*, 23, 312-323.

Gallais, A. & Guy, P. (1970). Breeding for heterosis in autotetraploïds. *Report of the Meeting of the Eucarpia Fodder Crops Section*. Ed. S.E.I., Versailles, 105-118.

Gallais, A. (1975). The use of heterosis in autotetraploïd cross fertilized plants with some application to lucerne and cocksfoot. *Proc. Meeting of the Eucarpia Fodder Crops Section*. Ed. Nuesch, B., Zurich, 50-56.

Gallais, A. (1976 a). Sur la signification de l'aptitude générale à la combinaison. *Ann. Amélior. Plantes*, 26, 1-13.

Gallais, A. (1976 b). A general approach to dependence relationships among genes with some applications. International Conference on Quantitative Genetics. Iowa State University.

Gillois, M. (1964). La relation d'identité en génétique. Thèse Fac. Sci. Paris.

Harris, D.L. (1964). Genotypic covariances between inbred relatives. *Genetics*, 50, 1319-1348.

Hill, M.O. (1973). Reciprocal averaging : an eigenvector method of ordination. *J. Ecol.*, 61, 237-249.

Hill. M.O. (1974). Correspondence Analysis : a neglected multivariate method. *Appl. Stat.*, 23, 340-354.

Hill, R.R., Jr. (1975). Parental inbreeding and performance of alfalfa single-crosses. *Crop Sci.*, 15, 373-375.

Horner, T.W. (1956). Theoretical and experimental study of self fertilized populations. *Biometrics*, 12, 404-414.

Kidwell, F.F. & Kempthorne, O. (1966). An experimental test of quantitative genetic theory. *Der Züchter, Genet. Breed. Res.*, 36, 163-167.

Marquardt, D.W. (1970). Generalized inverses, ridge regression, biased linear estimation, and non linear estimation. *Technometrics*, 12, 3, 591-613.

Rotili, P.(1976). Performance of diallel crosses and second generation synthetics of alfalfa derived from partly inbred parents. I. Forage yield. *Crop Sci.* 16, 247-251.

Taneja, G.C. & Negi, S. (1963). Estimates of covariances between full sibs and between half sibs for bristle number and wing length in samples of *Drosophila* bred to varying levels of inbreeding. *Jour. Genetics*, 58, 347-357.

Tomassone, R. (1967). Une méthode d'investigation : la regression orthogonale. *Ann. Sci. forest.*, 24, 233-258.

Appendix

States of Identity of Rank 2, for 2 to 8 homologous genes, such that there is no gene independent of the others

1	i|i	16	iij|iij	31	iii|jjjj	46	ijjj|ikkk
2	i|ii	17	ijk|ijk	32	iii|jjkk	47	iijj|ijjj
3	ii|ii	18	iii|jjj	33	iii|ijjj	48	iijj|ijkk
4	ij|ij	19	iii|ijj	34	iii|iijj	49	ijjk|iijk
5	ii|jj	20	iij|ijj	35	iij|jjjj	50	iiii|iijj
6	i|iii	21	iij|jkk	36	iij|iiij	51	iijj|iikk
7	i|ijj	22	ii|iiii	37	ijj|iikk	52	iiij|ijkk
8	ii|iii	23	ii|jjjj	38	ijj|ikkk	53	ijkk|ijkk
9	ii|jjj	24	ii|jjkk	39	ijk|ijkk	54	iiii|jjjj
10	ii|ijj	25	ii|iijj	40	iij|ijkk	55	iiij|ijjj
11	ij|ijj	26	ii|ijjj	41	ijj|iijj	56	iijj|iijj
12	i|iiii	27	ij|ijjj	42	ijj|iiij	57	iiii|ijjj
13	i|iijj	28	ij|iijj	43	iijj|kkll	58	iiij|iijj
14	i|ijjj	29	ij|ijkk	44	ijkl|ijkl	59	iiii|iijj
15	iii|iii	30	iii|iiii	45	ijkk|ijll	60	ijjj|ijjj
						61	iiii|iiii

VII

Results and Theory with Animals

Theoretical and actual genetic progress in dairy cattle

L. D. Van Vleck
DEPARTMENT OF ANIMAL SCIENCE
CORNELL UNIVERSITY, ITHACA, NY 14853

INTRODUCTION

Most farm animals have not been very useful for comparing theoretical and actual genetic progress from selection for numerous reasons. Changing selection goals, low selection intensities, variable and inefficient evaluation procedures, and crossbreeding have all contributed to the problem. The dairy cow may be the major exception. Widespread use of artificial insemination has made possible intense selection and theoretically high correlations between predicted and true values as well as data sets suitable for estimating fixed effects and variance components which are necessary for accurate genetic evaluation. In addition, such data sets can be used to monitor genetic change. The dairy cow also has a singular function, to produce milk, although other traits are sometimes considered in selection. Crossbreeding has also not been extensively practiced. Performance of about 25% of cows in the United States is recorded, and a large fraction of these have their sires identified.

The purpose of this paper is to compare theoretical genetic progress for milk yield as computed for rather ideal circumstances with estimates of that actually observed. Evidence and opinions as to the reasons for attaining less gain than that possible with optimum breeding schemes will be presented.

SOME CHARACTERISTICS OF DAIRY CATTLE

The characteristics of performance in the dairy cow although not unique offer both simplification and complexity to the problems of predicting response from selection and in estimating actual response. Naturally, milk yield is limited to the female. Artificial insemination is widely and successfully used to find and disseminate the genetic superiority of the opposite sex, the bull--as many or more than 50,000 services per year. A cow completes a lactation about once a year, beginning the first one in the United States at about two years of age. Repeated records may be both a complication and an advantage. The size of the record is related to age and may also be used in selecting which cows are retained for future lactations. Most parturitions yield single births although up to 5% may result in twins.

Dairymen are also individuals. Some through cooperative testing associations identify their cows and record the production of milk. Others may do neither, which is one of the causes of differences among recommendations as to optimum breeding plans. Should improvement be optimized for the whole population or for the recorded and identified population which supplies the information necessary for genetic progress?

THEORETICAL GAIN

Dickerson and Hazel (1944) first derived an expression for predicting genetic progress per year given the genetic selection differentials and generation intervals for males and females. A modification of this by Rendel and Robertson (1950) or some variation of their formulation has been widely used (Skjervold, 1963,

1966; Skjervold and Langholz, 1965; Specht and McGilliard, 1960; Hunt et al., 1972, 1974; Vaccaro, 1974). Searle (1961), VanVleck (1964), and Miller (1969) have used recursive formulas to predict genetic gain which have partially considered the fact that generations are not discrete but overlap and that progress in the first years of a program is not uniform. Hinks (1970, 1971, 1972) and Bichard et al. (1973) have considered the problem of overlapping generations, and Hill (1972, 1974) has established matrix procedures to predict response in such cases. After the crude attempt of VanVleck (1964) to relate the cost of genetic gain and expected genetic gain, some rather sophisticated discounting procedures have evolved (Hill, 1971; Hinks, 1971; Hunt et al., 1972, 1974; Brascamp, 1973; Cunningham and McClintock, 1974; McClintock and Cunningham, 1974; Vinson and Freeman, 1974).

For the purposes of this discussion, the development of Rendel and Robertson (1950) should be sufficient to describe the four sources of genetic gain and the relative importance of those sources in dairy cattle breeding.

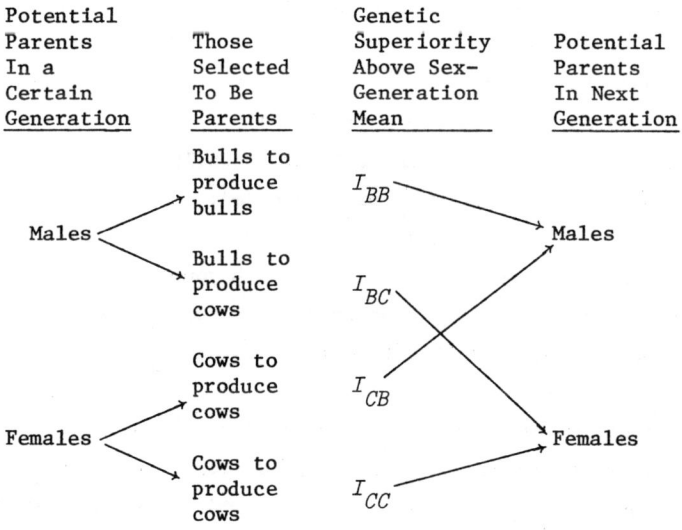

Then,

$$\Delta G/\text{yr} = \frac{I_{BB} + I_{BC} + I_{CB} + I_{CC}}{L_{BB} + L_{BC} + L_{CB} + L_{CC}}$$

where the L's are mean generation intervals for the four groups. As is well known, with the assumption of normality,

$$I_{jk} = r_{\hat{G}_{jk} G_{jk}} i \sigma_G$$

where $r_{\hat{G}G}$ is the correlation between predicted and true genetic value, i is the standardized selection differential corresponding in the normal distribution to z/p where z is the height of the ordinate of the standard normal distribution at the truncation point for selection and p is fraction selected, and σ_G is the genetic standard deviation. The principles of predicting G from records of relatives come from selection index procedures developed by Lush (see, in particular, Lush, 1931, 1933, 1947).

There are many factors which influence the I's and L's which make straightforward optimization difficult. These have been listed in increasing detail by VanVleck (1964), Skjervold and Langholz (1964), Hunt et al. (1972), and Brascamp (1973) and include:

> heritability of the trait (usually about .25 for milk yield);
> population size;
> selection intensity on dams of bulls;
> number of sires of sons replaced annually;
> proportion of cows with milk recorded;
> proportion of recorded cows bred to young bulls;
> number of daughters for testing each young bull;
> number of services to obtain a recorded daughter;
> semen output of tested bulls;
> fertility rate;
> proportion of proved bulls replaced each year.

Predictions of maximum gain have ranged from about 1% per year in a natural service situation with somewhat less than usual generation intervals where all gain comes from I_{CB} and I_{CC} (Rendel and Robertson, 1950), leading them to predict a probable gain of about .6% per year in typical pedigree herds. For small populations in an AI situation, Robertson and Rendel (1950) projected a gain of 1.69% for a 2000 cow unit and 2.05% for a 10,000 cow unit, suggesting somewhat more progress was possible for larger units. Most predictions fall in these ranges. For a 2000 cow unit (all recorded) and an optimum program, the must quoted proportions of gain from the four sources are: I_{CC}, 6%; I_{BC}, 18%; I_{CB}, 33%; I_{BB}, 43%. With many more inseminations now possible to proved bulls than then, the fraction attributable to I_{BC} may be somewhat greater. For example, Schmidt and VanVleck (1974) suggested that with attainable selection intensities that the proportions of gain could be: I_{CC}, 2%; I_{BC}, 26%; I_{CB}, 32%; I_{BB}, 39%.

These calculations show that little improvement (2-6%) would be expected from selection of cows to leave heifers and that most gain would be due to matings to produce young bulls to test in the next generation: $I_{CB} + I_{BB}$, 71-76%.

These calculations make some assumptions which may not be as unrealistic as they first appear.

Expected responses are based on the assumption of normality. Certainly the distribution of milk records approximates the normal distribution except on the tails. Optimum programs, however, would involve selection of the top 1-5% in many instances.

Another assumption is that selection is essentially for milk yield and that selection is based entirely on predicted genetic value for milk yield and by truncation. As will be shown later, other traits besides milk yield often are given more emphasis in practice than milk yield. Selection is almost never of a truncation nature and may vary according to biological necessities.

Theory suggests that expected responses can be projected for only one generation since the variances and distributions will change depending on the intensity of selection (Cochran, 1951). Most projections have assumed no change in variation. All evidence to date suggests that no decrease in variation has occurred in dairy cattle for milk yield. If anything, total variation has increased with increasing mean level of yield (VanVleck, 1966). Perhaps the implication is that measurements on the usual scale do not have an identical distribution throughout the range of measurement. If the scale of measurement is inappropriate, then the ratio of genetic to phenotypic variation (heritability) should not be expected to be the same for all levels of production. Several papers (Mason and Robertson, 1956; Robertson, O'Connor, and Edwards, 1960; VanVleck, 1963; VanVleck and Bradford, 1964) reported a somewhat higher fraction of genetic variance at higher levels of production as estimated from paternal half-sib correlations. Daughter-dam regression did not show the same pattern (Bradford and VanVleck, 1964). Bradford and VanVleck (1964) also reported the regression of daughter on dam to be linear for the range of dam production, which implies expected response with intense selection and with little selection to be in agreement with that expected from a common heritability value.

ACTUAL PROGRESS

There appear to be two important aspects in evaluating genetic progress. The more important from a practical viewpoint may be in how well genetic evaluations predict what will happen. If that agreement is poor, then there is not much justification in recommending such procedures to dairymen. The simplest demonstration is whether the daughter performance is what is predicted. If it is, the program can be pushed with great vigor. Data from several institutions have provided decisive empirical tests of both male and female evaluation. Results reported by Freeman (1976) indicate that evaluation methods predict differences well. Open heifers were selected from Iowa herds for high and low performance from the estimated transmitting ability (one-half additive genetic value) of the dam and the predicted difference (also one-half additive genetic value) of the sire. The expected difference for the heifers was 1810 lb of milk, and the actual difference in the Iowa State University herd was 1856 lb. In Tennessee, Richardson and Bearden (1976) selected bulls to improve production for use with one group of cows and to improve both production and udder conformation with another group of cows. The average difference in predicted difference was 210 kg milk. The daughter production was different by 224 kg. Wilk, Legates, and McDaniel (1976) compared daughters of high predicted difference Jersey bulls with those of control bulls in a North Carolina State University project. The average predicted difference between the two groups of bulls over two generations was 580 kg. In the first generation, the actual daughter difference was 618 kg and in the second generation 1017 kg. These demonstrations illustrate the opportunity for selection using the tools of quantitative genetics and statistics for evaluating bulls and cows. Results in larger populations of animals over a longer time period have not been as dramatic, probably for several reasons. In addition, methods of measuring change are inherently more difficult than differences in one generation of selection.

Until recently, the most widely used method of estimating genetic gain has been that of Smith (1962) or some modification of his method. The method is based on differences in regression coefficients and generally involves a doubling to give equal credit to male and female parents. Numerous estimates appear in the literature and are somewhat variable (e.g., Harville and Henderson, 1967; Burnside and Legates, 1967; Powell and Freeman, 1974). A strong inverse relationship also appears to exist between the genetic and environmental trend estimates. Often an estimate of a high genetic trend is accompanied by an estimate of a small or negative environmental trend.

The most accurate method of estimating genetic trend would be to predict the genetic value of all animals in the population and obtain the average for each year. Earlier evaluation procedures with varying bases and unspecified properties were not suitable. The advent of the best linear unbiased prediction (BLUP) procedures (Henderson, 1966, 1973, 1974, 1975) for sire evaluation made possible the evaluation of bulls used over long periods of time all compared to a constant base. Estimates of genetic trend due to the sire contribution could then be obtained by weighting for each year the sire values by either probable conceptions or number of daughters freshening in recorded herds. Such estimates have been made for several populations: Schaeffer, Freeman, and Burnside (1975) in Ontario; Kennedy and Moxley (1975) in Quebec; Olson and Jensen (1976) in Wisconsin; Powell and Freeman (1974) on Minnesota, Iowa, and Nebraska data; and Everett, Keown, and Clapp (1976) for the northeastern United States. Estimates of genetic trend were made by doubling the trend due to sire contribution.

Extension of the general BLUP method to cow evaluation would allow direct computation of genetic trend. Slanger, Everett, and Henderson (1976) have described a reasonable computing strategy for predicting genetic value of all cows in recorded herds making use of the rapid computation of the inverse of the numerator relationship among all animals as discovered by Henderson (1975, 1976). Hintz, Everett, and VanVleck (1977) have applied this procedure to northeastern United States data. As a byproduct, the trend in natural service sires was also computed. Rather than examine all estimates of sire contribution to genetic trend, only those for the Northeast will be discussed.

Figure 1 shows the overall trends in artificially and naturally sired cows by year of first freshening as well as the sire contribution to each trend. When the sire was unknown, it was assumed to be a natural service sire and one-half of the cow's estimated genetic value was assigned to the natural service average for that year of freshening. Some so-called natural service daughters may have been unidentified AI daughters and some natural

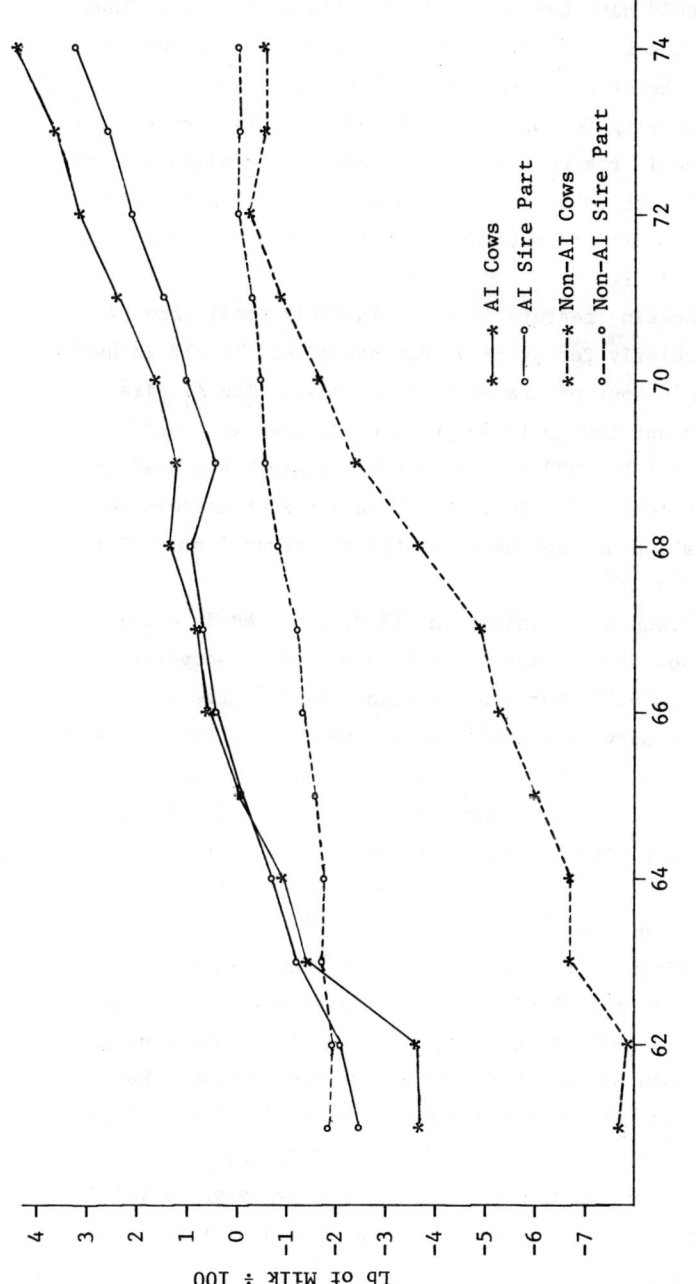

FIG. 1

Genetic Trend of Cows and the Sire Contribution in AI and non-AI Populations

service sires could have been sons of AI bulls or AI dams. Thus, the natural service trends may not be indicative of what would have happened in herds using only natural service sires.

The most striking feature of the trend shown in Figure 1 for the AI population is the reliance of the overall trend on the sire contribution for each year. Doubling the sire trend appears to be of doubtful validity in estimating overall trend. The reason, however, is not clear.

The most shocking feature is the relatively small rate of progress, particularly for an area that has prided itself in having one of the world's most progressive AI programs. The AI sire trend was only about 600 lb in 14 yr, and the overall trend was only a little more than 800 lb. Although the gap between AI and NS sires appears to be widening, the recent rate of improvement in the NS population has not been greatly different from that in the AI population.

Everett (personal communication, 1976) supplied data which allowed comparison of the sire trend from the major cooperative stud and that from all other studs selling semen in the area. These and natural service (non-AI) sire trends are shown in Figure 2. If the merit of all studs had equalled that of the best stud, the average AI daughter would have produced 200-250 lb more per lactation. In fact, the average "natural service" sire used during that period was superior to the average of the AI sires used in other studs. As stated previously, such a result may be due to some artifact of the data. The trend due to sires from the best stud accounted for about 800 lb improvement in 14 yr. Even this if doubled is only about 1% per year and certainly less than the 2% which should be possible. Figure 1 clearly shows, however, that genetic trend has not been twice the trend in sires. Reasons for the disparity need to be determined although the number of generations involved and the selection in the early years may not be great enough to result in genetic trend equalling twice the sire contributions.

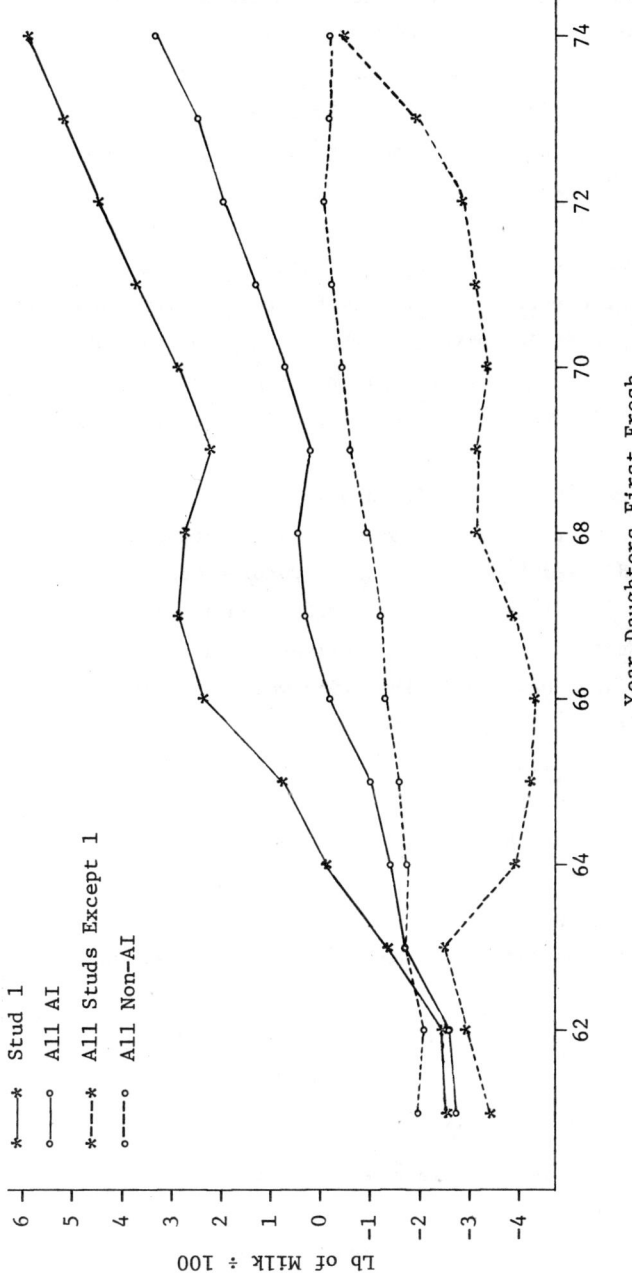

FIG. 2

Comparison of Sire Contribution to Yearly Genetic Value for AI Stud 1, All AI Studs, All AI Studs Except Stud 1, and All non-AI Sires

The average yearly value of the merit of bulls purchased based on pedigree selection and planned mating is shown in Figure 3 for AI bulls in stud 1 in the Northeast (Henderson et al., 1973). If the superiority of bulls selected at this stage is a constant each year, then that trend is also a measure of the improvement in the population. Figure 3 shows, however, that the superiority in initial pedigree selection is not constant. The large jump from 1965-1966 was due to a revision in the stud's selection policy, but once that occurred, the policy had to be maintained to prevent a drop. Any major effect in the whole population due to the gain from pedigree selection would not begin for about 5 yr and then would phase in gradually, depending on how quickly older bulls were culled.

Figure 3 also compares the difference in trends between the mean of young bulls entering stud 1 for sampling and the mean of all AI bulls in stud 1 weighted by number of recorded daughters (the year fresh trend of Figure 2 for all stud 1 bulls plotted 3 yr earlier to correspond to when the services would have occurred). The bulls used heavily in 1960-1965 were much superior to the young bulls sampled in those years. Intense pedigree selection of the young bulls beginning in 1965-1966 made the young bulls essentially equal to the proved bull stud. Selection from those superior young bulls should then begin to increase the overall trend a few years later as appears to be occurring.

Although such trends indicate substantial and permanent progress can and has been made, why have the gains not been more substantial?

REASONS FOR LESS THAN MAXIMUM PROGRESS

Obviously, progress has been less than the maximum possible as predicted from optimum breeding plans. Examination of some of the reasons may increase the chances of making more gain in the

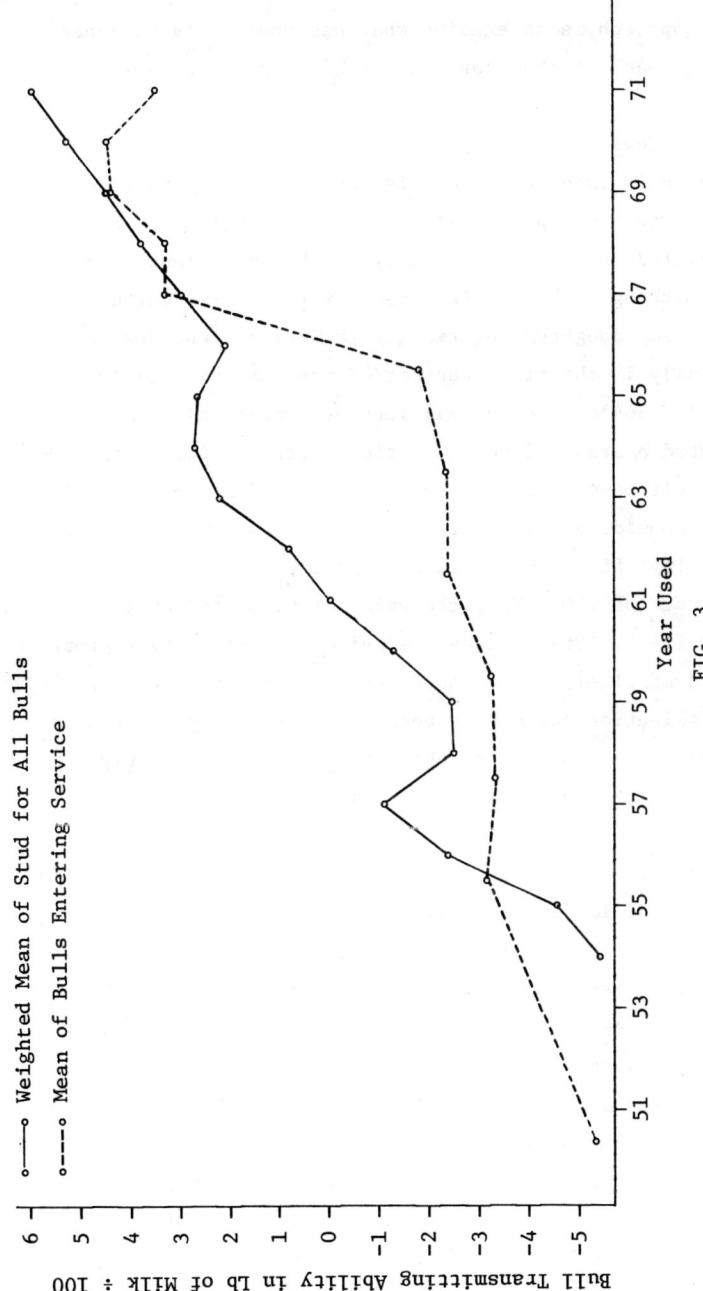

FIG. 3

Comparison Within Stud 1 Between Bulls First Entering Service and the Weighted Mean of All Bulls Used by Year

future. One approach is to examine what has apparently happened in practice for each of the four sources of progress.

Cows to Produce Cows

Little gain is expected from this source despite the efforts of some dairy extension specialists to encourage selection on the basis of estimated transmitting ability or the equivalent estimated average transmitting ability. Pedigree information is quite accurate in predicting daughter superiority (McGilliard and Freeman, 1976), especially if the first records of the dam are used (Deaton and McGilliard, 1964). The calculations of Rendel and Robertson (1950) suggested a standardized selection intensity factor of .49-.63 based on selection of replacements from dams in the top 60-70% of the herd. Schmidt and VanVleck (1974) may have been more realistic in assuming effective selection from the top 90% and a selection factor of .20. Yet, the evidence presented by Allaire and Henderson (1966, 1967) suggests a selection intensity factor for milk yield of .11-.14 even though the factors possible for single trait selection could have been from .36-.51--quite similar to those proposed by Rendel and Robertson (1950). Apparently, many traits other than milk yield are considered in choosing replacement heifers.

The literature also considers the other within-herd selection problem--that of deciding which cows stay in the herd to have another lactation. That decision process, which may concern the dairyman more than genetic selection, may indirectly play a large role in which cows leave replacements. Evidence is clear that in retrospect the terminal lactation receives nearly all the weight in the decision to save or cull a cow (see, e.g., Allaire and Henderson, 1966). Although the biological reasons are not apparent, the relative weights for various traits in making the selection decision were: milk, 1.00; depth of body, .53; dairy character, .53; breeding trouble, .32; depth of udder, .34; strength of fore udder attachment, .34; strength of rear udder attachment,

.25. Berger, Harvey, and Rader (1973) found that for every set of records considered in retrospect at least one type trait received more weight than production in the selection decision.

The obvious conclusion is that the proportion of genetic progress due to cows to produce heifer replacements is much less than 6% or even 2%. Whether more effort should be made in this area is also doubtful because of the relatively small gain which could result. Undoubtedly, some culling of cows for other than production is necessary and not too harmful to overall genetic progress for milk yield.

Bulls to Produce Cows

This path appears to have contributed the most to past progress through the widespread use of proved bulls (Henderson et al., 1973) since this is the largest part of observed trends. This source of progress, however, has been much less important than is possible because of the relatively high fraction of sampled bulls that are returned to service. Vinson and Freeman (1972) as part of a study of the selection practices of seven bull studs between 1960 and 1970 reported that 25% of tested bulls were returned to service but about 30% were lost for various nonselection reasons. Culling for low production accounted for 35% and own and daughter type for another 10% of the young bulls. Thus, between 1 in 2 and 1 in 3 of survivors were returned to service. Most optimum plans would return 1 in 5 to 1 in 20.

The impact of those returned to service after sampling has probably been lessened also by difficulties in genetic evaluation. Daughters of a few popular pedigree bulls may have received special treatment and hence a biased evaluation.

Failure to adjust proofs for mates which are better or poorer than average would also decrease genetic progress because of improper evaluation. Most research, however, shows that the problem of selection of mates is not very important (Slanger, 1972; Hillers and Freeman, 1966; Miller and Corley, 1965).

Although the herdmate and contemporary comparison methods of Henderson and Robertson revolutionized sire evaluation and led to much genetic progress, the methods allowed for no genetic trend and assumed sires of herdmates averaged the same for daughters of all sires. The success of the herdmate evaluation began to create problems in evaluating sires and must have reduced the rate of genetic progress from what would have been predicted. This problem may have been especially critical in selection of sires of sons when the choice was between bulls from many years apart. The new BLUP-based procedures now in effect in the northeastern United States, Canada, and Britain should correct this problem in the future.

Differential semen pricing may also have reduced possible progress for yield by discouraging use of highly evaluated and priced bulls even though pricing on the basis of probable profitability appears justified (Everett, 1975).

The changeover from fresh semen to frozen semen reduced the number of breeding units for the best and most used bulls in the United States for Holsteins and has resulted in the maintenance of larger numbers of proved bulls in the AI studs with a resulting decrease in selection intensity. The large number of bulls now available to dairymen because of frozen semen, no doubt, has also led to more than necessary consideration of traits other than production in the choice of bulls to use.

Cows to Produce Bulls

This source of progress appears important in theory but may have been one of the weak links in the past. A large selection differential is easily possible but has not been attained (Vinson, 1975). VanVleck and Carter (1972) and Freeman (1970) have shown that a cow's records have not predicted her son's genetic value very well although Butcher and Legates (1976) found closer agreement.

VanVleck (1969) found the retrospective index for pedigree selection of young bulls to be 70-90% as efficient as the theoretical index for predicting milk value mostly because of overemphasis on records of their dams.

Records other than for the first lactation on a cow have been especially less useful in prediction than theory would indicate. There has also been strong overemphasis on other traits in selection of bull dams. In retrospect, the index used by the average bull stud was only 74% and 57% as efficient as the theoretical index giving 1:3 and 1:20 emphasis on type and milk yield (Vinson and Freeman, 1972). The reason was primarily due to too much weight for type classification.

The generation interval for cows to produce bulls was 7-8 yr in the study of Vinson and Freeman (1972) as compared to the minimum of 5 yr thought possible by Rendel and Robertson (1950). In general, generation intervals in practice are much larger than optimum. For example, the average age of sires that produced sons was 11.3 yr compared to a theoretical 7 yr. These increases in the sum of generation intervals could easily reduce progress by 30% from the optimum. The major reason for the increased generation intervals would seem to be the desire of bull studs to be very accurate in evaluating bull dams and bull sires before mating them. Unfortunately, the added records on cows that start well in the first lactation and are treated well to help in making them acceptable to bull studs may actually have a negative influence on genetic progress because of the bias due to the preferential treatment in addition to lengthening the generation interval.

Sires to Produce Sons

Theoretical projections point to the importance of this source of genetic progress. Recommendations for most optimum programs suggest using a very limited number of the best proved bulls to sire sons--often only the top 2-4 out of every tested batch (Skjervold, 1966; Hunt et al., 1972). Bull studs have not been that selective as indicated by personal observation and by the long generation interval reported by Vinson and Freeman (1972). Sires of sons in their study were also selected about equally for production and type score of their daughters although the production emphasis was split between fat and milk yield--more on fat yield than milk yield. The retrospective indexes for sires of sons were only 68-84% efficient as compared to the theoretical indexes for 1:3 and 1:20 emphasis on type and production.

The regression coefficient to estimate a young bull's estimated transmitting ability from his pedigree index including sire's proof and dam's pedigree has been .6-.7 (VanVleck and Carter, 1972; Butcher and Legates, 1976). The intercept has been negative, indicating positive genetic trend. No doubt, genetic trend and overlapping generations have partially caused the lower than expected regression. The regression equation, however, does indicate that pedigree information is quite valuable.

CONCLUSIONS

Genetic progress in dairy cattle has been considerable, although much slower than theoretically possible. The results under controlled conditions and with fixed selection goals in institutional herds where results are close to those predicted for cow and bull selection as compared to results under field conditions indicate one major reason why theoretical progress exceeds actual progress. Selection in practice places considerable emphasis on nonproduction traits. Studies of past practices of bull studs also suggest too much reliance on later records of cows which may be biased by preferential treatment and which increase generation intervals substantially. Selection intensities for bulls to produce cows and particularly for bulls to produce sons are not nearly as great in practice as are theoretially optimum.

THEORETICAL AND ACTUAL PROGRESS IN DAIRY CATTLE 561

The unknown and changing base of herdmate evaluations has undoubtedly reduced progress.

New evaluation procedures using BLUP procedures and relationships among animals together with elimination or proper use of later records should lead to more accurate evaluation. In fact, such procedures could be used to estimate yearly trends in the four sources of genetic progress to monitor more adequately what is happening in practice. The solution to the problem of how to increase actual progress to levels closer to optimum will require strict adherence to the basic rules of selection--use of the most accurate methods of genetic evaluation and intense selection of bull dams, bull sires, and bulls to produce heifers with more correct economic emphasis on milk yield and other traits.

BIBLIOGRAPHY

[1] Allaire, F.R. and C.R. Henderson (1966). Selection practiced among dairy cows. I. Single lactation traits. *J. Dairy Sci. 49*, 1426.

[2] Allaire, F.R. and C.R. Henderson (1966). Selection practiced among dairy cows. II. Total production over a sequence of lactations. *J. Dairy Sci. 49*, 1435.

[3] Allaire, F.R. and C.R. Henderson (1967). Selection practiced among dairy cows. III. Type appraisal and lactation traits. *J. Dairy Sci. 50*, 194.

[4] Berger, P.J., W.R. Harvey, and E.R. Rader (1973). Selection for type and production and influence on herdlife of Holstein cows. *J. Dairy Sci. 56*, 805.

[5] Bichard, M., A.H.R. Pease, P.H. Swales, and K. Özkütük (1973). Selection in a population with overlapping generations. *Anim. Prod. 17*, 215-227.

[6] Bradford, G.E. and L.D. VanVleck (1964). Heritability in relation to selection differential in cattle. *Genetics 49*, 819.

[7] Brascamp, E.W. (1973). Model calculations concerning economic optimalization of AI-breeding with cattle. I. The economic value of genetic improvement in milk yield. *Z. Tierz. Züchtungsbiol. 90*, 1-15.

[8] Brascamp, E.W. (1973). Model calculations concerning economic optimalization of AI-breeding with cattle. II. Effect of costs on the optimum breeding plan. *Z. Tierz. Züchtungsbiol. 90*, 126-140.

[9] Brascamp, E.W. (1974). Model calculations concerning economic optimalization of AI-breeding with cattle. III. Profitability of performance testing in a dual-purpose breed according to meat production and the effect of beef crossing. *Z. Tierz. Züchtungsbiol. 91*, 176-187.

[10] Burnside, E.B. and J.E. Legates (1967). Estimation of genetic trend in dairy cattle populations. *J. Dairy Sci. 50*, 1448.

[11] Butcher, K.R. and J.E. Legates (1976). Estimating son's progeny test from his pedigree information. *J. Dairy Sci. 59*, 137.

[12] Cochran, W.G. (1951). Improvement by selection. *Proc. 2nd Berkeley Symp. Math. Statist. and Prob.*, 449.

[13] Cunningham, E.P. and A.E. McClintock (1974). Selection in dual-purpose cattle populations. Effect of beef crossing and cow replacement rates. *Ann. Génét. Sél. Anim. 6*, 227.

[14] Deaton, O.W. and L.D. McGilliard (1964). First, second, and third records of a cow to estimate superiority of her daughters. *J. Dairy Sci. 47*, 1004.

[15] Dickerson, G.E. and L.N. Hazel (1944). Effectiveness of selection on progeny performance as a supplement to earlier culling in livestock. *J. Agr. Res. 49*, 459-476.

[16] Everett, R.W. (1975). Income over investment in semen. *J. Dairy Sci. 58*, 1717.

[17] Everett, R.W., J.F. Keown, and E.E. Clapp (1976). Production and stayability trends in dairy cattle. *J. Dairy Sci. 59*, in press.

[18] Freeman, A.E. (1976). Recommended genetics management program. *Proc. Large Herd Symp.* Gainesville: Univ. of Florida Press.

[19] Freeman, M.G. (1970). What has been realized from pedigree selection of dairy bulls. Paper presented at the 65th Ann. Meeting of the ADSA, Univ. of Florida, Gainesville.

[20] Harville, D.A. and C.R. Henderson (1967). Environmental and genetic trends in production and their effects on sire evaluation. *J. Dairy Sci. 50*, 870-875.

[21] Henderson, C.R. (1966). A sire evaluation method which accounts for unknown genetic and environmental trends, herd differences, season, age effects, and differential culling. *Proc. of Symp. on Estimating Breeding Values of Dairy Sires and Cows.* Washington, D.C.

[22] Henderson, C.R. (1973). Sire evaluation and genetic trend. *Proc. of the Anim. Breeding and Genet. Symp. in Honor of Dr. Jay L. Lush.* July 29, 1972, Blacksburg, Va.

[23] Henderson, C.R. (1974). General flexibility of linear model techniques for sire evaluation. *J. Dairy Sci. 57*, 963.

[24] Henderson, C.R. (1975). Best linear unbiased estimation and prediction under a selection model. *Biometrics 31*, 68.

[25] Henderson, C.R. (1975). Use of relationships among sires to increase accuracy of sire evaluation. *J. Dairy Sci. 58*, 1731-8.

[26] Henderson, C.R. (1975). Use of all relatives in intraherd prediction of breeding values and producing abilities. *J. Dairy Sci. 58*, 1910.

[27] Henderson, C.R. (1976). A simple method for computing the inverse of a numerator relationship matrix used in prediction of breeding values. *Biometrics 32*, 69.

[28] Henderson, C.R., H.W. Carter, and J.T. Godfrey (1954). Use of contemporary herd average in appraising progeny tests of dairy bulls. *J. Anim. Sci. 13*, 959.

[29] Henderson, C.R., R.W. Everett, G.R. Ufford, and L.R. Schaeffer (1973). *Genetics Research Report to Eastern Artificial Insemination Cooperative*. Ithaca, N.Y.

[30] Hill, W.G. (1971). Investment appraisal for national breeding programmes. *Anim. Prod. 13*, 37-50.

[31] Hill, W.G. (1972). Effective size of populations with overlapping generations. *Theor. Pop. Biol. 3*, 278-89.

[32] Hill, W.G. (1974). Prediction and evaluation of response to selection with overlapping generations. *Anim. Prod. 18*, 117-139.

[33] Hillers, J.K. and A.E. Freeman (1966). Two sources of genetic error in sire proofs. *J. Dairy Sci. 49*, 1245.

[34] Hinks, C.J.M. (1970). The selection of dairy bulls for artificial insemination. *Anim. Prod. 12*, 569.

[35] Hinks, C.J.M. (1971). The genetic and financial consequences of selection amongst dairy bulls in artificial insemination. *Anim. Prod. 13*, 209.

[36] Hinks, C.J.M. (1972). The effects of continuous sire selection on the structure and age composition of dairy cattle populaions. *Anim. Prod. 15*, 103.

[37] Hintz, R.L., R.W. Everett, and L.D. VanVleck (1977). Estimation of genetic trend from cow and sire evaluations. *J. Dairy Sci.*, in preparation.

[38] Hunt, M.S., E.B. Burnside, M.G. Freeman, and J.W. Wilton (1972). Impact of selection, testing, and operational procedures on genetic progress in a progeny testing artificial insemination stud. *J. Dairy Sci. 55*, 829.

[39] Hunt, M.S., E.B. Burnside, M.G. Freeman, and J.W. Wilton (1974). Genetic gain when sire sampling and proving programs vary in different artificial insemination population sizes. *J. Dairy Sci. 57*, 251.

[40] Lush, J.L. (1931). The number of daughters necessary to prove a sire. *J. Dairy Sci. 14*, 209.

[41] Lush, J.L. (1933). The bull index problem in the light of modern genetics. *J. Dairy Sci. 16*, 501.

[42] Lush, J.L. (1947). Family merit and individual merit as bases for selection. *Amer. Nat. 81*, 241.

[43] Kennedy, B.W. and J.E. Moxley (1975). Genetic trends among artificially bred Holsteins in Quebec. *J. Dairy Sci. 58*, 1871.

[44] Mason, I.L. and A. Robertson (1956). The progeny testing of dairy bulls at different levels of production. *J. Agr. Sci. 47*, 367.

[45] McClintock, A.E. and E.P. Cunningham (1974). Selection in dual purpose cattle populations: Defining the breeding objective. *Anim. Prod. 18*, 237-248.

[46] McGilliard, M.L. and A.E. Freeman (1976). Predicting daughter milk production from dam index. *J. Dairy Sci. 59*, 1140.

[47] Miller, P.D. (1969). Relating progeny superiority to genetic trend in cattle. *J. Anim. Sci. 28*, 577.

[48] Miller, R.H. and E.L. Corley (1965). Usefulness of information on mates of sires in artificial insemination. *J. Dairy Sci. 48*, 580.

[49] Olson, K.E. and E.L. Jensen (1976). Estimation of genetic trend in Wisconsin Holsteins. Paper presented at the 71st Annu. Meeting of the ADSA, Raleigh, N.C.

[50] Powell, R.L. and A.E. Freeman (1974). Genetic trend estimators. *J. Dairy Sci. 57*, 1067.

[51] Rendel, J.M. and A. Robertson (1950). Estimation of genetic gain in milk yield by selection in a closed herd of dairy cattle. *J. Genetics 50*, 1.

[52] Richardson, D.P. and B. Bearden (1976). Response from two systems of selecting sires. Paper presented at the 71st Annu. Meeting of the ADSA, Raleigh, N.C.

[53] Robertson, A., L.K. O'Connor, and J. Edwards (1960). Progeny testing dairy bulls at different management levels. *Anim. Prod. 2*, 141.

[54] Robertson, A. and J.M. Rendel (1950). The use of progeny testing with artificial insemination in dairy cattle. *J. Genetics 50*, 21.

[55] Schaeffer, L.R., M.G. Freeman, and E.B. Burnside (1975). Evaluation of Ontario Holstein dairy sires for milk and fat production. *J. Dairy Sci. 58*, 109.

[56] Schmidt, G.H. and L.D. VanVleck (1974). *Principles of Dairy Science*, 181. San Francisco: W.H. Freeman and Co.

[57] Searle, S.R. (1961). Estimating herd improvement from selection programmes. *J. Dairy Sci. 44*, 1103-12.

[58] Skjervold, H. (1963). The optimum size of progeny groups and optimum use of young bulls in AI breeding. *Acta Agr. Scand. XIII*, 131.

[59] Skjervold, H. (1966). Selection schemes in relation to artificial insemination. *Proc. 9th Int. Congress of Anim. Prod., Edinburgh*, 250-61.

[60] Skjervold, H. and H.J. Langholz (1964). Factors affecting the optimum structure of AI breeding in dairy cattle. *Z. Tierz. Züchtungsbiol. 80*, 25-40.

[61] Slanger, W.D. (1972). The influence of production differences between mates on the evaluation of AI sires. M.S. Thesis, Cornell University, Ithaca, N.Y.

[62] Slanger, W.D., E.L. Jensen, R.W. Everett, and C.R. Henderson (1976). Programming cow evaluation. *J. Dairy Sci. 59*, in press.

[63] Smith, C. (1962). Estimation of genetic change in farm livestock using field records. *Anim. Prod. 4*, 239.

[64] Specht, L.W. and L.D. McGilliard (1960). Rates of improvement by progeny testing in dairy herds of various sizes. *J. Dairy Sci. 43*, 63.

[65] Vaccaro, R. (1974). Economics of different alternatives to progeny test dairy bulls through artificial insemination. Ph.D. Thesis, Cornell University, Ithaca, N.Y.

[66] VanVleck, L.D. (1964). Sampling the young sire in artificial insemination. *J. Dairy Sci. 47*, 441.

[67] VanVleck, L.D. (1966). Change in variance components associated with milk records with time and increase in mean production. *J. Dairy Sci. 49*, 36.

[68] VanVleck, L.D. (1969). Relative selection efficiency in retrospect of selected young sires. *J. Dairy Sci. 52*, 768.

[69] VanVleck, L.D. and G.E. Bradford (1964). Heritability of milk yield at different environmental levels. *Anim. Prod. 6*, 285.

[70] VanVleck, L.D. and H.W. Carter (1972). Comparison of estimated daughter superiority from pedigree records with daughter evaluation. *J. Dairy Sci. 55*, 214.

[71] Vinson, W.E. (1975). Symposium: Choosing and sampling young bulls. Selection differentials. *J. Dairy Sci. 58*, 1071.

[72] Vinson, W.E. and A.E. Freeman (1972). Selection of Holstein sires for future use in artificial insemination. *J. Dairy Sci. 55*, 1621.

[73] Vinson, W.E. and A.E. Freeman (1974). Pedigree selection and semen banking of young dairy sires for artificial insemination. *J. Dairy Sci. 57*, 105.

[74] Wilk, J.C., J.E. Legates, and B.T. McDaniel (1976). Comparison of daughters of Jersey bulls with high predicted differences for milk with progeny of unrelated control bulls. Paper presented at the 71st Annu. Meeting of the ADSA, Raleigh, N.C.

Success and failure of quantitative genetic theory in poultry

A. W. Nordskog
DEPARTMENT OF ANIMAL SCIENCE
IOWA STATE UNIVERSITY, AMES, IA 50011

INTRODUCTION

Let me say to begin with that the "successes" in poultry breeding may well outweigh the "failures". Improvement in nonreproductive traits, most of which are highly heritable -- body size, egg size, growth rate, etc. -- pretty well follows theoretical expectation. I have no quarrel with these. On the other hand, failures seem to be restricted to the reproductive traits that generally are not highly heritable. In poultry, as well as other domestic species of animals, reproductive performance is economically very important. Hence, we have a problem. What I mean by "failure" is that response to selection, say, for rate of egg production, is almost always far less than expectation. Furthermore, in almost all special studies reported in the literature the low response is not the result of exhaustion of genetic variance.

In reviewing several breeding experiments, I will try to explain recalcitrant responses to selection in conventional quantitative genetic terms (i.e., as generating negative genetic correlations) but also in terms of nonconventional concepts as antagonistic selection and incompatible traits. Finally, I wish to describe some work we have done on nonadditive genetic components of variance and on sex-linkage, maternal effects and major gene effects.

EGG PRODUCTION AND BROILER PRODUCTION TODAY

Figure 1 shows where we are today in the United States in egg production and in broiler meat production. These data came from <u>Agricultural Statistics</u>, 1973, and are discussed in a 1975 publication by the National Academy of Sciences called <u>Agriculture Production Efficiency</u>. The implication is that egg production has plateaued but efficiency of broiler production continues making improvement. For the egg breeder, this raises the question: what does this plateau mean? It looks as if we are approaching about 230 eggs and leveling off there. Interestingly, as far back as the early 50's, in discussions at the National Poultry Breeders' Roundtable, we talked about plateaus. Yet during the last 2 decades or more, egg production has steadily improved on a national basis. Of course, we are not sure how much of the improvement is due to nongenetic causes. Furthermore, we are not sure of how much of the genetic improvement has resulted simply from discarding poor strains. We can't determine from Figure 1 how effective selection within a population has been. Whether we are up against a genetic plateau within a selected population remains an unanswered question.

I think that concentrating on controlling diseases, through vaccination, prophylactics or genetic improvement might automatically give us a further lift in egg production. If we were able to hold body size and egg size at some constant level, I wonder how much "genetic slippage," to use Dickerson's term, occurs for a trait like rate of egg production. I am still rather

QUANTITATIVE GENETIC THEORY IN POULTRY

impressed with the fairly thorough analysis of Kimber data that Dr. Dickerson reported a number of years ago on the subject of genetic slippage (Dickerson, 1955). His study failed to show any real change in rate of egg production over a 23-year period 1931-53.

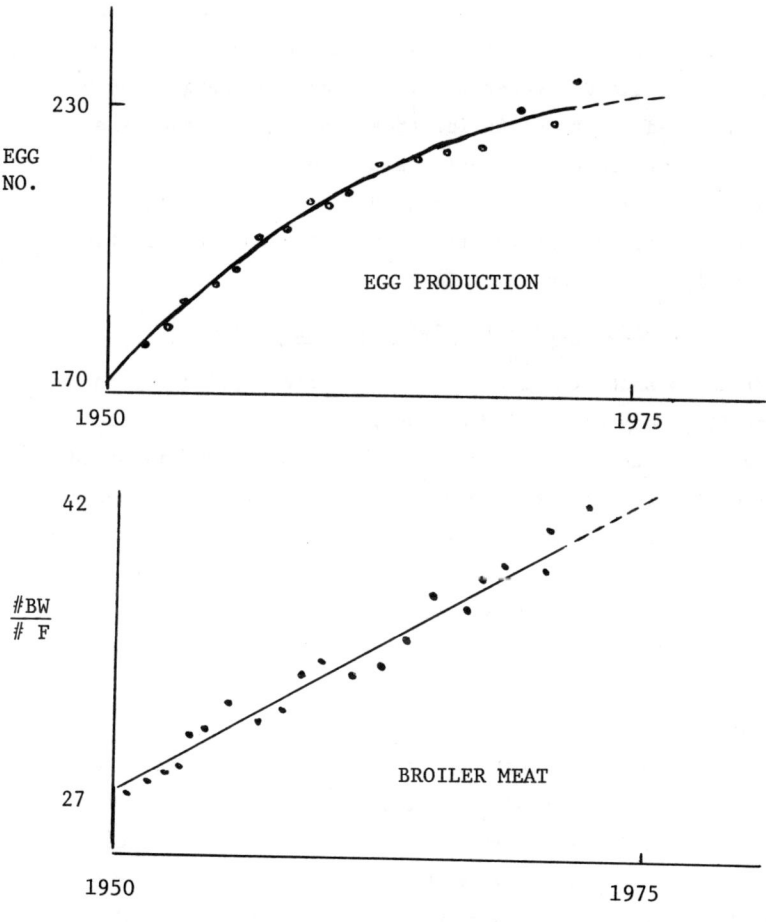

AGRICULTURAL STATISTICS, 1973

FIG. 1

Trends in egg production and broiler meat. (From Agriculture Production Efficiency 1975, ARC, NAS).

Figure 2 gives the results of an 11-year selection experiment for early rate of egg production that we carried out at Iowa State as part of the NC Regional project involving 2 lines: Leghorn A and Fayoumi J. For the first 8 generations we seemed not to go anywhere; at the 7th generation we dropped the Fayoumi J line. The concomitant increase in inbreeding due to small population size does not fully account for the failure. For reasons that we cannot fully explain, we started making progress after the 9th generation, but both egg weight and body weight declined. Correction for these correlated responses would have again reduced genetic gain in egg rate to near zero. The full-record more or less paralled the progress obtained with the part-record, but the heritability of the part-record was considerably lower than the full-record.

SINGLE-TRAIT AND TWO-TRAIT SELECTION

Figure 3 shows the results from divergent selection in each of 2 traits, body weight and egg weight, and the correlated responses with them. The direct responses are shown in solid lines and the correlated responses in dotted lines. In general, the direct responses are strong, and as expected, the correlated responses are lower than the direct responses. For body weight, the direct and correlated responses are symmetrical, but not for egg weight. Selecting for high egg weight in the line D did not result in as strong a correlated response in body weight as expected.

I now turn to the question of intentional two-trait selection in contrast to single-trait selection.

Figure 4 illustrates the results expected in 2 traits. The 2 traits are body weight (X) and egg weight (Y). Both axes are scaled in standard deviation units. There are 4 quadrants, I(+,+), II(+,-), III(-,-) and IV(-,+) where the signs indicate the direction of response to the X and Y traits, respectively.

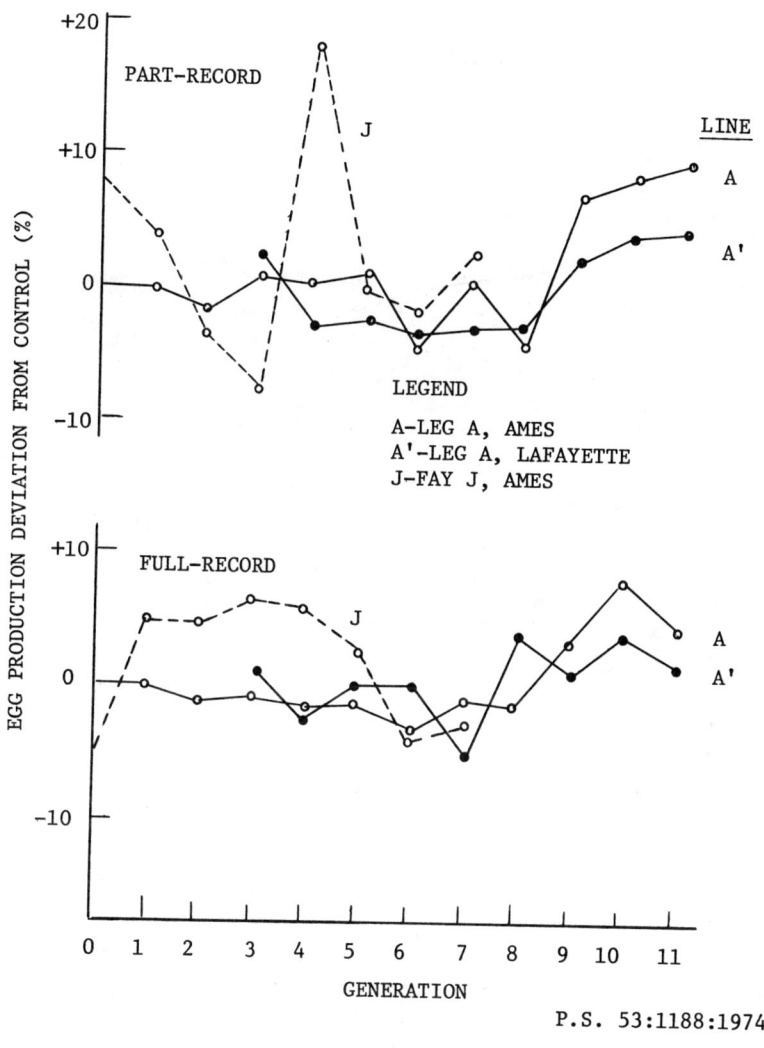

FIG. 2

Response to selection for rate of egg production in a Leghorn line, A, and a Fayoumi line, J.

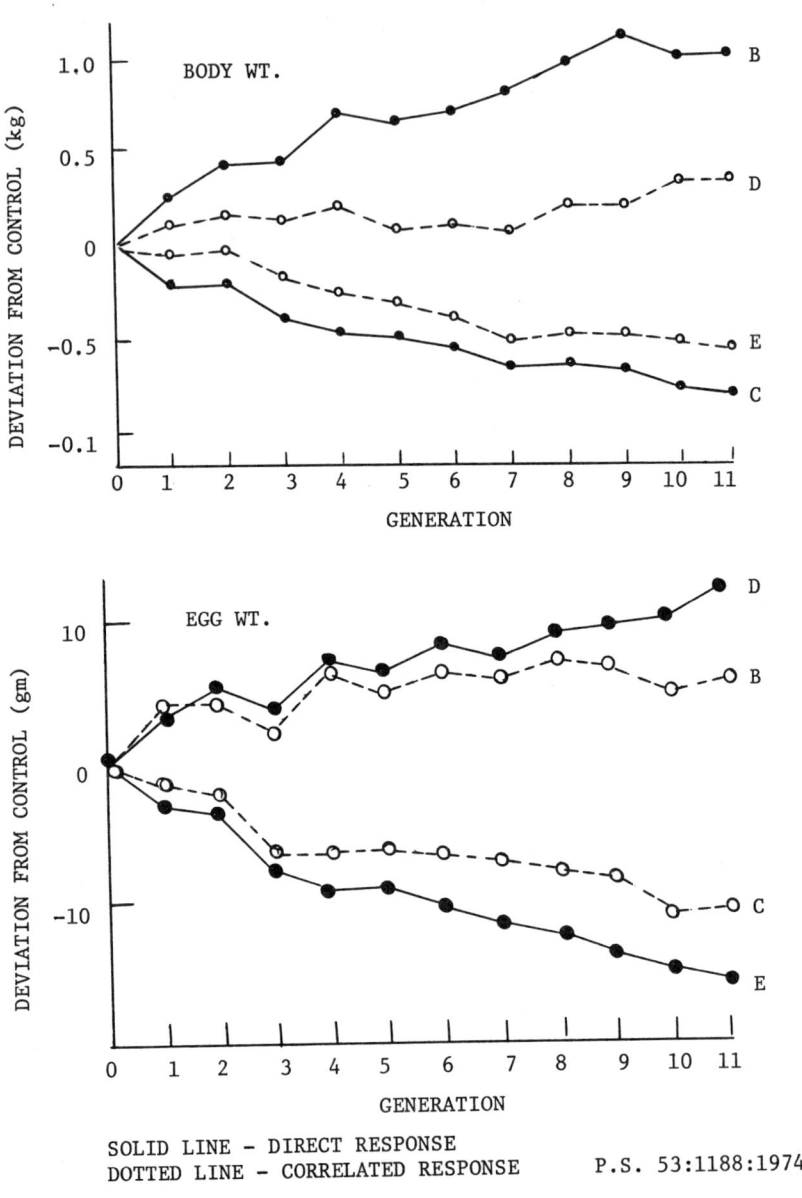

FIG. 3

Direct and correlated response to selection for egg weight.

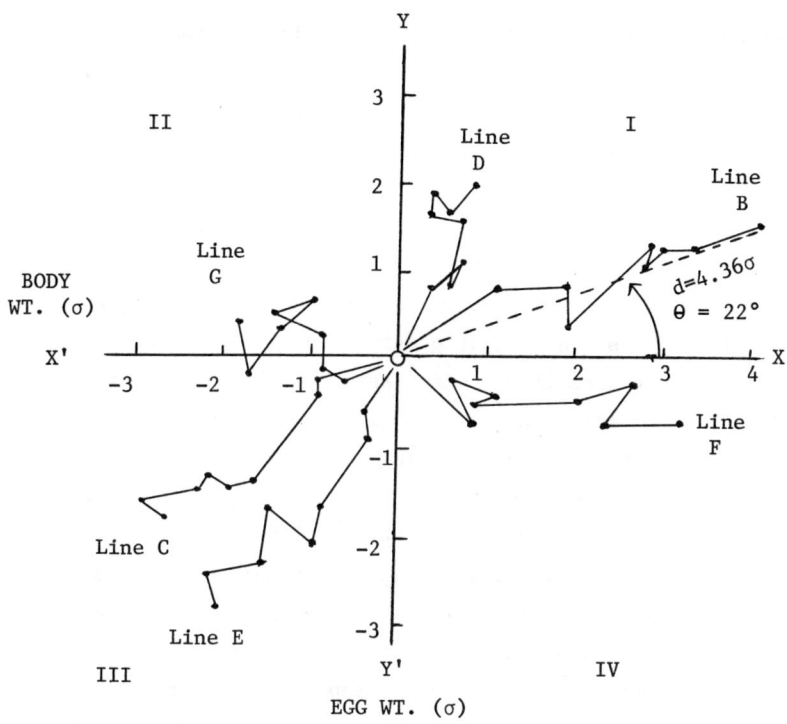

SINGLE-TRAIT SELECTION (QUAD. I & III)
TWO-TRAIT ANTAGONISTIC SELECTION (QUAD. II & IV)

FIG. 4

Response to selection that compares "normal" selection (+ +, or
- -) in quadrants I & III with antagonistic selection (- +, or
+ -) in quadrants II and IV. (P.S. 53:1188:1974).

The responses to single-trait selection are directed into quadrants I and III, but only deliberate two-trait selection can direct responses into quadrants II and IV. B is the response curve after 8 generations of selection for body weight alone. D is the response after 8 generations of selection for egg weight alone. Thus, the results of a single or 2-trait selection experiment can be described in terms of just two parameters: d, the vector distance, and θ, the angle that it makes with the X axis.

Four kinds of lines can be selected for using just two traits. Each line is unique to a particular quadrant I, II, III or IV as the case may be. I call selection that gets you into quadrants II or IV as antagonistic selection because, for the two traits considered, we are selecting against the genetic correlation.

For line B we can describe the selection result obtained in 8 generations as the vector distance, 4.36 standard deviations with an angle of $\theta = 22°$. Likewise we can map the distance and angle for the other lines. Note that progress in Line F, selected for large body size and small egg size, seemed to be only moderately successful. The direction (θ) more closely followed the body-weight axis than the egg-weight axis. This means that we didn't have quite the right kind of selection index originally hoped for. Note also that the selection path of line G was very erratic. Success in driving the selection response into quadrant II was poor. Line G was selected for large egg size and small body size. I call this kind of selection not only antagonistic, because it is contrary to the sign of the genetic correlation, but also incompatible in the sense that very small chickens don't naturally lay very large eggs; current quantitative genetics theory seems not to be of help in this case.

Table I gives the genetic correlations between body weight, egg weight and rate of egg production. For lines B and D the estimated correlations with rate are strongly negative. For lines C and E, with one exception, the estimates are positive. These results suggest that extreme selection for highly heritable traits may indeed change the sign of a genetic correlation.

NONADDITIVE GENETIC VARIANCE

Back about 25 years ago, one of Dr. Lush's favorite questions put to young staff and graduate students was "If there was one thing that you most wanted to know about the genetics of a population, what would it be?" As I recall, the most-wanted

TABLE I.

Genetic Correlations From Different Leghorn Lines

Line	Selected for	Genetic Correlations		
		BW·EW	BW·EP	EW·EP
A	High egg prod.	0.70	0.18	0.05
B	High body wt.	0.06	-0.67	-1.78
C	Low body wt.	0.00	0.22	-0.42
D	High egg wt.	0.15	-1.36	-1.08
E	Low egg wt.	0.93	0.21	0.43

From Nordskog et al., 1974, Poult. Sci. 53:1188-1219

information was on nonadditive genetic variance. The reason, I suppose, was that it seemed difficult to get a handle on it, especially in contrast to additive genetic variance. It seemed axiomatic, in terms of quantitative genetics, that if we knew what σ_A^2 and σ_{nonA}^2 were along with the relevant r_G's, the determination of the optimum breeding system for any population with the required parameter estimates would be strictly routine. The next most important thing we thought we want to know was, with a given rate of selection, how would the parameters change? This kind of information, we thought, should allow us to decide what breeding system to follow and, as the parameters changed, we could easily decide at what time we should switch to a more efficient breeding system.

Looking back now we see that animal breeders haven't learned much about the nature of the genetic variance of a population other than the additive portion. I expect the same holds true as well for plant breeders. The difficulty probably lies in the complexity of the problem, but with the enormous rise the last few years in computer capability, one would think that more progress in identifying nonadditive variance might be forthcoming.

We have tried two approaches to getting information on the problem in chickens. The first was to estimate the nonadditive variance as a "salvage operation" in populations undergoing selection (Silva et al., 1976). The second was based on a designed experiment to estimate sire x dam interaction (Sato, 1975).

For the Silva study we tried two models: Bohidar, 1964, and Willham, 1965 (Table II). The covariances between different relatives were expressed in terms of additive, nonadditive and environmental components in variance. Data from 10 generations of Leghorn line A, selected for high rate of egg production, and 3 generations of a new population (line Q) were used to estimate covariances. The Bohidar model provides for sex-linkage, and the Willham model has, instead, components for a maternal effect and an interaction effect involving additive and maternal effects. Each model has a term for a common environment effect between full-sib and the variance, σ_W^2, within full-sib groups. The 3/4 sibs have a sire in common, but their dams are full sisters. By solving the entire matrix of covariance relationships, the variance components were estimated.

TABLE II

Expected Composition of Covariance

Relatives	σ_A^2	σ_D^2	* σ_L^2	** σ_M^2	σ_{AM}^2	σ_C^2	σ_W^2
Dam-Daughter	1/2	0	0	1/2	5/4	0	0
G." G. "	1/4	0	0	1/4	5/8	0	0
Aunt-Niece	1/4	0	0	1/2	3/4	0	0
Full Sibs	1/2	1/4	1/2	1	1	0	0
3/4 Sibs	3/8	1/8	0	1/2	1/2	0	0
1/2 Sibs	1/2	0	1/2	0	0	0	0
W/Full Sibs	1/2	3/8	1/2	0	0	0	1

*Bohidar (1964) model
**Willham (1972) model

Table III gives the results of the two models. Only 4 of 12 traits studied are presented: body weight, egg weight, part-record and full-record. The results shown are a composite of the two lines analyzed, which altogether involved over 9,000 pedigree records. The estimates of additive effects are generally in agreement with the literature, although the values for egg weight seem unusually high.

TABLE III

Percentage Covariance Between Relatives

	BW	EW	Egg Prod. Part Rec.	Egg Prod. Full Rec.
Bohidar:				
Additive	72	98	5	21
Dominance	-8	-56	-4	23
Sex Linkage	6	-17	2	6
Environ. C	3	6	-2	-3
Environ. Random	27	69	99	53
	100-	100-	100-	100-
Willham:				
Additive	84	67	9	10
Dominance	12	-125	24	79
Maternal	-5	-11	3	21
Mx	0	44	-11	35
Environ. C	0	5	0	0
Environ. Random	9	120	75	25

From Silva et al., 1976, Br. Poult. Sci. 17:525-538

The estimates of sex linkage and dominance (Bohidar model) are disappointly erratic, and several are negative.

The Willham model gave similar results and is not much more informative. Here the covariance between relatives is partitioned into additive, dominance, maternal effects, maternal effects x additive effects, environment C and random environment. Again, the estimates of nonadditive effects are erratic, and the only reasonable conclusion, other than to say that both models yield consistent estimates of additive effects, is that the nonadditive parameters are probably small, and the sampling errors of them are high.

I turn now to the Sato study on sire x dam interaction.

Figure 5 shows the diallel mating plan. The idea was to produce progeny from 2 sires in the same hatch and thereby to avoid sire x hatch interaction effects. This was accomplished by blood typing sires and their progeny. Sires were paired according to blood typing contrasts at the B locus.

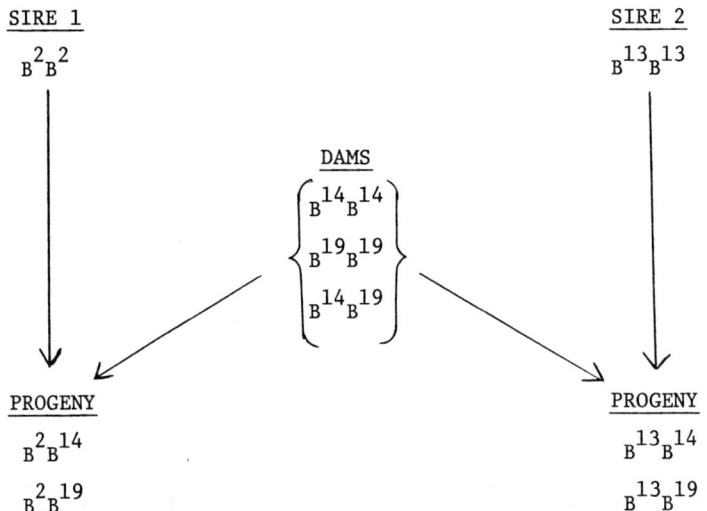

FIG. 5

Diallel mating plan using blood group markers

The theoretical variance components are given in Table IV, and the mean square expectations are given in Tabe V.

TABLE IV

Theoretical Fractions of Genetic Variance

Variance Components

	A	D	AA	AD	DD	SL	M
Sire	1/4		1/16			1/2	
Dam	1/4		1/16				1
Sire x Dam		1/4	1/8	1/8	1/16		
W/Full Sibs	1/2	3/4	3/4	7/8	15/16	1/2	
Total	1	1	1	1	1	1	1

A = Additive   D = Dominance   SL = Sex-linkage   M = Maternal

TABLE V

Expected Mean Squares Within Year

Pairs of Sires $\sigma^2_e + k_6 \sigma^2_{SD/P} + k_7 \sigma^2_{D/P} + k_8 \sigma^2_{S/P} + k_9 \sigma^2_P$

Sire/Pair $(") + k_4 \; (\;") + k_5 \; (\;")$

Dam/Pair $(") + k_2 \; (\;") + k_3 \; (\;")$

Sire x Dams/Pair $(") + k_1 \; (\;")$

Residual $(")$

Table VI gives the estimates of variance components for several traits. It seems that there is little σ^2_{SD} variance for physical traits, egg weight and age at maturity. For rate of egg production, although the results are not statistically significant, the σ^2_{SD} components were larger than the estimates of sire variance. Thus, it seems that nonadditive variance, presumably dominance, may be of special importance for egg production.

TABLE VI

Percentage Variance Components

	P	S	D	SD	E	T
Shank length	4.67	8.40	13.60	2.12	71.21	100
Body wt. 55 wk.	8.45	6.84	20.48	1.12	63.06	100
Egg wt. 55 wk.	12.76	9.22	20.38	-9.24	66.88	100
Part-record Egg Prod.	-1.44	1.94	5.54	7.25[a]	86.71	100
Full-record Egg Prod.	-0.32	2.67	6.04	5.05[b]	86.56	100
Age at Maturity	4.51	5.51	3.35	-0.45	87.08	100

a = Prob. <0.10   b = Prob. <0.16

P = Pens   S = Sires/P   D = Dams/P   SD = SD/P   E = W/SD/P
T = Total

The corresponding heritability estimates are given in Table VII. Note that the ratio of dominance to additive variance is 2-3 times for rate of production but much less for the more highly heritable traits of shank length, body weight and egg weight.

Table VII

Percentages of Additive (h_A^2) and Dominance (h_D^2) Variance and Their Ratio

	h_A^2	h_D^2	h_D^2/h_A^2
Shank length	34	8	0.25
BW (55)	28	4	0.17
EW (55)	37	-37	-
Rate (early)	8	29[a]	3.62
Rate (full)	11	20[b]	1.89

a P = 0.10, b P = 0.16

Evidence for the importance of nonadditive effects or nicking, as they bear on performance in commercial poultry, is still in short supply. Uncovering such evidence by statistical salvage from routine performance data as was done by the Silva study doesn't seem too fruitful because of large sampling effects and error biases that were not removed from the estimates. Specially designed experiments, such as producing diallel mating used in the Sato study, should be more useful. Tentatively, nicking seems not to be of importance for highly heritable traits but may be for reproductive traits.

SEX-LINKAGE, MATERNAL EFFECTS AND MAJOR GENES

One of the most important diseases of the fowl is avian leukosis. This occurs in two forms: lymphoid leukosis (LL) and Marek's disease. The latter is now well controlled by vaccination. However, no successful vaccines have yet been prepared to control LL. For that reason, breeders rely on a selection program for genetic resistance. LL is caused by several different strains of RNA viruses. It is now known that genetic resistance to virus infection is simply inherited: one gene determines cellular resistance to infection by a given virus strain subgroup. The most widely spread subgroup of virus among commercial flocks is subgroup A. Subgroups B, C, D and E also have been identified, but are less important. Most of this work has been done at the U.S. Regional Poultry Laboratory at East Lansing.

In a recently completed study (Nordskog and Pevzner, 1976), we have found that a heavy-breed Rhode Island Red is genetically more resistant to LL than a Leghorn strain. When reciprocal crosses are made between these two breeds, the cross from the Leghorn male mated to the Red female usually gives twice as high mortality as the reciprocal cross. The classical interpretation for this has been based on the assumption of sex linkage: the male parent always transmits its Z sex-chromosome to its female progeny. Our studies, however, now support the recent contention [U.S. Regional Poultry Laboratory, East Lansing], that genetically resistant hens don't produce maternal antibodies. When Leghorn hens, genetically susceptible to LL, are used as breeders they have acquired immunity to LL and transmit these antibodies to their genetically susceptible progeny, which protects them for the first 3-4 weeks of life. Thus, differences in LL mortality between reciprocal cross progeny of the Leghorn and Red breeds may be due to either genetic resistance or protection from transmission of maternal antibodies from genetically susceptible mothers. Also, it is possible that both sex chromosomes and maternal antibody transmission (under separate genetic control) are important resistance mechanisms against lymphoid leukosis. Thus, what has seemed to be sex-linked control of a complex quantitative trait (lymphoid leukosis) now turns out to be simply inherited by a single locus but operating as a maternal antibody transmission effect.

CONCLUSIONS

1. In chickens response to selection for highly heritable traits such as body size and egg weight follows theoretical prediction very well, at least for short-term periods of 10-15 generations.

2. Prediction of selection response to rate of egg production almost always is overestimated. Part but not all of this "nonresponse" can be accounted for by inbreeding depression. Because egg size and body weight are economically important and usually correlated negatively to production, the commercial breeder's task of improving production is more complicated. We have evidence that extreme selection for high body weight or egg weight makes the genetic correlations with egg rate negative, but selection for low body weight or egg weight has an opposite effect on the correlations.

3. The commercial breeder favors small-bodied birds that

lay large eggs. This kind of selection is antagonistic in the sense that selection is against the sign of the genetic correlation. Also, there is a physical incompatibility for extremely small birds to lay large eggs. Current quantitative genetics theory does not provide for these phenomena.

4. Little is known of the importance of nonadditive genetic variance in chickens. An attempt to obtain estimates as a "salvage" operation using least squares procedures from a rather large amount of routine-collected performance data gave inconsistent and generally unsatisfactory results. On the other hand, in a specially designed experiment to measure sire by dam interaction, nonadditive variance (dominance) was 2 to 3 times as large as additive fraction.

5. Maternal effects for viability may be very important in chickens. In the case of lymphoid leukosis, the genetic control may be at very few loci. The mechanism of control can be either genetic resistance or passive immunity acquired by the chickens by maternal antibody transmission. Thus, variation in certain quantitative traits that for years have been assumed to be under polygenic control, may turn out to be controlled by a relatively few major genes.

ACKNOWLEDGMENT

Journal Paper No. J-8670 of the Iowa Agriculture and Home Economics Experiment Station, Ames, Iowa. Project 1711. In cooperation with the North Central Regional Poultry Breeding Project, NC-89.

BIBLIOGRAPHY

Bohidar, N. R. (1964). Derivation and estimation of variance and covariance components associated with covariance between relatives under sex-linked transmission. Biometrics 20, 505-516.

Dickerson, G. E. (1955). Genetic slippage in response to selection for multiple objectives. Cold Spring Harbor Symp. Quant. Biol. Vol. XX:213-224.

Nordskog, A. W., H. S. Tolman, D. W. Casey and C. Y. Lin (1974). Selection in small populations of chickens. Poult. Sci. 53, 1188-1219.

Nordskog, A. W., and I. Y. Pevzner (1976). Sex-linkage versus maternal antibodies in the genetic control of disease. World's Poult. Sci. J. (in press).

Sato, M. (1975). The influence of dominance and gene interaction on egg production and other traits in laying hens. M.S. thesis. 51 pp. Iowa State University.

Silva, M. A., P. J. Berger and A. W. Nordskog (1976). On estimating non-additive genetic parameters in chickens. Br. Poult. Sci. 17, 525-538.

Willham, R. L. (1963). The covariance between relatives for characters composed of components contributed by related individuals. Biometrics 19, 18-27.

Past, present and potential contributions of quantitative genetics to applied animal breeding

Dewey L. Harris
USDA-ARS
PURDUE UNIVERSITY, WEST LAFAYETTE, IN 47906

ABSTRACT

Aspects of quantitative genetic "theory" relevant to applied animal breeding is reviewed. The focal point for this review is the ability of the breeder to predict genetic response from alternative schemes. Phenomena influencing long-term response are seen as being troublesome and the potential to predict these occurrences, so as to allow more effective planning of breeding programs, is assessed.

INTRODUCTION

As I suggested in the introductory session, the key need of the applied animal breeder is the ability to predict the response to alternative selection schemes. This ability allows him to make the correct decisions concerning the optimum scheme to make rapid changes toward his selection objective for his particular populations. In this paper, I intend to explore the degree to which the animal breeder has been able to reliably use predictions of selection response in his planning in the past, to explore some of the known difficulties in the interpretation of response to selection, and to explore the degree to which it will be possible in the future to augment the breeder's ability to make these predictions and, thus, to make the proper decisions. The term prediction as we are using it here can mean (and hopefully does mean in many cases) a reliable prediction of the absolute expected response. Sometimes, however, he can only predict the relative expected response. In instances where this is not possible, the breeder sometimes has only a suspicion of possible occurrences of phenomena in his population due to observations in other populations. This, of course, is not a satisfactory situation for the breeder.

There have been several references during this conference to the fact that many of the tools of the practicing animal breeder are more statistical than genetic. It is well to remember that animal breeding was a serious activity of many stockmen prior to the discovery of the Mendelian basis of inheritance, and the principle guiding the practice of breeding was the observation that "like tends to beget like". Thus, much of present day animal breeding involves the use of more sophisticated statistical techniques to amplify this primitive but still relevant principle and could be done without any knowledge of the genetic basis for inheritance. In a fundamental sense, the response to selection is the coefficient of regression of offspring on parents multiplied times the selection differential for the parents. Of course, the Mendelian basis for inheritance leads us to expect that the coefficient of regression of offspring on male parents will be equal to the coefficient of regression of offspring on female parents for traits influenced by only autosomal loci. But this could have been determined statistically without any knowledge of basic genetics. Many times in practice the covariance of parent and offspring is indirectly estimated from the observed covariance between full sibs or twice the covariance of half sibs. This latter point involves making use of the knowledge of genetics to augment the statistical tools. Because the regression of offspring on parents is a part of the basic description of response to selection and because this regression is composed of variances and covariances between parent and offspring, animal breeders have been very interested in the genetic parameterization of the genotypic variance and of the covariance between relatives. A series of papers by Fisher (1918), Cockerham (1954), and Kempthorne (1954, 1957) have led to the following general formulae:

$$V(G_X) = \sum_{s,t=0}^{n} \sigma^2_{A^s D^t} \quad \text{for } 1 \leq s+t \leq n$$

$$\text{Cov}(G_X, G_Y) = \sum_{s,t=0}^{n} (2r_{xy})^s u_{xy}^t \sigma^2_{A^s D^t} \quad \text{for } 1 \leq s+t \leq n$$

If one assumed that the parameters in this parameterization did not change from generation to generation and that the responses to each generation of selection were cumulative, the total response to selection could then be expected to be a linear function of the number of generations of selection and would continue to increase as long as the generation number increased. These assumptions would be equivalent to assuming that an infinite number of loci with infinitesimal effects were influencing the trait under selection. In addition to these assumptions, the assumptions behind the development of the covariance between relatives as related to selection theory are that populations are of infinite size, have regular autosomal diploid inheritance, are random mating (Hardy-Weinberg) populations, and have no linkage between loci.

Of course, most of these assumptions are not realized in specific populations with which animal breeders are concerned. He is concerned with populations of finite size and with traits influenced by a finite (but unknown) number of loci with finite (but unknown) effects. Also, he is concerned with traits that are influenced by the loci on the sex chromosomes, and he is concerned with linkage between the autosomal loci, as well as between the sex-linked loci. In many instances, the breeder is concerned with populations where the mating structure departs from that described by the familiar Hardy-Weinberg formula. Many people have attempted to extend the formula for the covariance between relatives for situations in which one or more of these assumptions does not hold. These include Schnell (1963), Van Aarde (1963), Bohidar (1960), Gillois (1964), Harris (1964), and Weir and Cockerham (1976). However, no one has been able to develop a satisfactory general formula for the covariance between relatives when several of these assumptions are relaxed. All of this provides the quantitative geneticist with plenty of ex-

QUANTITATIVE GENETICS IN ANIMAL BREEDING

cuses as to why the parameterization of the covariance between relatives and the selection theory derived therefrom might not be completely authenticated. However, in many instances, this theory seems to be relatively reliable for the description of the response to selection for several generations. The most noteworthy example of this would be the first 50 generations of the selection experiment for pupa weight described by Enfield (1972 and 1976). The response was relatively linear and in agreement with expectations from evaluation of the genetic parameters of the population. Of course, after 50 generations Enfield's response from selection began to change. There are also many examples of selection experiments in which the response to selection is not in agreement with that expected. Examples of the latter case are the papers earlier this week by Nordskog (1976) and by Eisen (1976). However, in both of these cases, the response to selection was observed in an unreplicated experiment. Thus, it is not possible to evaluate the amount of variability in the response to selection due to random genetic drift and other random factors peculiar to the specific selection population. This form of departure from expectation does not represent an invalidation of our quantitative genetic and selection theory. It just means that our predictions do not account for everything that influences our results in a finite population. Factors such as random genetic drift are never going to be "predictable", and the breeder has to be content with the occurrence of this phenomena due to the basic random nature of the genetic mechanism.

NON-LINEARITY OF LONG-TERM RESPONSE

Now let us examine some of the departures from the

"straight-line" response to selection and the explanations for their occurrence. Al-Murrani (1974) has reviewed many of these same phenomena as they relate to limits of selection, but it seems worthwhile to look at them again with respect to how they affect "predictability" of response. Most selection experiments in which it has been possible to precisely evaluate longterm improvement have shown a concave downward curve if extended over a long enough period of time. Such "plateauing" is illustrated in the Tribolium experiment of Enfield. This is not surprising if one considers that in an initial population with a certain number of loci with segregating alleles, there are numerous potential genotypes, and one or more of these is associated with a phenotype that is superior to the rest. The superior phenotype associated with this genotype is undoubtedly going to be finite on some appropriate scale of merit. Thus, if time, as measured in generations, is extended indefinitely, the response to selection will necessarily diminish so the cumulative response will not exceed this upper maximum genotypic value. Let us term this maximum genotypic value the potential for the population. This terminology is consistent with that of Griffing (1963). Equivalently, an asymptotic approach to this limit is expected due to the exhaustion of genetic variance as selection continues long enough to fix the alleles at many of the loci. Of course, if this maximum genotype involves heterozygosity at some of the loci, the maximum genotype is not possible to attain in a segregating population but can be obtained in a F_1 cross between selected populations. However, in addition to the maximum for an F_1 cross, there is also a maximum average genotypic value for segregating populations and a maximum genotypic value for homozygous populations. Thus there are three potentials possible when the maximum genotypic value involves heterozygosity. The potential for the F_1 cross could also be achieved in a reciprocal recurrent selection procedure as proposed by Comstock, Robinson and Harvey (1949) if it were to be carried out in two infinite populations. The potential for segregating populations could be achieved in a conceptual infinite random mating population under selection over an infinite amount of time. Similarly, the homozygous potential might be achieved in an infinite population with inbreeding. In order that these potentials are achievable, it has to be assumed that there is no "multi-peak epistasis" that would trap the populations prior to achieving these potentials. This latter assumption does not preclude the existence of epistasis but assumes that it is of limited magnitude.

Another phenomenon that has been observed in certain experimental work has been asymmetry of response to selection. Falconer (1960) offered several explanations for this occurrence. These were as follows:

1. Asymmetry in selection differentials.
2. Genetic asymmetry.
3. Selection for heterozygotes.
4. Inbreeding depression effects.
5. Maternal effects.

Falconer offered three possible ways in which asymmetry of selection differentials could occur:

1. Natural selection in one direction.
2. Correlation between fitness and the metric trait under selection.
3. Variance of the trait related to the mean.

For genetic asymmetry to occur, it is necessary that the dominance of gene action be in one direction for the majority of loci and for gene frequencies to be consistently above or below .5. The conditions of genetic asymmetry imply that there is a finite number of loci with major effects upon the trait. However, in Dr. Falconer's presentation this week (Falconer, 1976) he suggested that random genetic variation and lack of experimental replication may be accounting for much of the observed asymmetry when up and down selection is carried out in an unreplicated experiment.

Dickerson (1965) has presented evidence showing that there is an effect when selection is relaxed such that the mean of the population regresses toward the original mean prior to selection. Dickerson termed this effect a "recombination loss", implying that it is due to the recombination of selected epistatic effects. Griffing (1960a, 1960b) clearly demonstrated the theoretical basis for this showing that additive by additive epistatic effects in the population would give a temporary response to selection that is lost in later generations due to recombination between the two loci involved. In practice, it has been difficult to evaluate the magnitude of this phenomena due to the difficulty in obtaining reliable estimates of the magnitude of additive by additive epistatis.

Another occurrence which would lead to a relaxed selection effect is antagonistic natural selection. The best demonstration of the occurrence of this phenomena is in the paper by Enfield (1976). Of course, observations on this can be made in a selection experiment by comparing the intended selection differential with the achieved selection differential.

Both of these explanations of a relaxed selection effect would also lead to a decline in the response to continued selection.

To clarify the terminology and the description of response and improvement from selection in the presence of a relaxed selection effect, I would propose that the term "genetic response" be used to describe the difference between the mean of a population undergoing selection and the mean that it would have if selection had not been carried out. I would propose that the term "genetic improvement" be used to describe the difference between the mean of a population undergoing selection and the mean of the same selected population in some prior generation after adjustment for environmental differences. Thus genetic response is the sum of "genetic improvement" and the magnitude of the "relaxed selection effect".

Another possible occurrence in selection experiments is that, with the presence of genotype-environment interactions, the response in a different environment is somewhat less than the response in the environment where the testing and selection was carried out. The magnitude to which the response in another environment is less depends on the magnitude of the genotype-environment interaction or, alternatively, upon the magnitude by which the genetic correlation for the trait in the two different environments is less than one. This has been described by Dickerson (1962).

FINITE POPULATIONS

Up to now we have been considering phenomena that can occur in the "long term" in conceptual "infinite" populations. Of course, these phenomena also occur in finite populations. However, there are additional phenomena that occurs in finite populations that are worthy of consideration and which lead to departures from the response expected in conceptual "infinite" populations. The primary manifestation of finite population size is that, due to the random sampling of parents producing gametes to produce offspring and due to the random sampling of the phenotypic selection of parents, there will be variability of response. This variability of response is especially true at individual loci where the changes in gene frequency may fluctuate considerably due to the presence or absence of genes in the finite number of individuals selected. There are two consequences resulting from this random sampling. One of these is that there will be random fluctuations in observed response. This is part of the reason for the jagged nature of plots of generation means in a long term selection experiment. Of course, the other reason for this jaggedness is the presence of environmental differences from generation to generation. The random fluctuations in the response are also manifested in the differences between replicates of the same population undergoing precisely the same selection scheme and criteria. These differences are a reflection of the chance processes involved in selecting in a finite population due to the random basic nature of the genetic mechanism. However, there is a second manifestation of the variability of response in a finite population. In a conceptual infinite population where chance fluctuations do not occur, all loci continue to segregate until they are fixed by selection. However, in a finite population, the chance fluctuations may lead to fixation of alleles by chance. This fixation may be in either the favorable or unfavorable direction. The occurrence of inbreeding depression is due to this chance fixation, and the magnitude of the inbreeding depression effect is a function of the average dominance deviations of the homozygotes in the population. However, when the chance fixation leads to the elimination of desirable alleles, that is, elimination of alleles present in the genotype with maximum potential genotypic value, there is necessarily a reduction in the magnitude of the potential since this previously maximum genotypic value is no longer possible. This reduction in potential will be greater in earlier generations of a selection program due to the fact that more of the desirable alleles will be at low frequency and thus have greater probabilities of being lost by chance. Figure 1 gives an illustration of this phenomena. The conceptual infinite random mating population would change by selection such that the mean asymptotically approaches the original potential. When selection is carried out in a finite random mating population, the population mean is expected to increase and converge towards a potential that is declining as the desirable alleles in the earlier potential genotype are eliminated by chance. The magnitude of this reduction in potential is related to the probability of fixation of alleles which has been described by Kimura (1957) and by Robertson (1960). In Figure 1, note that the decline in the early generations of selection of the finite population relative to the infinite population is the inbreeding depression effect. However, the inbreeding depression effect may differ in magnitude from the reduction in potential. The inbreeding depression effect is a function of the mean dominance deviation of homozygotes (Kempthorne, 1957). However, the reduction in potential involves the difference

FIG. 1

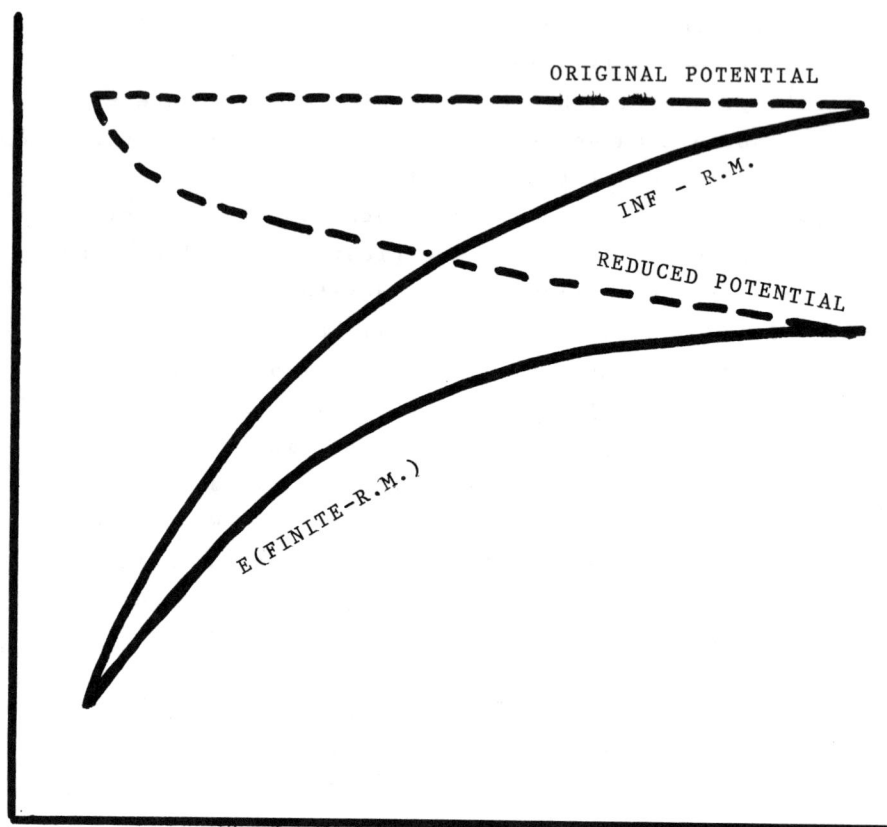

Schematic diagram showing the expected relative decline in response of selection in a finite population relative to that expected in a conceptual infinite population. This greater decline is associated with a reduction in potential due to chance loss of desirable alleles.

between the homozygote for all the desirable alleles (or the heterozygote when there is overdominance) and the values for the homozygote of the less desirable alleles which becomes fixed.

It is relatively easy to conceive and draw the representative diagram presented in Figure 1. However, up to now it has not been possible to develop algebraic formulae to precisely describe these phenomena. Thus, it is not practically possible at this point for the breeder to exactly describe or to predict the response curve for a finite population undergoing selection. Our infinite population theory allows us (if we know the population parameters) to predict the slope at the beginning of the infinite population curve. Of course, as the parameters change, we can change our prediction of the slope if we have new estimates of the parameters. Also, it is possible to experimentally evaluate, and thus predict, the inbreeding depression effect as a function of the coefficient of inbreeding (Kempthorne, 1957). The coefficient of inbreeding may be calculated for finite populations from the formula by Wright (1921). However, this formula is appropriate only for finite populations undergoing random mating. Whenever selection is being carried out, there will be an increase in inbreeding (as calculated from pedigree) due to the tendency to select several individuals from good families. On the other hand, with overdominance or pseudo-overdominance, there could be a tendency to select for heterozygotes, and thus the decline in heterozygocity could be less than that implied by Wright's formula.

The work of Kimura (1957) and Robertson (1960) indicates that the probability of fixation of the wrong allele is a function of $N \times \bar{i}$, where N is effective population size for a monoecious population and $\bar{i}$ is the selection differential in standard deviation units. Thus, the difference between the original potential and the reduced potential at the limit of a large number of generations will also be a function of $N\bar{i}$. Even though we cannot precisely predict the magnitude of the difference between the original potential and the reduced potential, we can obtain a relative evaluation from knowledge of effective population size and standardized selection differentials for monoecious populations. Unfortunately, none of the animal species is monoecious. It is not clear, to this author at least, precisely what is the dioecious analog to $N\bar{i}$. In many populations undergoing selection, there will be a lesser number of males from the number of females due to the capacity of males of most species to fertilize several females. This allows a greater selection differential among the males. Thus the N for males may differ from the N for females, and the $\bar{i}$ for males may differ from the $\bar{i}$ for females. In species such as dairy cattle where milk production is expressed only in the females and in chickens where egg production is only expressed in the females, there will be different selection criteria for the two sexes. Is the following formula the dioecious analog of $N\bar{i}$:

$$\frac{N_m \bar{i}_m r_{GI_m} + N_f \bar{i}_f r_{GI_f}}{2} ?$$

If this function or some other function of these same quantities can be established as being monotonically related to the probability of fixation of a less desirable allele, it would be possible for a breeder to obtain a relative comparison of the decline in potential for alternative breeding schemes. A precise comparison is not going to be possible in general, because this probability is also a function of the distribution of gene frequencies of desirable and less desirable alleles as well as the relative magnitude of the allele effects at different loci. Since these detailed aspects of the genetics are generally not known to the breeder, it seems doubtful that a precise estimate of the reduction in potential is going to be possible.

The computer simulation work of Qureshi, Kempthorne, and Hazel (1968) and Qureshi (1968) and Latter (1965), as well as the theoretical work of Hill and Robertson (1966), shows that linkage in a finite population will further reduce the potential relative to free recombination because of the increase of chance fixation of less desirable alleles when they are linked with desirable alleles that are being increased by selection.

Please note that even when overdominance is not present as a part of the gene action in the population, crossing may be desirable as the last step in the selection program so as to provide recovery from the chance fixation of undesirable alleles due to limited population size and linkage. When the probability of fixation of an undesirable allele is small, the probability of chance fixation in two populations would be quite small, and thus if dominance is near complete and in the desired direction, the heterozygotes produced by crossing are likely to be superior to either one of the pure populations. The cross seems even likely in most cases to be superior to a pure population of twice the size.

The indication that long-term response is a function of effective breeding population size as well as selection intensity leads the breeder to the dilemma illustrated in Figure 2. When the breeder has the choice between two breeding schemes, with short and long-term improvement as depicted here, which does he choose? Since short-term response to selection is proportional to $\bar{i}$, consideration only of short-term response would lead to selection of the scheme with the greater $\bar{i}$. However, that greater $\bar{i}$ may be associated with a smaller effective population size of breeders (N) such that the long-term expected potential of the selection scheme in a finite population may be less than for another scheme with smaller $\bar{i}$ but larger N. James (1972) has suggested the comparison of cumulative discounted improvement for large finite number of generations as the appropriate basis for comparison. James' work is the best effort in this area but still does not represent a complete mathematical description of all the relevant concerns of the breeder in designing a breeding program because James evaluated the worth of genetic improvement as being proportional to the mean of the population. For a commercial breeder the value of genetic improvement is also influenced by the potential for producing improved breeding stock for sale or distribution and this capacity will be a function of the size of the population. Thus increased size of the population is advantageous up until the capacity to produce breeding stock exceeds the potential market for the breeder. After that, increased population size has no increase of value and is actually detrimental because of the anticipated increased cost of maintaining larger population sizes.

FIG. 2

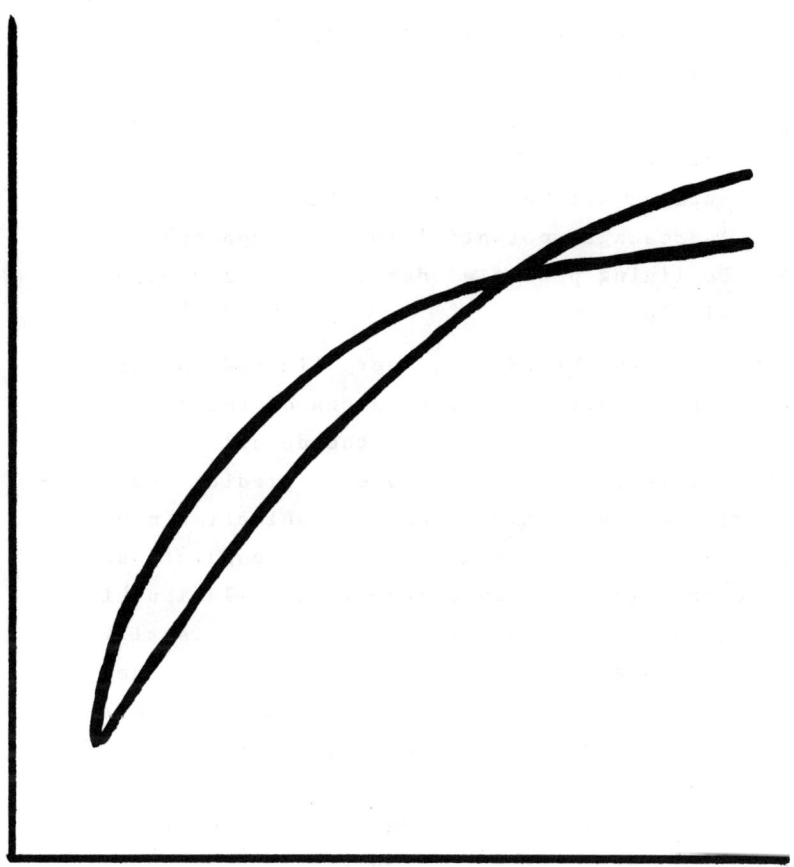

TIME

Schematic diagram showing alternative long-term improvement from two selection schemes - one with greater earlier improvement and the other with greater long-term improvement.

SUMMARY AND CONCLUSIONS

In summary, let us state that there are 4 reasons for the curvilinearity of improvement from selection. These are as follows:

1. Recombination loss of selected epistatic effects.
2. Antagonistic natural selection.
3. Approach to potential (maximum genotype).
4. Declining potential due to loss of desirable alleles.

Unfortunately, the breeder has very limited ability to predict or anticipate the occurrences of these four phenomena. Without knowledge of the details of the loci and alleles, it is impossible to predict the maximum potential. Antagonistic natural selection may be anticipated due to experiences in other populations, but since this is possibly a non-linear relationship to the change in the population mean, it is usually not possible to anticipate completely prior to the occurrence of the change in population mean. The recombination loss of selected epistatic effects is of limited predictability by the breeder because of the relative difficulty of obtaining estimates of the magnitude of epistatic effects or epistatic components of variance. The declining potential due to loss of desirable alleles can be predicted relatively as being a function of population size and intensity of selection.

In summary, there are nine groups of genetic characteristics influencing long-term response to selection. These are as follows:

1. Intended selection proportion, available genet-

ic variability, and accuracy of selection criteria.
2. Departure of achieved selection proportion from the intended
   a. Due to inbreeding effects upon reproduction.
   b. Due to changes in fitness associated with changes in metric traits.
3. Departures of selection differential from that expected
   a. Due to departures from normality.
   b. Due to chance in finite populations.
4. Inbreeding depression effects upon primary traits.
5. Magnitude of temporary response due to selection of epistatic effects and recombination loss due to the same.
6. Relationship between pureline and crossline performance and its implication upon type of tests and response.
7. Potential associated with maximum genotype; thus, the approach to this potential would be asymptotic.
8. Decline in potential due to fixation of undesirable alleles.
9. Random genetic drift.

Of course, the first item in this list includes the basic quantities used to predict short-term response to selection when the regression of response on criterion is linear and the criterion is normally distributed. This involves quantities that are possible for the breeder to know or to estimate.

Items 2a, 3a, 4, 6 and 8 are such that the breeder has a limited ability to estimate or anticipate in his particular selection program.

Items 2b, 3b, 5, 7 and 9 are phenomena that are impossible for the breeder to evaluate and are likely to continue to be so. In particular, item number 9, random genetic drift, which has seemingly led to considerable confusion in the interpretation of unreplicated selection experiments, is naturally unpredictable. However, the breeder does have some control over random genetic drift. The larger the population undergoing selection, the less is going to be the magnitude of departures from expectation due to random genetic drift. Thus, there seems to be 5 fundamental reasons for conducting a selection program in a large population. These are as follows:

1. The breeder can obtain more reliable estimates of population parameters.
2. The breeder can obtain more stability of expected response.
3. There will be less inbreeding depression.
4. There will be a lower probability of fixing less desirable alleles.
5. There will be a greater capacity for producing breeding stock for sale with limited opportunity for relaxed selection effect.

However, there is one reason for conducting selection programs in small populations, and that is the cost of conducting the selection program. Of course, this latter reason can predominate in many instances. The competitive business challenge for the breeder is to develop a breeding program that will give him a breeding stock production capacity near his market potential and thus he will be able to recover costs and a fair profit. Smaller populations can have the capacity to produce larger quantities of breeding stock if there are more intervening generations of multiplication. But these intervening generations of multiplication can lead to an increase in the relaxed selection effect. The breeder therefore wants to have a population large enough to accomplish the first four objectives and large enough to satisfactorily meet his capacity for producing breeding stock for sale at a profit relative to his cost of production.

With the present state of knowledge, the breeder should consider the following 4 factors in choosing and implementing a breeding plan:

1. Choose a basic plan (including selection proportions but not actual numbers) to maximize short-term response.
2. Implement the plan in scope sufficient to substantially reduce probability of fixation of undesirable alleles, but, for which cost can be supported.
3. Monitor parameter estimates each generation.
4. As evidence of reduction of rate of response or decline in genetic variability is observed, plans may be changed or sources of introducing new variability can be explored.

Unfortunately, our knowledge at the present time is relatively incomplete to give us the ability to reliably set lower limits on population sizes necessary to accomplish step No. 2. Quantitative geneticists need to develop the understanding and knowledge to be able to assess the relative probabilities of fixation of undesirable alleles such that the breeder can design breeding programs for their long-term potential as well as for their short-term potential. This latter objective is difficult since this probability, and, thus, the reduction in potential of selection response, are a function of gene frequencies and average effects of individual genes. Thus, precise knowledge is not possible. Hopefully, relative comparisons coming from experimentation or computer simulation can lead to an understanding that will support the breeder's need for a long-term evaluation (relative, at least) as he designs his breeding program.

BIBLIOGRAPHY

Al-Murrani, W. K. (1974). The limits to artificial selection. Anim. Brdg. Abstr. 42: 587-592.

Bohidar, N. R. (1960). Role of sexlinked genes in quantitative inheritance. Unpublished Ph.D. thesis. Library, Iowa State University, Ames, Iowa.

Cockerham, C. C. (1954). An extension of the concept of partitioning hereditary variance for analysis of covariance among relatives when epistosis is present. Genetics 39: 859-882.

Comstock, R. E., H. F. Robinson and P. H. Harvey. (1949(. A breeding procedure designed to make maximum use of both general and specific combining ability. Agron. J. 41: 360-367.

Dickerson, G. E. (1962). Implications of genetic-environmental interaction in animal breeding. Anim. Prod. 4: 47-63.

Dickerson, G. E. (1965). Experimental evaluation of selection theory in poultry. GENETICS TODAY (Proc. XI Int'l. Cong. Genetics, The Hague, 1963.) 3: 747-761.

Eisen, E. J. (1976). Selection indices and results with mice. Proc., Int'l. Conf. on Quant. Gen.

Enfield, F. D. (1972). Patterns of response to 70 generations of selection for pupa weight in Tribolium. Proc. Nat'l. Breeders Roundtable, Kansas City 21: 52-70.

Enfield, F. D. (1976). Selection experiments in Tribolium designed to look at gene action issues. Proc. Int'l. Conf. on Quant. Gen.

Falconer, D. S. (1960). <u>Introduction to quantitative genetics</u>. Oliver and Boyd, Edinburgh and London.

Falconer, D. S. (1976). Some results of the Edinburgh selection experiments with mice. Proc. Int'l. Conf. on Quant. Gen.

Fisher, R. A. (1918). The correlation between relatives on the supposition of Mendelian inheritance. Trans. Royal Soc. Edinburgh 52: 399-433.

Gillois, M. (1964). La relation d'identité en génétique. Thèse, Fac. Sciences, Paris, 294 p.

Griffing, B. (1960a). Theoretical consequences of truncation selection based on the individual phenotype. Aust. J. Biol. Sci. 13: 307-343.

Griffing, B. (1960b). Accommodation of linkage in mass selection theory. Aust. J. Biol. Sci. 13: 501-526.

Griffing, B. (1963). Comparisons of potentials for general combining ability selection methods utilizing one or two random-mating populations. Aust. J. Biol. Sci. 16: 838-862.

Harris, D. L. (1964). Genotypic covariances between inbred relatives. Genetics 50: 1319-1348.

Hill, W. G., and Alan Robertson. (1966). The effect of linkage on limits to artificial selection.

James, J. W. (1972). Optimum selection intensity in breeding programmes. Anim. Prod. 14: 1-9.

Kempthorne, O. (1954). The correlations between relatives in a random mating population. Proc. Royal Soc. London B, 143: 103-117.

Kempthorne, O. (1957). An introduction to genetic statistics. ISU Press, Ames, Iowa.

Kimura, M. (1957). Some problems of stochastic processes in genetics. Ann. Math. Stat. 28: 883-901.

Latter, B. H. D. (1965). The response to artificial selection due to antosomal genes of large effect. II. The effects of linkage on limits to selection in finite populations. Aust. J. Biol. Sci. 18: 1009-1023.

Nordskog, A. W. (1976). Success and failure of quantitative genetic theory in poultry. Proc., Int'l. Conf. on Quant. Gen.

Qureshi, A. W., O. Kempthorne, and L. N. Hazel. (1968). The role of finite population size and linkage in response to continued truncation selection. I. Additive gene action. Theor. and Appl. Gen. 38: 256-263.

Qureshi, A. W. (1968). The role of finite population size and linkage in response to continued truncation selection. II. Dominance and Overdominance Theor. and Appl. Gen. 38: 264-270.

Robertson, A. (1960). A theory of limits in artificial selection. Proc. Royal Soc., B, 153: 234-249.

Schnell, F. W. (1963). The covariance between relatives in the presence of linkage. Statistical Genetics and Plant Breeding Edited by W. O. Hanson and H. F. Robinson. Nat'l. Acad. Sci. - Nat'l. Res. Council Publ. 982: 463-483.

Van Aarde, I. M. R. (1963). Covariances of relatives in random mating populations with linkage. Unpublished Ph.D. thesis. Library, Iowa State University, Ames, Iowa.

Weir, B. S., and C. C. Cockerham. (1976). Two-locus theory in quantitative genetics. Proc., Int'l. Conf. on Quant. Gen.

Wright, S. (1921). Systems of Matings I-V Genetics 6: 111-178.

VIII

Mixed Model Theory in Quantitative Genetics

Prediction of future records

C. R. Henderson
DEPARTMENT OF ANIMAL SCIENCE
CORNELL UNIVERSITY, ITHACA, NY 14853

1. INTRODUCTION

The problem discussed in this paper is the following. The breeder has available a set of records, y, from which he wishes to predict the relative magnitude of future records. These predictions are then used to make selection decisions. For example, y might be milk records of the progeny of a set of dairy bulls from which the production of future progeny of these bulls is to be predicted. As a second example, y could represent milk records on all cows in a herd from the time that milk recording began until the present. The dairyman wishes to predict future milk yields on those cows still surviving that have records as well as the yields on heifers that have not yet reached producing age. As a third example, y represents a record on each of three traits on a set of animals. The breeder wishes to predict some function (not necessarily linear) of the records on future progeny of these animals.

Now, of course, the method of prediction should be dictated by what we want to do with the predictions, by computational costs, and by how well the results are understood by potential users. From the viewpoint of what we want to do with the predictions, two alternative and not necessarily incompatible goals are:

1. Maximizing the expected value of the functions of future records when individuals are selected upon the basis of their predictions.
2. Finding the most accurate "estimates" of functions of future records.

I tend to consider the second of these the more important since my responsibility as a researcher is not to make selections but rather to assist breeders in their selections by providing them with the best possible tools for selection. One must realize that the best selection for one breeder may not necessarily be the best for some other breeder, for their goals may differ, particularly with respect to relative emphasis to be placed on different traits. Consequently, we should provide "best" evaluations for each trait and then tell the breeder how to utilize this information to meet his goals. Further, the question of cost complicates the selection problem. The best selection from the standpoint of genetic improvement is not necessarily the best from an economic viewpoint. For example, the cost of an insemination from one dairy bull may be very different from that for another one. With the above considerations in mind, my efforts have been and will continue to be to emphasize most accurate evaluation possible within the limitations of computing costs and the costs of obtaining records suitable for evaluation. Accordingly, this paper deals with evaluation methods, which I choose to call prediction.

The problem of predicting future records is a special case of prediction of a nonobservable random vector, say g, that is distributed jointly with y. In the case of prediction of records, g is some function of future records. Also, g could be some variable, the prediction of which leads to a prediction of future records. Examples are given later. (An underscored lower case letter denotes a column vector, an underscored upper case letter a matrix, and an italicized lower case letter a scalar. Any of these can have one or more subscripts.)

2. BEST PREDICTION

We wish to predict g_i from a function of $\underline{y}$, say $f_i(\underline{y}) = \hat{g}_i$. How should this be done? It seems logical to try to minimize the prediction error, $\hat{g}_i - g_i$, or some function of error. The most commonly used notion is to minimize, if possible, $E(\hat{g}_i - g_i)^2$. Such a prediction method is called "best prediction" (BP), or the predictor itself is called a "best predictor" (also BP). It turns out that

$$\hat{g}_i = E(g_i | \underline{y}) \; ;$$

that is, the minimizing value is the conditional mean of g_i given $\underline{y}$. Some additional properties of BP are:

1. $E(\hat{\underline{g}}) = E(\underline{g})$. Thus, the predictor is unbiased.
2. $E[(\hat{\underline{g}} - \underline{g}),(\hat{\underline{g}} - \underline{g})'] = $ variance-covariance matrix of the prediction errors $= E_y[\text{Var}(\underline{g}|\underline{y})]$, that is, the mean in repeated sampling of the conditional variance-covariance matrix of $\underline{g}$ given $\underline{y}$.
3. $\text{Cov}(\hat{\underline{g}},\underline{g}') = \text{Var}(\hat{\underline{g}})$.
4. $\text{Cov}(\hat{\underline{g}},\underline{y}') = \text{Cov}(\underline{g},\underline{y}')$.
5. Correlation between $\hat{g}_i$ and g_i is maximum and is $\sigma_{\hat{g}_i}/\sigma_{g_i}$.
6. If $\underline{y}'$ can be partitioned into subvectors $[\underline{y}_1' \; \ldots \; \underline{y}_q']$ each of the same length and with the corresponding elements of $\underline{g}$ as $[g_1 \; \ldots \; g_q]$ such that the density function $h(\underline{y}_i, g_i)$ is the same for all $i = 1, \ldots, q$ and such that $(\underline{y}_i, g_i)$ is distributed independently of $(\underline{y}_j, g_j)$ for all $i \neq j$, the following property proved by Cochran (1951) is true. If truncation is carried out on $\hat{g}_i$, the resulting expected mean of the selected g_i is maximum for *any* function of $\underline{y}$.

Proofs of properties 1-5 above can be found in Searle (1973).

Now, it would appear to be impossible to find a more desirable prediction method than the one described above. Unfortunately, however, best predictors can seldom be found. Note that the form of distribution must be known, we must have the mathematical tools required for finding the conditional mean, and finally the numerical values of the parameters of the distribution must be known. Is the last requirement ever fulfilled? Further, in order for property 6 to be true, minimum requirements would be that the candidates for selection are unrelated (so that g_i are independent) and every candidate must have the same type of information and number of records (so that every y_i has the same length and the joint distribution of every $[y_i, g_i]$ is the same).

Fortunately, if the distribution is multivariate normal, best prediction simplifies greatly, for then the conditional mean of g is linear in y and the only parameter values needed are the first and second moments. Best prediction in this case is identical computationally to selection index discussed in the next section.

3. BEST LINEAR PREDICTION (SELECTION INDEX)

Suppose we are willing to restrict ourselves to predictions that are linear in y. That is,

$$\hat{g}_i = a_i + \underline{b}_i \underline{y}$$

where a_i is some scalar and $\underline{b}_i$ is a vector of weights chosen so that $E(\hat{g}_i - g_i)^2$ is minimum. It is found that

$$\hat{\underline{g}} = E(\underline{g}) + \underline{C}'\underline{V}^{-1}[\underline{y} - E(\underline{y})]$$

where $\underline{C}' = \text{Cov}(\underline{g}, \underline{y})$ and $\underline{V} = \text{Var}(\underline{y})$. Some useful properties of best linear prediction (BLP) are:

1. $E(\hat{\underline{g}}) = E(\underline{g})$. Thus, the predictor is unbiased.
2. $\text{Var}(\hat{\underline{g}} - \underline{g}) = \text{Var}(\underline{g}) - \underline{C}'\underline{V}^{-1}\underline{C}$.
3. $\text{Var}(\hat{\underline{g}}) = \text{Cov}(\hat{\underline{g}}, \underline{g}') = \underline{C}'\underline{V}^{-1}\underline{C}$.
4. The correlation between $\hat{g}_i$ and g_i is maximum in the class of linear functions of the form $a_i + \underline{b}_i'\underline{y}$.
5. When the distribution is multivariate normal, $\hat{\underline{g}}$ is the Bayesian estimate of $\underline{g}$ under a quadratic loss function (Solomon, 1971).
6. When the distribution is multivariate normal, the probability of a correct choice between g_i and g_j is maximized

for any linear function of $\underline{y}$ (see Henderson, 1963).

7. BP of $\underline{m}'\underline{g}$, a linear function of $\underline{g}$, is $\underline{m}'\hat{\underline{g}}$, that is, the same linear function of the BP of $\underline{g}$.

Best linear prediction is easy to compute when the number of records is small and is well known by animal breeders, but it does have the serious deficiency that $E(\underline{y})$ is usually if not always unknown. In contrast, $Var(\underline{y})$ and $Cov(\underline{g},\underline{y}')$ are often assumed to have been estimated quite accurately, and consequently most animal breeders seem willing to substitute these estimates for the parameter values and then to proceed as though these are the true values. With regard to the mean of $\underline{y}$, the assumption is usually made that it has the form, $\underline{X}\underline{\beta}$, where $\underline{X}$ is a known matrix and $\underline{\beta}$ is a fixed, unknown vector. Often, however, certain of the elements of $\underline{\beta}$ have been estimated quite accurately from prior data and can be used accordingly, as for example effects of age on milk production. Let $\underline{\beta}' = [\underline{\beta}_1'\vdots\underline{\beta}_2']$ and assume that $\underline{\beta}_1$ is known. Then, the record vector can be defined as $\underline{y}-\underline{X}_1\underline{\beta}_1$.

It is most unlikely, however, that all elements of $\underline{\beta}$ have been well estimated. In fact, in some cases, the sample at hand provides the only information. Consequently, the usual procedure used by breeders is to estimate $\underline{X}\underline{\beta}$ from the data, compute $\underline{y}-\underline{X}\hat{\underline{\beta}}$, and then proceed to apply selection index (BLP) as though $\hat{\underline{\beta}} = \underline{\beta}$. The next section deals with the question of the optimum choice of an estimate of $\underline{\beta}$.

4. BEST LINEAR UNBIASED PREDICTION

It was pointed out in the preceding section that the best linear prediction contains the numerical values of the means of $\underline{g}$ and of the records. It has also been assumed, naively I think, that the animals to be evaluated are all members of the same population. Consequently, their means have been assumed to be zero. Of course, any other constant would do equally well since its choice would not affect the ranking of predictions. Having made the assumption of equal means, the practice has then been to estimate $\underline{X}\beta$ by some method, sometimes by arithmetic means of levels of factors but often by regular least squares. The crucial question is what is the best estimate to use to substitute in selection index. Henderson (1963) solved a more general problem; namely, he found the linear function of $\underline{y}$ with mean $\underline{0}$ which minimizes the prediction error variance. That is, he minimized $E(\hat{g}_i - g_i)^2$ subject to $E(\hat{g}_i) = 0$. The result was $\hat{\underline{g}} = \underline{C}'\underline{V}^{-1}(\underline{y}-\underline{X}\hat{\underline{\beta}})$ where $\hat{\underline{\beta}}$ is a solution to the generalized least squares equations, $\underline{X}'\underline{V}^{-1}\underline{X}\hat{\underline{\beta}} = \underline{X}'\underline{V}^{-1}\underline{y}$. This method has been called best linear unbiased prediction (BLUP). Note that this is the selection index result with $\underline{X}\hat{\underline{\beta}}$ substituted for $\underline{X}\beta$.

It is probably not surprising to find that of *all* estimates of $\underline{X}\beta$, the GLS (also best linear unbiased) estimate is best to use in selection index. What is not so intuitively obvious is that of *all* linear functions of $\underline{y}$ the best one is in fact selection index using a GLS estimate of $\underline{X}\beta$ for the parameter value. Some useful properties of BLUP are:

1. $E(\hat{\underline{g}}) = E(\underline{g})$. This, of course, was required in the method.
2. $\text{Var}(\hat{\underline{g}}-\underline{g}) = \text{Var}(\underline{g}) - \underline{C}'\underline{V}^{-1}\underline{C} + \underline{C}'\underline{V}^{-1}\underline{X}(\underline{X}'\underline{V}^{-1}\underline{X})^{-}\underline{X}'\underline{V}^{-1}\underline{C}$ where $(\underline{X}'\underline{V}^{-1}\underline{X})^{-}$ denotes any generalized inverse of $\underline{X}'\underline{V}^{-1}\underline{X}$.
3. $\text{Var}(\hat{\underline{g}}) = \text{Cov}(\hat{\underline{g}},\underline{g}') = \underline{C}'\underline{V}^{-1}\underline{C} - \underline{C}'\underline{V}^{-1}\underline{X}(\underline{X}'\underline{V}^{-1}\underline{X})^{-}\underline{X}'\underline{V}^{-1}\underline{C}$.
4. The correlation between $\hat{g}_i$ and g_i is maximum in the class of linear functions of $\underline{y}$ that have zero expectations.
5. When the distribution is multivariate normal, $\hat{\underline{g}}$ is BLUE and MLE of $E(\underline{g}|\underline{y})$, the conditional mean of $\underline{g}$ for fixed $\underline{y}$.
6. When the distribution is multivariate normal, the linear function in the class of such functions with mean zero that maximizes the probability of selecting the better of g_i and g_j is $\hat{g}_i - \hat{g}_j$.
7. BLUP of $\underline{m}'\underline{g}$ is $\underline{m}'\hat{\underline{g}}$.

PREDICTION OF FUTURE RECORDS

4.1. Different and Unknown Means of Variables to Be Predicted

Suppose now that we cannot assume validly that all animals come from the same population. Rather, we assume that

$$g_i = k'_i \beta + m'_i u$$

where u is assumed to have null means. It seems logical in this case to predict g_i by the sum of the GLS estimate of $k'_i\beta$ plus BLUP of $m'_i u$. Henderson (1975c) proved that this minimizes $E(\hat{g}_i - g_i)^2$ in the class of linear functions of y that are unbiased, that is, for which $E(\hat{g}_i) = E(g_i) = k'_i\beta$.

When choices must be made between animals from different populations with different means and these means are unknown, the use of the method of this section has not been proved to have optimum properties with regard to ranking, as is the case with ranking normally distributed variables with known or with equal means. This is one of the unsolved problems in selection.

4.2. An Efficient Method for Computing BLP and BLUP

The foregoing methods described for BLP (selection index) and BLUP are difficult computationally if there are many elements in y, that is, if there are many records. The difficulty arises in that the large nondiagonal matrix V must be inverted. A much easier method was described by Henderson (1950) for certain prediction problems.

Suppose now that a linear model for y can be written as

$$y = X\beta + Zu + e$$

where X and Z are known matrices; β is a fixed vector, usually unknown; u and e are nonobservable random vectors with $\text{Var}(u) = G\sigma^2$, $\text{Var}(e) = R\sigma^2$, and $\text{Cov}(u,e') = 0$. G and R are known, nonsingular matrices; σ^2 is a scalar, the value of which is not needed in prediction. Now write the following equations:

$$\begin{pmatrix} X'R^{-1}X & X'R^{-1}Z \\ Z'R^{-1}X & Z'R^{-1}Z + G^{-1} \end{pmatrix} \begin{pmatrix} \hat{\beta} \\ \hat{u} \end{pmatrix} = \begin{pmatrix} X'R^{-1}y \\ Z'R^{-1}y \end{pmatrix}.$$

Henderson et al. (1959) proved that $\hat{\beta}$ is a GLS solution for β, and Henderson (1963) proved that $\hat{u}$ is BLUP of u. Then, of course, because of the invariance property of BLUP, if $g_i = m_i'u$, BLUP of g_i is $m_i'\hat{u}$.

Also, if β is known, $\hat{u}$, the solution to the following equations, is BLP (selection index);

$$(Z'R^{-1}Z + G^{-1})\hat{u} = Z'R^{-1}(y - X\beta) \; .$$

In most applications, $R = I$, and G has a form that is easy to invert. This method has been applied to some very large problems, for example, the evaluation of approximately 3000 sires (the u vector) with progeny in at least 240,000 herd-year-seasons comprising most of the elements of the β vector.

The method has the further advantage in that the sampling variances derive directly from a g-inverse of the coefficient matrix. Let a g-inverse be

$$\begin{pmatrix} C_{11} & C_{12} \\ C_{12}' & C_{22} \end{pmatrix} .$$

Then, if $K'\beta$ is estimable

$$\text{Var}(K'\hat{\beta}) = K'C_{11}K\sigma^2 \; ;$$
$$\text{Cov}(K'\hat{\beta}, \hat{u}' - u') = K'C_{12}\sigma^2 \; ;$$
$$\text{Var}(\hat{u} - u) = C_{22}\sigma^2 \; .$$

5. PREDICTION WHEN VARIANCES AND COVARIANCES ARE UNKNOWN

It has been assumed in the preceding sections that the needed variances and covariances, V and C (or G and R in the mixed linear model) are known.

How do we proceed if these parameters are unknown? Geneticists commonly use a combination of genetic theory and estimates of variances from whatever data have been published to derive estimates of these parameters. If these prior estimates have small sampling variances and if the data to be used for prediction can legitimately be regarded as a random sample from the same population as the data from which the prior estimates were obtained, the predictions should be about as good as if we knew the variances. Some work has been done on the sensitivity of the predictions to errors in these variances. In general, it appears that accuracy of ranking of single traits is quite insensitive to errors in estimates of variances, but multiple trait selection is much more sensitive to errors.

A particularly difficult problem is that of prediction from a body of data that provides the only information available on the variances. Econometricians have been concerned with this problem with respect to estimation of $\underline{\beta}$. If I understand the literature correctly, they commonly estimate the variance-covariance matrix of the error vector by some method that yields consistent estimators, regard these estimates as the parameter values, and then apply generalized least squares. This seems somewhat strange to me since consistency is a large sample property and many of their samples are not large. In contrast, animal breeders, often with very large samples, seem to prefer to use unbiased estimates of variances in their selection index methods and more recently in their BLUP equations. The question as to which is the best procedure to follow if the variances are unknown indeed may be an unsolvable problem.

One possibility is to estimate variances by maximum likelihood. Then, if these estimates are substituted for the corresponding parameter values in the mixed model equations, the resulting solution to $\underline{\beta}$ is maximum likelihood, and the solution to $\underline{u}$ is the maximum likelihood estimate of the conditional mean of $\underline{u}$ given $\underline{y}$. The mixed model equations solution can be used conveniently to estimate the variances as pointed out by Henderson (1973). Whether or not the resulting predictions are better or worse than those resulting from substitution of variances estimated by easier methods such as Henderson's (1953) methods is not known. Of course, it is known that maximum likelihood does have certain optimum properties in large samples.

The variance components model is

$$y = X\beta + Z_1 u_1 + \ldots + Z_c u_c + e$$

where $\mathrm{Var}(u_i) = I\sigma_i^2$, $\mathrm{Var}(e) = I\sigma_e^2$, $\mathrm{Cov}(u_i, u_j') = 0$ for all $i \neq j$, and $\mathrm{Cov}(u_i, e') = 0$ for all i. Then, $Z'R^{-1}Z + G^{-1}$ of the mixed model equations becomes

$$\begin{pmatrix} Z_1'Z_1 + \dfrac{I\sigma_e^2}{\sigma_1^2} & Z_1'Z_2 & \cdots & Z_1'Z_c \\ Z_2'Z_2 & Z_2'Z_2 + \dfrac{I\sigma_e^2}{\sigma_2^2} & \cdots & Z_2'Z_c \\ \vdots & \vdots & & \vdots \\ Z_c'Z_1 & Z_c'Z_2 & \cdots & Z_c'Z_c + \dfrac{I\sigma_e^2}{\sigma_c^2} \end{pmatrix}.$$

Let the inverse of this matrix be

$$\begin{pmatrix} T_{11} & T_{12} & \cdots & T_{1c} \\ T_{12}' & T_{22} & \cdots & T_{2c} \\ \vdots & \vdots & & \vdots \\ T_{1c}' & T_{2c}' & \cdots & T_{cc} \end{pmatrix}.$$

The estimate of σ_e^2 is $(y'y - \beta'X'y - u_1'Z_1'y - \ldots - u_c'Z_c'y)/n$ where β and u are solutions to mixed model equations and n is the number of elements in y. The estimate of σ_i^2 is $(u_i'u_i + \sigma_e^2 \mathrm{tr} T_{ii})/q_i$ where q_i is the number of elements in u_i. The solution to β is MLE of β, and the solution to u is the MLE of the conditional mean of u. These equations require, of course, an iterative solution. The proof of these results is given in a paper submitted to *Biometrics* by Henderson, Ufford, and Schaeffer.

6. APPLICATIONS

This section and the remainder of this paper will deal with some applications of the principles presented above and in particular BLUP.

PREDICTION OF FUTURE RECORDS

6.1. Prediction of Future Records in an Additive Genetic Model Using Single Records

Suppose that we wish to predict future records of a set of animals, some of which have one record and others none. An additive genetic model is assumed. Let the model for the records be

$$\underline{y} = \underline{X}\underline{\beta} + \underline{Z}\underline{a} + \underline{e}$$

where $\underline{a}$ represents the vector of additive genetic merits of the animals to be evaluated, and $\underline{Z}$ is a matrix with as many rows as there are records and as many columns as animals to be evaluated. The columns pertaining to animals with no records have all null elements. For example, if there are 5 animals, the 2nd, 3rd, and 4th only having records,

$$\underline{Z} = \begin{pmatrix} 0 & 1 & 0 & 0 & 0 \\ 0 & 0 & 1 & 0 & 0 \\ 0 & 0 & 0 & 1 & 0 \end{pmatrix}.$$

$$\text{Var}(\underline{a}) = \underline{A}h^2\sigma_y^2 ;$$
$$\text{Var}(\underline{e}) = \underline{I}(1-h^2)\sigma_y^2 ;$$
$$\text{Cov}(\underline{a},\underline{e}') = \underline{0}$$

where $\underline{A}$ is the numerator relationship matrix of the animals to be evaluated, h^2 = heritability of the trait in an unselected, noninbred population, and σ_y^2 is the variance of records in this population. Then, the variance of the record on the ith animal is $(1 + f_i h^2)\sigma_y^2$, and the covariance between records on the ith and jth animals is $a_{ij} h^2 \sigma_y^2$ where f_i is the inbreeding coefficient of the ith animal and a_{ij} is the numerator relationship between the ith and jth animal. Then, the BLUP equations for predicting $\underline{a}$ are

$$\begin{pmatrix} \underline{X}'\underline{X} & \underline{X}'\underline{Z} \\ \underline{Z}'\underline{X} & \underline{Z}'\underline{Z}+\underline{A}^{-1}\frac{1-h^2}{h^2} \end{pmatrix} \begin{pmatrix} \hat{\underline{\beta}} \\ \hat{\underline{a}} \end{pmatrix} = \begin{pmatrix} \underline{X}'\underline{y} \\ \underline{Z}'\underline{y} \end{pmatrix}.$$

The prediction of future records on the animals evaluated in these equations is $\hat{\underline{a}}$.

The additive genetic merit (and predicted record) of an animal not appearing in the $\underline{a}$ vector above is

$$\underline{c}'\underline{A}^{-1}\hat{\underline{a}}$$

where $\underline{c}'$ is the row vector of numerator relationships between this animal and those in the equations above. In the special case of an animal with no records and no progeny with records but with evaluated sire and dam, $\underline{c}'\underline{A}^{-1}\hat{\underline{a}}$ simplifies to the mean of the $\hat{a}_i$ of its parents.

If the number of animals in $\underline{A}$ is large, the rapid methods described by Henderson (1975a, 1975b, 1976a, 1976b) with a modification of one of them by Quaas (1976) can be used to compute $\underline{A}^{-1}$.

In this application using $\underline{A}$ and in subsequent ones, it is essential that all records be included that were used in making selection decisions.

6.2. Prediction of Future Records in an Additive Model Using Multiple Records

In this section, we deal with traits on which more than one record can be obtained, lactation yields in dairy cows, for example. Let the model for all records be

$$\underline{y} = \underline{X}\underline{\beta} + \underline{Z}\underline{a} + \underline{Z}\underline{p} + \underline{e}$$

where $\underline{a}$ is the vector of additive genetic merits of all animals, $\underline{p}$ is the vector of permanent environmental effects, and $\underline{Z}$ is a matrix of 1's and 0's relating a particular record to a particular individual. If the individual has no record, the column of $\underline{Z}$ relating to that animal is null. For example, if there are 3 animals to be evaluated, the first with 2 records, the second with 1 record, and the third with none,

$$\underline{Z} = \begin{pmatrix} 1 & 0 & 0 \\ 1 & 0 & 0 \\ 0 & 1 & 0 \end{pmatrix}.$$

$$\text{Var}(\underline{a}) = \underline{A}h^2\sigma_y^2 ;$$
$$\text{Var}(\underline{p}) = \underline{I}(r-h^2)\sigma_y^2 ;$$
$$\text{Var}(\underline{e}) = \underline{I}(1-r)\sigma_y^2$$

where r = repeatability and σ_y^2 = variance of single records in a noninbred population. $\text{Cov}(\underline{a},\underline{p}')$, $\text{Cov}(\underline{a},\underline{e}')$, and $\text{Cov}(\underline{p},\underline{e}')$ are null. The BLUP equations are

PREDICTION OF FUTURE RECORDS

$$\begin{pmatrix} X'X & X'Z & X'Z \\ Z'X & Z'Z+A^{-1}\frac{1-r}{h^2} & Z'Z \\ Z'X & Z'Z & Z'Z+I\frac{1-r}{r-h^2} \end{pmatrix} \begin{pmatrix} \hat{\beta} \\ \hat{a} \\ \hat{p} \end{pmatrix} = \begin{pmatrix} X'y \\ Z'y \\ Z'y \end{pmatrix}.$$

Note that all elements of $\hat{p}$ pertaining to animals with no records all equal 0. Consequently, the equations and coefficients corresponding to $\underline{p}$ of these animals can be deleted.

The prediction of the future production of the ith animal is $\hat{a}_i + \hat{p}_i$. The prediction of production of animals without records is the same as in the preceding section.

6.3. Prediction of Future Production Under an Additive Plus Dominance Model

The model is the same as in Section 6.2 except now a dominance term has been added and the method is restricted to a noninbred population.

The model is

$$\underline{y} = X\underline{\beta} + Z\underline{a} + Z\underline{d} + Z\underline{p} + \underline{e}$$

where $\underline{d}$ refers to dominance value with partitioning as described by Cockerham (1963). The covariances between $\underline{d}$ and $\underline{a}$, $\underline{p}$, and $\underline{e}$ are null.

$$\text{Var}(\underline{a}) = \underline{A}\sigma_a^2 ;$$
$$\text{Var}(\underline{d}) = \underline{D}\sigma_d^2$$

where $\underline{D}$ is the dominance relationship matrix;

$$\text{Var}(\underline{p}) = I\sigma_p^2 ;$$
$$\text{Var}(\underline{e}) = I\sigma_e^2 .$$

Then the variance of a record is $\sigma_a^2 + \sigma_d^2 + \sigma_p^2 + \sigma_e^2$. Covariance between two records on the same individual is $\sigma_a^2 + \sigma_d^2 + \sigma_p^2$. Covariance between records on the ith and jth animal is $a_{ij}\sigma_a^2 + d_{ij}\sigma_d^2$. The BLUP equations are

$$\begin{pmatrix} \underline{X}'\underline{X} & \underline{X}'\underline{Z} & \underline{X}'\underline{Z} & \underline{X}'\underline{Z} \\ \underline{Z}'\underline{X} & \underline{Z}'\underline{Z}+\underline{A}^{-1}\frac{\sigma_e^2}{\sigma_a^2} & \underline{Z}'\underline{Z} & \underline{Z}'\underline{Z} \\ \underline{Z}'\underline{X} & \underline{Z}'\underline{Z} & \underline{Z}'\underline{Z}+\underline{D}^{-1}\frac{\sigma_e^2}{\sigma_d^2} & \underline{Z}'\underline{Z} \\ \underline{Z}'\underline{X} & \underline{Z}'\underline{Z} & \underline{Z}'\underline{Z} & \underline{Z}'\underline{Z}+\underline{I}\frac{\sigma_e^2}{\sigma_p^2} \end{pmatrix} \begin{pmatrix} \hat{\underline{\beta}} \\ \hat{\underline{a}} \\ \hat{\underline{d}} \\ \hat{\underline{p}} \end{pmatrix} = \begin{pmatrix} \underline{X}'\underline{y} \\ \underline{Z}'\underline{y} \\ \underline{Z}'\underline{y} \\ \underline{Z}'\underline{y} \end{pmatrix}.$$

The prediction of a future record (expressed as a deviation from 0) on the ith animal included in the equations above is

$$\hat{a}_i + \hat{d}_i + \hat{p}_i .$$

The predicted production of an animal not appearing in these equations is

$$\underline{c}'\underline{A}^{-1}\hat{\underline{a}} + \underline{d}'\underline{D}^{-1}\hat{\underline{d}}$$

where $\underline{c}'$ is the row vector of the numerator relationship between the animal and those above and $\underline{d}'$ is a similar vector for the dominance relationship.

The methods of this section can be extended to other components of genetic variation, e.g., additive × additive.

6.4. Evaluation of Multiple Traits in an Additive Genetic Model

This section deals with the problem of evaluating several correlated traits from records on the individual and its relatives. Suppose that the additive genetic variance matrix of traits in the same animal is $\underline{G}_0$, the corresponding environmental variance-covariance matrix is $\underline{R}_0$, the genetic and environmental variables are uncorrelated, and consequently the phenotypic variance-covariance matrix is $\underline{G}_0 + \underline{R}_0 = \underline{V}_0$, say. Suppose that the ith animal has a record on each of the traits and the model for the vector of records is denoted by $\underline{y}_i$, the model for which is

$$\underline{y}_i = \underline{X}_i\underline{\beta} + \underline{a}_i + \underline{e}_i ;$$
$$\text{Var}(\underline{a}_i) = \underline{G}_0 ;$$
$$\text{Var}(\underline{e}_i) = \underline{R}_0 ;$$
$$\text{Cov}(\underline{a}_i, \underline{e}_i') = \underline{0} .$$

Now, if $\underline{\beta}$ is known and we do not wish to use records on relatives of the ith animal, the regular selection index evaluation of

PREDICTION OF FUTURE RECORDS

$$\underline{a}_i = \underline{G}_0 \underline{V}_0^{-1}(\underline{y}_i - \underline{X}_i \underline{\beta}) \ .$$

Then, if the net merit is defined as $\underline{m}'\underline{a}_i$, the solution for this animal is $\underline{m}'\underline{\hat{a}}_i$.

If certain records are missing, we substitute for $\underline{G}_0$ and $\underline{V}_0$, these matrices reduced according to records missing.

Now suppose we want to use records on relatives and we cannot assume that $\underline{\beta}$ is known. We first consider the case in which we have n animals each with a record on all q of the traits. Then, if we order the records traits within animals, we have for our general BLUP equations $\underline{X}' = [\underline{X}_1' \quad \underline{X}_2' \quad \ldots \quad \underline{X}_n']$, $\underline{Z} = \underline{I}$, and

$$\underline{G} = \begin{pmatrix} a_{11}\underline{G}_0 & a_{12}\underline{G}_0 & \ldots & a_{1n}\underline{G}_0 \\ a_{12}\underline{G}_0 & a_{22}\underline{G}_0 & \ldots & a_{2n}\underline{G}_0 \\ \vdots & \vdots & & \vdots \\ a_{1n}\underline{G}_0 & a_{2n}\underline{G}_0 & & a_{nn}\underline{G}_0 \end{pmatrix}$$

where a_{ij} are elements of $\underline{A}$, the numerator relationship matrix. $\underline{G}$ can be written as $\underline{A}*\underline{G}_0$ where * denotes the direct product operation (Searle, 1966). Similarly,

$$\underline{R} = \begin{pmatrix} \underline{R}_0 & 0 & \ldots & 0 \\ 0 & \underline{R}_0 & \ldots & 0 \\ \vdots & \vdots & & \vdots \\ 0 & 0 & \ldots & \underline{R}_0 \end{pmatrix} = \underline{I}*\underline{R}_0 \ .$$

Now, because of the form of $\underline{G}$ and $\underline{R}$, these can be inverted easily;

$$\underline{G}^{-1} = \underline{A}^{-1}*\underline{G}_0^{-1}$$

where $\underline{A}^{-1}$ is easy to compute by the rapid methods cited above and $\underline{G}_0^{-1}$ is a small matrix. Similarly,

$$\underline{R}^{-1} = \underline{I}*\underline{R}_0^{-1} = \begin{pmatrix} \underline{R}_0^{-1} & & 0 \\ & \ddots & \\ 0 & & \underline{R}_0^{-1} \end{pmatrix} \ .$$

Then, the BLUP equations can be written, realizing that because $\underline{Z} = \underline{I}$,

$$\underline{X}'\underline{Z} = \underline{X}' \ ;$$
$$\underline{Z}'\underline{Z} = \underline{I} \ ;$$
$$\underline{Z}'\underline{y} = \underline{y} \ .$$

A more realistic situation is one in which certain of the animals to be evaluated have one or more missing records. In fact, some may have no records. In that case, $\underline{Z} \neq \underline{I}$ but rather is written by deleting from an $nq \times nq$ identity matrix all rows pertaining to missing records. For example, if there are three animals and two traits of interest and the first animal has records on both traits, the second has a record on only the first trait, and the third animal has no records,

$$\underline{Z} = \begin{pmatrix} 1 & 0 & 0 & 0 & 0 & 0 \\ 0 & 1 & 0 & 0 & 0 & 0 \\ 0 & 0 & 1 & 0 & 0 & 0 \end{pmatrix}.$$

Also, $\underline{R}$ is modified by deleting both rows and columns pertaining to missing records. In the above example,

$$\underline{R} = \begin{pmatrix} r_{11} & r_{12} & 0 \\ r_{12} & r_{22} & 0 \\ 0 & 0 & r_{11} \end{pmatrix}$$

where r_{ij} are elements of $\underline{R}_0$. $\underline{G}$ and $\underline{a}$ are the same as in the full information case, and consequently all traits on all animals can be predicted.

6.5. Evaluation from Progeny with One Record Each

In this section, we are concerned with the evaluation of animals using only progeny records. Each progeny has only one record used in the analysis. The vector of progeny records, $\underline{y}$, has the model

$$\underline{y} = \underline{X}\underline{\beta} + \underline{Z}\underline{s} + \underline{e}.$$

The most commonly assumed elements in $\underline{\beta}$ by those evaluating dairy sires in artificial insemination programs are herd-year-season effects and group effects (generation and stud, usually). $\underline{s}$ refers to a vector of sire values, with possibly some of the sires having no progeny. $\underline{Z}$ relates records to sires. For example, suppose there are three sires, the first with 4 progeny, the second with 2, and the third with none. Then, if records are ordered within sire,

PREDICTION OF FUTURE RECORDS

$$\underline{Z} = \begin{pmatrix} 1 & 0 & 0 \\ 1 & 0 & 0 \\ 1 & 0 & 0 \\ 1 & 0 & 0 \\ 0 & 1 & 0 \\ 0 & 1 & 0 \end{pmatrix}.$$

The assumptions about $\underline{s}$ and $\underline{e}$ are as follows in addition to the assumption of null means:

$$\text{Var}(\underline{s}) = \tfrac{1}{4}h^2 \underline{A}\sigma_y^2 ;$$
$$\text{Var}(\underline{e}) = (1-\tfrac{1}{4}h^2)\underline{I}\sigma_y^2 ;$$
$$\text{Cov}(\underline{s},\underline{e}') = \underline{0} .$$

Then the BLUP equations are

$$\begin{pmatrix} \underline{X}'\underline{X} & \underline{X}'\underline{Z} \\ \underline{Z}'\underline{X} & \underline{Z}'\underline{Z}+\underline{A}^{-1}\frac{4-h^2}{h^2} \end{pmatrix} \begin{pmatrix} \hat{\underline{\beta}} \\ \hat{\underline{s}} \end{pmatrix} = \begin{pmatrix} \underline{X}'\underline{y} \\ \underline{Z}'\underline{y} \end{pmatrix} .$$

If all sires are assumed to be members of the same population, the relative merit of their future progeny is predicted by differences in the elements of $\hat{\underline{s}}$. If groups are in the model, the prediction of the ith sire is $\hat{s}_i$ + an element of $\hat{\underline{\beta}}$ corresponding to the group in which the sire falls.

6.6. Evaluation from Progeny with Multiple Records

The problem in this section is the same as in 6.5 except that some or all of the progeny have more than one record. Now we write the model for the records as

$$\underline{y} = \underline{X}\beta + \underline{Z}_s \underline{s} + \underline{Z}_p \underline{p} + \underline{e}$$

where $\underline{p}$ refers to individual progeny. To illustrate $\underline{Z}_s$ and $\underline{Z}_p$, suppose there are three sires. The first has 3 progeny with 3, 2, and 1 records, respectively. The second sire has 2 progeny with 2 and 1 records, respectively, and the third sire has no progeny. Then, if records are ordered progeny within sires,

$$Z_s = \begin{pmatrix} 1 & 0 & 0 \\ 1 & 0 & 0 \\ 1 & 0 & 0 \\ 1 & 0 & 0 \\ 1 & 0 & 0 \\ 1 & 0 & 0 \\ 0 & 1 & 0 \\ 0 & 1 & 0 \\ 0 & 1 & 0 \end{pmatrix};$$

$$Z_p = \begin{pmatrix} 1 & 0 & 0 & 0 & 0 \\ 1 & 0 & 0 & 0 & 0 \\ 1 & 0 & 0 & 0 & 0 \\ 0 & 1 & 0 & 0 & 0 \\ 0 & 1 & 0 & 0 & 0 \\ 0 & 0 & 1 & 0 & 0 \\ 0 & 0 & 0 & 1 & 0 \\ 0 & 0 & 0 & 1 & 0 \\ 0 & 0 & 0 & 0 & 1 \end{pmatrix}.$$

s, p, and e are random vectors with null means.

$$\text{Var}(s) = \tfrac{1}{4}h^2 A \sigma_y^2 ;$$
$$\text{Var}(p) = (r - \tfrac{1}{4}h^2) I \sigma_y^2 ;$$
$$\text{Var}(e) = (1-r) I \sigma_y^2 .$$

Cov(s,p'), Cov(s,e'), and Cov(p,e') are all null. This, of course, is an additive genetic model, the mates of sires are assumed unrelated, and the usual simple repeatability model is assumed.

Then, in a noninbred population, it is seen that the variance of a record is $(\tfrac{1}{4}h^2 + r - \tfrac{1}{4}h^2 + 1 - r)\sigma_y^2 = \sigma_y^2$. The covariance between two records on the same progeny is $(\tfrac{1}{4}h^2 + r - \tfrac{1}{4}h^2)\sigma_y^2 = r\sigma_y^2$. The covariance between half-sibs is $\tfrac{1}{4}h^2 \sigma_y^2$. The covariance between a progeny of the ith sire and the progeny of the jth sire is $\tfrac{1}{4}h^2 a_{ij} \sigma_y^2$.

Then the BLUP equations are

PREDICTION OF FUTURE RECORDS

$$\begin{pmatrix} \underline{X'X} & \underline{X'Z}_s & \underline{X'Z}_p \\ \underline{Z'_sX} & \underline{Z'_sZ}_s + \frac{4(1-r)}{h^2}\underline{A}^{-1} & \underline{Z'_sZ}_p \\ \underline{Z'_pX} & \underline{Z'_pZ}_s & \underline{Z'_pZ}_p + \frac{4(1-r)}{4r-h^2}\underline{I} \end{pmatrix} \begin{pmatrix} \hat{\underline{\beta}} \\ \hat{\underline{s}} \\ \hat{\underline{p}} \end{pmatrix} = \begin{pmatrix} \underline{X'y} \\ \underline{Z'_sy} \\ \underline{Z'_py} \end{pmatrix}.$$

$\underline{Z'_sZ}_s$ is simply a diagonal matrix with a total number of progeny records for a sire in the diagonal. Similarly, $\underline{Z'_pZ}_p$ is a diagonal matrix with the diagonal elements being the number of records on an individual progeny. $\underline{Z'_sZ}_p$ contains in each row the number of records on the individual progeny of the sire and zeros. $\underline{Z'_sy}$ and $\underline{Z'_py}$ are totals of records by sire and progeny, respectively.

6.7. Multiple Trait Evaluation by Progeny Tests

If we wish to evaluate sires with respect to several correlated traits from records on their progeny, a combination of the methods of Sections 6.4 and 6.5 can be used. Details are presented by Henderson (1976c).

6.8. Nonlinear Merit Function

There are some practical situations in which the net merit of an individual is a nonlinear function of certain traits (see, for example, Wilton et al., 1968). As stated in Section 2, best prediction would be the logical method to use in this case, but it is unlikely that the required distribution and its parameters are known. A feasible method is available if $(\underline{y},\underline{u})$ are jointly normally distributed, if the first and second moments of the distribution are known, if the merit function is a polynomial in $\underline{u}$, and if we are willing to accept a prediction that is a polynomial in $\underline{y}$. In this case one can describe the problem as one of best linear prediction where the data vector available for prediction contains elements that are products and powers of the individual elements of $\underline{y}$.

For example, suppose that the merit function is

$$g = u_1 + 2u_2 + .5u_1^2 + .6u_1u_2 + .8u_2^2 .$$

Suppose that $\underline{y} = [y_1 \quad y_2]$ and we choose $\hat{g}$ of the form $\hat{g} = E(g) + b_1[y_1-E(y_1)] + b_2[y_2-E(y_2)] + b_3[y_1^2-E(y_1^2)] + b_4[y_1y_2-E(y_1y_2)] + b_5[y_2^2-E(y_2^2)]$. Then the prediction of this form that minimizes prediction error variance requires for its computation the parameters

$E[u_1 \quad u_2 \quad u_1^2 \quad u_1u_2 \quad u_2^2 \quad y_1 \quad y_2 \quad y_1^2 \quad y_1y_2 \quad y_2^2]$;

$E[y_1 \quad y_2 \quad y_1^2 \quad y_1y_2 \quad y_2^2][y_1 \quad y_2 \quad y_1^2 \quad y_1y_2 \quad y_2^2]'$;

$E[y_1 \quad y_2 \quad y_1^2 \quad y_1y_2 \quad y_2^2][u_1 \quad u_2 \quad u_1^2 \quad u_1u_2 \quad u_2^2]'$.

Fortunately, the expectations of products of normally distributed random variables can be written as functions of the means and of the variance-covariance matrix of variables comprising the product. These can be derived from the moment generating function. Let the vector $\underline{w}$ have a multivariate normal distribution with mean $\underline{\mu}$ and variance-covariance matrix $\underline{V}$. Then the moment generating function from Mood (1950) is

$$\exp(\underline{t}'\underline{\mu} + \underline{t}'\underline{V}\underline{t}) \ .$$

Using this function, some results are

$E(w_1w_2) = \mu_1\mu_2 + v_{12}$;

$E(w_1w_2w_3) = \mu_1\mu_2\mu_3 + \mu_1v_{23} + \mu_2v_{13} + \mu_3v_{12}$;

$E(w_1w_2w_3w_4) = \mu_1\mu_2\mu_3\mu_4$

$\qquad + \mu_1\mu_2v_{34} +$ all other terms of this type

$\qquad + v_{12}v_{34} +$ all other terms of this type ;

$E(w_1w_2w_3w_4w_5) = \mu_1\mu_2\mu_3\mu_4\mu_5$

$\qquad + \mu_1\mu_2\mu_3v_{45} +$ all other terms of this type

$\qquad + \mu_1v_{23}v_{45} +$ all other terms of this type ;

$E(w_1w_2w_3w_4w_5w_6) = \mu_1\mu_2\mu_3\mu_4\mu_5\mu_6$

$\qquad + \mu_1\mu_2\mu_3\mu_4v_{56} +$ other terms of this type

$\qquad + \mu_1\mu_2v_{34}v_{56} +$ other terms of this type

$\qquad + v_{12}v_{34}v_{56} +$ other terms of this type .

6.9. Restricted Best Linear Unbiased Prediction

Kempthorne and Nordskog (1959) solved the problem of constructing an index to select individuals such that the expected

PREDICTION OF FUTURE RECORDS

value of a set of linear functions of additive merits of several traits is null in the individuals selected for some defined merit function. Quaas and Henderson (1976) have generalized this method to solve the same problem but with the means of records unknown and using records on relatives. They call this method "restricted best linear unbiased prediction."

Let $\underline{u}_i$ be the subvector of additive genetic values on several traits on the ith animal. We want to predict $\underline{m}'\underline{u}_i$ so that $\underline{C}'\underline{u}_i$ is uncorrelated with the predictor $\underline{m}'\underline{\hat{u}}_i$, so that $\underline{m}'\underline{\hat{u}}_i$ has expectation zero, and with these 2 restrictions has minimum prediction error variance. The method for doing this is found to be identical to the method of Section 6.4 except that in place of $\underline{R}_0^{-1}$ a block diagonal matrix is used that is computed from $\underline{R}_0$, $\underline{G}_0$, $\underline{C}$, and $\underline{Z}$.

6.10. Prediction of the Merit of Single Crosses

We assume that a random set of lines has been derived from the same original population, that these lines all have the same inbreeding coefficients, and a random set of progeny is obtained from single crosses among these lines. The problem is to predict the merit of a single cross (including those on which there are no data). It is assumed that the "pure" line data will not be used. Let the model for an observation on the cross between the ith line used as the male line and the jth line used as the female line be

$$y_{ijk} = \underline{x}'_{ijk}\underline{\beta} + s_i + d_j + r_{ij} + e_{ijk}$$

where $\underline{x}'_{ijk}\underline{\beta}$ defines the fixed effects in the model.

$$\text{Var}(s_i) = \sigma_s^2 \; ;$$
$$\text{Var}(d_j) = \sigma_d^2 \; ;$$
$$\text{Cov}(s_i, d_i) = \sigma_{sd} \; ;$$
$$\text{Var}(r_{ij}) = \sigma_r^2 \; ;$$
$$\text{Cov}(r_{ij}, r_{ji}) = \sigma_{rr'} \text{ for } i \neq j \; ;$$
$$\text{Var}(e_{ijk}) = \sigma_e^2 \; .$$

All other covariances = 0.

Then the BLUP equations for predicting $\underline{s}$, $\underline{d}$, and $\underline{r}$ are written by first writing least squares equations as though $\underline{s}$, $\underline{d}$, and $\underline{r}$ are fixed. These equations have the form

$$\begin{pmatrix} \underline{C}_{11} & \underline{C}_{12} & \underline{C}_{13} & \underline{C}_{14} \\ \underline{C}'_{12} & \underline{C}_{22} & \underline{C}_{23} & \underline{C}_{24} \\ \underline{C}'_{13} & \underline{C}'_{23} & \underline{C}_{33} & \underline{C}_{34} \\ \underline{C}'_{14} & \underline{C}'_{24} & \underline{C}'_{34} & \underline{C}_{44} \end{pmatrix} \begin{pmatrix} \hat{\underline{g}} \\ \hat{\underline{s}} \\ \hat{\underline{d}} \\ \hat{\underline{r}} \end{pmatrix} = \begin{pmatrix} \underline{t}_1 \\ \underline{t}_2 \\ \underline{t}_3 \\ \underline{t}_4 \end{pmatrix}.$$

Now modify these equations as follows:

to $\begin{pmatrix} \underline{C}_{22} & \underline{C}_{23} \\ \underline{C}'_{23} & \underline{C}_{33} \end{pmatrix}$, add $\begin{pmatrix} \underline{I}\sigma_s^2 & \underline{I}\sigma_{sd} \\ \underline{I}\sigma_{sd} & \underline{I}\sigma_d^2 \end{pmatrix}^{-1} \sigma_e^2$;

to $\underline{C}_{44}$, add $\underline{V}^{-1}$ where $\underline{V}$ is a matrix with σ_r^2 in the diagonals and $\sigma_{rr'}$ in the positions corresponding to r_{ij} with r_{ji}; all other elements of $\underline{V}$ are zero.

From the solution to these equations, the prediction of the merit of the cross of line of sire i by line of dam j is

$$\hat{s}_i + \hat{d}_j + \hat{r}_{ij} .$$

The prediction of the merit of the ith line of sire in crosses with a random sample of dams from other random lines of the population is $\hat{s}_i$ and similarly for the prediction of the merit of the jth line of dam.

BIBLIOGRAPHY

[1] Cochran, W.G. (1951). Improvement by selection. *Proc. 2nd Berkeley Symp. Math. Statist. and Prob.*, 449.

[2] Cockerham, C.C. (1963). Estimation of genetic variance. *Statistical Genetics and Plant Breeding* (W.D. Hanson & H.F. Robinson, Eds.). Washington: National Academy of Sciences--National Research Council, 53-93.

[3] Henderson, C.R. (1950). Estimation of genetic parameters. *Biom. 6*, 186.

[4] Henderson, C.R. (1953). Estimation of variance and covariance components. *Biom. 9*, 226.

[5] Henderson, C.R. (1963). Selection index and expected genetic advance. *Statistical Genetics and Plant Breeding* (W.D. Hanson & H.F. Robinson, Eds.). Washington: National Academy of Sciences--National Research Council, 141-63.

[6] Henderson, C.R. (1973). Sire evaluation and genetic trends. *Proc. of the Anim. Breeding and Genet. Symp. in Honor of Dr. Jay L. Lush.* July 29, 1972, Blacksburg, Va.

[7] Henderson, C.R. (1975a). Rapid method for computing the inverse of a relationship matrix. *J. Dairy Sci. 58*, 1727.

[8] Henderson, C.R. (1975b). Inverse of a matrix of relationships due to sires and maternal grandsires. *J. Dairy Sci. 58*, 1917.

[9] Henderson, C.R. (1975c). Best linear unbiased estimation and prediction under a selection model. *Biom. 31*, 423.

[10] Henderson, C.R. (1976a). A simple method for computing the inverse of a numerator relationship matrix used for prediction of breeding values. *Biom. 32*, 69.

[11] Henderson, C.R. (1976b). Inverse of a matrix of relationships due to sire and maternal grandsires in an inbred population. *J. Dairy Sci.*, in press.

[12] Henderson, C.R. (1976c). Multiple trait sire evaluation using the relationship matrix. *J. Dairy Sci. 59*, 769.

[13] Henderson, C.R., Kempthorne, O., Searle, S.R., & von Krosigk, C.M. (1959). The estimation of environmental and genetic trends from records subject to culling. *Biom. 15*, 192.

[14] Kempthorne, O. & Nordskog, A.W. (1959). Restriced selection indices. *Biometrics 15*, 10.

[15] Mood, A.M. (1950). *Introduction to the Theory of Statistics.* New York: McGraw-Hill.

[16] Quaas, R.L. (1976). Computing the diagonal elements and inverse of a large numerator relationship matrix. *Biometrics*, in press.

[17] Quaas, R.L. & Henderson, C.R. (1976). Restricted best linear unbiased prediction of breeding values. *Biometrics*, submitted.

[18] Searle, S.R. (1966). *Matrix Algebra for the Biological Sciences (Including Applications in Statistics).* New York: John Wiley and Sons, Inc.

[19] Searle, S.R. (1973). Derivation of prediction formulae. BU-482-M, Biometrics Unit, Cornell Univ., Ithaca, New York.

[20] Solomon, D.L. (1971). A Bayesian interpretation of the genetic selection index. BU-362-M, Biometrics Unit, Cornell Univ., Ithaca, New York.

[21] Wilton, J.W., Evans, D.A., & VanVleck, L.D. (1968). Selection indices for quadratic models of total merit. *Biometrics 24*, 937.

Estimation of quantitative genetic parameters

Robin Thompson
A.R.C. UNIT OF STATISTICS
THE KING'S BUILDINGS, EDINBURGH EH9 3JZ, SCOTLAND

1. INTRODUCTION

The genetic parameters that will be considered are genetic variances and covariances. Standard references (for instance Cockerham [1963], Falconer [1960] and Kempthorne [1957]) have discussed the common methods and problems associated with estimating these parameters. Typically some system of mating is used to generate sets of relatives raised in one or more environments. Often an analysis of variance (for collateral relatives) or covariance (for non-collateral relatives) based on the mating and environmental design can be easily constructed. The resulting variance and covariance components are usually easily interpreted in terms of covariances between relatives. These covariances between relatives can be also interpreted in terms of genetic and environmental components and hence estimates of genetic variance can be derived. The key role of the analysis of variance is not surprising since it is a way of partitioning variance, and genetic variances arose out of partitioning phenotypic variance (Fisher [1918]).

In many cases (in fact most of those discussed in standard texts) this partition of variance is enough to make estimation simple and efficient. However there are cases when this is not so and in this paper we discuss maximum likelihood (ML) methods in some of these cases. There are two main cases. One case is the balanced designs where a partition of variance is possible but there are more covariances between relatives than parameters to estimate and hence for some parameters more than one estimate can be derived. In Section 2 ML estimation is discussed and in particular a simple estimation procedure very similar to weighted least squares is given.

The other case considered is the unbalanced designs, which can occur, for example, when it is impossible to raise families of equal size. In this situation many estimation procedures for variance components have been suggested (Searle [1971]). These are usually based on analogies with the analysis of variance for balanced data. In the past, ML estimation has been rarely attempted primarily because of the computational difficulties. It is argued in Section 3 that, in some cases at least, the computational difficulties are no worse than in many ad hoc schemes and that the terms in the ML estimating equations are often useful in animal breeding studies.

In Section 4 we discuss the modifications needed when animals used as parents are selected on their phenotypic performance. In Section 5 we discuss the case when data are available from more than 2 discrete generations.

2. ESTIMATION FROM BALANCED DESIGNS

The case when a balanced design generates more covariances between relatives than parameters is now discussed. For example consider a hierarchical structure in which there are s sires,

ESTIMATION OF QUANTITATIVE GENETIC PARAMETERS 641

d dams mated to each sire and n offspring raised from each dam, and data available on offspring and parents (Hill and Nicholas [1974], Thompson [1976]). This design generates five variances and covariances between relatives, namely, covariances between full sibs, σ_{fs}, between half sibs, σ_{hs}, between father and offspring, σ_{fo}, between mother and offspring, σ_{mo}, and phenotypic variance σ_p^2. In heritability estimation these structural parameters are often interpreted in terms of three environmental and genetic parameters: phenotypic variance, additive genetic variance, σ_A^2, and σ_K^2, the part of the full sib covariance not due to additive variance, which therefore contains dominance and common environmental terms. The relationship between the two sets of parameters is given in Table I.

TABLE I

Covariances among Relatives in terms of σ_A^2, σ_K^2 and σ_p^2.

Covariance	σ_A^2	σ_K^2	σ_p^2
σ_{fs}	1/2	1	0
σ_{hs}	1/4	0	0
σ_{fo}	1/2	0	0
σ_{mo}	1/2	0	0
σ_p^2	0	0	1

We will use this model for illustration even though there are several assumptions about genetic relationships, for instance no epistasis and no maternal effects, that might not be appropriate. We see there are 5 covariances between relatives and 3 parameters to estimate. Hill and Nicholas [1974] have shown how the parent-offspring and half sib estimates can be combined by evaluating the variances and covariances of the estimates. This is tedious and difficult to generalize. We now know how the ML estimates can be conveniently calculated.

Suppose X_i, Y_{ij} and Z_{ijm} represent observations on sires, dams and offspring and these are normally distributed about means μ_X, μ_Y and μ_Z, with covariances between observations given in

Table I. A convenient way of summarizing the data is to calculate 3 sum of squares and products matrices representing variation within dams ($\underset{\sim}{S}_1$) between dams within sires ($\underset{\sim}{S}_2$) and between sires ($\underset{\sim}{S}_3$). We let $\bar{X}_{.}$, $\bar{Y}_{i.}$ etc. denote means taken over the subscript replaced by a dot and

$$z_{ijm} = Z_{ijm} - \bar{Z}_{ij.}, \quad z_{ij.} = \bar{Z}_{ij.} - \bar{Z}_{i..}, \quad z_{i..} = \bar{Z}_{i..} - \bar{Z}_{...},$$

$$y_{ij} = Y_{ij} - \bar{Y}_{i.}, \quad y_{i.} = \bar{Y}_{i.} - \bar{Y}_{..}, \quad x_i = X_i - \bar{X}_{.}.$$

The sum of squares and products matrices are:

$$\underset{\sim}{S}_1 = (\Sigma \Sigma \Sigma z^2_{ijm}), \quad \underset{\sim}{S}_2 = \begin{pmatrix} n \Sigma \Sigma z^2_{ij.} & \Sigma \Sigma z_{ij.} y_{ij} \\ i j & i j \\ \Sigma \Sigma z_{ij.} y_{ij} & \Sigma \Sigma y^2_{ij} \\ i j & i j \end{pmatrix},$$

$$\underset{\sim}{S}_3 = \begin{pmatrix} nd \Sigma z^2_{i..} & d \Sigma z_{i..} y_{i.} & \Sigma z_{i..} x_i \\ i & i & i \\ d \Sigma z_{i..} y_{i.} & d \Sigma y^2_{i.} & \Sigma y_{i.} x_i \\ i & i & i \\ \Sigma z_{i..} x_i & \Sigma y_{i.} x_i & \Sigma x^2_i \\ i & i & i \end{pmatrix}. \quad (1)$$

We note that the z, y and x squared terms represent terms in the analysis of variance of offspring, dam and sire measurements respectively and the cross product terms represent terms in the analysis of covariance. The degrees of freedom associated with $\underset{\sim}{S}_1$, $\underset{\sim}{S}_2$ and $\underset{\sim}{S}_3$ are $\nu_1 = sd(n-1)$, $\nu_2 = s(d-1)$ and $\nu_3 = s-1$. The expected value of $\underset{\sim}{S}_h$ denoted by $\nu_h V_h$ can be written in terms of the covariances between relatives. We find

$$\underset{\sim}{V}_1 = (\sigma^2_p - \sigma_{fs}), \quad \underset{\sim}{V}_2 = \begin{pmatrix} \sigma^2_p - \sigma_{fs} + n(\sigma_{fs} - \sigma_{hs}) & \sigma_{mo} \\ \sigma_{mo} & \sigma_p \end{pmatrix},$$

$$\underset{\sim}{V}_3 = \begin{pmatrix} (\sigma^2_p - \sigma_{fs} + n(\sigma_{fs} - \sigma_{hs}) + nd\sigma_{hs} & \sigma_{mo} & \sigma_{fo} \\ \sigma_{mo} & \sigma^2_p & 0 \\ \sigma_{fo} & 0 & \sigma^2_p \end{pmatrix}. \quad (2)$$

The likelihood of all the data can be partitioned into two parts, one due to the fixed effects and one due to error contrasts i.e. contrasts with expectation independent of the fixed effects.

ESTIMATION OF QUANTITATIVE GENETIC PARAMETERS

We use this latter log-likelihood, $\mathcal{L}$, to estimate the variance parameters arguing that in the absence of knowledge about the fixed effects the former provide no information about the variance parameters (Patterson and Thompson [1971]). In this example the $\mathcal{L}$ is equivalent to the log-likelihood of $\underset{\sim}{S}_1, \underset{\sim}{S}_2$ and $\underset{\sim}{S}_3$ and can be written as

$$\mathcal{L} = \text{const} - \frac{1}{2} \sum_{h=1}^{3} \nu_h \left[\log |\underset{\sim}{V}_h| + \text{tr}(\underset{\sim}{M}_h \underset{\sim}{V}_h^{-1}) \right] \quad (3)$$

where $\underset{\sim}{M}_h = \underset{\sim}{S}_h / \nu_h$. In order to differentiate (3) with respect to the parameters we express the $\underset{\sim}{V}$'s as a linear function of the parameters, i.e.

$$\underset{\sim}{V}_h = \underset{\sim}{X}_{hA}\,\sigma_A^2 + \underset{\sim}{X}_{hK}\sigma_K^2 + \underset{\sim}{X}_{hp}\sigma_p^2 \quad (4)$$

where the $\underset{\sim}{X}$'s are known matrices. The matrices $\underset{\sim}{X}_{hi}$ (h=1,2,3; i=A,K,p) can be derived by replacing the covariances in $\underset{\sim}{V}_h$ by the corresponding coefficients for σ_i^2 in Table I, for instance $\underset{\sim}{X}_{1A} = (-\tfrac{1}{2})$, $\underset{\sim}{X}_{1K} = (-1)$ and $\underset{\sim}{X}_{1p} = 1$. The values of σ_i^2 that maximize (3) satisfy

$$\frac{\partial \mathcal{L}}{\partial \sigma_i^2} = \sum_{h=1}^{3} \nu_k \,\text{tr}(\underset{\sim}{V}_h^{-1}\underset{\sim}{M}_h\underset{\sim}{V}_h^{-1}\underset{\sim}{X}_{hi}) - \sum_{h=1}^{3} \nu_K \,\text{tr}(\underset{\sim}{V}_h^{-1}\underset{\sim}{X}_{hi}) = 0 \quad (5)$$

Usually (5) cannot be solved explicitly and an iterative solution is needed. One based on using the expected values of the second differentials that is very similar to weighted least squares is suggested by Anderson [1973]. In this scheme $\hat{\sigma}_i^2$ is estimated from

$$\sum_{j=1}^{3} \tilde{A}_{ij}\,\hat{\sigma}_j^2 = \tilde{B}_i \quad (6)$$

where $\tilde{A}_{ij} = \sum_{h=1}^{3} \nu_h\,\text{tr}(\tilde{\underset{\sim}{V}}_h^{-1}\underset{\sim}{X}_{hi}\tilde{\underset{\sim}{V}}_h^{-1}\underset{\sim}{X}_{hj})$, $\tilde{B}_i = \sum_{h=1}^{3} \nu_h\,\text{tr}(\tilde{\underset{\sim}{V}}_h^{-1}\underset{\sim}{X}_{hi}\tilde{\underset{\sim}{V}}_h^{-1}\underset{\sim}{M}_h)$ (7)

and $\tilde{\underset{\sim}{V}}_h$ is an initial estimate of $\underset{\sim}{V}_h$. The procedure is repeated using $\hat{\sigma}_i^2$ to give $\tilde{\underset{\sim}{V}}_h$ (from (4)) until the estimates converge. The relationship with weighted least squares becomes apparent if we consider the linear model

$$m_h = \sum_{i=1}^{q} x_{hi}\,\theta_i + e_h \quad (h = 1, \ldots, H) \text{ with the e's uncorrelated,}$$

with variances w_h. The weighted least squares estimates of θ_i satisfy

$$\sum_{j=1}^{q} A_{ij} \hat{\theta}_j = B_i \qquad (8)$$

where $A_{ij} = \sum_{h=1}^{H} w_h^{-1} x_{hi} x_{hj}$ and $B_i = \sum_{h=1}^{H} w_h^{-1} x_{hi} m_h$. (9)

Obviously the weight given in (6) to $\underset{\sim}{M}_h$ depends on ν_h and $\tilde{V}_h$. If the $\underset{\sim}{M}_h$ are scalars then the weights are inversely proportional to V_h^2/ν_h which is not surprising since then $\underset{\sim}{M}_h$ has a χ^2 distribution with mean $\underset{\sim}{V}_h$ and degrees of freedom ν_h. This procedure has been introduced using the hierarchical example but can be used whenever the data can be split into independent sum of squares and product matrices and their expectation is a linear function of variance parameters. Other analogies with least squares carry over; if $\underset{\sim}{A}$ is singular not all the parameters can be estimated, $2\hat{\underset{\sim}{A}}^{-1}$ gives the asymptotic variance-covariance matrix of the estimates which makes it relatively easy to compare alternative designs and the efficiency of the ML versus other estimation procedures.

Equations similar to (6) and (7) arise in other estimation procedures. For example they occur in minimum norm quadratic unbiased estimation (MINQUE) (Rao [1973]) if $\tilde{V}_h$ is chosen to correspond with the norm being minimized. Other methods (for instance Horn, Horn and Duncan [1975]) follow from replacing $\tilde{V}_h$ by V_h in part of (7) and manipulating (6). Another possibility is to use weighted least squares on the elements of $\underset{\sim}{M}_h$ or the covariances between relatives (Hayman [1960]). This leads to the same estimates as in the ML procedure but needs the derivation of variances of and covariances between the elements of $\underset{\sim}{M}_h$ or the covariances between relatives. Although, in theory, these can be found, in practice the calculations can be intractable. Eisen [1967] suggested a design for estimating maternal genetic variances that generated 13 covariances between relatives and so $13 \times 14/2 = 91$ variances and covariances would be needed to implement the weighted least squares procedure.

3. ESTIMATION IN UNBALANCED DESIGNS

We discuss in this section ML estimation in unbalanced designs. In unbalanced designs sensible partitions of the data are not as obvious as in balanced designs. Often linear models for the observations (as opposed to linear models for the variance parameters as in Section 2) are introduced to generate appropriate partitions. A simple two factor model will be used to illustrate the main points. Extensions to more general models follow naturally but need matrix algebra to express the results compactly.

We assume a linear model of the form

$$y_{ki\ell} = \alpha_k + b_i + e_{ki\ell} \quad . \tag{10}$$

In sire evaluation this model is often used and then $y_{ki\ell}$ is the yield of the ℓ-th daughter of sire i in herd-season k, α_k is the effect of herd-season k and $e_{ki\ell}$ is a random variable normally distributed with mean zero and variance σ^2. If no other assumptions are made about α_k and b_i then α_k and b_i are called fixed effects and (10) is a fixed effects model. Alternatively if we assume the b_i are normally distributed with mean zero and variance σ_b^2 then the b_i are called random effects and (10) is then a mixed effects model. This mixed model implies that $\text{var}(y_{ki\ell}) = \sigma^2 + \sigma_b^2$ and $\text{cov}(y_{ki\ell}, y_{k'i'\ell'}) = \sigma_b^2$ if $i = i'$ and $= 0$ if $i \neq i'$. So the model can be written

$$y_{ki\ell} = \alpha_k + e'_{ki\ell} \tag{11}$$

where $e'_{ki\ell}$ has variance $\sigma^2 + \sigma_b^2$ and $\text{cov}(e'_{ki\ell}, e'_{k'i'\ell'}) = \sigma_b^2$ if $i = i'$ and 0 if $i \neq i'$. Often (11) is a covenient way of thinking about genetic models and helps in formulating the linear model (10). In the sire evaluation case if the covariances between daughters of a bull are assumed to be $\sigma_A^2/4$ and the variance of an observation is σ_p^2 this is consistent with a mixed model with $\sigma_b^2 = \sigma_A^2/4$ and $\sigma^2 = \sigma_p^2 - \sigma_A^2/4 = \sigma_p^2(1-h^2/4)$, where $h^2 = \sigma_A^2/\sigma_p^2$.

Searle [1971] has reviewed methods for estimating σ_b^2 and σ^2. Most follow the simple recipe of equating two sums of squares to their expectations. One of the commonest (called the method of fitting constants or Henderson's method 3) is now outlined because the development is useful in understanding ML estimation. If α_k and b_i were fixed effects and were estimated by least squares they would satisfy

$$n_{k0} \hat{\alpha}_k + \sum_i n_{ki} \hat{b}_i = y_{k0}, \quad (12)$$

$$\sum_k n_{ki} \hat{\alpha}_k + n_{0i} \hat{b}_i = y_{0i}, \quad (13)$$

where n_{ki} is the number of daughters of sire i in herd-season k and 0 indicates summation over a suffix. An analysis of variance can be constructed:

Source	Sum of squares	
Herd-seasons	$\sum y_{k0}^2 / n_{k0}$	
Sires (adjusted for Herd-seasons)	$\sum \hat{b}_i y_{0i}$	(14)
Residual	$\sum y_{ki\ell}^2 - \sum \hat{\alpha}_k y_{k0} - \sum \hat{b}_i y_{0i}$	(15)

The sires sum of squares is the difference between fitting a model with α_k and b_i and with α_k. The residual sum of squares is the sum of squares of deviations $(y_{kij} - \hat{\alpha}_k - \hat{b}_i)$. In the method of fitting constants (14) and (15) are equated to their expectation, which are functions of σ^2 and σ_b^2, and hence σ^2 and σ_b^2 can be estimated. The efficiency of this procedure is in general unknown and depends on the degree of unbalance and relative magnitude of σ^2/σ_b^2. However using this method with some unbalanced designs more precise estimates can be obtained if some of the data are removed (Swiger, Harvey, Everson and Gregory [1964]). Another misgiving I feel is in the ambivalence in the role of b_i. They are first assumed to be fixed effects to generate the sum of squares (14) and (15) and then assumed to be random effects to calculate their expectations. Further, if the mixed model is interpreted as (11) i.e. as a model for α_k with correlated errors it can be argued that weighted least squares and not least squares should be used to estimate α_k. Weighted least squares would usually require the inversion of a matrix of size the number of observations. However in mixed models of this

ESTIMATION OF QUANTITATIVE GENETIC PARAMETERS 647

type Henderson (in Henderson, Kempthorne, Searle and Von Krosigk [1959]) has shown that this inversion can be eliminated and that the weighted least squares estimate of α_k satisfies

$$n_{k0} \tilde{\alpha}_k + \sum_i n_{ki} \tilde{\beta}_i = y_{k0} , \qquad (16)$$

$$\sum_k n_{ki} \tilde{\alpha}_k + (n_{0i} + \gamma^{-1}) \tilde{\beta}_i = y_{0i} , \qquad (17)$$

where $\gamma = \sigma_b^2/\sigma^2$. These equations are very similar to (12) and (13) except that the coefficient n_{0i} in (13) is replaced by $(n_{0i} + \gamma^{-1})$ in (17). Henderson [1973] has emphasized that the $\tilde{\beta}_i$ can be interpreted as the predicted breeding values of bull i. In the sire evaluation case

$$\tilde{\beta}_i = [n_{0i} h^2/(4 + (n_{0i} - 1)h^2)] (y_{0i} - \sum_k n_{ki} \tilde{\alpha}_k)/n_{0i} \quad \text{i.e. the mean}$$

daughter-yield corrected for herd seasons is regressed back by a factor $[n_{0i} h^2/(4 + (n_{0i} - 1)h^2)]$.

Further $\tilde{\beta}_i$ plays a key part in the ML estimation of σ^2 and σ_b^2. Patterson and Thompson [1971] have shown that the ML estimating equations are equivalent to equating

$$\Sigma y_{kij}^2 - \Sigma \tilde{\alpha}_k y_{k0} - \Sigma \tilde{\beta}_i y_{0i} \quad \text{and} \quad \Sigma \tilde{\beta}_i^2 \quad \text{to their expected value.}$$

The first term is similar to a sum of squares of residuals (cf (15)), and the second term is the sum of squares of bull's predicted values. Once again an iterative scheme is usually needed to estimate σ^2 and σ_b^2 since (17) depends on σ_b^2/σ^2. In practical cases I have found the iterative scheme outlined by Patterson and Thompson [1971] has converged in two or three iterations. The fitting constant method essentially gives equal weight to each observation, other methods give weights to the family means that are functions of their size. For the one-way classification i.e. only one herd-season $\Sigma \tilde{\beta}_i^2$ can be written as $\Sigma \gamma^2 [\gamma + n_{0i}^{-1}]^{-2} [(y_{0i} - n_{0i} \tilde{\alpha}_1)/n_{0i}]^2$ and $[\gamma + n_{0i}^{-1}]^{-2}$ is the weighting suggested by Robertson [1962].

The ML method can be extended to more complicated cases.

Thompson [1977] has considered the unbalanced version of the hierarchical design discussed in Section 2. Kempthorne and Tandon [1953] (for a single family classification) and Ollivier [1974] (for a hierarchical family classification) have suggested regression schemes weighting families according to size in order to make most use of the parent-offspring information. The ML method automatically does this and, where appropriate, uses the extra information from the sib covariances and enables fixed effects to be estimated.

The ML method can also be generalized to deal with q traits. In the two factor case we estimate $q \times q$ matrices $\underset{\sim}{\sigma^2}$ and $\underset{\sim}{\sigma^2_b}$. Useful equations are then (Thompson [1973])

$$n_{k0}\,\tilde{\alpha}_{km} + \sum_i n_{ki}\,\tilde{\beta}_{im} = y_{k0m} \qquad (18)$$

$$\sum_k n_{ki}\,\tilde{\alpha}_{km} + n_{0i}\,\tilde{\beta}_{im} + \sum_{k=1}^{q} \gamma^{-1}_{mk}\,\tilde{\beta}_{ik} = y_{0im} \qquad (19)$$

where $\underset{\sim}{\gamma}^{-1} = \underset{\sim}{\sigma^2}(\underset{\sim}{\sigma^2_b})^{-1}$ and the suffix m represents the m-th trait $(m = 1, \ldots, q)$. Again $\tilde{\beta}_{im}$ can be interpreted as the predicted value for sire i for trait m. It is equivalent to combining the data on all q traits, corrected for herd-season effects, by means of a selection index to give the predicted value for the m-th trait. The ML estimating equations are natural extensions of the univariate equations. For instance, the sum of squares and products of the values $\tilde{\beta}_{im}$ are used. If $\underset{\sim}{\gamma}^{-1}$ is diagonal, then (18) and (19) separate into q parts each like (16) and (17) and there is no connection between the q variates.

If $\underset{\sim}{\gamma}^{-1}$ is not diagonal a canonical transformation of the variates enables the equations to be solved in q parts. The q new derived variates are

$$y^*_{kijr} = \sum_m T_{rm}\,y_{kijm} \qquad (m = 1, \ldots, q)$$

where $\underset{\sim}{T}$, the matrix of coefficients T_{rm} satisfies

$$\underset{\sim}{T}\,\underset{\sim}{\sigma^2_b}\,\underset{\sim}{T}' = \underset{\sim}{I} \quad \text{and} \quad \underset{\sim}{T}\,\underset{\sim}{\sigma^2}\,\underset{\sim}{T}' = \underset{\sim}{D}$$

where $\underset{\sim}{D}$ is a diagonal matrix. These canonical variates sometimes have a genetic interpretation in terms of which linear combination

ESTIMATION OF QUANTITATIVE GENETIC PARAMETERS

of traits is most heritable (Rouvier [1969]). They might also be useful in interpreting results on the effects of errors in parameter estimates on the efficiency of selection indices (Harris [1964]).

4. SELECTION OF PARENTS

Sometimes, either through design or accident, the animals that are used as parents are chosen on their phenotypic performance. Then some of the usual methods of estimation are biased, for example heritability if estimated by sib covariances, genetic correlations if estimated by parent-offspring regression. In this section it is argued that these difficulties are removed if ML is used.

Suppose we have observations on $\nu_1 + \nu_2$ parents (y_i) and on ν_2 offspring (z_i). Suppose also y_{1i} and y_{2i} are normally distributed with mean zero and variance V_1 and V_{22} and covariances V_{12} between y_i and z_i and also that parents are chosen at random. Let S_1 be the sum of squares for the parental data and $\underline{S}_2$ the sum of squares and cross products matrix for parent and offspring data. Let $\nu_2 \underline{V}_2$ be the expected value of $\underline{S}_2$, then $\underline{V}_2$ and $\underline{S}_2$ can be partitioned as

$$\underline{V}_2 = \begin{pmatrix} V_{11} & V_{12} \\ V_{21} & V_{22} \end{pmatrix} \quad \underline{S}_2 = \begin{pmatrix} S_{11} & S_{12} \\ S_{21} & S_{22} \end{pmatrix}$$

where $V_{11} = V_1$ and $V_{21} = V_{12}$. The log-likelihood can be written as

$$\mathcal{L} = \text{const} - \frac{1}{2}(\nu_1 \log|V_1| + \text{tr}[(S_1 - S_{11})V_1^{-1}] + \nu_2 \log|\underline{V}_2| + \text{tr}(\underline{S}_2 \underline{V}_2^{-1})) \quad . \tag{20}$$

$\mathcal{L}$ is of the same form as (3) and hence (6) can be used if V_1 and $\underline{V}_2$ are linear functions of the unknown parameters. An alternative instructive form for (20) follows if we partition $\mathcal{L}$ into two independent parts, one part, $\mathcal{L}_1$, the log-likelihood of y_{1i}, the parental data and another part, $\mathcal{L}_2$ the log-likelihood of $z_i - V_{21}V_{11}^{-1}y_i$ which can be thought of as the offspring record given (or conditional on) the parental record. Defining $S_{22 \cdot}$ and

$V_{22.}$ as the sum of squares and the variance of $z_i - V_{21}V_{11}^{-1}y_i$, we find $\mathcal{L}_1$ and $\mathcal{L}_2$ can be written as

$$\mathcal{L}_1 = \text{const} - \frac{1}{2}((\nu_1 + \nu_2)\log|V_1| + \text{tr}(S_1 V_1^{-1})), \qquad (21)$$

$$\mathcal{L}_2 = \text{const} - \frac{1}{2}(\nu_2 \log|V_{22.}| + \text{tr}(S_{22.} V_{22.}^{-1})). \qquad (22)$$

We see ML essentially makes use of three pieces of information. The parental data gives information on V_1, regression of z_i on y_i gives information on $V_{21}V_{11}^{-1}$ and $z_i - V_{21}V_{11}^{-1}y_i$ gives information on $V_{22.}$.

Suppose parents are chosen on their parental values, then following Kempthorne and Von Krosigk (in Henderson et al [1959]) and Curnow [1961] we can write the log-likelihood as the log-likelihood of parental values plus the log-likelihood of offspring values given parental values. This log-likelihood is again $\mathcal{L}_1 + \mathcal{L}_2 = \mathcal{L}$ (20) and the iterative scheme in Section 2 can be used. One minor modification is needed, A_{ij}, depends on the expected values of the second moments $E(y_i z_i)$ and $E(z_i^2)$. Following Curnow [1961] we express these conditional on the selected parental values. Let $\nu_1 M_{11} = S_{11}$, then noting that $z_i - V_{21}V_{11}^{-1}y_i$ has variance $V_{22.}$ and is independent of y_i we find

$$E(y_i z_i) = M_{11} V_{11}^{-1} V_{12} \qquad (23)$$

$$E(z_i^2) = V_{22.} + V_{21}V_{11}^{-1} M_{11} V_{11}^{-1} V_{12} = V_{22} - V_{21}V_{11}^{-1}(V_{11} - M_{11}) V_{11}^{-1} V_{12} \qquad (24)$$

$$= V_{22} - (1-K) V_{21} V_{11}^{-1} V_{12} \quad \text{if } M_{11} = K V_{11}.$$

Using (23) and (24) A_{ij} can be found to be

$$A_{ij} = \sum_{h=1}^{2} \nu_h \, \text{tr}(V_h^{-1} X_{hi} V_h^{-1} X_{hj}) - 2\,\text{tr}\,(V_1^{-1}(S_{11} - \nu_1 V_1) V_1^{-1} X_{1i} V_1^{-1} X_{1j})$$

$$+ 2\,\text{tr}\left[\begin{pmatrix} V_1^{-1}(S_{11} - \nu_1 V_1) V_1^{-1} & 0 \\ 0 & 0 \end{pmatrix} X_{2i} V_2^{-1} X_{2j}\right]. \qquad (25)$$

Terms similar to (25) have been given by Curnow [1961] (for parent-offspring data) and Thompson [1973 and 1976] (for multivariate parent-offspring data and multivariate hierarchical structures). Using (23) and (24) it can be checked that (5) gives unbiased estimating equations for θ_i. Note that in effect we are estimating the variances in the unselected population. Covariances and variances in the selected population could be evaluated using formulae similar to (23) and (24). Equations (23) and (24) and natural extensions of them have been used to investigate the effect of selection of parents on several common estimation procedures. They can be used to justify parent-offspring regression to estimate heritability (Falconer [1960]), to give measures of biases and to suggest correction factors for sib-covariance estimates of heritability and parent-offspring estimates of genetic correlation (Reeve [1953], Brown and Turner [1968]).

The formulae (20) - (25) have been written so that they hold if y_i and z_i are vectors. Obviously V_1, V_{12}, V_{22} etc. will then be interpreted as matrices of the appropriate size. Sometimes data are only available on the selected parents and the offspring. Then maximizing the conditional likelihood of the offspring given the parents, $\mathcal{L}_2$, seems an obvious suggestion. If $\mathcal{L}_2$ is written as $\mathcal{L} - \mathcal{L}_1$ it is of the form of (3). This way of writing the log-likelihood is similar to writing, in a non-orthogonal analysis of variance, the sum of squares for factor B after adjusting for factor A as the sum of squares for factors A and B minus the sum of squares for A (Searle [1971]). Henderson [1975] has discussed estimating fixed effects and predicting random effects from unbalanced designs using a similar conditional approach.

5. MORE THAN TWO GENERATIONS

In this section we give a convenient form for the covariances between relatives in different generations in terms of the additive genetic variance. We assume for simplicity that generations are discrete so that an individual in generation t is the offspring of parents in generation $(t-1)$ and that there are N_t individuals in the t-th generation.

In each generation we order the individuals by sex with males first. Suppose we start in generation 0 and if we assume the N_0 individuals are unrelated then, the coefficients of parentage of the individuals can be represented by a $N_0 \times N_0$ matrix, R_{00}, equal to $\frac{1}{2} I$. We now define matrices Z_t of size $N_{t+1} \times N_t$ relating the individuals of generation (t+1) with those in generation t in order to calculate the other coefficients of parentage. The Z_t matrices can be written as

$$\begin{pmatrix} Z_{tmm} & Z_{tmf} \\ Z_{tfm} & Z_{tff} \end{pmatrix} \qquad (26)$$

where the elements of Z_t are either $\frac{1}{2}$ or 0. The (j, k) element of Z_{tmm} is $\frac{1}{2}$ only if the k-th male of generation t is the father of the j-th male in generation (t+1). The other blocks of Z are defined similarly. Hill [1974] used similar matrices and notes that the blocks of Z represent the alternative pathways of genes

$$\left(\begin{array}{c|c} \text{males from males} & \text{males from females} \\ \hline \text{females from males} & \text{females from females} \end{array} \right).$$

The relationship matrix for the first three generations (which indicates the general form) can now be written, if no individual is inbred, as (Thompson [1977])

$$R = \frac{1}{2} \begin{pmatrix} I_0 & 0 & 0 \\ Z_0 & I_1 & 0 \\ Z_1 Z_0 & Z_1 & I_2 \end{pmatrix} \begin{pmatrix} I_0 & 0 & 0 \\ 0 & \frac{1}{2}I_1 & 0 \\ 0 & 0 & \frac{1}{2}I_2 \end{pmatrix} \begin{pmatrix} I_0 & Z_0' & Z_0'Z_1' \\ 0 & I_1 & Z_1' \\ 0 & 0 & I_2 \end{pmatrix} \qquad (27)$$

This is the product of a lower triangular matrix with a diagonal matrix and with an upper triangular matrix. The inverse of R, sometimes used for predicting random effects (Henderson [1976]) has a simple form since the left inverse of the lower triangular matrix is

$$\begin{pmatrix} I_0 & 0 & 0 \\ -Z_0 & I_1 & 0 \\ 0 & -Z_1 & I_2 \end{pmatrix}.$$

The variance matrix of the observations assuming just an additive genetic component and an environmental component is $\underset{\sim}{V} = 2\underset{\sim}{R}\ \sigma_A^2 + \underset{\sim}{I}\ \sigma_e^2$. Suppose the records in the t-th generation are $\underset{\sim}{y}_t$ and, for simplicity, these are normally distributed about a mean of 0. ML estimation of σ_A^2 and σ_e^2 depends on calculating $\underset{\sim}{V}^{-1}$ and, except for some special cases (for example observations only available on one sex (Thompson [1977])) in most practical cases this is not feasible. One suggestion is to work with deviations from parental values $\underset{\sim}{y}_{t+1}^+ = \underset{\sim}{y}_{t+1} - \underset{\sim}{Z}_t\ \underset{\sim}{y}_t$, since the variance matrix of these deviations is tridiagonal and the covariance between deviations two or more generations apart are zero. When each dam is mated to only one sire the variance structure for $\underset{\sim}{y}_t^+$ corresponds to a hierarchical analysis of variance with sire and dam components σ_e^2 and covariance between full sibs $\sigma_e^2 + \sigma_A^2/2$.

Another possibility is to work with $\underset{\sim}{y}_t^*$, the t-th generation values conditional on, or given, the ancestors' records, which can be interpreted as deviations of actual from predicted values. For example, $\underset{\sim}{y}_1^* = \underset{\sim}{y}_1 - (\sigma_A^2/(\sigma_e^2 + \sigma_A^2))\ \underset{\sim}{Z}_0 \underset{\sim}{y}_0$, where $(\sigma_A^2/(\sigma_e^2 + \sigma_A^2))\underset{\sim}{Z}_0 \underset{\sim}{y}_0$ represents a vector of mid-parent values regressed back by a factor $(\sigma_A^2/(\sigma_e^2 + \sigma_A^2))$ and hence are predicted values of $\underset{\sim}{y}_1$. The terms for the next generations are more complicated, but if we approximate the variance of $\underset{\sim}{y}_1^*$ by $v_1\ \underset{\sim}{I}$, where v_1 is the variance of the elements of $\underset{\sim}{y}_1^*$, we can approximate $\underset{\sim}{y}_2^*$ by $\underset{\sim}{y}_2 - \underset{\sim}{Z}_1 \underset{\sim}{y}_1 + (\sigma_e^2/v_1)\underset{\sim}{Z}_1 \underset{\sim}{y}_1^*$. Using the same type of approximation and a recursive argument similar to that of Bulmer [1971] in succeeding generations we find that $\underset{\sim}{y}_{t+1}^*$ might be approximated by

$$\underset{\sim}{y}_{t+1}^* = \underset{\sim}{y}_{t+1} - \underset{\sim}{Z}_t \underset{\sim}{y}_t + (\sigma_e^2/v_t)\underset{\sim}{Z}_t \underset{\sim}{y}_t^* \qquad (28)$$

where $v_t = \sigma_p^2 + d_t$, $d_{t+1} = \tfrac{1}{2} d_t - \tfrac{1}{2} H_t v_t^{-1} H_t$, $d_0 = 0$, $\sigma_p^2 = \sigma_A^2 + \sigma_e^2$ and $H_i = \sigma_A^2 + d_i$. As in Bulmer's case the d_i quickly converge to a limiting value d^*.

$$d^* = (-2\sigma_A^2 - \sigma_p^2 + [\sigma_p^4 + 4\sigma_A^2 (\sigma_p^2 - \sigma_A^2)]^{\frac{1}{2}})/4 .$$

The values of d^*/σ_p^2 for various values of σ_A^2/σ_p^2 are given in Table II and we see approximately $d^*/\sigma_p^2 = -\frac{1}{2}(\sigma_A^2/\sigma_p^2)^2$. $\sigma_p^2 + d^*$ can be thought of as the variance between the actual value and the predicted value using all parental, grand parental etc. information.

Table II

Values of d^*/σ_p^2 for various values of σ_A^2/σ_p^2.

σ_A^2/σ_p^2	0.0	0.2	0.4	0.6	0.8	1.0
d^*/σ_p^2	-0.000	-0.030	-0.100	-0.200	-0.330	-0.500

Sometimes the regression of response on selection differential is used to estimate heritability (Falconer [1960], Hill [1971, 1972]). This is similar to putting $d_t = 0$ in (28) and working with the mean values of y_t^* rather than the individual values. The variance-covariance matrix of the mean values can be derived by arguments similar to the development of (28) which I find more appealing than Hill's [1971, 1972] intuitive genetic approach.

BIBLIOGRAPHY

ANDERSON, T.W. [1973]. Asymptotically efficient estimation of covariance matrices with linear structure. Ann. Statist. 1, 135-141.

BROWN, G.H. and TURNER, H.N. [1968]. Response to selection in Australian Merino sheep. II Estimates of phenotypic and genetic parameters for some production traits on Merino ewes and an analysis of the possible effects of selection on them. Aust. J. of Agric. Research 19, 303-22 (Corrigendum, 21, 182).

BULMER, M.G. [1971]. The effect of selection on genetic variability. Amer. Natur. 105, 201-211.

COCKERHAM, C.C. [1963]. Estimation of genetic variances. Statistical Genetics and Plant Breeding. Nat. Acad. Sci. Nat. Res. Council Publ. 982, 53-94.

CURNOW, R.N. [1961]. The estimation of repeatability and heritability from records subject to culling. Biometrics 17, 553-66.

EISEN, S.J. [1967]. Mating designs for estimating direct and maternal variances and direct-maternal covariances. Can. J. Genet. Cytol. 9, 13-22.

FALCONER, D.S. [1960]. Introduction to Quantitative Genetics. Ronald Press Co., New York, N.Y.

FISHER, R.A. [1918]. The correlation between relatives on the supposition of Mendelian inheritance. Trans. Royal Soc., Edinburgh 52, 399-433.

HARRIS, D.L. [1964]. Expected and predicted progress from index selection involving estimates of population parameters. Biometrics 20, 46-72.

HAYMAN, B.I. [1960]. Maximum likelihood estimation of genetic components of variation. Biometrics 16, 369-381.

HENDERSON, C.R. [1973]. Sire evaluation and genetic trends. In Proc. Anim. Breed. Genet. Symp. Blacksburg, Virginia pp 10-41. American Society of Animal Sciences, Champaign, Illinois.

HENDERSON, C.R. [1975]. Best linear unbiased estimation and prediction under a selection model. Biometrics 31, 423-447.

HENDERSON, C.R. [1976]. A simple method for computing the inverse of a numerator relationship matrix used in prediction of breeding values. Biometrics 32, 69-83.

HENDERSON, C.R., O. KEMPTHORNE, S.R. SEARLE and C.M. von KROSIGK. [1959]. The estimation of environmental and genetic trends from records subject to culling. Biometrics 15, 192-218.

HILL, W.G. [1971]. Design and efficiency of selection experiments for estimating genetic parameters. Biometrics 27, 293-311.

HILL, W.G. [1972]. Estimation of realized heritabilities from selection experiments. I Divergent selection. II Selection in one generation. Biometrics 29, 747-780.

HILL, W.G. [1974]. Prediction and evaluation of response to selection with overlapping generations. Anim. Prod. 18, 117-139.

HILL, W.G. and F.W. NICHOLAS. [1974]. Estimation of heritability by both regression of offspring on parent and intraclass correlation of sibs in one experiment. Biometrics 30, 447-468.

HORN, S.D., R.A. HORN and D.B. DUNCAN [1975]. Estimating variances in linear models. Journal of the American Statistical Association, 70, 380-385.

KEMPTHORNE, O. [1957]. An introduction to Genetic Statistics. John Wiley and Sons, New York, N.Y.

KEMPTHORNE, O. and O.B. TANDON. [1953]. The estimation of heritability by regression of offspring on parent. Biometrics 9, 90-100.

OLLIVIER, L. [1974]. La regression parent-descendant dans le cas de descendances subdivisees en familles de taille inegale. Biometrics 30, 59-66.

PATTERSON, H.D. and R. THOMPSON. [1971]. Recovery of interblock information when block sizes are unequal. Biometrika 58, 545-554.

RAO, C.R. [1973]. Linear Statistical Inference and its Applications (Second Edition) Wiley, New York, N.Y.

REEVE, E.C.R. [1953]. Studies in quantitative inheritance. III Heritability and genetic correlation in progeny tests using different mating systems. J. Genet. 51, 520-542.

ROBERTSON, A. [1962]. Weighting in estimation of variance components. Biometrics 18, 413-415.

ROUVIER, R. [1969]. Ponderation des valeurs genotypiques dans la selection par indice sur plusieurs caracteres. Biometrics 25, 295-308.

SEARLE, S.R. [1971]. Linear Models. Wiley, New York, N.Y.

SWIGER, L.A., W.R. HARVEY, D.O. EVERSON and K.E. GREGORY [1969]. The variance of intraclass correlation involving groups with one observation. Biometrics 20, 818-826.

THOMPSON, R. [1973]. The estimation of variance and covariance components with an application when records are subject to culling. Biometrics 29, 527-550.

THOMPSON, R. [1976]. Design of experiments to estimate heritability when observations are available on parents and offspring. Biometrics 32, 283-304.

THOMPSON, R. [1977]. The estimation of heritability with unbalanced data. To appear in Biometrics.

Diallel and multi-cross designs: What do they achieve?

Klaus Hinkelmann
DEPARTMENT OF STATISTICS
VIRGINIA POLYTECHNIC INSTITUTE AND STATE UNIVERSITY
BLACKSBURG, VA 24060

INTRODUCTION

Basic to experiments in quantitative genetics is the underlying mating design as it determines what kinds of genetic information can be obtained, at least theoretically, from such an experiment (e.g. Hinkelmann, 1975). Among the mating designs that have been used most often are diallel crosses. Derived from the Greek word "διάλληλος," Schmidt (1919) introduced the name "diallel" to denote all possible crosses among a collection of male and female animals. More generally now, a diallel cross refers to a set of all possible crosses among a collection of genetic entities, such as (monoecious or dioecious) individuals or (inbred) lines.

The various types of diallel crosses can be classified as follows:

(i) Diallel mating design type I (Hinkelmann and Stern, 1960), also referred to as factorial mating design (Cockerham, 1963) or North Carolina design II (Comstock and Robinson, 1952): Consists of all crosses among s males and d females;

(ii) diallel mating designs type II involving p monoecious individuals or (inbred) lines with the following subtypes (e.g. Griffing, 1956a):

D_1: consists of $p(p-1)/2$ crosses, $p(p-1)/2$ reciprocal crosses, and p parents;

D_2: consists of $p(p-1)/2$ crosses and p parents;

D_3: consists of $p(p-1)/2$ crosses and $p(p-1)/2$ reciprocal crosses;

D_4: consists of $p(p-1)/2$ crosses;

(iii) partial diallel crosses of types I and II (e.g. Hinkelmann, 1966, Curnow, 1963);

(iv) two-level diallel crosses (Hinkelmann, 1974); and

(v) variations of the pure forms mentioned above (e.g. topcross designs).

With regard to multi-cross experiments, the concept of the type II diallel cross has been extended to the

(i) three-way mating design consisting of all $p(p-1)(p-2)/6$ crosses among p lines (Rawlings and Cockerham, 1962a);

(ii) four-way mating design consisting of all $p(p-1)(p-2)(p-3)/12$ double crosses among p lines (Rawlings and Cockerham, 1962b); and

(iii) partial three-way and four-way mating designs (Hinkelmann, 1965, 1968).

From a practical point of view, these mating designs provide a very simple and convenient method of generating a number of crosses in one or two generations; but often this is where the simplicity ends. What comes after the crosses have been grown and quantitatively evaluated, is not always clear. When I say this I do not mean to imply that we do not know how to analyze the data from such an experiment (although this point can be argued; see below). In fact, several methods of analysis have been proposed over the years, giving rise to a certain amount of controversy and discussion. However, what is even more important is the ambiguity in the interpretation of the results and the inferences that can be drawn from them.

When we talk here about inferences from diallel and multi-cross experiments, we must really distinguish between two aspects: a theoretical and a practical. In terms of the title of this paper I can only try to discuss the question: What <u>can</u> be achieved with diallel and multi-cross experiments? This is a theoretical question concerned with the general nature and scope of such experiments and the validity of their results. The other question is: What <u>has been</u> achieved with diallel and multi-cross experiments? This is a more practical question although it is, of course, related to the first: Have practitioners, i.e. animal and plant breeders, found the theoretical developments helpful and to what extent has that shaped and influenced their breeding programs?

Although the results of many diallel experiments have been reported in the literature, using one or the other method of analysis, I have not been able to really find an answer to the last question. In other words, researchers have tended to report on the information aspects of diallel and multi-cross experiments but not on their decision aspects. This is unfortunate since this latter aspect is of major importance from the point of view of this conference.

It is, of course, impossible to discuss here everything that has been said and written about diallel and multi-cross analyses. Therefore, I would first like to make some general statements, after that consider a particular situation bringing out and discussing some aspects and types of information associated with diallel analysis, and then relate this more generally to other cross experiments.

GENERAL STATEMENTS ABOUT DIALLEL ANALYSIS

In order to discuss what type of inference can be made from a diallel experiment, one must know first of all whether one is dealing with, what I have referred to at another occasion (Hinkelmann, 1975), a comparative diallel experiment or an exploratory diallel experiment. In the first case we usually wish to make comparisons of some sort among the lines included in the experiment, such as comparisons among their average performance in crosses, or comparisons among the crosses themselves. This is fairly straight-forward and concepts of quantitative genetics do not really enter at this point except such statistical-genetical concepts as general and specific combining ability, concepts very much related to main effects and interactions in factorial experiments.

In the exploratory diallel experiment we generally wish to make much wider inferences and obtain genetic information of a different type, namely information about various types of gene action like additive effects, dominance deviations, and different epistatic effects. Now, gene action effects are defined in the context of a well specified population. Various reference populations have been proposed and considered. These are:

(i) A random mating population from which the crosses are considered to be a random sample. This is equivalent to assuming that the set of inbred lines used in the diallel experiment is a random sample from a population of inbred lines which in turn were derived from a parent population by means of an inbreeding system free from forces which change gene frequencies (Kempthorne, 1956). [It has been shown (Griffing, 1956b) that a self-mating system of a random mating population in equilibrium leads to homozygous lines and that the gene frequencies in the original population can be obtained by diallel crossing among those inbred lines];

(ii) the set of parental lines utilized in the diallel cross, or more precisely, the equilibrium population obtained by random mating the specific set of inbred lines (e.g. Jinks, 1954, Hayman, 1954a,b);

(iii) a derived population which is a random mating population defined by the gene frequencies of the parents utilized in the diallel cross (Kuehl, Rawlings and Cockerham, 1968).

It does not seem to be at all clear which of these reference populations should be used as arguments for or against each one of them have been advanced. The important point to keep in mind is that the reference population, together with additional genetical and statistical assumptions, determines what type of genetic information can be obtained, how it can be obtained, and how it should be interpreted. To put it in a different way, this is presumably what Mather and Jinks (1971) mean when they say (although in a slightly different context): "...unless the breeding history of the individuals is known, no worthwhile biometrical genetical analysis or interpretation is possible".

The assumptions referred to above that have to be made for analyzing a diallel experiment, are generally of three different kinds:

(i) Those that are of a general nature, such as a.) normal diploid segregation, b.) the phenotypic value is equal to the sum of genotypic values and environmental effects, the latter being associated at random with the genotype. Both these assumptions have been relaxed (Savchenko, 1966, Matzinger and Kempthorne, 1956);

(ii) those that are necessitated by the choice of the reference population, such as a.) the parents are homozygous, b.) the genes are independently distributed among the parents, the latter assumption being crucial for reference population (ii) where, however, it may or may not be satisfied; and

(iii) those that are determined partly by the type of diallel to be analyzed and partly by the type of information desired, such as a.) there are only additive genetic and dominance effects (diallel D_4), b.) there are no differences among reciprocal crosses.

The point that I am trying to make here is that some of the assumptions are absolutely necessary while others can be relaxed depending on the particular situation. The researcher must determine which assumptions fall into which category based on the information he has about the genetic material and the type of information desired, before he "chooses" the parental lines and embarks upon a particular type of diallel experiment.

In any case, any information obtained from a diallel experiment pertains to the chosen reference population. This information is usually in the form of genetic variance and covariance components estimated from second order statistics such as sums of squares from an analysis of variance table or sums of products.

ANALYSES OF A DIALLEL EXPERIMENT OF TYPE D_3

So far I have discussed diallel analysis in very general terms. It is, of course, impossible to treat all conceivable situations in detail. Instead I shall consider the type D_3 diallel, consisting of $p(p-1)/2$ crosses among p inbred lines and the $p(p-1)/2$ reciprocal crosses, and describe various suitable methods of analysis, how they are related and what information can be obtained. At the same time this will bring out the limitations of diallel experiments.

I shall make the following general assumptions with others to be added as the need arises for the specific models to be discussed:

(i) Normal diploid segregation;
(ii) completely inbred parental lines (F=1);
(iii) reference population (i) applicable;
(iv) no genotype-environment interaction.

If for a specific cross i denotes the parental line and j denotes the female line, then the yield of the k-th offspring of this cross shall be denoted by Y_{ijk}.

It is possible and useful to formally express this yield in terms of different models (to be given below) representing different facets or aspects of looking at data from such a diallel experiment. Generally speaking, model 1 (M1) is a structural-statistical model as it relates to the structure of the data in statistical terms but in the genetic context; model 2 (M2) is a global genetic model as it expresses types of gene actions (direct

DIALLEL AND MULTI-CROSS DESIGNS

and maternal) assumed to be present; and model 3 (M3) represents a functional genetic model as it relates more specifically to the chromosomal functions and effects. It seems to me that these models together and the relationships among them shed some light upon the question: What can be achieved with this type of diallel experiment?

Let me now state the three models more explicitly and comment on them:

(M1) $Y_{ijk} = \mu + g_i + g_j + s_{ij} - m_i + m_j + r_{ij} + \varepsilon_{ijk}$,

where g_i is the general combining ability of the i-th line, s_{ij} is the specific combining ability associated with the crosses $i \times j$ and $j \times i$, m_i is the differential maternal effect of the i-th line if it is used as maternal line ($- m_i$ could be considered as the corresponding paternal effect), and r_{ij} is the specific reciprocal effect of the cross $i \times j$ (with $r_{ij} = - r_{ji}$). All terms in M1, except μ, are considered to be random variables with means zero and variances σ_g^2, σ_s^2, σ_m^2, σ_r^2, and σ_ε^2, respectively, and no covariances among them, except between g_i and m_i which is denoted by σ_{gm} (this covariance has been noted by Henderson, 1948, but it is not used by Schaffer and Usanis, 1969).

(M2) $Y_{ijk} = \mu + A_{(ij)} + D_{(ij)} + A_{m(j)} + D_{m(j)} + \varepsilon_{ijk}$,

where $A_{(ij)}$ and $D_{(ij)}$ represent the direct additive and dominance effects associated with the genotype of the cross $i \times j$, $A_{m(j)}$ and $D_{m(j)}$ represent the maternal additive and dominance effects associated with the j-th maternal line. All effects in M2, except μ, are random variables with mean zero and variances and covariances σ_A^2, σ_D^2, $\sigma_{A_m}^2$, $\sigma_{D_m}^2$, σ_{AA_m}, σ_{DD_m}, and σ_ε^2, respectively (for definitions see Kempthorne, 1957, and Willham, 1963).

For model 3 it is necessary to specify which sex is homogametic and which sex of the progeny is being considered. For purposes of illustration I shall consider here the case of a homogametic male species and female progeny (for other cases see Eisen, Bohren, and McKean, 1966):

(M3) $Y_{ijk} = \mu + A_i + L_i + A_j + M_j + S_{ij} + \varepsilon_{ijk}$,

where A_i is the cumulative additive effect of the autosomal genes for line i, L_i is the cumulative effect of the sex-linked genes for line i, M_i is the maternal effect of line i, and S_{ij} is the cumulative non-additive genetic effect specific to the cross i × j. The effects in M3, except μ, are considered to be independent variables with means zero and variances σ^2_{Aut} (to distinguish it from σ^2_A in M2), σ^2_L, σ^2_M, σ^2_S, and σ^2_ε, respectively.

These are models which have been proposed and have been used for diallel analysis. It seems important to me to consider how these models relate to each other and how the various variance and covariance components can be estimated. This is achieved through the intermediary of covariances among relatives. For the present situation there are five such covariances:

$C_1 = \text{Cov}(Y_{ijk}, Y_{ijk'}) = \text{Cov (Full sibs)}$
$C_2 = \text{Cov}(Y_{ijk}, Y_{jik'}) = \text{Cov (Reciprocal full sibs)}$
$C_3 = \text{Cov}(Y_{ijk}, Y_{ij'k'}) = \text{Cov (Paternal half sibs)}$
$C_4 = \text{Cov}(Y_{ijk}, Y_{i'jk'}) = \text{Cov (Maternal half sibs)}$
$C_5 = \text{Cov}(Y_{ijk}, Y_{j'ik'}) = \text{Cov (Reciprocal half sibs)}$.

For the models given above and under the assumptions stated earlier these covariances among relatives can be expressed in terms of variance and covariance components (for M1,2,3) as given in Table I.

Equating corresponding covariances and solving yields the relationships among the variance and covariance components as presented in Table II. The results speak for themselves. However, it may be worthwhile to point out some details:

(i) σ^2_g contains in addition to the additive genetic variance also maternal effects components, which has implications with regard to the role σ^2_g usually plays in estimating heritability;

(ii) the fact that $\sigma^2_r = 0$ in terms of the components of M2 and M3 might imply that it is perhaps a measure of non-additivity of direct genetic and maternal effects;

TABLE I

Covariances Among Relatives

Model 1

$C_1 = 2\sigma_g^2 + \sigma_s^2 + 2\sigma_m^2 + \sigma_r^2$

$C_2 = 2\sigma_g^2 + \sigma_s^2 - 2\sigma_m^2 - \sigma_r^2$

$C_3 = \sigma_g^2 + \sigma_m^2 - 2\sigma_{gm}$

$C_4 = \sigma_g^2 + \sigma_m^2 + 2\sigma_{gm}$

$C_5 = \sigma_g^2 - \sigma_m^2$

Model 2

$C_1 = \sigma_A^2 + \sigma_D^2 + 2\sigma_{AA_m} + \sigma_{A_m}^2 + \sigma_{D_m}^2$

$C_2 = \sigma_A^2 + \sigma_D^2 + 2\sigma_{AA_m}$

$C_3 = 1/2\sigma_A^2$

$C_4 = 1/2\sigma_A^2 + 2\sigma_{AA_m} + \sigma_{A_m}^2 + \sigma_{D_m}^2$

$C_5 = 1/2\sigma_A^2 + \sigma_{AA_m}$

Model 3

$C_1 = 2\sigma_{Aut}^2 + \sigma_L^2 + \sigma_M^2 + \sigma_S^2$

$C_2 = 2\sigma_{Aut}^2 + \sigma_S^2$

$C_3 = \sigma_{Aut}^2 + \sigma_L^2$

$C_4 = \sigma_{Aut}^2 + \sigma_M^2$

$C_5 = \sigma_{Aut}^2$

TABLE II

Relationships Among Parameters

Models 1/2

$\sigma_g^2 = 1/2\sigma_A^2 + \sigma_{AA_m} + 1/4\sigma_{A_m}^2 + 1/4\sigma_{D_m}^2$

$\sigma_s^2 = \sigma_D^2$

$\sigma_m^2 = 1/4\sigma_{A_m}^2 + 1/4\sigma_{D_m}^2$

$\sigma_r^2 = 0$

$\sigma_{gm} = 1/2\sigma_{AA_m} + 1/4\sigma_{A_m}^2 + 1/4\sigma_{D_m}^2$

$\sigma_A^2 = 2\sigma_g^2 + 2\sigma_m^2 - 4\sigma_{gm}$

$\sigma_D^2 = \sigma_s^2$

$\sigma_{A_m}^2 + \sigma_{D_m}^2 = 4\sigma_m^2$

$\sigma_{AA_m} = 2\sigma_{gm} - 2\sigma_m^2$

Models 1/3

$\sigma_g^2 = \sigma_{Aut}^2 + 1/4\sigma_L^2 + 1/4\sigma_M^2$

$\sigma_s^2 = \sigma_S^2$

$\sigma_m^2 = 1/4\sigma_L^2 + 1/4\sigma_M^2$

$\sigma_r^2 = 0$

$\sigma_{gm} = 1/4\sigma_M^2 - 1/4\sigma_L^2$

$\sigma_{Aut}^2 = \sigma_g^2 - \sigma_m^2$

$\sigma_S^2 = \sigma_s^2$

$\sigma_M^2 = 2\sigma_m^2 + 2\sigma_{gm}$

$\sigma_L^2 = 2\sigma_m^2 - 2\sigma_{gm}$

Models 2/3

$\sigma_A^2 = 2\sigma_{Aut}^2 + 2\sigma_L^2$

$\sigma_D^2 = \sigma_S^2$

$\sigma_{A_m}^2 + \sigma_{D_m}^2 = \sigma_L^2 + \sigma_M^2$

$\sigma_{AA_m} = -\sigma_L^2$

$\sigma_{Aut}^2 = 1/2\sigma_A^2 + \sigma_{AA_m}$

$\sigma_S^2 = \sigma_D^2$

$\sigma_M^2 = \sigma_{A_m}^2 + \sigma_{D_m}^2 + \sigma_{AA_m}$

$\sigma_L^2 = -\sigma_{AA_m}$

DIALLEL AND MULTI-CROSS DESIGNS 669

(iii) σ_A^2 is determined by the additive effects of autosomal and sex-linked genes; and

(iv) the expression for σ_L^2 indicates that σ_{AA_m} is non-negative (in terms of estimates this is reflected in the results of Godwin, 1972).

The relationships in Table II are important in two respects: theoretically, in that it enables us to better relate the various parameters to each other and translate results from one situation to another, and practically, in that it enables us to estimate the parameters once we know how to estimate the parameters associated with one model. This estimation process can be achieved for the parameters of M1 through the quadratic and bilinear forms associated with its least squares analysis, e.g. the analysis of variance table as given in Table III. The E(MS) in Table III show how the variance and covariance components are estimated and how tests of hypotheses can be done (under the assumption of normality). It is then an easy matter to obtain estimates of the variance and covariance components associated with M2 and M3.

COMMENTS ABOUT THE ANALYSIS

I regard the procedure as outlined in the previous section as a viable and useful approach to diallel analysis. In describing it I have already stressed its strong points, namely the various types of information that become available. Still, certain difficulties and problems remain which need to be mentioned in order that one can make a realistic assessment of the situation and understand the limitations of diallel experiments more generally. Here then is a list.

1. The relationships among parameters of the three models, as given earlier, assume of course that the models are correct. In particular for M2 this means that epistatic effects (direct and maternal) are assumed to be absent. They could have been included in the model but then no "clear" estimates can be obtained because of the limited number of covariances among relatives.

TABLE III

Analysis of Variance for Model 1

Source	d.f.	S.S.*)	E(M.S.)**)
g	$p-1$	$\frac{1}{2(p-2)} \sum_i (X_{i.} + X_{.i})^2 - \frac{2}{p(p-2)} X_{..}^2$	$\sigma_e^2 + 2\sigma_s^2 + 2(p-2)\sigma_g^2$
s	$p(p-3)/2$	$\frac{1}{2} \sum_{\substack{i,j \\ i<j}} (x_{ij} + x_{ji})^2 - \frac{1}{2(p-2)} \sum_i (X_{i.} + X_{.i})^2 + \frac{X_{..}^2}{(p-1)(p-2)}$	$\sigma_e^2 + 2\sigma_s^2$
m	$p-1$	$\frac{1}{2p} \sum_i (X_{i.} - X_{.i})^2$	$\sigma_e^2 + 2\sigma_r^2 + 2p\,\sigma_m^2$
r	$(p-1)(p-2)/2$	$\frac{1}{2} \sum_{\substack{i,j \\ i<j}} (x_{ij} - x_{ji})^2 - \frac{1}{2p} \sum_i (X_{i.} - X_{.i})^2$	$\sigma_e^2 + 2\sigma_r^2$
ε	$p(p-1)(n-1)$	$\sum_{ijk} (Y_{ijk} - \bar{y}_{ij.})^2$	σ_ε^2
		S.P.	E(M.P.)
gm	$p-1$	$\sum_i \hat{g}_i \hat{m}_i = \frac{1}{2p(p-2)} \sum_i (X_{.i}^2 - X_{i.}^2)$	$\frac{p-1}{p-2} \sigma_{gm}$

*) $x_{ij} = \bar{y}_{ij.}$, $X_{i.} = \sum_{j \neq i} x_{ij}$, $X_{.i} = \sum_{j \neq i} x_{ji}$, $X_{..} = \sum_{\substack{i,j \\ (i \neq j)}} x_{ij}$

**) $\sigma_e^2 = \sigma_\varepsilon^2/n$, n = number of progeny per cross

DIALLEL AND MULTI-CROSS DESIGNS 671

2. In M1 one could have distinguished more generally between paternal and maternal effects. Again this would have led to problems of estimation.

3. The standard errors of the estimators for the variance components can be evaluated using a method given by Rao (1968). Consideration must be given to the fact that the sums of squares for general combining ability and maternal effects (see Table III) are correlated. The precision of the estimators is most likely not very good, unless one has a fairly large diallel experiment. Hence the information obtained is usually rather vague and inconclusive.

4. Tests of hypotheses concerning the parameters of M2 and M3 are not at all clear and perhaps for the most part not easily available.

5. There does not seem to be a way of testing for model adequacy, i.e. to test whether a model with additive and dominance effects is satisfactory. One could fit a model with additive and additive × additive epistatic effects but that generally would appear to be a less logical choice to the model considered here.

6. With regard to M3 only female progeny have been considered. If male progeny can also be used their observations would have to be expressed by a different model (see Eisen, Bohren and McKean, 1966) and both models combined should be used for the relationships with parameters of M1 and M2.

7. The assumption of complete inbreeding (F=1) of the parental lines can be relaxed at least for M1 and M2 (Matzinger and Kempthorne, 1956), but this leads to difficulties with M3 in that one would have to redefine what is meant by the cumulative additive effect of autosomal genes of an individual line and how that term enters into the model for an offspring's yield, etc.

OTHER EXPERIMENTS

In the previous sections I have considered one type of diallel experiment in much detail as an illustration of what can be done with experimental data from cross experiments. It is, of course, clear that appropriate modifications have to be made for each situation. This goes for the other types of diallel experiments as well as for other multi-cross experiments. In addition to considering the analysis only for reference population (i), it should also be considered for the other two reference populations, except that I do not know yet how to do it.

Of the diallel experiments, the type D_1 has perhaps been discussed most often. It is intimately connected with the names Mather-Jinks-Hayman (see Mather and Jinks, 1971, for references) after a first, but different analysis was provided by Yates (1947). It seems in order then to comment briefly on their analysis (to the extent that I can). First, it should be noted that their analysis is based on reference population (ii). Secondly, additional assumptions are employed such as no multiple allelism, the genes are independently distributed among the parental lines, and there are no genetic differences among reciprocal crosses. These assumptions may be rather restrictive, in particular the last one, since the main reason for doing reciprocal crosses seems to be to detect and assess expected differences among them. With the stated assumptions their analysis evolves mainly around variances and covariances of offspring and parental arrays which are used to (i) check the adequacy of the additive-dominance model, (ii) provide a measure of the average level of dominance, (iii) obtain an indication of the distribution of dominant and recessive genes among the parents and (iv) estimate genetic variance components. In that these procedures are based to some extent on conditional arguments and in that the first test depends heavily on the assumption of no epistasis, it does not seem to be clear at all just how reliable the analysis and its conclusions are.

It would seem that the type of analysis proposed in this paper for diallel analysis might also work for other multi-cross designs although I have not tried it. Rawlings and Cockerham (1962a,b) have given models for three-way and four-way cross experiments,

with alternative models indicated by Hinkelmann (1963) which
are extensions of M1. If no reciprocal crosses are made, i.e.
if maternal and reciprocal effects are assumed to be absent, then
M2 presents no problems, otherwise care must be taken to reflect
properly at which stage reciprocal crosses have been made, as
there now are various possibilities. The same goes for M3. In
any case, with multi-cross experiments there exist many more types
of relatives, covariances among which can be used to obtain information on more parameters than were possible with the diallel
analysis.

FINAL REMARKS

I hope that this discussion has provided a balanced picture
of what can be achieved with diallel and multi-cross experiments.
Depending on one's expectations this picture is neither entirely
bright nor entirely bleak.

From some points of view the information obtained from a
diallel is somewhat limited and could have been obtained perhaps
from other types of genetic experiments. This, however, is no
argument against diallel experiments as they present a very convenient (from the researcher's point of view) method of generating
this type of data. Using in addition other types of crosses, e.g.
backcrosses, etc., may increase the amount of information. At the
same time this may lead to other difficulties since generally more
generations will be involved and so, for example, the problem of
genotype-environment interaction cannot be neglected. Also deciding what exactly the relevant reference population is may not
be entirely clear.

Multi-cross experiments share to some extent the problems just
mentioned, although they are immediate extensions of diallel experiments and hence share the same reference population. Experimentally they are, of course, more difficult than diallel experiments and the added information that one may gain theoretically
may be hard to recover.

Even if one can, the final and crucial question is: How can one utilize this information to the maximum possible extent in breeding work? That ultimately determines the usefulness of these experiments.

BIBLIOGRAPHY

Cockerham, C. C. (1963). Estimation of genetic variance components. Statistical Genetics and Plant Breeding. Natl. Acad. Sci. Natl. Res. Council Publ. 982, 53-94.

Comstock, R. E. and Robinson, H. F. (1952). Estimation of average dominance of genes. Heterosis, Iowa State College Press, 494-516.

Curnow, R. N. (1963). Sampling the diallel cross. Biometrics 19, 287-306.

Eisen, E. J., Bohren, B. B. and McKean, H. E. (1966). Sex-linked and maternal effects in the diallel cross. Austral. J. Biol. Sci. 19, 1061-1071.

Godwin, M. (1972). Genetic and maternal effects on lactation and growth in mice. M. Sc. Thesis, Univ. of Florida.

Griffing, B. (1956a). Concept of general and specific combining ability in relation to diallel crossing systems. Austral. J. Biol. Sci. 9, 463-493.

Griffing, B. (1956b). A generalized treatment of the use of diallel crosses in quantitative inheritance. Heredity 10, 31-50.

Hayman, B. I. (1954a). The analysis of variance of diallel tables. Biometrics 10, 235-244.

Hayman, B. I. (1954b). The theory and analysis of diallel crosses. Genetics 39, 789-809.

Henderson, C. R. (1948). Estimation of general, specific and maternal combining abilities in crosses among inbred lines of swine. Ph.D. Thesis, Iowa State College.

Hinkelmann, K. (1963). Design and analysis of multi-way genetic cross experiments. Ph.D. Thesis, Iowa State Univ.

Hinkelmann, K. (1965). Partial triallel crosses. Sankhyā (A) 27, 173-196.

Hinkelmann, K. (1966). Unvollständige diallele Kreuzungspläne.
Biom. Zeit. 8, 242-265.

Hinkelmann, K. (1968). Partial tetra-allel crosses. Theor. Appl.
Genetics 38, 85-89.

Hinkelmann, K. (1974). Two-level diallel cross experiments.
Silv. Genet. 23, 18-22.

Hinkelmann, K. (1975). Design of genetical experiments. A Survey
of Statistical Design and Linear Models. J. N. Srivastava,
Ed. North-Holland.

Hinkelmann, K. and Stern, K. (1960). Kreuzungspläne zur
Selektionszüchtung bei Waldbäumen. Silv. Genet. 9, 121-133.

Jinks, J. L. (1954). The analysis of continuous variation in a
diallel cross of nicotiana rustica. Genetics 39, 767-788.

Kempthorne, O. (1956). The theory of the diallel cross. Genetics
41, 451-459.

Kempthorne, O. (1957). An Introduction to Genetic Statistics.
Wiley.

Kuehl, R. O., Rawlings, J. O. and Cockerham, C. C. (1968).
Reference populations for diallel experiments. Biometrics
24, 881-901.

Mather, K. and Jinks, J. L. (1971). Biometrical Genetics.
Cornell Univ. Press.

Matzinger, D. G. and Kempthorne, O. (1956). The modified diallel
table with partial inbreeding and interaction with environment. Genetics 41, 822-833.

Rao, J. N. K. (1968). On expectations, variances, and covariances
of ANOVA mean squares by 'synthesis'. Biometrics 24, 963-978.

Rawlings, J. O. and Cockerham, C. C. (1962a). Triallel analysis.
Crop Sci. 2, 228-231.

Rawlings, J. O. and Cockerham, C. C. (1962b). Analysis of double
cross hybrid populations. Biometrics 18, 229-244.

Savchenko, V. K. (1966). Estimation of the general and specific
combining abilities of polyploid forms in diallel crossing
systems. Genetika 2 (No.1), 17-22.

Schaffer, H. E. and Usanis, R. A. (1969). General least squares analysis of diallel experiments: A computer program. Diall. Res. Rep., Genet. Dept., North Carolina State Univ. No. 1.

Schmidt, J. (1919). La valeur de l'individu à titre de générateur appréciée suivant la méthode du croisement diallèle. Compt. Rend. Lab. Carlsberg 14, No. 633.

Willham, R. L. (1963). The covariance between relatives for characters composed of components contributed by related individuals. Biometrics 19, 18-27.

Yates, F. (1947). The analysis of data from all possible reciprocal crosses between a set of parental lines. Heredity 1, 287-301.

IX

Special Invited Paper

Modes of evolutionary change of characters

Sewall Wright
UNIVERSITY OF WISCONSIN, MADISON, WI 53706

After listening to the discussion of the differences between quantitative genetics and population genetics, I realized that the title, "Modes of evolution," that I had submitted for my talk was not altogether appropriate for a conference on the former. I had, however, come to realize that this title was much too inclusive for the time at my disposal. Evolution includes both the transformation of characters and the splitting of species. These occur simultaneously, indeed, under the mutation theories of de Vries, Goldschmidt and Willis, but are largely distinct under the theories generally accepted now. I decided that I had better restrict myself to transformation and thus consider my subject to be "Modes of evolutionary change of characters." This has the fortunate consequence this is the aspect of evolution most closely related to quantitative genetics so that I may hope that what I say will not be wholly inappropriate.

I will begin with a brief historical account of evolutionary change under the assumption of Mendelian heredity, with special reference to my own approach to the subject.

Most geneticists in the period immediately after the rediscovery accepted some more or less attenuated version of de Vries' mutation theory. Species were assumed to be homallelic with respect to the "wild type" alleles at all loci except for deleterious mutations, kept at low frequencies, and the much rarer favorable ones that replace the wild type alleles, usually one at a time. Bateson (1909) held this view and so did Morgan (1932), though with some qualifications.

There were, however, Mendelians in this period who took a view more like that of Darwin in his statement, "without variability nothing can be effected; slight individual differences, however, suffice, and are probably the chief or sole means in the production of species."

Castle early challenged the common belief, tracing to de Vries, that selection of quantitative variability could produce no permanent change. He proved the contrary in experiments with hooded rats, pushing the amount of white from about 50% nearly to self white in one direction, nearly to self black in the other, before increasingly low vigor stopped the experiment (1916). He thought at first that his selection was operating on minor mutations of the spotting factor itself, a view supported by one clearly demonstrated intermediate allele in the plus series, but after carrying through a crucial test (1919) he accepted the hypothesis of multiple independent modifiers. East, in the same laboratory, was meanwhile obtaining massive support for the multiple factor hypothesis in experiments with maize and species of Nicotiana (1910, 16).

As Castle's assistant in his rat experiments, 1912-15, and from my contacts with East, I started strongly in favor of the view that species are significantly heterallelic at most loci, but with respect to alleles that differ only slightly in effect.

My own primary interest was, however, in physiological

genetics, in particular in the complicated interaction systems that I found in the cases of coat color and hair direction of guinea pigs (Wright 1916, 17). I accepted the hypothesis that genes act largely by determining the specificity of enzymes. It would be a quarter of a century before Beadle and Tatum (1941) provided massive support for this thesis, but there was some support then from Garrod (1908), Durham (1904) and Onslow (1915) and apart from this, it seemed almost a logical necessity. The biochemists (e.g. L.G. Henderson, with whom I took a course) held that what distinguishes the capabilities of different cells are their arrays of enzymes. Presumably organisms differed primarily in the properties of the enzymes of corresponding cells. Geneticists, however, attributed the differences to their genotypes. These views could be reconciled only if the genes determined the enzymes.

The complex network of biochemical and developmental processes intervening between action of primary gene products and phenotypic characters implied that many loci affect each character in a complex interaction system and that each gene affects many characters.

With respect to evolution, I was so impressed by the extraordinary nature of an organism from the physico-chemical standpoint, that it seemed to me that natural selection must somehow be operating among the interaction systems as wholes to be sufficiently effective. There was, however, the obvious dilemma that gene combinations are broken up so rapidly under biparental reproduction that the slow process of natural selection could operate only on the net effects of the separate genes, except where linkage is so close that combinations behave almost as if alleles.

A solution of this dilemma emerged while I was in the Animal Husbandry Division of the U.S. Bureau of Animal Industry during the period 1915-25. The chief of the Division, Mr. G.M. Rommel, had started 23 lines of brother-sister mating with guinea pigs in 1906. Careful records had been kept but no one had analyzed the data until I was brought in for this purpose.

Comparisons of inbreds with contemporary controls revealed the usual marked decline on the average in viability, fecundity and weight (Wright 1922a), with recovery in the crosses (1922c), which I made, but what interested me especially was the profound differentiation among the inbred strains (Wright 1922b). This applied separately to all aspects of vigor, to pattern, quality and intensity of color, in part in respects too slight for Mendelian analysis, in morphology (lengths of legs relative to body, shapes of nose and of ears, carriage of back). I could easily recognize the strain of any loose guinea pig on the floor. The strains also differed in sizes and shapes of internal organs, temperament and physiological traits such as resistance to tuberculosis (Wright & Lewis 1921) and histocompatibility (Loeb & Wright 1927). This suggested that if species in nature tend to be sufficiently subdivided to become differentiated at random with respect to their arrays of gene frequencies at the thousands of heterallelic loci, there could be material for natural selection by means of selective diffusion from those localities in which the more favorable interaction systems happened to have occurred. Evolution would proceed by the selection among somewhat inbred lines.

The application of the inbreeding coefficient, that I had arrived at, to the herd book records at the Shorthorn breed of cattle indicated that artificial selection must have operated largely in this way (Wright 1923a,b, McPhee & Wright 1925). No doubt breeders had practiced individual selection within their own herds to the best of their abilities and some had been luckier than others, but the course of change in the breed as a whole had clearly been guided by selective diffusion, here taking the form of selection of sires overwhelmingly from the herds that were perceived as the best: those of the Colling brothers in the foundation period in the late 18th century, those of Bates and a few others in the early 19th century and, in the late 19th century that of Cruickshank, who had developed a very different type from Bates.

This study, of course, merely put in quantitative terms what was well known to the livestock breeders. It suggested, however, a theory of evolution quite different from simple mass selection, one that I attempted to express in mathematical form in 1925, my last year in Washington. It did not reach publication, however, until 1931 (abstract 1929).

Turning to England, the debate between the mutationists led by Bateson and the Darwinians led by Pearson and Weldon was exacerbated by the nonacceptance of Mendelian heredity by the latter group. The feelings developed (which I encountered in a conversation with Bateson about 1921) probably delayed experimental population genetics in England for several decades. This was, in spite of papers by the biometrician Yule in 1902 and 1906 in which he showed that Galton's law of ancestral heredity, viewed as a multiple regression equation based on observed correlations of individuals with ancestors, was quite reconcilable with the Mendelian mechanism, especially in conjunction with the multiple factor hypothesis which he proposed. This was really implied by Pearson's mathematical results (1904) though he himself would not recognize it.

The geneticists seem to have paid no attention, at least until Fisher elaborated the reconciliation in 1918 and brought it to bear on evolutionary theory in 1930. Experimental population genetics did not get well under way in England until taken up by Fisher's student, Mather, in 1941.

Castle (Castle and Allen) had been the first in America to present Mendelian results. He was not a mathematician but in the same year (1903), stimulated by a misunderstanding of Yule's 1902 paper, he worked out for the first time the course of change of zygote frequencies after a cross, under complete elimination of the recessive segregants in each generation. He also showed that the binomial square distributions, which he found, remained unchanged, if the selection was discontinued at any point. This was the first presentation of what came to be called the Hardy-Weinberg principle, except for application by Yule to randombred descendents of a cross.

The theory of the changes of zygotic frequencies under selection was somewhat extended by H.T.J. Norton (1915) in England but it was reserved for Haldane (1924) to make a systematic study of the courses under a great variety of conditions. This put the evolutionary concept of Bateson (and Morgan) into mathematical form. Haldane (1957) later demonstrated that the cost of fixation of a mutation to the reproductive excess of the species is the same, irrespective of the intensity of selection. He estimated that some 300 generations are used up in any substitution. According to this, the more rapid evolutionary changes could only be accounted for under mass selection by a succession of substitutions with major effects.

Fisher's (1930) "fundamental theorem of natural selection," under which the rate of progress in fitness equals the additive genetic variance, put the Darwinian concept of evolution based on quantitative variability into mathematical form.

It has been asserted (Mayr 1959) that the mathematical formulations of Haldane, Fisher and myself in about 1930 all applied to the same oversimplified "bean bag" theory of evolution. Haldane accepted this in a paper in 1964. There were indeed no mathematical inconsistencies among the three theories but the biological postulates were actually about as far apart as possible (Wright 1960). This will become apparent from the classification of modes of evolutionary transferaction under Mendelian heredity, which I will take up next.

TABLE I

Modes of Evolutionary Transformation of Species According to Mode of Reproduction and Controlling Factors

Evolution under Uniparental Reproduction

I Exclusive Uniparental Reproduction
II Predominant Uniparental Reproduction, Occasional Crosses

Evolution under Biparental Reproduction

I Evolution not under selective control
   A. Control by mutation pressure (Haldane 1933)
   B. Control by random drift (Kimura 1968)

II Control by mass selection. Population panmictic.
   A. On the basis of rare major mutations (Haldane 1924, 32)
      1. According to separate net effects
         a. Under unchanging conditions
         b. Under changing conditions
         c. After a genetic change
      2. According to combination effects under strong linkage

   B. On the basis of multiple, largely minor mutations
      1. According to additive genetic variances (Fisher 1930)
         a. Under unchanging conditions
         b. Under changing conditions
         c. After genetic change
      2. According to effects of allelic chromosome arrangements

   C. Jointly with random drift

III Control by selective diffusion. Population subdivided.
   A. On the basis of local adaptive differention
   B. On the basis of local random drift. Single gene differences
      1. "Altruistic" genes (Haldane 1932, Wright 1945)
      2. "Criminal" genes
   C. On the basis of local multifactorial, peak shifts (Wright 1929, 1931, 1932)
      1. Major factor and modifiers (Wright 1951)
      2. Systems of minor factors
         a. Throughout species
         b. Within groups of small transient colonies (Wright 1940)(often associated with fixation of translocation (Wright 1941))

This table is not intended to be a list of mutually exclusive theories. Each entry may hold in particular cases and, indeed, for different genes or groups of interacting genes, in the same species. As I have noted, it applies only to the transformation, not to the splitting of species, or to the evolution of the higher categories. I have presented classifications of processes at these levels (Wright 1949) but there is not time for discussion here. Thus it only touches on cytologic changes, which undoubtedly play a major role in speciation.

I will say only a little about evolution under uniparental heredity. Under exclusive uniparental reproduction, selection necessarily applies to the genotype as a whole and thus is fully effective with respect to interaction effects. All clones but the best are, however, rapidly eliminated within any homogeneous environment. Thereafter evolution is restricted to very rare favorable mutations.

If, however, crossing occurs often enough to prevent reduction to a single clone, the field of variability is maintained. The process is exceedingly well adapted to providing clones, adapted to take most advantage of novel ecologic opportunities. It also is more favorable for a cumulative process than exclusive uniparental reproduction but there are serious limitations. A wide cross breaks down whatever complex adaptations have been built up, while a cross between clones that differ little, leads merely to improvement along a narrow line.

The primary categories are according to presence or absence of effective selection and to mode, if present. I will take up first control of evolution where selection is so weak that either mutation pressure or mere accidents of sampling controls the course of change.

Letting N be the effective population number, v and u the rates of mutation per generation to and from the leading allele (with gene frequency q) and s the selective advantage of this allele over the next, mutation pressure controls the course of change if

$4Nv > 4Nu > 1 > 4Ns$. The equilibrium frequency of the gene is given by $\hat{q} = v/(u + v)$ with only second order disturbance from selection.

Control by mutation pressure has been most often suggested for the evolutionary degeneration of organs that have become useless since most mutations are observed to act in this direction. It is likely, however, that two kinds of selection are usually more important: selection against the useless organ as an encumbrance and increased selection for alleles that further the development of other characters. (Wright 1929, 64). On this hypothesis, vestigial organs should degenerate in proportion to the amount of evolutionary change in physiologically correlated characters. Haldane (1933) gave an excellent discussion of the role of mutation pressure in which he took a somewhat more favorable view of its importance.

Turning to control by random drift in the species as a whole, (IB), this tends to prevail only if $4Nv$ and $4Nu$ are less than 1, with $4Ns$ still smaller. Gene substitution by this means has been advocated in recent years by Kimura (1968) and by King and Jukes (1969) with respect to amino acid substitutions in the course of phylogeny in such protein molecules as hemoglobin and cytochrome c. They suggest that the amino acids in considerable portions of the molecule serve no purpose other than filling space and thus are completely neutral. The rate of nucleotide substitution (and of mutation) at this level is only about 10^{-9} per year, several orders of magnitude less than observed for mutations distinguished by phenotypic affects. The latter are probably always composites (Wright 1931). The criterion $4Nv < 1$, for control by random drift instead of by mutation pressure, in the absence of selection, is thus much more plausible in this case than for mutations with phenotypically recognizable effects which I rejected in 1929 and later, except for species on the verge of extinction.

Category II, under biparental reproduction, control by mass

selection, is widely accepted as the only controlling process. The primary subcategories are essentially those considered by Darwin: selection of relatively rare "sports" (major mutations) or of the genetic component of the omnipresent quantitative variability.

A second mode of subdivision is according to whether mass selection operates merely on the net effects of genes as must usually be the case, or according to interaction effects. The latter is possible only under such strong linkage that combinations behave almost as if alleles. This seems clearly to have occurred in the evolution of complex polymorphisms. The most important cases seem to be the chromosome polymorphisms based on the absence of crossing over in inversion heterozygotes (Dobzhansky, 1970). These have been studied on a grand scale in species of Drosophila by Dobzhansky and others: The phenomena has also been studied in grasshoppers (M.J.D. White) and may well be important in many groups not yet adequately studied. Such polymorphisms permit the building up of segregating complex interaction systems, by which the species can adapt to different environmental conditions almost as if several species, rolled into one.

The evolutionary process in this case is remarkably similar to that under predominant uniparental reproduction. Selection in heterozygotes corresponds to interclonal selection while the recombination in homozygous arrangements corresponds to that of F_2 of crosses between clones. The interaction systems that are built up may involve both major and minor gene differences.

A third set of alternatives under mass selection has to do with whether conditions remain constant or change. Under long continued constancy, genetic variability tends to be exhausted and recurrent mutation supplies nothing new. Further evolution depends on the exceedingly rare occurrence of novel favorable mutations.

With changing environmental conditions, however, the

previously most favorable allele at a locus may become less favorable than another, already present. There is usually material for rapid readaptation. This process is, however, somewhat of the nature of a treadmill.

Any gene substitution, however, opens up the way for changes at other loci that have a favorable modifying effect (IIA1c, IIB1c). Mass selection may thus build up interaction systems that already have a start. This is a very important evolutionary process, but is not one that initiates new interaction systems. The impossibility of the latter is the most serious limitation of mass selection.

This brings us to category III, evolution controlled by selective diffusion from the more favorable local populations within the species. This, of course, depends on the occurrence of local differentiation.

The most obvious cause of such differentiation is adaptation to different regional environmental conditions. This, however, is a divisive process, likely to lead to the formation of subspecies and occasionally to splitting of the species, not evolution of the species as a whole. It is possible, however, that a chain of regional genetic changes from this cause may lead to an adaptation of general value that could hardly have been arrived at otherwise; which then spreads through the species by selective diffusion (Wright 1940).

Fine scaled subdivision of a species in the absence of environmental differentiation gives the possibility for evolution of single gene traits that are advantageous to groups of individuals without necessarily being advantageous to individuals in competition with each other. If neutral in the latter respect, genes giving group advantage will obviously tend to be fixed throughout the species by selective diffusion from the groups in which they happen to have reached high frequencies by random drift, there being no opposing force.

If individually disadvantageous, there must be some critical degree of disadvantage that just balances the above tendency toward fixation of the genes giving group advantage. There is thus a possibility for the establishment of "altruistic" genes. Similarly mutations that are advantageous to individuals, but are damaging to the group ("criminal" genes) may be kept at low frequences by selective diffusion from those groups with low frequences by random drift. The conditions for the overbalancing of the pressure from mass selection of either sort by extreme random drift are, however, rather severe, (Haldane 1932, Wright 1945), though not impossible as some have maintained. This sort of selection has been advocated most vigorously in recent years by Wynne-Edwards (1962).

Subcategory IIIC is the one, already discussed briefly, that I have considered most favorable for adaptive evolution wherever population structure makes it possible: Simultaneous random drift in all sufficiently small local populations at all loci that are strongly heterallelic with respect to nearly neutral alleles, gives a broad basis for the local appearance of novel favorable interaction systems, by peak-shifts. These tend to be stabilized by local mass selection and to spread throughout the species by selective diffusion (Wright 1929, 31, 32 and later). It is to be noted as favorable to this view, that all but a few of the heterallelic loci are kept at near-neutrality by consideration of cost to the reproductive excess. With respect to strong heterallelism, this may usually be the consequence of the balance between mutation pressure and sufficiently weak selective differences between the leading alleles. In contrast with mass selection, considerations of cost do not limit the rate of evolution of species as a whole by this process in which cost may be apportioned among many localities, and interaction systems instead of single genes are the units of selection.

While the conditions for any particular peak-shift are somewhat severe, there is no tendency to reversal after a shift has occurred. This is in contrast with the necessity for continuously greater pressure of selective diffusion from localities with extreme random drift than from the opposed individual selection in the cases of "altruistic" or "criminal" genes: Moreoever, the number of potentially altruistic or criminal genes is presumably rather limited while the number of potentially favorable interaction systems where random drift is occurring simultaneously among thousands of nearly neutral loci within each locality; and in doing this more or less independently in a great many localities, is virtually unlimited.

The requirement of near-neutrality of the alleles involved makes it relatively unlikely that interaction systems of two or more major mutations will be established by peak-shifts. This does not apply, however, to a single major mutation that offers the possibility of a major advance if the usual deleterious side effects can be eliminated. If recurrent at a usual rate, the system of major mutation and an array of modifiers that remove the side effects, can readily be established by a local peak-shift (IIIC1).

The most favorable situation for a peak-shift occurs where there is a region within the species in which colonies are continually becoming extinct to be restored by stray migrants, perhaps single fertilized females, from the more successful colonies (Wright 1940).

The same condition is almost necessary for the establishment of a translocation which exhibits the usual semisterility of heterozygotes (Wright 1941). There is thus likely to be a correlation between speciation initiated by fixation of a translocation, and the occurrence of multiple peak-shifts. In the main, I have assumed that the transformation of species which I have discussed, and the splitting of species which I have not, are independent processes in contrast with the mutation theories of deVries, Goldschmidt and Willis. There is, however, likely to be a correlation where new species arise on the basis of a translocation, and transformation occurs simultaneously by peak-shifts.

It has been shown by Ayala and his associates (1974) by electrophoretic studies of proteins, that two subspecies or related species tend to show either strong predominance of the same allele or nearly complete substitution, rarely an intermediate situation. We must look to comparisons between local populations of the same subspecies for indications on how such substitutions come about.

The same allele tends to predominate at most loci within subspecies, but there are usually one or more loci in which gene frequencies of localities differ by as much as 0.50, suggesting at least incipient substitution.

Let me compare two cases: First, a study of the allozymes of 16 loci of Drosophila pavani in collections from 14 localities in Chile by Kojima and associates (1972). The localities were chosen as being as diverse as possible in altitude and latitude. Only three of the 16, however, showed gene frequency differences of more than 0.50. These are shown in the upper part of Table II.

TABLE II

The distribution among local populations of frequencies of leading alleles in cases in which there is a difference of at least 0.50, in Drosophila pavani with large population numbers and extreme environmental differences (in Chile) and in Peromyscus polionotus albifrons with small population numbers and negligible environmental differences (in the Florida panhandle and adjacent Alabama).

	Locus	F	.05	.15	.25	.35	.45	.55	.65	.75	.85	.95	no. pop.
Drosophila pavani (16 loci)	Pgi	.17					1	2	3	2	1	5	14
	Idh	.15		2	1	2	3		2	3	1		
	Aph	.13			5	3	2	1	1	1	1		
Peromyscus polionotus albifrons (32 loci)	Ldh1	.66	1		1			1				7	10
	Est1	.50		1						1	2	6	
	Pgm3	.41			1		2					7	
	α-Gpd	.30	1	1				1		4	1	2	
	Est2	.26				1			2		2	5	

Neighborhood numbers were not estimated but were presumably too great for appreciable random drift, judging from studies in other Drosophila species such as D. pseudoobscura and D. willistoni. It would thus seem probable that these three substitutions are due to selective adaptation to different conditions.

Consider next comparisons of allozyme frequencies among 10 localities in the range of the old-field mouse, Peromyscus polionotus albifrons from the data of Selander et al (1971). Five of the 32 loci showed gene frequency differences of more than 0.50.

There was considerably more differentiation than in the case of D. pavani as shown by the values of the fixation index F (which takes all alleles into account), yet there can be little environment difference among localities in the coastal panhandle of Florida and adjacent Alabama. Studies of neighborhood size in species of Peromyscus from the data of Dice and Howard (1951) and of Blair (1951) indicate probable effective numbers small enough to permit much random drift at times of population minimum and thus provide the conditions for peak-shifts. Thus it is unlikely that the differences were due to adaptations to different environments but probable that they were due to peak-shifts. Other examples could be given that indicate one or other of these situations. A survey of species differing in range, density, and population structure, in environmental diversity and amounts of secular environmental change, and consideration of the differences among loci in various respects, makes it unlikely that all evolutionary transformation is controlled in the same way.

BIBLIOGRAPHY

[1] Ayala, F.J., Tracey, M.L., Barr, L.G., McDonald, J.F. & Perez-Salas, S.(1974). Genetic variation in natural populations of five Drosophila species and the hypothesis of selective neutrality of protein polymorphisms. Genetics 77,343-64.

[2] Bateson, W.(1909). Mendel's Principles of Heredity. Cambridge: Cambridge University Press.

[3] Beadle, G.W. & Tatum, E.L.(1941). Genetic control of biochemcial reactions in Neurospora. Prof. Nat. Acad. Sci., USA 27,499-506.

[4] Blair, W.F.(1951). Population structure, social behavior and environmental selection in a natural population of the beach mouse Peromyscus polionotus leucocephalus. Contrib. Lab. Vert. Biol., Univ. Mich. 48,1-47.

[5] Castle, W.E.(1903). The laws of heredity of Galton and Mendel, and some laws governing race improvement by selection. Proc. Am. Acad. Arts & Sci. 39.

[6] _____(1916). Further studies of piebald rats and selection with observations on gametic coupling. Carnegie Institution of Washington. Publ. no. 241,163-87.

[7] _____(1919). Piebald rats and selection. A correction. Amer. Nat. 53,370-76.

[8] Castle, W.E. & Allen, G.M.(1903). The heredity of albinism. Proc. Am. Soc. Arts & Sci. 38.

[9] Darwin, C. (1968). The variation of animals and plants under domestication. 2nd ed. New York: D. Appleton, 1883.

[10] Dice, L.R. & Howard, W.E. (1951). Distance of dispersal by prairie deer mice from birth places to breeding sites. Contrib. Lab. Vert. Biol., Univ. Mich. Publ. no. 50, 1-15.

[11] Dobzhansky, Th. (1970). Genetics of the evolutionary process. New York: The Columbia Univ. Press.

[12] Durham, F.M. (1904). On the presence of tyrosinases in the skins of some pigmented vertebrates. Proc. Roy. Soc. London B. 74, 310-13.

[13] East, E.M. (1910). A Mendelian interpretation of variation that is apparently continuous. Am. Nat. 44, 65-82.

[14] ____(1916). Studies on size inheritance in Nicotiana. Genetics 1, 164-76.

[15] Fisher, R.A. (1918). The correlations between relatives on the supposition of Mendelian inheritance. Trans. Roy. Soc. Edinburgh 52, 399-433.

[16] ____(1930). The Genetical Theory of Natural Selection. Oxford: Clarendon Press.

[17] Garrod, A.E. (1908). The inborn errors of metabolism. Lancet Jan. 4, 11, 18, 25.

[18] Goldschmidt, R. (1940). The Material Basis of Evolution. New Haven: Yale Univ. Press.

[19] Haldane, J.B.S. (1924). A mathematical theory of natural and artificial selection. Part 1, Cambridge Phil. Soc. Trans. 23, 19-41

[20] ____(1932). The Causes of Evolution. London: Longmens Green.

[21] ____(1933). The part played by recurrent mutation in evolution. Am. Nat. 67, 5-19.

[22] ____(1957). The cost of natural selection. J. Genet. 55, 511-24.

[23] ____(1964]. A defense of bean bag genetics. Persp. in Biol. & Med. 7, 343-59.

[24] Kimura, M. (1958). Evolutionary rate at the molecular level. Nature 217,624-26.

[25] King, J.L. & Jukes, T.H. (1969). Non-Darwinian evolution. Science 164,788-98.

[26] Kojima, K., Smouse, P., Yang, S., Neir, P.S. & Brncic, D. (1972). Isozyme frequency patterns in Drosophila pavani associated with geographical and seasonal variables. Genetics 72,721-32.

[27] Loeb, L. & Wright, S. (1927). Transplantation and individuality differentials in inbred strains of guinea pigs. Amer. J. Pathol. 3,251-83.

[28] Mather, K. (1941). Variation and selection of polygenic characters. J. Genet. 41,159-93.

[29] McPhee, H.C. & Wright, S. (1925). Mendelian analysis of of the pure breeds of livestock. III The Shorthorns. J. Hered. 16,205-15.

[30] Mayr, E. (1959). Where are we? Cold Spring Harbor Symp. Quant. Biol. 24,1-14.

[31] Morgan, T.H. (1932). The Scientific Basis of Evolution. Philadelphia: J.B. Lippincott Co.

[32] Norton, H.T.J. (1915) in Punnett, R.C. (1915). Mimicry in butterflies. Cambridge: Cambridge Univ. Press.

[33] Onslow, H. (1915). A contribution to our knowledge of the chemistry of coat colors in animals and of dominant and recessive whiteness. Proc. Roy. Soc. London B89, 36-58.

[34] Pearson, K. (1904). On a generalized theory of alternative inheritance with special reference to Mendel's laws. Phil. Trans. Roy. Soc. London A203,53-86.

[35] _____ (1909). On the ancestral correlations of a Mendelian population mating at random. Proc. Roy. Soc. London B81,219-25.

[36] _____ Selander, R.K., Smith, M.H., Yang, S.Y., Johnson, W.E. & Gentry, J.B. (1971). Biochemical polymorphism and systematics in the genus Peromyscus. I Variation in the old-field mouse Peromyscus polionotus. In Studies in genetics VI. Univ. Texas Publ. no. 7103, 49-90. Ed. by M.R. Wheeler.

[37] de Vries, H.(1901). Die Mutationstheorie. Leipzig: Veit.

[38] White, M.J.D.(1973). Animal Cytology and Evolution. Cambridge: Cambridge Univ. Press (3rd ed).

[39] Willis, J.C.(1922). Age and Area. Cambridge: Cambridge Univ. Press.

[40] Wright, S.(1916). An intensive study of the inheritance of color and of other coat characters in guinea pigs with especial reference to graded variation. Carnegie Institution of Washington. Publ. no. 241,59-160.

[41] _____(1917). Color inheritance in mammals. J. Hered. 8, 224-35.

[42] _____(1922a). The effects of inbreeding and crossbreeding on guinea pigs. I Decline in vigor. Bull. no. 1090 U.S. Dept. Agric. 1-36.

[43] _____(1922b). ibid II Differentiation among inbred families. Bull. no. 1090 U.S. Dept. Agric. 37-63.

[44] _____(1922c). ibid III Crosses between highly inbred families. Bull. no. 1121 U.S. Dept. Agric. 1-59.

[45] _____(1923a). Mendelian analysis of the pure breeds of livestock. I. The measurement of inbreeding and relationship. J. Hered. 14,339-48.

[46] _____(1923b). ibid II The Duchess family of Shorthorns as bred by Thomas Bates. J. Hered. 14,405-22.

[47] _____(1927). The effects in combination of the major color factors of the guinea pig. Genetics 12:530-69.

[48] _____(1929a). Fisher's theory of dominance. Am. Nat. 63:274-79.

[49] _____(1929a). Evolution in a Mendelian population. Anat. Rec. 44,287.

[50] _____(1931). Evolution in Mendelian populations. Genetics 10,97-159.

[51] _____(1932). The roles of mutation, inbreeding, crossbreeding, and selection in evolution. Proc. 6th Internat. Congress Genetics 1,356-66.

[52] _____ (1940). Breeding structure of populations in relation to speciation. Am. Nat. 74,232-248.

[53] _____ (1941). On the probability of fixation of reciprocal translocations. Am. Nat. 75,513-22.

[54] _____ (1945). Tempo and mode in evolution: a critical review. Ecology 26,415-19.

[55] _____ (1949). Population structure in evolution. Proc. Amer. Phil. Soc. 93,471-78.

[56] _____ (1951). Fisher and Ford on the Sewall Wright effect. Am. Scientist 39,452-58.

[57] _____ (1960). Genetics and twentieth century Darwinism: a review and discussion. Am. J. Human Genetics 12:365-72.

[58] _____ (1964). Pleiotropy in the evolution of structural reduction and of dominance. Am. Nat. 98,65-69.

[59] Wright, S. & Lewis, P.A.(1921). Factors in the resistance of guinea pigs to tuberculosis with especial regard to inbreeding and heredity. Am. Nat. 55,20-50.

[60] Wynne-Edwards, V.C.(1962). Animal dispersion in relation to social behavior. New York: Harper Publ. Co.

[61] Yule, G.U.(1902). Mendel's laws and their probable relation to interracial heredity. New Phytol. 1,192-207,222-28.

[62] _____ (1906). On the theory of inheritance of quantitative compound characters and the basis of Mendel's law. A preliminary note. 3rd Intern. Cong. Genetics 140-42.

MODES OF EVOLUTIONARY CHANGE OF CHARACTERS

COMMENTS BY PAUL L. CORNELIUS

It has been suggested at least a couple of times this week that our quantitative genetic models may not be very descriptive of the real world. Let me suggest that the same may be true for evolutionary genetics. Now, I like building models. But I try to warn my students about the pitfalls to be encountered in the extrapolation of a model beyond the range of the data to which it was fitted. Evolutionary theory has given us some nice models for what happens, or possibly could happen, in biological populations. When we try to claim that these models offer an adequate explanation for our own existence, we are guilty of an unjustified extrapolation. Unfortunately, it is difficult to divorce this issue from our emotions.

Dobzhansky, in his text, <u>Genetics of the Evolutionary Process</u>, states, "Life can be understood without recourse to the assumption of any transcendental powers". He, of course, could just as easily have stated, "Life can be understood <u>with</u> the assumption of transcendental powers". However, since life is not now very well understood at all, both statements are quite possibly false. But the amazing thing is that an intelligent and inquiring mind would not do one or the other of two things: (1) To examine the evidence concerning the underlying assumption itself, bringing in, perhaps, evidence from history. What, for example, is the significance of the occurrence of a mysteriously empty tomb on a Sunday morning in approximately A. D. 30? and (2) To seek to answer the question, "How well can life be understood <u>with</u> and <u>without</u> the assumption of so-called 'transcendental powers'?"

Darwinian evolution cannot be a fact of history if there has not been sufficient time for it to occur. The rocks bearing the fossils of multi-cellular life supposedly cover a time span of 500 million years. The dating of this geologic column is based on two things: (1) A uniformitarian concept of geology, and (2) An evolutionary concept of biology. Either or both could be false. The argument is something like this: Dinosaurs lived

70 million years ago because they are found in Cretaceous rock. But how do we know Cretaceous rock is 70 million years old? Because it contains dinosaur fossils, of course. Here we have a case of circular reasoning where each half of the circle, perhaps innocently, assumes the other half has proved its premise.

Now to get more specific. Fossilized pine tree pollen has been found in pre-Cambrian rocks from the Grand Canyon, rocks supposedly 250 million years older than the first occurrence of pine trees. Human footprints have been found in both Cretaceous and Carboniferous rocks. A human skull has been found in a coal seam. These finds strongly suggest that there is something wrong with the way the geologic column is dated.

So then let us consider the question, "How old is the earth"? Now one of the amazing things here is that the maximum likelihood estimate seems to be the oldest, or nearly oldest, measurement we can obtain. A number of measures suggest that the earth is much too young to accommodate the evolutionary theory. Among them are:

1. Rate of decay of the earth's magnetic field.
2. Rates of helium accumulation in the earth's atmosphere.
3. Rates of mineral accumulation in the ocean. Nickel is of particular interest, since it derives primarily from meteoritic dust, the influx of which is probably fairly constant.
4. Absence of large accumulations of meteoritic dust, either on the earth or the moon.
5. If man has existed for one million years, his annual population growth rate in prehistoric times must have been as small as 0.00002. This is based on an estimated population at the time of Christ of 300 million and projecting back 1 million years. Could man's fitness ever have been that poor? (The present rate is about 2% per year.)

Now let me get to the point of this whole issue. Aside from the fact that our academic discipline is building on sand if it is building on underlying assumptions which are false, I think we are very careless about the sociological, psychological, and if you will, spiritual, impact on students of what we teach. We live in a world where large numbers of people, most notably young adults and teenagers, apparently lack any reason or purpose for existing, drifting aimlessly, without any belief in a Creator who loves them and has plans for them, in despair, on drugs or alcohol, involved in irresponsible sexual behavior, teenage crime, suicide. Such I believe are the fruits of what they are being taught concerning their origin.

REPLY BY SEWALL WRIGHT

It is a gross misrepresentation to allege that paleontologists have based their estimates of geologic time on circular reasoning. Strata have always been identified by all available evidence: the similarity of the total arrays of minerals and fossils in neighboring regions, the spatial relations among these strata within regions and the continuity of the systems from region to region. Such comparisons, however, merely establish an order of succession. The dating rests now on quantitative determination of the products of decay of certain of the elements, the rates of which are determinable in the laboratory from the amounts of radiation. These rates have been found to be unaffected by any known environmental conditions. They differ enormously among the elements: that of carbon-14, for example, is so rapid that it is usable for dating back only to about 40,000 years ago. Others are available for dating back for about a million years or for a few billion years ago.

Acceptable evidences for the current human species (Homo sapiens) have so far been found back only for some 250,000 years; for the genus Homo back for some 1.3 million years, and for the family Hominidae back some 14 million years or only 5.5 million years, depending on whether or not Ramapithecus is considered to diverge significantly in the human direction from generalized anthropoids.

During more than 99 percent of his history, man (in the sense of genus Homo) must have lived by hunting and gathering, a mode of life under which population density would necessarily have been extremely limited in comparison with the present situation. Total numbers presumably fluctuated in accordance with fluctuating environmental conditions, under control of famine and disease in periods of unfavorable conditions. There is no reason to postulate continual population growth such as that initiated by the invention of agriculture some 10,000 years ago or in recent years by modern technology. The current rapid rate will inevitably be reversed soon (in terms of total human history) by the exhaustion of essential resources, if not still sooner by massive release of nuclear energy.

Dr. Cornelius' final paragraph indicates that he is not primarily concerned with objective evaluation of evidence on the history of life on the earth, but rather with what, in his opinion, it would be best sociologically, psychologically and spiritually to have people believe. Others think that the truth is best in the long run.

X
General Papers

Quantitative genetics and the design of breeding programs

Ralph E. Comstock
DEPARTMENT OF GENETICS AND CELL BIOLOGY
UNIVERSITY OF MINNESOTA, ST. PAUL, MN 55108

INTRODUCTION

Barring major catastrophies the human population of the world appears destined, at minimum, to double before it stops growing. Moreover, this doubling must be anticipated within the next 35-40 years. Possibly population growth will never stop before food is limiting but unless and until that is established there will be compelling reasons for keeping food production abreast with population. And in view of problems posed by transportation and by trade balances among nations, the geographic distribution of food production should, to the extent possible, parallel that of population.

Genetic improvement is by no means the only avenue through which increases in food production can be achieved but, assuming its potential contribution to be substantial, it will be my point of departure that we must now become very serious concerning maximum exploitation of the allele resources of our economic species in minimum time. By maximum exploitation of the allele resource of a species, I mean synthesis of ideal genotypes,

genotypes that possess the most useful allele (or alleles, in the case of overdominance) of each gene that has any bearing on value of the genotype. Obviously, in view of genotype-environment interaction, the allele composition of ideal genotypes will vary depending on the population of environments in which the genotypes are to be used.

Optimum procedure for synthesis of such ideal genotypes is not obvious in all details. However, because phenotypic value in agricultural species depends on various traits of which most are multifactorial and incompletely heritable, it appears certain that recurrent selection (defined to embrace all cyclic systems in which the selections of any one cycle are employed as the parents or gene source for the first generation of the following cycle) will be required. It is worth noting and emphasizing that the theory and a substantial portion of the empirical evidence that justify this conclusion are contributions from quantitative genetics.

Assuming that recurrent selection will be necessary, I shall discuss the present state of quantitative genetics, as it relates to breeding, in terms of its contributions to the breeders bases for making the prime decisions required in the design of adequate recurrent selection programs. What has quantitative genetics accomplished and what has it failed to accomplish that is relevant to those decisions? In view of present urgency I will focus on programs aimed at close approach to ideal genotypes without any outcrossing (allele introduction) after the initial foundation population has been put together.

Before going further, let me say that what follows is presented with functionally diploid species in mind. You are all aware that quantitative genetics has given little attention to other species.

CONDITIONS FOR SUCCESS VIA RECURRENT SELECTION

In the preceding remarks "success" has been identified as the

DESIGN OF BREEDING PROGRAMS

development of a genotype having the best possible allele at each locus. The required conditions are:
(1) that locus by locus the rank order values of alleles are not conditioned by allele frequencies at other loci,
(2) that the selection criterion is such that the most useful allele of every segregating gene is favored, and
(3) that effective population size is sufficient to trivialize the ultimate consequence of genetic drift.

The first of these has to do with epistasis. It reflects the important fact that simple epistasis (rank order values of alleles not conditioned by background genotype) does not interfere with success via recurrent selection. In contrast, given multiple peak epistasis, the allele frequency changes fostered by recurrent selection may be such that the limit approached is associated with a lesser peak (a genotype that is not ideal); i.e., the end result will be conditioned by the matrix of allele frequencies in the initial foundation population.

Condition (2) poses the following pair of questions:
(a) Given a specific definition of the ideal phenotype, how should the selection criterion be constructed?
(b) Can the phenotype that will be ideal in the future (e.g., 30 or 40 years from now) be foreseen with sufficient accuracy?

The first of these is a genetic issue that will be discussed later. The second is a nongenetic issue but critical in relation to the genetic consequences of recurrent selection. Given the wrong goal, the selection criterion will doubtless be wrong with the result that alleles other than the most useful will be increased in frequency, perhaps fixed, before the proper goal correction is made. The possibility that phenotypic need will not be correctly perceived 10 or more years in advance poses a flaw in the concept of synthesizing ideal genotypes by recurrent selection. However, there seems no way out of the dilemma other than continuing realistic review of phenotypic goals in the light of whatever socioeconomic logic, facts, or trends are relevant.

MAJOR DECISIONS IN THE DESIGN OF A RECURRENT SELECTION PROGRAM

In my judgment these are the decisions relative to the issues in the following list.

(1) The goal of the program
    (a) Phenotypic goal
    (b) Target population of environments
(2) Foundation population(s)
(3) Selection system
(4) Selection criterion
(5) Effective population size

Most of what follows will be devoted to these issues with specific attention to the underlying genetic issues and to whether those in the realm of quantitative genetics have or have not been resolved.

Phenotypic Value

The phenotypic goal of a program is conveyed by statements concerning the measurement of phenotypic value. While this in itself is not a genetic issue, it is important in relation to much that is genetic that we remind ourselves that, in the real world of the breeder, the value of a breed, variety, or cultivar always depends on a number of traits. If any of us is ever inclined to minimize that fact, we have only to reflect briefly on such a list of variables as yield, quality of product, production efficiency, aspects of reproduction, disease resistances, and adaptations to environment.

Before leaving this subject, I feel bound to warn that unjustified convictions concerning what is genetically feasible can have unfortunate effects on goal identification. As an example consider "litter size" in beef cattle. Many have believed, and probably still do, that the allele resource of the bovine species is too limited to support the development of cattle with an average "litter size" closer to two than to one. But this has not been proven beyond doubt and such an increase would bring a major reduction in the production cost of beef. It appears

DESIGN OF BREEDING PROGRAMS

axiomatic that if a close approach to the full genetic potential of a species is to be realized, all changes that would have substantial value should be attempted. In support is one of the major contributions of quantitative genetics, the experimental demonstration that a very great variety of quantitative traits, including ones strongly canalized, responds to selection in well-conceived programs.

Target Population of Environments and Number of Programs

Environment varies in many ways; some highly predictable, others less so with predictability ranging downward from moderate to perhaps zero. Because the rank order values of genotypes are not constant over the vast array of environments in which a species may have utility, decisions are required relative to the populations of environments for which high genotypic value will be sought in a single variety, cultivar, or breed. For example, corn breeders might identify such a <u>target population of environments</u> (TPE) with a geographical region (consisting perhaps of two or more disconnected portions) and within that region with particular soil types, plant densities, levels of fertilization, etc.

It is readily apparent that the subdivision of the totality of environments into TPE's has conflicting dimensions. Homogeneity within populations with respect to predictable aspects of environment can be increased by finer subdivision. But because the number of recurrent selection programs must be at least equal to the number of TPE's the total cost for programs of adequate size is thereby increased. The TPE's of an optimum set would be structured to minimize the joint costs of (a) the recurrent selection programs required and (b) the rank order changes in genotype value associated with intrapopulation variation in predictable aspects of environment.

This problem area has received minimal attention from quantitative genetics. Procedure for estimation of genotype-environment interaction variances has been outlined and numerous estimates have actually been obtained and reported. In general, however, these have been interpreted with reference to genotype-environment interaction as a source of nongenetic variance among selection units of one kind or another.

In contrast, quantitative genetics has not shown how such estimates can logically be employed in decisions concerning TPE's. For example, I do not know how large a component of genotype-environment interaction variance can be, as a fraction or multiple of genetic variance, when there is no variation among environments in the rank order values of genotypes (or family groups of genotypes) nor am I aware that this has been discussed in the literature. Perhaps the situation is best summarized by asserting that quantitative genetics might reasonably be expected to develop a description of the kind of data and analysis that would provide an objective basis for delineation of TPE's but that to date this has not been done.

Selection System

This is the decision area in which quantitative genetics has made its greatest contribution.

The major question is whether the ideal genotype sought can be synthesized by recurrent selection in a one population program or whether a two (or more) population program with selection for combining ability is required. The underlying genetic issue is overdominance and the question stated above can be restated as follows. Can the ideal genotype be homozygous (for the best allele) at nearly all loci or is it one that must be heterozygous at a substantial number of loci?

The overdominance issue is not completely resolved but important evidence, mostly from two types of work, has been produced. Studies of genetic variance components indicate that overdominance is not a major feature in the genetics of single quantitative traits, not excepting such highly heterotic ones as grain yield of maize. The second major source of evidence has

DESIGN OF BREEDING PROGRAMS 711

been long-term recurrent selection for single traits in pilot species. Correlated responses (normally negative in the case of reproduction), residual genetic variance at plateaus, and the changes in levels of the selected trait and reproduction following relaxation or reversal of selection have provided strong evidence for a substantial amount of pleiotropy. More specifically to the point, the totality of this evidence indicates that alleles favorable relative to one trait are often unfavorable relative to another. The obvious corollary, considering the multi-trait nature of total phenotypic value in economic species, is that overdominance can be (probably is) important at the level of total value without being important in the genetics of single traits. In saying this, I think I am echoing ideas expressed by Dr. Falconer on Monday.

The elements of a theoretical base for comparing recurrent selection systems within either of the major classes of systems implied at the beginning of this section are the following.

(1) Expressions that approximate $E(\Delta \bar{Y})$, the expectation of change in the mean value of genotype per cycle of selection. In their usual form, these are premised on a no-epistasis genetic model and have been or can be obtained for every selection system that I have thought about. These always involve a genetic variance or covariance. Bases for estimating these parameters have been or can be provided but often are not required because in many systems comparisons the parameter is the same for both systems and therefore cancels out.

(2) The probability of fixation of an allele has been shown to increase as the product, Ns, increases. Here N = effective population size, or more precisely, the number such that variance of allele frequency during one selection program cycle is $q(1 - q)/2N$ and $\underline{s}$ = the "selective value" of the allele in question.

(3) A general procedure for approximating $\underline{s}$ as a function of $\underline{k}$, u/σ_x, and $\underline{f}$. Here $\underline{k}$ = the selection differential as a

multiple of σ_x, σ_x = the standard deviation of the selection criterion (x), u = one-half the effect on x of substituting, at the locus in question, the best homozygous genotype for the other homozygous genotype, and f, when it is not zero, is a coefficient of inbreeding common to all selection units (or the parents of those units). As implied by the definition of u, this procedure assumes only two alleles per gene. It probably could be generalized but probably then would have no greater value relative to selection system comparisons.

Selection systems have ordinarily been compared in terms of (1) $E(\Delta \bar{Y})$ per unit of time and (2) their costs for operation, given that number of selections is sufficient to make inbreeding per generation quite small; e.g., ≤ 0.03. In the present context; i.e., recurrent selection aimed at achieving ideal genotypes, I propose that the bases for comparison should be

(1) $E(\Delta \bar{Y})$ per unit of time,

(2) operation costs, and

(3) the product, Ns. While the quantity, u, in s cannot ordinarily be evaluated, it is invariant relative to selection systems and hence cancels from comparisons.

Because effective empirical comparisons of recurrent selection systems are ruled out by considerations of time and costs, the theoretical base for comparisons provided by quantitative genetics (and sketched above) must be viewed as a contribution of very special value despite the degree of approximation introduced by the enabling assumptions involved.

Selection Criterion

The multi-trait composition of phenotypic value in economic species poses the problem of how variations in the contributing traits should be reflected in the selection criterion employed in a recurrent selection program. As a base for thinking about this problem, we should recall that if the s-value for an allele is zero, the probability of fixation of the allele in the final product of recurrent selection will be equal to the original frequency of the allele. The probability will be higher or lower than that depending on whether the $\underline{s}$-value is positive or negative. It follows that the selection criterion in recurrent selection programs from which the breeder hopes to obtain ideal genotypes should make $\underline{s} > 0$ for each useful allele.

The best known solution to the trait-weighting problem is the one embodied in the criterion usually denoted as <u>the selection index</u> which I will refer to as the Smith-Hazel (S-H) index. It must be viewed as a major contribution of quantitative genetics. At the same time, we need to keep in mind that it is designed to maximize the effect of selection on total phenotypic value in the cycle immediately ahead, not the total change in phenotypic value that can be achieved by recurrent selection. Stated differently and more specifically, the procedure for construction of the S-H index does not insure $\underline{s} > 0$ for all useful alleles. On the contrary, even if all traits contributing to phenotypic value were included in an S-H index and the parameter estimates employed were all accurate, some of the consequent $\underline{s}$-values could be negative.

The problem appears to deserve renewed attention. One way to confer positive $\underline{s}$-values on all useful alleles is by use of the system of weights inherent in the specification of total phenotypic value; i.e., by using total phenotypic value as the selection criterion. I suspect that this may be the only way. It would require measurement of all traits contributing to phenotypic value and would shift comparative values of selection systems toward those in which there is high genetic variance among selection units (to minimize measurement costs per unit of response) and in which the selection units are families of sufficient potential size to allow optimizing measurement heritabilities.

Clearly there is more work for quantitative genetics in this area.

Effective Population Size

Among conditions, listed early in this paper, for success via recurrent selection was effective population size (N) sufficient to trivialize the ultimate consequence of genetic drift. The theory developed relative to probability of fixation is a major contribution of quantitative genetics (broadly defined) that provides very useful insights to required values of N. Consider the expression

$$P = \frac{1 - e^{-2Nsq}}{1 - e^{-2Ns}} \quad (1)$$

that has become more and more familiar to us over the past two decades. Here P = the probability of fixation of an allele, q = the initial frequency of the allele, e is the base of the natural system of logarithms, and s has meaning specified earlier. We have learned that equation (1) is an excellent approximation when N and s are constant through time and the gene in question segregates independently from all others with effects on the selection criterion. Thus, assuming knowledge of s and q and ignoring possible linkage effects, this expression, used iteratively, gives us the value of N required to achieve any chosen value of P. Our problem is to supply appropriate values of s and q.

It was noted earlier that an expression giving s as a function of k, u/σ_x, and f can be obtained for any selection system. The most familiar example is $s = 2ku/\sigma_x$ for the case of selection among individuals on the basis of their own phenotype. Unfortunately, we have almost no direct information concerning u-values when x is total phenotypic value or whatever other selection criterion that is appropriate in the kinds of recurrent selection program being discussed. In this circumstance I find it useful to note that

$$\frac{u}{\sigma_x} = \frac{u}{\bar{R}} \frac{\bar{R}}{\sigma_x}$$

DESIGN OF BREEDING PROGRAMS

where $\bar{R}$ = mean phenotypic value. Once the selection criterion (x) has been decided and a selection program is underway, data providing good estimates of $\bar{R}$ and σ_x should soon be available. Then, using equation (1), one can answer questions of the following nature: "What N is required to make P = 0.98 if $u/\bar{R}$ = 0.01 and q = 0.10?" Out of this, useful insights to guide decisions concerning effective population size can be obtained. Crucial, of course, are the lowest values of $u/\bar{R}$ and $\underline{q}$ that should be considered and the possible magnitudes of linkage effects.

Present information concerning linkage effects can be summarized in a rough way as follows. In general, linkage decreases fixation probabilities so the value of N must be increased to compensate. This unfavorable effect will be greatest when there is linkage disequilibrium of the type that must be anticipated when the foundation population is structured with the objective of aggregating useful alleles from diverse sources; e.g., when attempting to develop ideal genotypes (as the latter have been defined for the purposes of this paper). Insights concerning increases in N required to compensate for linkage effects have been provided by computer simulation studies but conclusions, even rather loose conclusions, depend on the assumptions made about linkage intensities.

We are also in serious need of better information concerning values of $u/\bar{R}$. Specifically, our need is for answers to questions such as these. How much of the potential for genetic improvement is associated with alleles for which $0.01 \leq u/\bar{R} < 0.02$, how much with alleles for which $0.005\ u/\bar{R} < 0.01$, etc.? I cannot suggest good or obvious ways for obtaining such information but it is obvious that information on numbers of genes that affect value of the organism is relevant. The greater the number of these, the less the average value of $u/\bar{R}$ must be. It should be added that information on gene number contributes as well to our base for reasonable assumptions concerning linkage intensities.

The most important thing to be said about frequencies of

favorable alleles (q-values) is that our information is inadequate. Another point worth making is the obvious one, that if any "strain" is used as the source of a fraction, p, of the original genes of a foundation population in the belief that useful alleles absent from all other components of the foundation may be contributed by that "strain," then q-values as low as p should be assumed when the decision concerning N is made.

To summarize, the now available theory relating probability of fixation to effective population size has special value despite our need for more complete information concerning allele frequencies, u-values, and linkage effects. Applications of that theory, using assumptions concerning q, s, and linkage that are clearly too favorable, demonstrate that effective population size required in recurrent selection programs aimed at synthesis of ideal genotypes is a lot greater than many of us had realized.

Foundation Population(s)

Ideal genotypes cannot be synthesized by recurrent selection unless all the required alleles are present ($q > 0$) in the foundation material or arise at some point by mutation. However, it can be shown that the expectation of generation time required to obtain a favorable allele in homozygous state if it must arise in the population by mutation is much greater than required if the allele is present originally in a frequency even as low as 0.05 or 0.1. Therefore, in the context of attempts to produce ideal genotypes in 100 generations or less, mutations can be ignored as a source of useful alleles.

How can the probability of $q > 0$ for each useful allele be maximized in a foundation population? It appears that this would be accomplished by a stratified sampling of the entire species so that all subpopulations would contribute genes to the foundation population. The major counterargument is that the frequencies of alleles superior relative to a specific TPE would then be lower on average than if the foundation population were constructed from the sources reasonably adapted to the TPE. Thus the achievement of high genotypic value will take longer when the broad sample approach is employed but the final level achieved may (not certainly will) be higher if effective population size has been adequate.

The genetic issue is the distribution of alleles and allele frequencies within species. Opinion among quantitative and agricultural geneticists ranges from the view that each subpopulation (race, strain, local population of pure lines, etc.) probably possesses one or more useful alleles that is absent in all or almost all of the rest to the view that all or almost all useful alleles are present ($q > 0$) in every subpopulation. The latter view is predicated on sufficient effective population size of subpopulations and the migration among them.

Instead of attempting further discussion, let me re-emphasize the existing divergence of opinions and summarize by saying that we need either (1) more information concerning allele distributions within species or (2) better synthesis and exposition of what is already known.

SUMMARY

The preceding discussion has been centered around the problems of synthesizing ideal genotypes for all of the environment populations in which an agricultural species has utility and doing it in minimum time. Quantitative genetics has made major contributions to the bases for the major design decisions involved in the performance of that task. At the same time there are significant issues in the realm of quantitative genetics that have not been resolved. Some of these appear tractable, others relatively intractable.

Finally, it is obvious that a complete set of ideal genotypes represents a goal that can be approached but not fully achieved because for that infinite investment would be required. It follows that optimum utilization of resources actually available will pose various problems. What, for example, is the optimum balance in worldwide maize breeding among number of recurrent selection programs (closely related to number of target populations of environment to be delineated), effective population size in single programs, and breadth of genetic base in foundation populations?

It appears from the perspective of the breeder that quantitative genetics still has a challenging future.

Status of quantitative genetic theory

Oscar Kempthorne
STATISTICAL LABORATORY
IOWA STATE UNIVERSITY, AMES, IA 50011

What Do We Want From Quantitative Genetics?

I consider that this is the basic question and that the answer is quite clear from some viewpoints and very unclear from other viewpoints.

Let me first consider the question from a "pure science" viewpoint. Mather and Jinks (1971) in their prefaces give one answer, which I quote:

> "to show the kind of evidence upon which the genetical theory of continuous variation is based, to bring out the special problems which it raises, to see how the familiar genetical concepts must be adapted to this new use, and to outline an analytical approach which can help us to understand our experimental results."

I find it very difficult to understand this statement. I wonder if I am alone. I shall return to the Birmingham school later. Falconer (1972) in his review of the Mather-Jinks book mentions the genetic architecture of metric traits. This seems to

---

[1] This paper was presented in part at the International Conference on Quantitative Genetics, held at Ames, Iowa, on August 16 to 21, 1976, and also contains what may be termed an "Epilogue" view of the conference.

Journal Paper No. J-8686 of the Iowa Agriculture and Home Economics Experiment Station, Ames, Iowa. Project 1669. This work was supported by Grants GM13827-10 and GM23339-01 from the National Institutes of Health.

me to convey something. We might wonder, for instance, if a trait is determined by additive gene action, or whether there is complete dominance or overdominance or epistasis. I think this question is meaningful. We might ask if a metric trait in a population exhibits variability because of a few genetic factors of large effect or an unknown but presumably very large number of genetic factors with very small effects. If we could answer this definitely, we would have a useful result, which would suggest further activities of various sorts. In spite of this "verbiage," I still find that I have perplexity about what the "pure" quantitative geneticist is seeking.

The contrasting outlook of the applied geneticist is very simple and direct. The corn breeder would like a genetic stock that produces consistently, say, a yield of 300 bushels or more per acre under average Iowa conditions. The poultry breeder wants a stock of laying hens that has considerably greater production than the present, and so on. The applied geneticist is interested in a science and a technology that will produce superior stocks. He needs desperately some sort of theory, though in fact, a quite remarkable amount of progress was achieved over the centuries without any theory apart from "use the best." Interestingly, this prescription is still the main basis of animal breeding, but with considerable statistical sophistication in determining which animals are best.

It is almost a cliché, I suppose, that in applied quantitative genetics there is a spectrum of conditions with two extremes. One extreme is that of the animal breeder who is and has to be concerned with the upgrading of a single existent population by selective breeding. At the other extreme is the plant breeder who has a fairly wide variety of stocks, who can make up new populations very easily and who has the twofold problem of making up populations within which to select and then of having processes for selection within chosen manufactured populations. In applied quantitative genetics there is also the technological role and impact of basic biological features of the species, particularly with respect to multiplication of stocks. Corn breeding via inbred lines is viable as an industry, whereas beef breeding is much less so.

So we have science, technology, science of technology and application of chosen technology, all of which must be integrated in practical uses.

Fisher's Theory of An Equilibrium Population

Without doubt, the basic and seminal paper in the theory of quantitative genetics is that of Fisher (1918). The paper itself gives a short account of the background of knowledge, opinion and controversy that was Fisher's basis. A definitive account would involve extensive study and would be a fascinating story in the history of science, which we may hope will be developed in the future. A review I wrote (Kempthorne, 1974) has stimulated some exploration (Norton and Pearson, 1976), and the reproduction of an hitherto unpublished paper of Fisher written in 1911 when he was a second-year undergraduate at Cambridge, a paper which contains ideas that have been brought forward independently in the subsequent decades. It is desirable, however, to give a very brief statement of the background, namely, that there was a deep and intense controversy as to whether the obvious and unquestionable fact of continuous variation of metric traits could be reconciled with the discrete processes based on Mendelian factors and Mendelian laws of inheritance. That there should have been controversy about this seems very strange, even by, say, the year 1910, and surely so by the 1920's.

The initial point of Fisher was perhaps strange to the conventional wisdom of the time but surely consonant with all that has been discovered since, including modern molecular biology, in that he took as a genetic model an indefinitely large number of segregating Mendelian loci with the possibility also of an arbitrary number of alleles at each locus. This is strongly relevant to the discoveries in the last two decades of a huge amount of genetic segregation in populations. Also, in contrast to classical Mendelism in which the role of environment was essentially negligible and could be ignored, it was obviously necessary to allow for environmental effects on metric traits. So Fisher gave the essential elements of modelling in the study of quantitative genetics, which have been used ever since. Indeed, I shall take the viewpoint that to some extent the stance taken by Fisher has dominated thought ever since, and has perhaps limited the approaches of subsequent workers. It is a tribute, of course, to Fisher's intellect that it has proved very difficult to move beyond the basis he set.

What then were Fisher's ingredients? I have to be brief and almost cryptic for reasons of space and time. There is an indefinitely large number of segregating loci; these are segregating "more or less" independently; their effects are essentially additive; hence we may call on the Central Limit Theorem to infer normality of distributions of total genotypic effect; environment is additive to genotype, and is randomly associated with genotype. Fisher had become interested in human populations from an early age (i.e., < 20 years) and received a considerable stimulus from the massive effort of Karl Pearson and his associates of applying statistical or biometrical procedures to human data. These workers had accumulated much human data, which in consonance with the "biometric school," were reduced to correlations between relatives. So the question was whether the observed correlations were consonant with what was at that time a very sophisticated Mendelian model. The stage was still further restricted to the case of an existent population that had arisen by "natural" processes and could not be experimentally manipulated by the various possible activities of planned crosses, planned inbreeding, planned selection, and so on. This history is relevant, I think, because it gives perhaps an indication of why Fisher did not get involved appreciably in the theory of selection for a metric trait, even though he gave, as will be seen, nearly all the bases that subsequent workers have used. It is also relevant to the nature of more recent work on humans, particularly with respect to mental abilities.

Fisher's work was complicated by the obvious necessity to take care of assortative mating, to which I will refer in a later section. In the area of random mating populations, Fisher established the covariances for the relationships commonly observable (i.e., without inbreeding) in human populations by a model with a single segregating locus, and then merely added over loci. It is natural that he was then concerned with the role of dominance in these covariances, which he developed with dominance generally. He was concerned a little with two-factor interactions, but after seeing how these entered into "simple" covariances for a random mating population, did not pursue the problem. He disposed of multiple allelomorphism, getting the same formal result as with two alleles. He tackled the role of linkage and, after some development, concluded that it was not important in his context; a pair of linked factors could, he thought, be replaced analytically by a single factor.

Perhaps the most curious logical fact that Fisher found in 1918 with respect to a <u>metric</u> trait and with no variability in fitness is the following. In a random mating single locus population with no selection, the Hardy-Weinberg law holds, and in this case one can apply the processes of linear models. With $P = G + E$, one can develop by least squares the decomposition

$$G = A + D$$

and in the population A and D are uncorrelated random variables with variances σ_A^2 and σ_D^2. The Fisherian recognition of this is not surprising, I suppose, because Fisher had developed the <u>modern</u> ideas of linear models and the analysis of variance essentially single-handedly, but withal we may wonder why a process of fitting such a linear model would bear a relationship to components of variance that give a simple structure to covariances of relatives.

Wright's Systems of Mating

Over somewhat the same period, more or less, Sewall Wright had been pursuing somewhat the same sort of problem, but with what I think should be characterized as a different thrust. Wright was concerned with the role of what I wish to call neutral genetic processes, and in particular inbreeding. By a neutral genetic process I mean a process in which the genetic operations of crossing and inbreeding are performed with no selection. For this, which is only a branch of Wright's total effort, he used his method of path coefficients, which is original and in general statistics highly important. This method, however, does not handle non-linear determination of a variable by prior variables except by very forced arguments (c.f., the treatment of the role of dominance, Wright, 1963). I shall abbreviate a huge history by stating two points: (a) by attaching a score to a particular gene and zero to all other genes, Wright obtained so-called genetic correlations, and (b) by supposing that phenotype is determined additively by randomly occurring genes and by randomly occurring environment that are additive in their effects, Wright determined covariances of relatives under a wide variety of circumstances and not just for an equilibrium random mating population as Fisher had done.

This process was used, after all, by Fisher in the first instance in his "proof" that normality of distribution of the metric trait would hold. In tacitly expressing criticism of this type of thinking, I have also to recognize that workers have done "what they can do," and to develop results for a wider model, involving even two loci, let alone one involving, e.g., 1000 loci, is very hard and, indeed, has not yet been accomplished. Parenthetically, the whole of theoretical population genetics is confronted with the same sort of problem. Nearly all theoretical population work uses single locus fitness, which is itself an obscure and quite inadequately thought-out and exposited concept.

QUANTITATIVE GENETIC THEORY

Malécot's Contribution to Quantitative Genetics

Here I shall abbreviate sharply. Malécot (1948 and earlier) formulated the notion of identity by descent and gave us a calculus based solely on probability ideas. He then obtained, with the use of probabilities of identity by descent, a general formula for covariances of relatives in a population under neutral processes, i.e., random mating or pure inbreeding, with a single locus segregating and with no selection. With hindsight, we see that the problem was rather simple, but, as throughout the saga, the recognition of inherent simplicity is the truly creative act.

General Work on Covariances of Relatives

The gaps in desired knowledge at this stage are rather obvious. We now have a theory of covariances of relatives, assuming a single segregating locus and purely random environmental effects, which are additive, based on the equation

$$P = A + D + E$$

with obvious connotations for the symbols.

It is appropriate to focus on this because so much work in the theory of applied quantitative genetics uses this model or even the simpler model

$$P = A + E$$

with assumptions of lack of correlation and so on. To make matters even more unsatisfying, it is assumed in much theory that one needs merely to work out what happens with a single segregating locus, treating all other loci as merely contributing to random error, and then to accommodate the obviously essential ingredient of there being many loci to "add over loci."

Fisher in 1918 had observed a very curious fact for a single locus random mating population. He had obtained simple covariances and found that they involve parameters that we now denote by σ_A^2 and σ_D^2. A Hardy-Weinberg population is representable as

	A	a
A	p^2	pq
a	pq	q^2

and this 2x2 table has proportional frequencies. Did Fisher know that the analysis of variance with proportional frequencies is clear and easy? The world did not know until perhaps two decades later. The upshot of this is that Fisher became highly enamored with the additive genetic variance, defined as the variance accounted for by a linear model consisting additively of gene effects. Parenthetically, I state my opinion that attributing deep significance to this parameter for the case of an arbitrary population may well be one of the intellectual errors of Fisher. I merely state the fact that I find the section "The genetic element in variance," pages 30-37 of "The Genetical Theory of Natural Selection" (Fisher, 1930, 1958), rather incoherent, as also Fisher (1941), even though I have understood and exposited the purely linear model aspects (Kempthorne, 1957, Chapter 16) without errors, I believe. To exposit my views is impossible here. My remarks have implications, clearly, to the status of Fisher's Fundamental Theorem of Natural Selection, which has bothered, and I think correctly, very many workers.

We may surmise that Fisher knew very early, and Yates by the 30's, that the analysis of variance of a multiple cross classified population of numbers is a rather straightforward matter under one and only one circumstance, namely when the frequencies of observations in the ultimate cells are proportional. The linear additive genetic model is useful for non-Hardy-Weinberg populations only with purely additive gene action, which leads to a trivial variance analysis. Otherwise, the analysis of variance of a cross-classified structure is very difficult and not unique. Proportionality of frequences was the clue, I believe, for Cockerham (1954) for the case of two alleles at a locus and certainly the clue for myself for the general case. The outcome is that if the population with an arbitrary number of loci and arbitrary numbers of alleles at each locus is in generalized Hardy-Weinberg equilibrium there is a simple analysis of variance; and if also there is no linkage, the linear model associated with the analysis variance gives terms that behave in a rather simple manner with respect to covariances of relatives. The upshot of this is two formulae:

$$\sigma_G^2 = \sigma_A^2 + \sigma_D^2 + \sigma_{AA}^2 + \sigma_{AD}^2 + \sigma_{DD}^2 + \ldots$$

and

$$\text{Cov}(X,Y) = \Sigma (2r_{XY})^u (u_{XY})^v \sigma_{A^u D^v}^2 .$$

This is described correctly, I hope, in my book (Kempthorne, 1957). This general result seemed at the time, I surmise, to be rather neat, as I suppose it is. I still find it remarkable that such pleasing formulae occur.

Let us suppose that, somehow or other, we know that the conditions necessary for these formulae hold. Should we feel that we have stripped out this situation? We have solved nicely a particular interesting problem, but having solved it we must immediately face problems of implementing the formulae and of assessing the value of what we have achieved. How good are these formulae? We shall observe covariances of relatives, and we shall then try to estimate the components of variance σ_A^2, σ_D^2, and so on, but we encounter extremely unpleasant problems. If we accept the formulae literally we have an unknown number of parameters, and to estimate the components we must necessarily have a number of different relationships between X and Y equal at least to the number of components. Fortunately, we are saved to some extent by the powering that occurs in the coefficients of the covariance formula, so that if we wish to get a good approximate idea of covariances for an arbitrary (noninbred) relationship, we can eliminate all but a few of the terms. Even here, however, the associated statistical problems are difficult as a result of the relationships between coefficients and the imprecision of sample covariances, which require large amounts of data for even moderate precision. It would be a huge task to review the statistical literature on this. The "simple" way out is merely to delete all terms except σ_A^2, as occurs in uses of Wright's path coefficients, or all terms except σ_A^2 and σ_D^2, as with Fisher, and in the simple quantitative genetic designs. As we shall see later, this is not at all bad for some purposes.

Let us pass over these technical problems and consider the situation if indeed all our assumptions hold and we are able to estimate all the components of genetic variance that are involved. I think we have to have a bifurcated view. On the one hand, we are able to say what every covariance not involving inbreeding is. We know how the variance of our population is made up. But on the other hand, this is from many viewpoints a pitifully weak picture of the genetic population. We have a remarkable result because the variability in the population depends on a huge number of genotypic frequencies and genotypic effects. We have achieved a description of variance and of covariances of relatives that depends only on genetic components of variance, which, however, depend on the gene frequencies and the genotypic effects. We have in fact managed to submerge almost all the interesting facts about our population. We can, however, use this description or characterization of the population for the prediction of short-term response to weak selection under special circumstances, and this has been used very widely in the choice of breeding plans.

It is interesting that the results on the variances and covariances do not depend at all on the nature of the trait except that it be representable by an arithmetic number. So, for instance, if our trait were presence or absence of some attribute, we could score presence by 1 and absence by 0, and our description in terms of components of variance would be correct. With the model assumptions, including arbitrary genotypic effects, the results are not approximate.

The next easy step is to suppose that in fact the metric traits of individuals of any specified relationship can be represented by a multivariate normal distribution. We are, then, clearly in business, because the multivariate normal distribution is specified totally by the mean vector and the variance-covariance matrix. Also, this distribution has the property that the distribution of any set of variances conditional on other variables is also multivariate normal, and the conditional expectation function given the values of some variables is linear. So by adjoining multivariate normality we have a huge simplification, which is widely used in animal breeding. The question of linearity of relationships between relatives assumed with normality was raised by Kempthorne (1960), and it was interesting that the question is raised by A. Robertson in this volume.

Complications of the Preceding

1. Genetic Mechanism

It is obvious that even if we have a generalized Hardy-Weinberg population at equilibrium, the phenomenon of linkage must be accommodated. I shall not give any review of this. Definitive work is by Schnell (1961, 1963) and van Aarde (1975). The upshot is that formulae can be developed, but they depend on the whole panorama of segregation parameters. This very difficult work has had very little impact on our quantitative genetic thinking, and in saying this, I am not being at all critical of it. We have to develop larger outlooks so that we can contemplate the formulae and give them some intuitive meaning. I expect this to be done by some fresh mind.

2. Population Structure

What results we have are for a generalized Hardy-Weinberg population. What can we do for populations that do not have this structure or some other very special structures to be discussed later? I suppose we have been completely stopped. We can only observe our populations and attempt a purely empirical statistical description. But perhaps someone will have ideas.

At this point it is appropriate to say a few words about Fisher (1918) again. In a remarkable effort, Fisher obtained some theoretical ideas about a population with assortative mating which is at equilibrium. Fisher developed a theory of covariances of relatives under a model with additive and dominance effects but with no epistasis for an equilibrium assortative mating population, in which the number of loci is "very large," the effects associated with any locus are "very small," and environment acts additively and at random. In spite of efforts to establish the validity of the whole development, obscurities remain (I have seen unpublished work by Vetta). It is necessary at least to mention Wright's work (1921) on assortative mating, which perforce, because of the method, deals with only additive gene action. It seems that neither Fisher's nor Wright's work has been extended in any essential way.

It would be remiss not to comment on the use of this work. Except by psychologists, the work has been almost (but not quite) ignored. I shall not give references, but will merely confine myself to the statement that the ideas were taken up, notably, by C. Burt and Arthur Jensen. In the latter case, a huge amount of controversy, often very bitter, has arisen. I have only a brief, perhaps cryptic, comment. The model is highly naive genetically, and, more critically, environmentally. In my opinion, there have been errors in statement and in logic on all sides. The whole episode is most unfortunate, and many of the unfortunate aspects are, I opine, due to the participants. The topic of "race" and IQ should be buried because there is in the foreseeable future no possibility of eliminating extreme naiveté of genetical, environmental and statistical modelling. The only viable approach to educability and education is by way of very carefully designed (with randomization) experiments on environmental factors. This has been the path followed by the best psychological and educational research of this century, as in the Milwaukee experiment (Heber et al, 1972).

3. Consanguineous Mating

An obvious potentially useful approach to understanding the

nature of quantitative inheritance is to make use of consanguineous mating, which leads, of course, to inbreeding. And, of course, inbreeding has been used extensively in applied quantitative genetics even before the time and ideas of Mendel. What then can be said, following the traditional route, about covariances of relatives with inbreeding?

It seems that if we obtain a population of generalized Hardy-Weinberg inbred structure, and then use random mating, the basic formulae remain valid, with the coefficients $2r_{XY}$ and u_{XY} properly defined to take account of the initial inbreeding (see Section 20.6 of Kempthorne, 1957). This work is restricted, however, to cases in which the relatives considered are not inbred.

The initial attack on this, for a base population of Hardy-Weinberg structure, was that by Wright, already referred to in which a single locus is segregating and the genes have additive effects. In that case, the covariance of two relatives X and Y is simply equal to $2r_{XY}\sigma_A^2$, where σ_A^2 is the genotypic and the additive variance in the base population. Furthermore, the coefficient $2r_{XY}$ is very simple to obtain by using a simple calculus on identity by descent, exposited, for example, by Kempthorne (1957, Chapter 5). Essentially all the results for simple cases had been derived at around 1920 by Wright. It is perhaps a comment on the state of animal breeding theory that there has been essentially no use of any theory more general than this very naive one.

The problem is, as one would expect, that a general theory of covariances of relatives, which may be inbred, is extremely difficult. The basic work in this area deals with a single locus and was done by Gillois (1964) and Harris (1964). The fact that comes to light immediately is that there are fifteen different possibilities with regard to identity by descent of four genes at a single locus is found to involve parameters of genotypic variation and covariance and five functions of relationship. A review of this work is impossible here. Extension of this to, say, the case of several loci has been pursued by Harris (1964) and Gallais (see his paper and references in this volume). An alternative attack by means of identity-by-descent is described in this volume by Weir and Cockerham. It seems that this work generates a huge number of parameters that will not be estimable with the usual amount of resources and I wonder if it will aid understanding.

An alternative route was considered by Kempthorne (1957). Here the game, confined to a single segregating locus, is to consider an inbreeding process, e.g., parent-offspring mating or full sibbing starting from a Hardy-Weinberg population, and to obtain covariances of relatives of particular relationships, e.g., parent in generation n and offspring in generation (n+1), or full-sibs in generation n. The virtue of this work is that, over and above the results of Wright, it does take account of dominance. This line of work has been largely ignored, perhaps because it has not been examined, or perhaps because it is not fertile, though this seems doubtful. The topic is treated, with references to related work, in Kempthorne (1957, Chapter 17). Clarification and extension of that work might be useful. The case of continued selfing lends itself to special treatment; in population genetics theory it behaves as a particular case of mutation with large mutation rates. Each individual has one parent, and all the complexities owing to an individual having two parents, which are usually "swept under the rug" by the use of haploid models, do not arise. The case of continued selfing with arbitrary epistasis but without linkage is treated by Kempthorne (1957, Chapter 20), but the results have not been found stimulating.

To reinforce an opinion stated earlier, the results obtained in these cases do not have any relation to additive variance defined in a least squares sense. This is part of my basis for rejecting completely any general force to this Fisherian concept, except in the case of random mating populations, and, of course, in the case of additive gene action when the notion of partition of variance is unnecessary.

QUANTITATIVE GENETIC THEORY   733

A Partial Evaluation of the Above

I have already made some comments in the above exposition, to which I add the following. The situation we are trying to model, and to model correctly and effectively, is incredibly complex. That we have been able to develop some theory, no matter how simplified, is remarkable, and as we shall see later some of the theory has been useful to obtain approximate ideas about selection. My task, however, is not to praise or condemn but to try to evaluate.

Let us put aside all the genetic complexities and consider the general nature of the modelling. The basis throughout the above has been that we have primitive zygotes containing nothing but the genotype as biochemical "stuff," of course, and that the primitive zygote is placed in a random environment. When the modelling is described in these harshly critical terms, its potential naiveté is displayed. It is obvious in mammals that there is cytoplasm and extra-nuclear material. It is obvious that the fertilized egg lives its early life in a special environment--that given by its mother. It is obvious that that environment is itself determined partially by nuclear forces, and it is obvious that purely environmental forces can influence the mother and the envrionment of the fetus. Various small efforts have been made on maternal effects, e.g., Kempthorne (1957, Section 15.11) and Willham (1963). One might think that this sort of phenomenon is confined to mammals, but flowering plant species, e.g., corn, have a facet that is somewhat similar logically. The endosperm of a plant surely contributes to the phenotypic expression and is part of the environment of the supposed diploid nucleus. In some respects, then, it is necessary to consider a special sort of triploid effect, and this has been done by van Aarde (1976, Variability attributable to direct effects of endosperm genotype, unpublished). The phenomenon is documented from classical Mendelian work by van Aarde, and associated biometric theory is worked out. Whether this will have impact, I do not know, but we can surely see that attaching a score to the nuclear sperm and one to the nuclear ovum is but a naive, if necessary, beginning.

Yet another critical matter is that the environment of an individual is determined by the population in which that individual grows. Implications of this have been pursued by Griffing in several publications, including this volume.

The phenomena just alluded to enter, of course, in the matter of humans and of IQ, but have, I believe, been essentially ignored. To add to the naiveté of the work, this aspect has been swept under the rug. It is comforting that at least one study, the famous Milwaukee one (Heber et al, 1972), <u>which is experimental</u>, takes real cognizance of the role of the mother after birth.

Other deficiencies of the area of quantitative genetic theory relate to the existence of sex. On the one hand, one has the phenomenon of sex chromosomes. On the other hand, it seems rather clear that expression of a nonsex-limited character will in general depend on the status of the individual with regard to sex chromosomes.

The existence of these well-established biological phenomena presents a deep dilemma to workers in quantitative genetic theory. On the one hand, the theory that incorporates them is much more complicated, and naturally, the biologist abhors complexity. On the other hand, the elucidation of the effects, even very approximately, requires planned experiments that are impossible, or nearly so (e.g., transplantation of fertilized eggs), and also requires huge experiments and difficult technical statistical problems. But this should not be a source of surprise since, in the last resort, the task is to model correctly the whole panorama of biological life and this is not going to be achieved by observing, even if very accurately, a few correlation coefficients or a few components of variance. A final remark with reference to human IQ and with respect to behavioral genetics needs to be made. It is surely obvious that the model of the nuclear genotype plus random environment is so naive as to vitiate any substantive conclusions, but the

QUANTITATIVE GENETIC THEORY 735

use of such models can thicken our journals and lead in some cases to totally unnecessary deep societal conflict.

The Biometrical Genetics Approach

It is interesting that Fisher (Fisher, Immer and Tedin, 1932) also initiated a quite different approach to quantitative genetics. In the foregoing, gene frequencies and gene effects are hidden in the parameters, and this has unfortunate consequences. What is another approach? With the notion of completely inbred lines, a suggestion is immediate, namely, to work with inbred lines. If we take two inbred lines and form their F_2, we have got rid of the gene frequency problem because all gene frequencies are 0, 1 or ½. We have P_1 and P_2; we can get F_1, F_2, BC_1, BC_2, and so on. Furthermore, we can use bulked populations, and we can look at population means. This was pursued by several, and a report up to the late 40's is given by Mather (1949); a summary of this work with brief evaluation is given by Kempthorne (1957, Chapter 21). The simple aspect of this is the nature of means of derived populations given in a highly related line of work by Anderson and Kempthorne (1954). This line of thought has been continued, e.g., by Eberhardt and others. The general line of this basic work is very important because it raises the question of scaling, which somehow "falls between the cracks" in the lines of work discussed earlier. The problem is a general one, associated with factorial experiments and error; one wants error to be additive and from a single distribution; one wants zero-interaction of the factors. Obviously, one can work with any 1-1 function of the observations. Rather obviously too, it is unlikely that a transformation that will achieve one of the desiderata will achieve the other.

I shall pass over any discussion and merely state that the results reported by Mather and Jinks (1971) and in various other works are quite emphatic, in my opinion, in indicating that non-removable epistasis is a general phenomenon (as one would expect). This poses an interesting question; attempts to show epistasis by analysis of second-degree statistics have been unsuccessful. This is a direct contrast. The reason I believe is simply that components of variance involve gene frequencies, and epistatic effects are smeared with small weights so as to be lost in the general variance characterization.

The next step in the biometrical approach, of which the basic ideas were presented by Fisher, Immer and Tedin (1932), uses selfing and the analysis of variance for hierarchical structures. But then comes the problem of developing a genetic interpretation of the components of variance. These are, of course, related to covariances of relatives, but with special non-Hardy-Weinberg populations. With no epistasis and no linkage, this is fairly easy, but otherwise not at all so. One may refer to Mather and Jinks (1971).

An extensive experimental effort with maize was made on somewhat the same lines in North Carolina (Comstock et al, 1958). A critical evaluation of that effort would be useful.

But supposing one has gone through the suggestions in this direction. What precisely has one achieved? I am not alone, I know, in having considerable doubts. This should be debated very seriously in the literature, but it is not. It is very easy to conduct the experiments of biometrical genetics; they provided Ph.D. thesis topics by the dozen, which are no more than imitative, and are, when questioned seriously, as uninformative as the original ones. A very popular experimental design uses some type of diallel cross. One view on the nature of the information these give is presented in this volume by K. Hinkelmann. Another view with, it seems, the different background that the inbred are a definite fixed set and not a random sample from a population, was developed by Hayman and Jinks, and is presented by Mather and Jinks (1971). I am not alone in finding this analysis very opaque. Also, as D. S. Falconer has said, the information it supposedly leads to has, it seems, no implication with regard to selection.

QUANTITATIVE GENETIC THEORY

The punch line in the Mather-Jinks theory is the notion of effective factors and the interpretation of effective factors. The notion itself is essentially, but with minor modifications, the Castle-Wright formula of the distant past, a formula based on purely additive gene action, which requires very strong and, it would seem, unwarrantable assumptions. The associated genetic "entities" are called by them "super-genes." I have the view that the whole concept here, rather like that of polygene (which is, I think, a mistaken one), is not viable nor useful. I suggest to the critical reader Chapter 11 of Mather and Jinks. I now give a few quotations, which motivate my doubts:

p. 325   "Since..., the super-genes which the chiasmata distinguish will not be of constant content. They will be variable even within a generation,...." "Finally, since chiasmata vary in position, a further breakdown of the effective factors must occur in later generations. The total number of factors found in these later generations will generally be greater than the first estimate. The effective unit of inheritance is thus a unit only for one generation, and even within this period it may well be a statistical rather than a physical unit."

p. 360   "The demonstrations of polygenic linkage,..., carries with it the demonstration that no permanent system of effective factors can be derived." "A second consequence of the breakdown of the factors is that they can be used as units only in a temporary sense."

finally,

p. 360   "They enable us to force behavior in the near future, to predict minimal limits and rates of advance under selection for one or two generations."

I am mystified because a lack of quasi-permanence in the so-called "super-genes" would seem to preclude any role in selection theory and because I do not see in the presentation any suggestions on how the results of the biometric analysis are to be used in deciding on a selection scheme. Falconer (1972, pp. 416-417) says:

> "The book disappointed me in two respects. First, one feels that the detailed and precise genetical knowledge gained from the analyses must have some practical use in application to plant breeding. Yet applications are hardly ever mentioned and are nowhere discussed in any detail. Second, there is very little comment by the authors on the results obtained.... I can only urge the authors to write another book to give us the benefit of their knowledge and experience of these more general problems."

I close with a general imprecise comment. It is quite clear, I think, that a model for quantitative inheritance must be statistical with respect to the Mendelian loci as well as with respect to individuals. A specification of genetic structure by single gene effects seems to be quite inviable. Can the Mather-Jinks ideas be fertile in this type of direction? We were very sorry that the Birmingham School was not represented at the conference to react to questions of this sort.

The Theory of Genetic Selection

From the viewpoint of human needs, the theory of quantitative genetics must be judged on the basis of whether it suggests schemes of selection and predicts successfully the results of selection schemes applied to real populations of importance to mankind. In the preceding parts of this essay, I have discussed our available theory for a population and the biometric relationships of relatives that may arise or be obtained by random mating or inbreeding. There are considerable gaps, as we have seen, but we should not blind our sight to the theoretical knowledge we do have. So now we turn to the theory of selection.

It is of some interest that the basic step was given by Fisher in 1918 (p. 403), though Fisher's interest was in the correlation of relatives, and in selection only insofar as assortative mating would induce changes in genotypic frequencies. I shall use Fisher's notation so that the reader may verify my account. It is curious that the "gem" lay hidden for a very long time. Consider a population with genotypic array $\bar{P}AA + 2\bar{Q}Aa + \bar{R}aa$, at a particular

QUANTITATIVE GENETIC THEORY

locus. As I have indicated, the Fisher model has a "large number" of loci with "very small" effects. Consider the subgroup of the population for which the metric trait has value x measured from the population mean. Let the effects of the AA, Aa and aa subgenotypes on the metric trait be a, d, -a, respectively, with mean m. Then using the normal distribution, Fisher shows that the frequency of the AA, Aa, aa subgenotypes in this subgroup of the population are given respectively by P, 2Q and R, where

$$P \doteq \overline{P} \{1 + \frac{x}{\sigma^2} (a-m) \}$$

$$Q \doteq \overline{Q} \{1 + \frac{x}{\sigma^2} (d-m) \}$$

$$P \doteq \overline{R} \{1 + \frac{x}{\sigma^2} (-a-m)\} \ .$$

Let us now change the notation a bit, and take a special case of Hardy-Weinberg structure. Write $\overline{P} = p^2$, $\overline{Q} = pq$, $\overline{R} = q^2$ with $P = p'^2$, $Q = p'q'$, $R = q'^2$, so that the prime indicates the frequencies for the chosen or selected subgroup. Then

$$p'^2 \doteq p^2 \{1 + \frac{x}{\sigma^2} (a-m) \}$$

$$p'q' \doteq pq \{1 + \frac{x}{\sigma^2} (d-m) \}$$

$$q'^2 \doteq q^2 \{1 + \frac{x}{\sigma^2} (-a-m)\} \ .$$

We see then that the gene frequencies in the chosen group are

$$p' \doteq p \{1 + \frac{x}{\sigma^2} [p(a-m) + q(d-m)] \}$$

$$q' \doteq q \{1 + \frac{x}{\sigma^2} [p(d-m) + q(-a-m)] \} \ .$$

For later purposes, it is useful to denote these by $p'(x)$, $q'(x)$, respectively. With a diploid model, AA, Aa, aa, relative "selective (viability) values" $1 + s_2$, $1 + s_1$, and $1 + s_0$, respectively, population $p^2 AA + 2pq Aa + q^2 aa$, then

$$p'^2 = \frac{p^2(1+s_2)}{p^2(1+s_2) + 2pq(1+s_1) + q^2(1+s_0)}, \text{ etc.},$$

$$= p^2\{(1+s_2) [1 - s_2 p^2 - 2s_1 pq - s_0 q^2]\}$$

$$\doteq p^2\{1 + s_2 - s_2 p^2 - 2s_1 pq - s_0 q^2\}$$

$$= p^2\{1 + s_2(1-p^2) - 2s_1 pq - s_0 q^2\}.$$

Above, we have

$$p'^2 = p^2\{1 + \frac{x}{\sigma^2} [a - p^2 a - 2pqd + q^2 a]\}$$

$$= p^2\{1 + \frac{x}{\sigma^2} [a(1-p^2) - 2pqd + q^2 a]\}.$$

We may verify that there is a correspondence between the representations if we take

$$s_2 = \frac{x}{\sigma^2} a, \quad s_1 = \frac{x}{\sigma^2} d, \quad s_0 = \frac{x}{\sigma^2}(-a).$$

We may then consider a group of the population, selected on the basis of the metric trait, and we accommodate this if we replace x by the average x, which we may write as $i\sigma$ where i is the standardized differential. Hence quantitative selection merges with qualitative selection by using

$$s_2 = i\frac{a}{\sigma}, \quad s_1 = i\frac{d}{\sigma}, \text{ and } s_0 = i\frac{(-a)}{\sigma}.$$

So we see that if we may assume that effects at a single locus are small, so that σ is essentially constant, and the standardized differential is kept constant, selection for a metric trait may be regarded as constant genotypic **viability** selection with regard to this locus. This line of thought is sometimes attributed to Haldane.

We may also note for the single locus case that

$$p'-p = i\frac{p\alpha}{\sigma}$$

where α is the least-squares "effect" of A versus a. If we consider a haploid population $pA + qa$ with weak viability selective values $1+s$ and 1 for A and a, respectively, then

$$p'-p \doteq spq \, .$$

We see that the haploid model corresponds in effect to the diploid one if

$$\frac{i\alpha}{\sigma} = sq$$

or

$$i \left[\frac{p(a-m) + q(d-m)}{\sigma} \right] = sq \, .$$

If $d = 0$, $m = (p-q)a$, then we have

$$i \frac{qa}{\sigma} = sq \quad \text{or} \quad s = i \frac{a}{\sigma} \, .$$

So diploid metric trait selection with respect to a single locus of small effect may be approximated with respect to gene frequency, if there is no dominance, by a haploid model with constant genotypic viability selection. Obviously, the case with dominance cannot be so approximated. The haploid representation has been used extensively, particularly by A. Robertson.

It is not my intention in this essay to make a eulogy of R. A. Fisher, and indeed the reader will have noted some critical remarks. But it seems that the field of quantitative genetics has spent decades pursuing lines of development Fisher set down in 1918 (actually mainly before 1916) and often has failed to read and understand the first steps he made.

In the 1918 paper Fisher gave (page 403) the regression of offspring on parent in the form

$$\frac{\sigma_A^2}{\sigma_P^2}$$

in modern terms. He did not feel it necessary, one supposes, to state specifically what other regressions were and what the implications are. It is a trivial consequence of this that if we select individuals on the basis of their value for the metric trait, and we mate these to the whole population, then the offspring mean is equal to

$$\tfrac{1}{2} \frac{\sigma_A^2}{\sigma_P^2} \text{Ave}(x)$$

which with normality of the trait and selection of the upper portion p of the population so that $\text{Ave}(x) = (z/p)\sigma_P$, comes out to be

$$\tfrac{1}{2} \frac{z}{p} \left(\frac{\sigma_A^2}{\sigma_P^2} \right) \sigma_P$$

which is recognized as the "work horse" formula of quantitative selection.

Suppose we mate the group measuring x_1 at random with the group measuring x_2. Then, working with the case of 2 alleles for simplicity of exposition (but not of development) we are mating a subpopulation having gene frequency $p'(x_1)$ with a subpopulation having gene frequence $p'(x_2)$. Then the offspring (sub)population has array

$$[\, p'(x_1)A + q'(x_1)a \,]\ [\, p'(x_2)A + q'(x_2)a \,]\,.$$

The argument can be pursued.

We see that we can do a number of interesting computations on this Fisherian basis. We can consider perfect phenotypic positive assortative mating, perfect phenotypic negative assortative mating, and intermediate cases. We can consider a wide variety of matings based on phenotype and on genetic operations of inbreeding. I hope to present an integrated account elsewhere. I shall, here, merely make the comment that the additive genetic variance defined in a least squares sense, which Fisher was so fond of, is not a useful parameter. And, to give what I regard as the 'coup de grace,' even if it were useful, it cannot be estimated from obtainable data.

The above argument is strongly dependent, it is clear, on there being (a) a single locus segregating and (b) genotypic effects small relative to $\sigma(=\sigma_P)$, the phenotypic standard deviation, (c) a constant normal distribution of environmental effects, which are additive, and (d) the approximation (which <u>cannot</u> hold perfectly) that a mixture of normal distributions can be taken to be a normal distribution. Latter (1965) gives very useful ideas on part of the approximation process. It seems rather definite that Fisher thought one could get a reliable result by treating each locus singly and regarding all other loci as contributing noise, which would behave as random environment, and then by adding the results for the separate loci. This supposition pervades the literature of quantitative genetics (and of population genetics with fitnesses multiplicative over loci). Surely it should be questioned.

The Fisherian argument may be generalized to an arbitrary number of loci, with a similar approximative process, by taking the qualitative viability selective value of a genotype to be

$$s(\text{genotype}) = \frac{i}{\sigma} [\text{genotypic effect - mean}] .$$

This has been exploited by Griffing in a fine series of papers, for which the Felsenstein-Taylor bibliography may be consulted. Griffing considered only upper truncation selection, with genotypic expression not dependent on sex, and with random mating of the selected group. I cannot give here an appropriate review of this line of work. The requirement in the mathematics is that the selection be very weak, so that changes in population due to selection do not affect variance properties. I find it very remarkable, though some of the results can be derived from other considerations as I shall note later, that the parameters that characterize the successive means depend not only on the recombination probabilities, as they obviously should, but also on the parameters σ_A^2, σ_D^2, σ_{AA}^2, σ_{AD}^2, and so on, that I developed (Kempthorne, 1954). The upshot is then that one has a formula for the successive population means under <u>weak</u> upper truncation selection with random mating of the selected subpopulation.

To some extent, the results of Fisher and of Griffing are not at all unexpected. Suppose we have a large number of loci that are to some extent nonepistatic, then with random mating, we may expect that the joint distribution of sire (S), dam (D) and offspring (O), for instance, with respect to the metric trait will be trivariate normal. This distribution is very special because conditional expectations of single variables are linear functions of the values of conditioning variables. Hence we have

$$O = \mu + \beta_{OS}(S-\mu) + \beta_{OD}(D-\mu) + e$$

with $E(e) = 0$ and with β_{OS}, β_{OD} determined from the variance matrix of S, D and O. And with random mating,

$$\beta_{OS} = \beta_{OD} = \frac{\text{Cov}(P,O)}{\sigma_P^2}.$$

Then we use the fact that

$$\text{Cov}(P,O) = \frac{1}{2}\sigma_A^2 + \frac{1}{4}\sigma_{AA}^2 + \ldots,$$

in the case of no linkage, and we have part of Griffing's results. It is interesting that one can obtain some results without invoking Mendelism at all, but merely use purely statistical ideas of correlation and regression. One can go further, I believe. The whole area of selection can be approximated by purely statistical ideas of correlation and regression. The ideas of Mendelism merge with these ideas, as Fisher showed (more or less), and the fact that the theory does not need Mendelism in some respects, and one can almost say, does not use Mendelism intimately is, I think, a reason for it having a moderate degree of robustness in relation to assumptions. Apart from a difficulty I shall mention later, one could proceed as follows. Let there be a population; let rules of forming mating couples be defined in terms of metric traits of individuals and/or in terms of relationship; let there be selection of individuals on the basis of metric traits or metric traits of related individuals; and finally let the offspring be measured. Then without an atom of formal Mendelism and with a large data set, the joint distribution of offspring and parents can be determined. One can examine this distribution and determine a prediction

QUANTITATIVE GENETIC THEORY   745

equation, which one can then apply for a few generations. The only flies in the ointment for this proposal are that every covariance would have to be determined from data and not inferred from, say, a coefficient of relationship and a heritability, and large data sets would be needed to control sampling error. So one could have a completely empirical selection procedure and a purely empirical process of obtaining a prediction of the result of continued selection. I suggest that this type of thinking should not be dismissed as a cranky idea. The reason that some predictions of the results of selection theory seem to work is that they are based on a process rather close to what I have sketched.

I should note before passing on that I do not have space to discuss the very important idea of reciprocal recurrent selection. A critical review of Comstock, Robinson and Harvey (1949) would be valuable.

Apart from potential deficiencies of the theory discussed with respect to aspects such as epistacy, linkage and environment, there is one very critical aspect. This is that the population is infinite. What this means in simple upper truncation selection is that the population size is some very big number, say, N, from which the best $mN(0 < m < 1)$ members are chosen; these mN individuals are mated at random to produce a very large population, say of the same size N, as the base for the next cycle. This is surely a very artificial structure. It does, however, provide some basis for predicting gain from selection. Before turning to the role of finite population sizes, it may be useful to discuss some deficiencies of infinite population theory. I have already mentioned that our infinite theory does have the deep deficiency that it merely predicts the immediate gain from 1 cycle of selection. Just how many cycles of selection may be predicted reasonably is quite unknown. To get an idea of what is involved, consider a single segregating locus with A at frequency p, with no dominance and effect α for the substitution of A for a. With weak selection, usual simple assumptions,

$$\Delta p = \frac{i\ p(1-p)\alpha}{\sqrt{2p(1-p)\alpha^2 + \sigma^2}} \ .$$

This equation can be iterated, of course, to obtain the whole selection response curve, and one can, of course, convert this, as Haldane did for qualitative selection, into a differential equation and solve. Even in this simple case, however, the only attributes determinable by conventional methods, it seems, are $2p(1-p)\alpha^2$ and σ^2. So even in this case we cannot determine $2(1-p)\alpha$ which is the total achievable gain. Perhaps under simple assumptions we can do this if we determine higher moments, a comment that recalls the thrust of the Fisher-Immer-Tedin (1932) paper, which seems to have had little impact, and perhaps justifiably. If now we consider the case of k loci we have

$$\Delta p_i = \frac{i\ p_i(1-p_i)\alpha_k}{\sqrt{2 \sum_j p_j(1-p_j)\alpha_j^2 + \sigma^2}} \ , \ i = 1,2,\ldots,k \ .$$

In this case, if we are given a set of pairs $\{(p_j, \alpha_j)\}$, where p_j, α_j are the initial frequencies and loci effects, we can compute easily by recursion the progress over cycles of each p_j and of the population mean. I have not been successful so far in my attempts to see how any set of observations on the initial population of the usual type enables a determination of the nature of the response curve over cycles to selection and of the limit that would be achieved. The same problem arises, of course, with finite populations.

There are, I think, some points of interest in this arena. We may note, for instance, that the change in mean under the very ideal circumstances is $\Delta\mu = i\sigma_A^2/(\sigma_A^2 + \sigma^2)$, where $\sigma_A^2 = 2p(1-p)\alpha^2$. It is clear, then, that a priori, there is no logical requirement that the response curve be concave, i.e., have decreasing 'slope.' It seems to be rare for the response in a second cycle to be greater that that in the first cycle of selection for attributes of usual interest in economic species; though, of course, one could get such a curve by transformation of attribute. On the question of limits, it is interesting that at the present conference we saw cases in which, apparently, one can make progress indefinitely (apart from natural limits such as 0%, 100%). The matter of being able to make a prediction of the genetic potential in different genetic populations is most obscure. For instance, we may have 6 inbred lines from which we generate 15 F_1 and then 15 F_2 populations. How are we to form a judgment on which population we should attack by quantitative selection? If the F_2 populations are equally variable, we should, presumably, take the population with the higher mean, but it seems obvious that we should examine a large sample of each F_2 population to aid the judgment. Should we take, for instance, the F_2 with the largest σ_A^2 (assumed measurable)? This one will give the largest immediate response, but will it give the highest mean after a number of cycles of selection? We do not know. The fact is, I believe, that we have no idea of the limits to selection in infinite (i.e., very large) populations with infinite (i.e., very many) observations on covariances of relatives. We cannot tell whether lizards can become dinosaurs or not under selection for size. At a more useful level, can we develop 300-bushels-per-acre corn, or hens that lay three eggs in two days? Is it important in theory to be able to do something about the questions I have raised? I think so, but perhaps we have to accept unanswerability.

Finiteness of Populations

There has been a huge intellectual effort in the past 20 years by workers, in works far too numerous to list, directed towards the role of the sampling variability that occurs with finite sampling from finite or infinite Mendelian populations. The logic of the game is fairly straightforward in terms of stochastic processes, but the mathematics are rather complex, even in simple cases, and very difficult with anything but the case of a single segregating locus. Just how this very extensive theory is relevant to theory of quantitative genetics merits lengthy presentation. I hope someone will give a definitive discussion of the question.

One aspect of this whole endeavor has been used by A. Robertson (1960) in his well-known theory of limits. Because this theory is relevant to quantitative genetic selection, it is perhaps useful to give a very elementary exposition. Consider a particle that moves over a set of states, which we index by $1, 2, \ldots, S, S+1, S+2, \ldots, S+k$, where states $S+1, S+2, \ldots, S+k$ are absorbing in the sense that once the particle reaches any one of the states it does not move. The first S states are not absorbing in that there is a nonzero probability of leaving any of them. Let u_i be the probability that the particle will reach state $S+1$, say, eventually, given that it is initially in state i. Let T_{ij} be the probability that the particle goes from state i to state j in one step. Then, clearly

$$u_i = \sum_{j=1}^{S} T_{ij} u_j + T_{i,S+1}$$

or, in matrix terms, with obvious definitions,

$$u = Tu + v .$$

Hence

$$(I-T)u = v, \text{ and } u = (I-T)^{-1} v.$$

One may be able sometimes to work from this base.

One interesting case is considered in this volume (Bailey). Suppose we have a selfing species; we start with an individual, obtain M offspring by selfing; we measure each offspring and select the best to begin a new cycle. If we have a single locus with 2 alleles A, a, then the states are Aa, AA and aa, where AA and aa are absorbing states. Then let

p_2 = Prob{Aa gives a selected offspring which is AA},

p_1 = Prob{Aa gives a selected offspring which is Aa},

and

p_0 = Prob{Aa gives a selected offspring which is aa}.

QUANTITATIVE GENETIC THEORY

If u = Prob{a line starting from Aa reaches the state AA}, then

$$u = p_2 + p_1 u ,$$

or

$$u = \frac{p_2}{1-p_1} = \frac{p_2}{p_2 + p_0} .$$

So to determine the probability u, we need only to obtain the ratio p_2/p_0 for a single cycle of selection, this being, however, not a small task.

If we have a population with gene frequency p, and if the state of the population at the beginning of any cycle of selection is given by the gene frequency, with p = 1 or 0 being absorbing states, and u(p) is the probability that a population starting from the state p eventually reaches the state 1, then

$$u(p) = \sum_{p'} T_{pp'} u(p') ,$$

where $T_{pp'}$ is the probability that in any one cycle the gene frequency changes from p to p', assumed to be the same for all cycles. This equation can be solved with a small number of states when one has obtained $T_{pp'}$ for all p, p', this indeed being no small task in general without highly heuristic arguments. Also direct attack with most partly realistic models involves a large number of states and $\{T_{pp'}\}$ depends on one or more parameters.

The special case used by Robertson is that of a diploid population with a single diallelic locus of size N, characterizable by gene frequency, which will be of the form i/2N, where i is the number of, say, A genes. Then

$$u(i) = \sum_{i'} t_{ii'} u(i') .$$

We may use the Euler summation formula

$$\sum_{x=1}^{X} f(x) = \int_{1}^{X} f(w)dw + \frac{1}{2} F(X) + R ,$$

or some variant, with remainder R small. This process gives us the integral equation

$$u(p) = \int T(p,p+x) \, u(p+x) \, dx \, ,$$

where $T(p,p+x)$ is the probability of a shift from p to $p+x$ in one cycle. Then by expanding $u(p+x)$ and ignoring remainders, we obtain the standard differential equation and, e.g., the solution for the case of no dominance, as

$$u(p) = \frac{1 - e^{-2NSp}}{1 - e^{-2Ns}} \, .$$

Work by several has shown that this remarkably simple formula gives very good approximation for the case of a single locus. A critical review of the voluminous literature would be valuable but is impossible here.

Attack primarily by W. G. Hill and A. Robertson on the case of 2 or more loci is extremely difficult and, with suggestions for parameterization from diffusion approximation, has depended on very extensive use of simulation and computers.

Apart from the great complexity of results obtained and the problems of extension to a large number of loci, for which see Robertson (1970), the utility of this approach to applied quantitative genetics is quite obscure to me. The behavior of an infinite population under quantitative selection depends, even with additive gene action, on gene frequencies, on gene effects and on recombination values. I was interested, but not surprised, that in the present conference the old questions of number of loci and size of effects were being raised again, and appeal was being made to an estimation procedure, Castle-Wright, that had been discarded (except by the Birmingham School) as requiring too many assumptions. In the case of a single locus with two alleles and AA-aa = α, say, we need to determine p and α, as in the case of the infinite population theory, to determine the fixation probability.

QUANTITATIVE GENETIC THEORY 751

The foregoing also brings sharply into focus a point that should have been obvious to us all decades ago, but was not. The step from a finite population π_1 to a finite population π_2, involving selection or not is, because of Mendelian processes and environmental variation, a random variable of very complicated structure. The result of a single line of successive populations is then a very complex vector random variable. A single "replicate" of a selection experiment is then a random variable of unknown properties, and is nearly useless, except that it indicates, surely, one possibility that occurred. Extensive verbal theorizing from one such line is obviously unjustified. Under a simple model, some theory applicable to observations on individuals in a single selected "line" has been developed by W. G. Hill (1970). The complexities of inference from a realization of a stochastic process have yet to be dealt with.

How To Design a Quantitative Genetic Study?

All the preceding, and considerations not mentioned, raise for the quantitative geneticist the very basic question of how to design an experimental quantitative genetic study, with, say, a laboratory species, such as Drosophila or mice or Tribolium. Obviously, one can conduct an upper truncation selection experiment for a metric trait. Obviously, too, such an experiment must be replicated, or one will be lost in a cloud of mere speculation, which may well be about random features of a particular realization. Every process that one applies must be replicated. The simplest replication is to make a number of independent starts from the base population. In this case, however, the successive results from a given start are obviously correlated in a serial way. If one introduces, say, relaxed selection or selection in an opposite direction at some point in time, this must be replicated. Exactly the same considerations hold in an observational study, whether with a quantitative genetics or population genetics orientation. A single time series realization may easily be worth little more "than the paper it is written on." These considerations have the consequence that even a naive exploratory experiment involves a massive effort.

Suppose one takes cognizance of the need for replication of Mendelian sampling as well as of environmental variability, to what scientific directions should a quantitative genetics experiment be oriented? I believe that this question has not been addressed seriously. I have commented on the simple experiments of "biometrical genetics." They do little more than enable estimation of some simple covariances of relatives. The same seems true of diallel cross experiments. What do general and specific combining ability experiments tell us, apart from indicating which of a set of lines "combines well" with the whole set of lines? It is critically necessary that the questions posed here are addressed seriously and in a questioning way.

The basic dilemma is that an experiment design must be based on a hypothesized theory, even a naive one. Insofar as theory of quantitative genetics is deficient, good ideas for quantitative genetic experiments do not arise.

The Matter of Reproductive Fitness

The mathematics of the theory of quantitative genetics discussed clearly uses intimately in the modelling the assumption that there is no genotypic variability in fecundity or viability. All covariances of relatives are based on this assumption. In the naive selection theory we have, it is assumed that there is no "natural selection;" all selected individuals have only variability in fecundity and viability represented by purely random error, which may then be essentially ignored in almost all the basic mathematical computations. The blending of "natural selection" with directed selection via metric attributes is an outstanding theoretical and experimental problem. This is potentially a topic on which ideas of population genetics and quantitative genetics may merge with fruitful results. Is the conflict between natural selection and humanly directed selection at the root of the well-known difficulties in poultry breeding for egg production? Is natural selection the explanation for the inability to maintain some lines in some species under intense inbreeding? How should one attempt to examine the possibilities?

Related, perhaps, to this is the view that may be advanced by some that yield in a species like maize, or egg production in poultry, is very much like an attribute of "natural fitness." I think this analogy can easily be very overworked and may be strongly misleading. I say this because the economically superior populations generated would, I believe, have no future in a "natural" millieu. It is well-established, it seems, that a very good double cross maize hybrid will degenerate very rapidly under "natural" conditions. It follows, I believe, that any attempt to relate the hybrid vigor that we find, e.g. in maize, to what one might expect under natural evolutionary processes is most unlikely to be fruitful. Any appeal to equilibrium arguments for naturally evolving populations is unlikely, I opine, to be effective in the situation where man is making huge changes both genetically and environmentally. In the area of man-directed selection and modification of environment, we have obviously not reached equilibrium, or we would fail to make progress and cease our efforts. Indeed, the justified concern of the globally oriented ecologist is that we are not at equilibrium, but on a "primrose path" that will lead to the destruction of natural life as we now see it. It should be noted that I use the adjective "justified." I do not wish to imply that the concern should lead to a complete return to purely natural processes, which could lead to the death by starvation of billions of existing humans.

The Theory of Applied Quantitative Genetics

This topic, like many others I have mentioned, merits deeply critical examination, and I regret being able to make only a few remarks. The general approach as exposited excellently by Falconer (1960), for instance, uses the idea that genes act essentially additively, and recourse is then made to a concept that I find rather obscure--that of breeding value, which is presumably the mean of an infinite offspring array.

The general structure of the logic may be presented rather succinctly as follows. Suppose we have a random variable Z', which equals (Y',X'), where $Y' = (Y_1, Y_2, \ldots, Y_p)$ and $X' = (X_1, X_2, \ldots, X_q)$, the whole variable Z having a $(p+q)$-variate probability distribution. Consider then estimating or predicting a variable $a'Y$ by means of $b'X$, with the criterion that $E[(a'Y - b'X)^2]$ shall be a minimum. Here the vector a is given, and b is to be determined. Let the variables have zero means and variance-covariance matrices as follows:

$$E(XX') = F, \ E(XY') = G, \ E(YY') = H .$$

Then we wish to find the vector b which minimizes

$$Q = a'Ha - 2b'Ga + b'Fb.$$

Differentiation gives the equation

$$Fb = Ga .$$

If we write, taking F to be invertible, as it will be,

$$b = F^{-1}Ga + \delta ,$$

we have

$$Q = a'Ha - a'GF^{-1}Ga + \delta'F\delta .$$

This is minimized, clearly, with δ as the null vector; thus we have the fully proved formal solution, given for a particular frequently occurring situation expressed by the well-known equation $Fb = Ga$. The general procedure was given by Fairfield Smith (1936), though it goes back for decades in psychological testing. The situation stated is very general in that $X_1, X_2, \ldots, X_q$ are observations on individuals, and $Y_1, Y_2, \ldots, Y_p$ are so-called breeding values. It may be that there is just one Y, or that there is a Y associated with each X. The basic formulae given by Lush, which have found such wide use in animal breeding, may be derived from this standard basis. Some of the more recent work by Henderson (e.g., this volume) is derivable by the same route with a model

$$Y = Z\beta + \text{error}$$

with the vector $Z\beta$ representing fixed effects, the same minimization, but with the additional restriction that $a'Z = \emptyset$ (the null vector), or with an additional assumed model

$$X = A\beta + \text{error},$$

again with $A\beta$ representing fixed effects.

To say this is not to imply, however, that the overall problem is solved. One may readily adjoin ideas of restricted selection indices (Kempthorne and Nordskog, 1959). There are very difficult questions of statistics. There are very difficult questions resulting from the existence of selection in the "X" data, e.g., selection of bulls by daughter records. Also, underlying application to large animals are very difficult questions relating to the overlapping of generations which are discussed by Hill and Pollak in this volume.

It seems that, at root, the same formalization holds for selection in a plant population, though complexities arise at a basic level because of the use of intense inbreeding, which is avoided in animal populations.

Concluding Remarks

To develop a fully adequate theory of quantitative genetics is a fantastically difficult task. One has only to contemplate Mendelian processes, which while very simple for a single segregating locus, are simply not workable for, say, 1000 linked loci. Even in the case of a single segregating locus, there are simply formulated theoretical questions that we have been unable to answer. Attack on the field requires, we see, almost the whole panoply of mathematics, and particularly ideas of stochastic processes, an area of the 20th century. The field of quantitative genetics uses also a considerable portion of the available statistical theory, which is a 20th century creation. We are involved in applied mathematics rather different from the conventional physical science type and much more difficult. That the problems are huge is obvious because the aim is, in some respects, to explain quantitative variation in the whole biological world.

That we see great deficiencies in our presently available theory should not be a surprise. It seems likely, however, that the mere following of the routes of the past will not be really effective. There is deep need for that very rare human attribute, creativity, which our human society does not support as much as it should. Instead, support is given for "polishing the apples of past thought," and a dedicated creative effort that is unsuccessful receives short shrift. The importance of quantitative genetics to human affairs merits much greater support of free, unfettered, and potentially unsuccessful effort than it has heretofore received. Obviously, much larger selection experiments must be made to suggest testable hypotheses. A somewhat natural reaction to selection experiments of the past is "Interesting, but so what?". Obviously mathematical and statistical research of difficult order is needed. Obviously, because the "gestalt" is so complicated, the mathematical and statistical problems must be approached by the computer, the fastest that exists, with large financial support, because an hour or two can merely generate a cloud of noise.

In recent years we have heard, via Camus, of "The Myth of Sisyphus," who it may be recalled spends eternity rolling a boulder to the top of a hill only to have it roll to the bottom again. I believe this myth has relevance to all science. We develop a naive theory, and we are momentarily very pleased with it, but then we see that our progress has generated even more difficult problems. Having rolled our boulder to the top of the hill, we find nearly the same boulder at the bottom of an even steeper hill. We see this in physics and chemistry and in evolutionary genetics, as described by Lewontin (1974) for instance, which I see as a tale full of creative effort, but ending not with a "bang" but with a "whimper." We see it, of course, also in qualitative genetics. Our triumphs of yesterday lead to our anxieties of today. There should be very little "mutual back-slapping" for the progress. But also we should not write polemics attempting to pin-point our failures.

On the contrary, we should perhaps feel somewhat gratified that we do have some theory with some predictive power, as was shown by several contributors to our conference. It seems quite unarguable that applied quantitative geneticists have made huge genetic improvements in some economic species, and part of that progress is a result of use of our admittedly very naive quantitative genetic theory.

* * * * * * * *

It is critical to note that this review of the status of quantitative genetic theory does not treat all subareas with appropriate attention and depth. To do so was quite impossible. Furthermore, the bibliography below is very incomplete. I can only say that most of the writings of the founders, Fisher, Haldane, Wright, of every contributor to the conference, and of many individuals mentioned without reference are important in the area. The prime source is the bibliography of Felsenstein and Taylor (1973), a huge proportion of which is relevant.

BIBLIOGRAPHY

(1) Anderson, V. L. and Kempthorne, O. (1954). A model for the study of quantitative inheritance. Genetics 39, 883-898.

(2) Cockerham, C. C. (1954). An extension of the concept of partitioning hereditary variance for analysis of covariance among relatives when epistasis is present. Genetics 39, 859-884.

(3) Comstock, R. E., Robinson, H. F. and Harvey, P. H. (1949). A breeding procedure designed to make maximum use of both general and specific combining ability. Agronomy Journal 41, 360-367.

(4) Comstock, R. E., Robinson, H. F., and Cockerham, C. C. (1958). Quantitative Genetics Project Report 1947-1957. Institute of Statistics Mimeo Series Number 167, North Carolina State University.

(5) Falconer, D. S. (1960). Introduction to Quantitative Genetics. New York: Ronald Press Co.

(6) Falconer, D. S. (1972). Review of Biometrical Genetics. Heredity 28, 415-417.

(7) Felsenstein, J. and Taylor, B. (1973). A Bibliography of Theoretical Population Genetics. National Technical Information Service, U.S. Dept. Commerce, Springfield, Va.

(8) Fisher, R. A. (1918). The correlation between relatives on the supposition of Mendelian inheritance. Transactions of the Royal Society of Edinburgh 52, 399-433.

(9) Fisher, R. A. (1930). The Genetical Theory of Natural Selection. Oxford: Clarendon Press. Reprint, (1958), New York: Dover Inc.

(10) Fisher, R. A. (1941). Average excess and average effect of a gene substitution. Annals of Eugenics, London 11, 53-63.

(11) Fisher, R. A., Immer, F. R. and Tedin, O. (1932). The genetical interpretation of statistics of the third degree in the study of quantitative inheritance. Genetics 17, 107-124.

(12) Gillois, M. (1964). La relation d'identité en génétique. Unpublished thesis. Faculty of Sciences. University of Paris.

(13) Harris, D. L. (1964). Genotypic covariances between inbred relatives. Genetics 50, 1319-1348.

(14) Heber, R., Gaber, H., Harrington, S., Hoffman, C. and Falender, C. (1972). Rehabilitation of Families at Risk for Mental Retardation. Rehabilitation Research and Training Center in Mental Retardation. Madison: University of Wisconsin.

(15) Hill, W. G. (1970). Theory of limits to selection with line crossing. Mathematic Topics in Population Genetics. (Ed., K. Kojima). 210-245. New York: Springer-Verlag.

(16) Kempthorne, O. (1954). The correlations between relatives in a random mating population. Proceedings of the Royal Society of London, B, 143, 103-113.

(17) Kempthorne, O. (1957). *An Introduction to Genetic Statistics*. New York: J. Wiley and Sons. Reprinted (1969) Ames: Iowa State University Press.

(18) Kempthorne, O. (1960). The biometrical relations between relatives and selection theory. *Biometrical Genetics*. (Ed., O. Kempthorne) 12-23. New York: Pergamon Press.

(19) Kempthorne, O. (1974). Review of collected papers of R. A. Fisher. *Social Biology* 21, 98-101.

(20) Kempthorne, O. and Nordskog, A. W. (1959). Restricted selection indices. *Biometrics* 15, 10-19.

(21) Latter, B. D. H. (1965). The response to artificial selection due to autosomal genes of large effect. I. Changes in gene frequency at an additive locus. *Australian Journal of Biological Sciences* 18, 585-598.

(22) Lewontin, R. C. (1974). *The Genetic Basis of Evolutionary Change*. New York: Columbia University Press.

(23) Malécot, G. (1948). Les mathématiques de l'hérédité. Masson et Cie, Paris. Eng. trans. (D. M. Yermanos). *The Mathematics of Heredity*. San Francisco: Freeman and Co.

(24) Mather, K. (1949). *Biometrical Genetics*. 1st Ed. London: Methuen.

(25) Mather, K. and Jinks, J. L. (1971). *Biometrical Genetics*. 2nd Ed. Ithaca, New York: Cornell University Press.

(26) Norton, B. and Pearson, E. S. (1976). A note on the background to, and refereeing of, R. A. Fisher's 1918 paper 'On the correlation between relatives on the supposition of Mendelian inheritance.' *Notes and Records of the Royal Society of London* 31, 151-162.

(27) Robertson, A. (1960). A theory of limits in artificial selection. *Proceedings of the Royal Society of London*, B. 153, 234-249.

(28) Robertson, A. (1970). A theory of limits in artificial selection with many linked loci. *Mathematical Topics in Population Genetics*. (Ed. K. Kojima) 246-288. New York: Springer-Verlag.

(29) Schnell, F. W. (1961). Some general formulations of linkage effects in inbreeding. *Genetics* 46, 947-957.

(30) Schnell, F. W. (1963). The covariance between relatives in the presence of linkage. Statistical Genetics and Plant Breeding. (Eds., W. D. Hanson and H. F. Robinson). 468-483. Washington, D. C.: National Academy of Sciences - National Research Council Publication No. 982.

(31) Smith, H. F. (1936). A discriminant function for plant selection. Annals of Eugenics 7, 240-250.

(32) van Aarde, I. M. R. (1975). The covariance of relatives derived from a random mating population. Theoretical Population Biology 8, 166-183.

(33) Willham, R. L. (1963). The covariance between relatives for characters composed of components contributed by related individuals. Biometrics 19, 18-27.

(34) Wright, S. (1921). System of mating. Genetics 6, 111-178.

(35) Wright, S. (1963). Discussion - Plant and animal improvement in the presence of multiple selective peaks. Statistical Genetics and Plant Breeding. (Eds., W. D. Hanson and H. F. Robinson) 116-122. Washington, D. C.: National Academy of Sciences - National Research Council Publication No. 982.

Quantitative inheritance, stabilizing selection and cultural evolution

M. W. Feldman
DEPARTMENT OF BIOLOGICAL SCIENCES
STANFORD UNIVERSITY, STANFORD, CA 94305

L. L. Cavalli-Sforza
DEPARTMENT OF GENETICS
STANFORD UNIVERSITY MEDICAL CENTER,
STANFORD, CA 94305

ABSTRACT

The dynamics of the bivariate gaussian distribution of genotypic and phenotypic values under gaussian stabilizing selection on the phenotype is studied. Relative to single locus genetics, this involves approximations about the genetic transmission rule. Once made these approximations allow the incorporation of phenotypic assortative mating, and more general transmission laws for the phenotype and the environment. Evolution of discrete valued phenotypes under selection is also discussed in the context of more general transmission rules possibly applicable to cultural phenomena.

INTRODUCTION

The Fisher (1918) framework for the study of continuous variation forms the basis for present approaches to the study of quantitative inheritance. This statistical theory has become separated from the study of evolutionary population genetics. Perhaps the last barrier against complete bifurcation has been the Fundamental Theorem of Natural Selection (Fisher, 1930; Kimura, 1958; Ewens, 1976) relating the dynamics of average population fitness to the additive genetic variance.

Our recent studies have involved the construction of models for evolution at both the phenotypic and genotypic levels. In the first of these studies (Cavalli-Sforza and Feldman, 1973) the point of departure was the original Fisher (1918) model of three genotypes A_1A_1, A_1A_2, A_2A_2 which contribute on the scale of the phenotypic measure $-a, d, a$, respectively, to the phenotype. The Hardy-Weinberg law for the gene frequencies was assumed to hold, and there was no selection on the phenotype. The phenotypic values of individuals of the genotypes A_1A_1, A_1A_2 and A_2A_2 at generation t+1 were $X_{1,t+1}$, $X_{2,t+1}$, $X_{3,t+1}$ respectively. Nongenetic parental influence on the offspring was introduced by regressing the offspring phenotype on the midparental phenotype. Then $X_{1,t+1}$ is determined as

(1)
$$\begin{aligned}X_{1,t+1} &= -a + b_{1,11} X_{1,t} + \varepsilon \text{ with probability } p^4 \\ &= -a + b_{1,12} \frac{(X_{1,t}+X_{2,t})}{2} + \varepsilon \text{ with probability } 2p^3q \\ &= -a + b_{1,21} \frac{(X_{1,t}+X_{2,t})}{2} + \varepsilon \text{ with probability } 2p^3q \\ &= -a + b_{1,22} X_{2,t} + \varepsilon \text{ with probability } 4p^2q^2\end{aligned}$$

Here ε is an error term with mean zero and constant variance σ^2. The regression coefficients $b_{i,jk}$ in (1) depend on the genotype of the offspring and the genotypes of both parents. There would be 15 such numbers and, with the other parameters, the total number of parameters to be specified would be twenty. We chose therefore to omit the dependence of the $b_{i,jk}$ on the last two subscripts so that the weight of the parental phenotypes affecting the offsprings' was determined only by the genotype of the latter.

Then, from (1) and the analogous calculations for the other

genotypes, it is seen that the phenotype of the child is a linear function of the parental phenotype, the function being different for each genotype. This is tantamount to assuming a very specific form of genotype-environment interaction. The environment here is, in fact, the phenotypes of the parents; it seems quite plausible that this component of the environment would be important for many human behavioral traits.

At any point in time, one can in principle calculate from (1) and its analogues for the other genotypes, within family variances, total variance, parent offspring covariance, covariances between adopted offspring and biological or adoptive parent, etc. Again, in principle, these can be iterated over time. Unfortunately, even in the simplified case with $b_{i,jk}$ depending only on i, this iteration must be done with a computer. Various choices for the parameters $-a$, d, a, p, σ^2 and the initial conditions were made and the analysis reported in Cavalli-Sforza and Feldman (1973).

The main qualitative conclusions of this analysis can be outlined as follows:

1. Correlations between relatives arise even if there is no genetic variation. Thus if $-a = d = a$, correlation between relatives is established by the phenotypic transmission alone.

2. When there is purely cultural transmission ($-a = d = a$) correlations between adoptive relatives are comparable to those between biological relatives.

3. Correlations between MZ twins reared together by their biological parents are always higher than the heritability when there is cultural transmission.

4. Correlations between MZ twins reared apart depend on whether one twin is reared with its biological parent or both are adopted.

The parameters $b_{i,jk}$ cannot be estimated in the absence of sufficient unbiased adoption data. But they do provide a framework within which phenotypic plasticity can be conceptualized. The fact that they may play a key role in determining the usual statistics of biometrical genetics suggests that the classical Fisherian framework is inadequate to describe situations where cultural transmission is important.

Stabilizing Selection on the Pheno-genotype.

The incorporation of natural selection into models for the dynamics of continuously varying phenotypes has turned out to be difficult. This is especially so if both genotypic and phenotypic properties are of interest. It seems inevitable that a theory which aims to track the evolution under natural selection of both genotypes and continuously varying phenotypes must involve some approximation at least insofar as genetic rules are concerned. Many such approximations are possible. For example Lande (1975) uses the formulation of Slatkin (1970) for stabilizing selection on a continuous phenotype, and defines the genetic variation in terms of a heritability parameter which is assumed to remain constant over time. Kimura (1965) used a diffusion procedure to simulate the process and obtain the distribution of the genotypic value.

We have taken a different approach motivated by our interest in evolution under a variety of modes of transmission. Each adult individual in the population at generation t is an observation on a bivariate Gaussian random variable (f_t, g_t) which we call the pheno-genotype. g_t is the genotypic value after selection and f_t is the phenotypic value after selection. Optimizing selection on the juvenile phenotype takes place via the survival probability density

(2) $$W(f) \propto \exp\left[-(f-\mu)^2/2S\right]$$

where μ is the optimum and the strength of selection is measured by S; large S is weak selection, small S is strong selection. Slatkin (1970) used a phenotypic transmission function $L(x;y,z)$ giving the probability that a mating between parents of phenotypes y and z produces an offspring of phenotype x. We introduce the transmission function $T(f',g';h,k,h'k')$ giving the probability that parents of pheno-genotypes (h,k) and (h',k') produce an off-

STABILIZING SELECTION AND CULTURAL EVOLUTION

spring of pheno-genotype (f',g'). It is in the specification of T that approximation to the standard rules of inheritance for small numbers of genes without selection must be made. We have used (Cavalli-Sforza and Feldman , 1976) a model specified by the following relations between the parents (h,k) and (h',k') and the offspring (f',g') before selection:

(3)
$$g' = \frac{k+k'}{2} + N(m_g, G_t/2 + M)$$
$$f' = g' + N(0, E) .$$

Here m_g and M are the changes in the mean and variance of the genotypic value caused by mutation and E is a constant environmental variance. For comparison with Mendelian inheritance (3) corresponds to what is usually called the polygenic fully additive model with independent effects from each locus. There are two basic assumptions inherent in (3). The first is that the variance between sibs is the same in each family. The heuristic justification for this is the large number of loci involved. The second is that the mean variance within sibships before selection (averaged over all families) is $G_t/2$, namely half that in the parental generation after selection. The error involved in this assumption is currently under investigation.

After (3), selection, via the function (2), operates and the variables G_{t+1}, F_{t+1}, $\mu_{g_{t+1}}$, $\mu_{f_{t+1}}$, $r_{f,g}^{t+1}$ i.e. the genetic and phenotypic means and the correlation coefficient, determined. These completely specify the bivariate gaussian distribution of adult pheno-genotypes in generation t+1 recursively in terms of those in generation t. In (3) the phenotype is essentially genetically determined (modulo E) and all analysis boils down to a linear fractional iteration in G_t. This converges to

(4a)
$$\hat{G} = M \{\sqrt{1+4(E+S)/M} - 1\}/2$$

while the phenotypic variance at equilibrium is

(4b) $$\hat{F} = S(\hat{G} + E + M)/(\hat{G} + E + M + S).$$
Obviously if $M=0$ then $\hat{G}=0$ (c.f. the Fundamental Theorem of Natural Selection) and $\hat{F}=SE/(S+E)$. At equilibrium the heritability after selection defined as $\hat{G}/\hat{F}$ is increased by the factor $1+E/S$ over that before selection, and can be greater than one.

Inbreeding.

Classical theory of the effect of inbreeding in the absence of selection (e.g. Wright 1943, 1965; Falconer 1960) can be used to establish that with purely additive effects the average within family genetic variance is $\Gamma = G_0(1-F)/2$ where G_0 is the usual additive genetic variance in the absence of selection and with random mating, and F is the coefficient of inbreeding. The total genetic variance in the population is $G_0(1+F)$. Assuming all individuals are inbred to the same constant extent, the correlation between the genotypic values of the parents is $2F/(1+F)$. In the absence of selection, then, the genetic variance in the whole population remains at $G_0(1+F)$. The effects of various regular schemes of inbreeding are currently under investigation (manuscript in preparation).

Assortative Mating.

Fisher (1918) pointed out that when the number of genes is large, the change in heterozygosity due to assortative mating is not expected to be large. This justifies the assumption that the average variance within families produced by the adults at generation t is $G_t/2$. Suppose that the correlation between phenotypic values of mating parents, i.e. the extent of assorting is a. Then the correlation between their genotypic values is $a(r^t_{f,g})^2$ where $r^t_{f,g}$ is the correlation between phenotypic and genotypic values in the parental generation (Wright, 1921). Again returning to (3), in the offspring generation t+1 before selection, the total genotypic variance is then

(5) $$G_t\left[1 + a(r^t_{f,g})^2/2\right].$$

(See also Crow and Felsenstein, 1968). An analysis of this case has been made (Feldman and Cavalli-Sforza, 1977). Apart from the assumptions pointed out in the presentation of (3), (5) also depends on the assumption that assorting causes no change in the average within family variance from that expected in the random mating case. When the number of loci is large and there is no selection Wright (1921) shows that the correlation between homologous genes is inversely proportional to the number of loci involved (see also Crow and Felsenstein, 1968). In Table I we present the equilibrium values for the phenotypic and genotypic variances $\hat{F}$ and $\hat{G}$ and the correlation coefficient between phenotypic and genotypic values at equilibrium for values of a between 0 and 1.0. Of course as a approaches 1.0 the model becomes far from realistic. In the Table, E=1, S=10 and M is the mutation variance. A more detailed numerical analysis is presented in Cavalli-Sforza and Feldman (1977).

TABLE I

Variances and Covariances at Equilibrium under Assortative Mating: E = 1, S = 10; a is the Correlation between Mating Phenotypes.

	M = 0.001			M = 0.01			M = 0.1		
	$\hat{F}$	$\hat{G}$	$\hat{r}$	$\hat{F}$	$\hat{G}$	$\hat{r}$	$\hat{F}$	$\hat{G}$	$\hat{r}$
0.0	0.995	0.104	0.294	1.179	0.327	0.479	1.736	1.000	0.690
0.2	1.104	0.236	0.420	1.354	0.538	0.573	1.968	1.281	0.734
0.4	1.683	0.936	0.678	1.796	1.073	0.703	2.298	1.681	0.777
0.6	2.313	1.699	0.779	2.362	1.758	0.784	2.699	2.166	0.814
0.8	2.860	2.361	0.826	2.888	2.395	0.828	3.117	2.672	0.842
1.0	3.335	2.936	0.853	3.354	2.958	0.854	3.518	3.157	0.861

Correlations Between Relatives.

Estimates of heritability usually derive from correlations between such relatives as monozygous twins, sib-pairs, parents and offspring. Under selection these values will change. We present here some results for correlations between relatives under the system (3) with selection (2). For monozygous twins (reared together with the same environmental contribution E to the phenotypic variance before selection) and no selection, the expected value of the correlation is

(6) $$r^*_{MZ} = (G + M)/(G + M + E).$$

Using the joint distribution of (g, f_1, f_2) where g represents the common genetic value and f_1, f_2 the twins' phenotypic values, the correlation after selection is

(7) $$r^{t+1}_{MZ} = \frac{S(G_t+M)}{(E+G_t+M)(E+S)+E(G_t+M)} = \frac{r^*_{MZ}}{1+E(1+r^*_{MZ})/S}$$

Obviously $r^{t+1}_{MZ} \to r^*_{MZ}$ as $S \to \infty$. Selection reduces the correlation between identical twins.

The between sib correlation before selection is classically expected to be

(8) $$r^*_{SS} = \frac{G}{2(G+M+E)}$$

By considering the joint distribution of (g_1, f_1, g_2, f_2) where the parental pheno-genotypes of (g_1, f_1) and (g_2, f_2) are the same it can be shown that

(9) $$r^{t+1}_{SS} = \frac{r^*_{SS}}{1+\left[1-(r^*_{SS})^2\right](G_t+M+E)/S},$$

which again converges to r^*_{SS} as $S \to \infty$.

The parent-offspring correlation before selection in generation t+1 is

(10) $$r^*_{PO} = \frac{W^*_{hf}}{\sqrt{F_t(G_t+M+E)}}$$

where F_t is the phenotypic variance after selection in generation t and G_t is the corresponding genotypic variance. W^*_{hf} is the parent-offspring phenotypic covariance before selection. After selection

(11) $$r^{(t+1)}_{PO} = \frac{r^*_{PO}}{\sqrt{1+\frac{(G_t+M+E)(1-r^2_{PO})}{S}}}$$

In all cases selection reduces the correlation between relatives, and, as might be expected it also reduces the phenotypic variances of those individuals, whose relatives (sib or twin) have survived, below that of random survivors.

Complex Transmission.

In our earlier writings on the transmission of cultural phenotypes we have considered several different modes of transmission. Phenotypic transmission is obviously possible from parent to child via either or both of the parents. Other relatives in the parental generation or earlier generations, or the peer generation may influence an offspring's phenotype. Teachers, political leaders and social hierarchies all transmit cultural information. The transmission envisaged in these examples is non-biological but can surely have biological ramifications if it leads, for example, to assortative mating or population subdivision. One model of phenotypic transmission (actually involving a genotype-phenotype interaction) was discussed earlier. In the present context phenogenotypic transmission is introduced most simply by rewriting equations (3), for instance, as

(12a) $\quad\quad\quad g = (k+k')/2 + N(m_g, G_t/2 + M)$

(12b) $\quad\quad\quad f = \gamma g + (1-\gamma)(\alpha_1 h + \alpha_2 h') + N(0, E)$.

Here the offspring phenotype before selection is made up of contributions γ from the parental genotypic values and $1-\gamma$ from their phenotypic values. α_1 and α_2 are the relative amounts contributed by the two separate parental phenotypes. This type of transmission involving both genotype and phenotype we call **complex**.

When $\gamma=0$ the model reduces to one of blending inheritance. When $\gamma=1$ there is purely genotypic transmission as in (3). In these simple cases complete iteration of the distribution of phenogenotypes is completely tractable. In intermediate cases, however, the mathematical structure of the time dependent behavior of the system is in terms of a matrix linear fractional transformation. Karlin (1977) has proven that in the present case the recursion system converges to a unique equilibrium. It is possible to show that this equilibrium can be expressed in terms of the admissible root of a seventh degree polynomial.

It is important to note that as γ becomes small, so that the parents' phenotypes are more important than their genotypes in determining the offspring phenotype, (12a) and (12b) become less dependent on each other. The selection acting on the phenotype has less feedback on the genotypic value. The latter then comes under the influence of M. As a result the genotypic variance can be greater than the phenotypic. We regard this as of particular relevance to the theory of canalization whose evolution can be viewed as the evolution of the transmission law.

Our most recent work concerning complex transmission with continuous phenotypes introduces a third variable, the environment, into the specification. Individuals are observations on a trivariate random variable (f,g,e). Selection acts on f, g obeys the "genetic" law (12a) and e is merely transmitted but participates with g in determining f. Assortative mating may act through either the phenotype or environment. Such an extension of (12) is important for cultural evolution because much of the transmission of phenotype from generation to generation, or within generations, may not involve relatives at all.

Discrete Phenotypes.

G. S. Watson (1960) proposed and analysed a model for the inheritance of cytoplasmic sex-ratio in <u>Drosophila</u>. The model involved matroclinal transmission and included a genetic resistance mechanism. Watson and Caspari (1960) and Caspari and Watson (1959) developed models for cytoplasmic sterility in plants and mosquitoes. In an epidemiology study, Fine (1975) has classified many such phenotypes in a wide variety of organisms. The primary mode of transmission of all of these phenotypes is non-genetic, although genetic influences on the phenotype are not precluded.

There are obvious analogues between cytoplasmic transmission of biologically active material and uniparental transmission, perhaps through direct teaching, of a cultural phenotype. In higher animals and man, however, there are no well established cases of genetic resistance to (or, for that matter, enhancement of) such cultural traits. We have recently constructed a series of models for a two-valued phenotype whose transmission between generations

STABILIZING SELECTION AND CULTURAL EVOLUTION

may be genetically modified (Feldman and Cavalli-Sforza, 1976).

In the terminology of the previous section, consider six pheno-genotypes $\overline{A_1A_1}$, A_1A_1, $\overline{A_1A_2}$, A_1A_2, $\overline{A_2A_2}$, A_2A_2. The presence or absence of the bar represents the presence or absence of the phenotype in the relevant genotype. The viability of the bar phenotype relative to the other is $1+S:1$. The frequency of A_1 is p and mating is random between all pheno-genotypes. One parent in the mating is assumed to be the transmitting parent (T.P.). If the offspring is A_1A_1 and the T.P. has the bar phenotype, which we henceforth call <u>skill</u>, the probability that the offspring also has the skill is $1-N_i$. If the TP did not have the skill then the probability that the offspring has it is $1-n_1$. Similarly $1-N_2$, $1-n_2$, $1-N_3$ and $1-n_3$ are defined for the genotypes A_1A_2 and A_2A_2 respectively. These parameters completely specify the model. Let k_1, k_2 and k_3 be the proportions of skill phenotype within A_1A_1, A_1A_2, A_2A_2 respectively. Then the evolution of the system is described by the recursion system.

(13) $\quad p' = p(1+SK_1)/(1+SK)$

(14) $\quad k_1' = (\alpha_1+\beta_1 K_1)/(1+SK_1)$

(15) $\quad k_2' = \tfrac{1}{2}\left[(\alpha_2+\beta_2 K_1)/(1+SK_1) + (\alpha_2+\beta_2 K_2)/(1+SK_2)\right]$

(16) $\quad k_3' = (\alpha_3+\beta_3)K_2/(1+SK_2)$

where

$$\alpha_i = 1-n_i \; ; \; \beta_i = (1+S)(1-N_i)-\alpha_i$$

and

$$K_1 = pk_1+qk_2, \; K_2 = pk_2+qk_1, \; K = pK_1+qK_2$$

it can be seen that the first recurrence is of exactly the same form as the classical genetic transformation with selection coefficients Sk_1, Sk_2 and Sk_3. But k_i depends on the previous evolution of both p and k_i. Thus there is a complicated frequency dependence in the selection.

Inspection of (13) produces the three gene frequency equilibria $\hat{p}=0$, $\hat{p}=1$ and

$$\hat{p} = (\hat{k}_2 - \hat{k}_3)/(2\hat{k}_2 - \hat{k}_1 - \hat{k}_3)$$

(17)
$$= \frac{(\alpha_2 - \alpha_3) + (\beta_2 - \beta_3)\hat{K}}{(2\alpha_2 - \alpha_1 - \alpha_3) + (2\beta_2 - \beta_1 - \beta_3)\hat{K}} .$$

Our previous analysis of these equilibria was made in the case $\alpha_i = 0$ for all i. Here we record some new results for the case $\alpha_i \neq 0$ for all i.

When $\hat{p} = 0$ then $\hat{k}_3$ is the unique solution, $\hat{k}_3^{(0)}$, of

(18)
$$S k_3^2 + k_3(1-\beta_3) - \alpha_3 = 0$$

and $\hat{k}_2$ is the unique solution, $\hat{k}_2^{(0)}$, of

(19)
$$S k_2^2 + k_2\{1-\beta_2/2 - SB/2\} - \alpha_2/2 - B/2 = 0$$

where

$$B = (\alpha_2 + \beta_2 k_3)/(1+Sk_3) .$$

The uniqueness of these roots is independent of the sign of S. There is then a unique $\hat{p} = 0$ equilibrium which we call $S^{(0)} = \{\hat{p}=0, \hat{k}_1^{(0)}, \hat{k}_2^{(0)}, \hat{k}_3^{(0)}\}$. Similarly at $\hat{p} = 1$ there is a unique equilibrium which we call $S^{(1)} = \{\hat{p}=1, \hat{k}_1^{(1)}, \hat{k}_2^{(1)}, \hat{k}_3^{(1)}\}$. Finally the polymorphic equilibrium (17) is completely specified by noting that $\hat{K}$ solves the cubic equation

(20)
$$\begin{aligned}&K^3 S(\beta_1+\beta_3-2\beta_2)\\&+ K^2\left[\beta_1+\beta_3-2\beta_2+S(\alpha_1+\alpha_3-2\alpha_2) - (\beta_1\beta_3-\beta_2^2)\right]\\&+ K\left[\alpha_1+\alpha_3-2\alpha_2-\alpha_1\beta_3-\alpha_3\beta_1+2\alpha_2\beta_2\right] + \alpha_2^2-\alpha_1\alpha_3 = 0 .\end{aligned}$$

Some results have been obtained in the general case but here we want to restrict attention to a class of special cases.

We pointed out earlier (Feldman and Cavalli-Sforza, 1976) that $\alpha_i = 0$ (all i) reduces (20) to a quadratic whose roots have peculiar properties. Indeed it is possible in this case that the boundary equilibria $S^{(0)}$ and $S^{(1)}$ be stable but (20) have no admissible roots. Convergence then occurs to a neutral curve of equilibria. When $\alpha_i \neq 0$ (all i) this behavior does not occur.

Suppose, in particular, that

(21) $$\frac{\alpha_2-\alpha_3}{2\alpha_2-\alpha_1-\alpha_3} = \frac{\beta_2-\beta_3}{2\beta_2-\beta_1-\beta_3} = \mu, \text{ say.}$$

In view of (17) the existence of a valid interior equilibrium requires that $\alpha_2-\alpha_3$ and $\alpha_2-\alpha_1$ be of the same sign, and similarly for $\beta_2-\beta_3$ and $\beta_2-\beta_1$. Then it is not difficult to see that (20) again reduces to a quadratic equation:

(22) $$K(1+SK) - \mu(\alpha_1+\beta_1 K) - (1-\mu)(\alpha_2+\beta_2 K) = 0.$$

(22) always has a single valid root $\hat{K}$ (i.e. $0 < \hat{K} < 1$). Under the condition (21) then, the pattern is similar to a classical one locus constant selection scheme with overdominance or heterozygote disadvantage. For ease of reference denote the genetically polymorphic equilibrium by $S^{(p)}$.

The stability analysis of $S^{(0)}$, $S^{(1)}$ and $S^{(p)}$ under the condition (21) is algebraically rather tedious but it turns out that the boundary equilibria $S^{(0)}$ and $S^{(1)}$ are locally unstable if and only if the interior equilibrium $S^{(p)}$ is stable. Again the situation parallels the classical one locus constant selection scheme. Global convergence has not been proven.

Further analysis of the cubic (20) without the restriction (21) has been made although it is not complete. In addition we have made a numerical study of the system (13)-(16) under iteration. The latter has produced some interesting empirical observations. One question of topical interest in ecological theory concerns the possibility of cycling. In no case have we observed an equilibrium cycle in either gene frequency or phenotype proportion in the population. However, in the initial stages of the evolution it is possible for K to move in a damped oscillation before moving into a monotonic trajectory. In the final stages of evolution we have found no counter examples to the increase of (1+SK), the analogue of the usual mean fitness function. Nor have we found cases where more than one interior stable point (i.e. $S^{(p)}$) exists.

These remarks are of some mathematical interest since they

are not true for transmission rules which are non-genetic. Suppose that the transmission rule is specified by the probabilities b_3, b_2, b_1, b_0 that matings $\overline{M}x\overline{F}$, $\overline{M}xF$, $Mx\overline{F}$, MxF produce an offspring with the skill phenotype (M and F are mother and father, and genotype is ignored). Suppose as before, that the relative viabilities of skilled to unskilled are 1+S:1. Then by choosing the b_i appropriately, the mean frequency of skill in the population may end up in a two point cycle (Feldman and Cavalli-Sforza, 1976). Our numerical analysis so far seems to indicate that this cannot occur in the system (13)-(16). A possible reason is that (14)-(16) have a linear fractional structure which precludes the cycling found even in simpler models of gene frequency dependent selection. These are conjectures and analytical work is ongoing toward their resolution.

Extensions of the Transmission Scheme.

It has often been pointed out that one of the major differences between cultural and biological evolution is that the former often occurs quickly and the latter slowly. We may view culture as the result of a biological adaptation which allows fitness to increase by the use of information from sources presently or formerly in the population. Models such as (13)-(16) include transfer of information only from parents $(1-N_i)$ or the inate possession of the information $(1-n_i)$. This structure has now been extended (M. Uyenoyama, Feldman and Cavalli-Sforza, in preparation) by the incorporation of an epidemic factor. Thus, if instead of k_1' k_2' k_3' in (14)-(16) we denote the right hand sides of these relations by k_1^*, k_2^*, k_3^* and by K^* their population mean, we can write a new recursion system

(23) $$p' = \frac{p(1+SK_1)}{1+SK}$$

(24) $$k_1' = k_1^* + h_1 K^* (1-k_1^*)$$

(25) $$k_2' = k_2^* + h_2 K^* (1-k_2^*)$$

(26) $$k_3' = k_3^* + h_3 K^* (1-k_3^*)$$

where h_1, h_2 and h_3 are the rates of phenotypic transmission from those in the group with the skill, and those of genotypes A_1A_1, A_1A_2 and A_2A_2 who do not have the skill after biological transmission is complete. This simple model allows one "round" of epidemic transfer between the rounds of biological transfer. When all α_i are zero and h_i equal, our numerical analysis has confirmed that the rate of phenotypic evolution is accelerated by the epidemic transmission, whereas it slows the evolution of the gene frequency below the rate attained in the previous study by Feldman and Cavalli-Sforza (1976). The incorporation of the factor $h \neq 0$ allows the maintenance of phenotypes with a selective disadvantage. New equilibrium configurations are also possible. From our numerical analysis of this case it appears that the equilibrium cycles which characterize classical epidemiological models do not arise.

The relationship of these models to quantitative inheritance in its usual sense is indirect. However, for the student of evolution it is of some interest to assess the robustness of the theory to modification of the transmission law as opposed to changes in the selective values of the phenotypes. It may turn out that this can best be accomplished in terms of a discrete phenotypic scale. Also of interest to the evolutionist are ways in which the switch from a primarily biological mode of evolution to a more rapid cultural mode can be modelled. As we have shown appropriate modification of the transmission law can lead to the maintenance of characters at a disadvantage in the usual Darwinian sense. Thus this theory is relevant to the ongoing debate on the evolution of altruistic phenotypes.

ACKNOWLEDGMENTS

This research was supported in part by Grant USPHS 10452-11-12, U.S. Atomic Energy Commission Grant AT(04-03)-326, and National Science Foundation Grant GB 37835.

BIBLIOGRAPHY

(1) Caspari, E. & Watson, G.S. (1959). On the Evolutionary Importance of Cytoplasmic Sterility in Mosquitoes. Evolution 13, 568-570.

(2) Cavalli-Sforza, L.L. & Feldman, M.W. (1973). Cultural Versus Biological Inheritance: Phenotypic Transmission From Parents to Children (A Theory of the Effect of Parental Phenotypes on Children's Phenotypes). Am. J. Hum. Genet. 25, 618-637.

(3) Cavalli-Sforza, L.L. & Feldman, M.W. (1976). Evolution of Continuous Variation: Direct Approach Through Joint Distribution of Genotypes and Phenotypes. Proc. Nat. Acad. U.S. 73, 1689-1692.

(4) Crow, J.F. & Felsenstein, J. (1968). The Effect of Assortative Mating on the Genetic Composition of a Population. Eugen. Quarterly 15, 85-97.

(5) Ewens, W.J. (1976). Remarks on the Evolutionary Effect of Natural Selection. Genetics 83, 601-607.

(6) Falconer, D.S. (1960). Introduction to Quantitative Genetics. New York: Ronald Press Co., 266-267.

(7) Feldman, M.W. & Cavalli-Sforza, L.L. (1976). Cultural and Biological Evolutionary Selection for a Trait Under Complex Transmission. Theor. Pop. Biol. 9, 238-259.

(8) Feldman, M.W. & Cavalli-Sforza, L.L. (1977). The Evolution of Continuous Variation II. Complex Transmission and Assortative Mating. Theor. Pop. Biol., to appear.

(9) Fine, P.E.M. (1975). Vectors and Vertical Transmission: An Epidemiological Perspective. Ann. N.Y. Acad. Sci. 266, 173-194.

(10) Fisher, R.A. (1918). The Correlation Between Relatives on the Supposition of Mendelian Inheritance. Trans. Roy. Soc. Edinb. 52, 399-433.

(11) Fisher, R.A. (1930). The Genetical Theory of Natural Selection. Clarendon Press Oxford.

(12) Karlin, S. (1977). Manuscript.

(13) Kimura, M. (1958). On the Change in Population Fitness by Natural Selection. Heredity 12, 145-167.

(14) Kimura, M. (1965). A Stochastic Model Concerning the Maintenance of Genetic Variability in Quantitative Characters. Proc. Nat. Acad. Sci. U.S. 54, 731-736.

(15) Lande, R. (1976). The Maintenance of Genetic Variability by Mutation in a Polygenic Character with Linked Loci. Genet. Res. 26, 221-235.

(16) Slatkin, M. (1970). Selection and Polygenic Characters. Proc. Nat. Acad. Sci. U.S. 66, 87-93.

(17) Watson, G.S. (1960). The Cytoplasmic "Sex Ratio" Condition in Drosophila. Evolution 14, 256-265.

(18) Watson, G.S. & Caspari, E. (1960). The Behavior of Cytoplasmic Pollen Sterility in Populations. Evolution 14, 56-63.

(19) Wright, S. (1921). Assortative Mating Based on Somatic Resemblance. Genetics 6, 144-161.

(20) Wright, S. (1943). An Analysis of Local Variability of Flower Color in Linanthus Parryae. Genetics 28, 139-156.

(21) Wright, S. (1965). The Interpretation of Population Structure by F-Statistics With Special Regard to Systems of Mating. Evolution 19, 395-420.

XI
Invited Discussions

Somatic cell genetics and genetic improvement

M. Gillois
LABORATOIRE DE GÉNÉTIQUE CELLULAIRE
CENTRE DE RECHERCHES DE TOULOUSE, I.N.R.A.
B.P. 12, 31320 CASTANET-TOLOSAN, FRANCE

ABSTRACT

The somatic cell genetics of domestic animals is now operational because all tools required for its development are available.

Objective information concerning the genotypes of animals kept for breeding purposes could be obtained by hybridization of primary cells from the animals with cells from reference strains.

Procedures and methods of quantitative inheritance applied to animal selection will be supplemented in the future by hybridization, whereas mathematical models which describe quantitative features will be modified.

THE PRESENT SITUATION FOR DOMESTIC ANIMALS

Very little is known about gene function, localization and mapping in mammals, except in man. Research on human gene mapping is now rapidly progressing. At the Rotterdam Conference in 1974, and at the last Baltimore Conference in 1975, the locations of about 200 genes on human chromosomes were reported. One of our aims is to accomplish the same kind of work on domestic animals, but the difficulties are numerous.

First, for obvious economic reasons, defective animals are slaughtered and it is very difficult to have primary cultures with genetic markers for domestic animals, as is possible for man. So we began to work on established cell strains in order to obtain a great number of mutant clones, using mutagenesis techniques and selective procedures.

For the pig cell strain PK 15, Mulsant has isolated biochemical classes of mutants and Gillois and Echard (15) have isolated temperature sensitive mutants.

Secondly, cell hybridization is the most commonly used technique for gene mapping. For human cells, P.E.G. chemical agent and Sendai virus are used to induce cell fusion. This virus was found to be ineffective in inducing cell fusion in pig cells. So, Gillois and Gilbert (15) experimented on cell fusion with several viruses. Among these, the para influenzae virus induced fusion in pig cells. During the past three months many experiments have been carried out to obtain intraspecific and interspecific hybrid clones, i.e., pig (PK) x pig (PK), pig (PK) x hamster (CHO, CCL 39). The first hybrid clones have just been obtained, and we are now extending this hybridization to bovine cells.

Thirdly, to characterize hybrid clones and localize genes on chromosomes, it is important to have a precise knowledge of the karyotype of the cell strains used. The banded karyotype and chromosome abnormalities of cell strains of domestic animals has been studied by Echard (4, 5, 15) for three pig cell strains (PK 15, F, RP), a bovine cell strain (MDBK), a rabbit cell strain (RL) and for primary cells of domestic animals--bovine, porcine and rabbit.

The gap between humans and domestic animals in the hybridization field has now been closed. This is all the more true, since important works on mutagenesis and selection have allowed the study of biological, veterinary and physiological problems.

Isolation of pig strain PK 15 mutants, affected in purine and pyrimidine metabolism, by Mulsant (15), has allowed him to use these mutants to determine which enzymatic activities of this metabolic pathway have an important physiological part in pig cells.

In particular, the major adenosine use pathways begin at its deamination in inosine, followed by a phosphorylation in hypoxanthine, rather than its straight conversion into AMP by adenosine enzyme. Now, adenosine is toxic for pig cells, except for pig mutant cells that have lost this adenosine kinase enzyme (AK^-). This important adenosine deamination by pig cells probably plays a detoxification role.

These same techniques have allowed Caboche (3, 15) to study methionine metabolism. Biochemical and genetic analysis of isolated mutants have given a better knowledge of two enzymatic activity regulations, i.e., the feedback inhibition regulation of menadione reductase enzyme, and the synthesis control of methionine adenosyl transferase enzyme (MAT).

It is possible to make a genetic analysis of cell cycle and mitosis mechanism by means of temperature sensitive (ts) mutants.

Gillois and Echard (6, 15) used nonselective isolation techniques for mutants (BUdR suicide) and developed a new technique to isolate perimitotic ts mutants. After a BUdR treatment at the nonpermissive temperature, pig PK cells were incubated with BUdR at the permissive temperature for one generation. Temperature-sensitive mutants which recover growth are killed, except for perimitotic mutants arrested by colcemid addition before the following ADN synthesis period.

In this way, Gillois and Echard isolated thermosensitive mutant clones. The kinetic and biochemical analysis of five of these clones, selected by BUdR--colcemid technique, confirmed their specific nature as cell-cycle temperature sensitive mutants.

Only the study of this type of mutant will lead to an understanding of the genetic control of the cell cycle. Good progress in this knowledge is necessary for the comprehension of cellular growth, and of the genetic control of mitosis and meiosis.

Interspecific cellular hybrids have been used in order to localize the genes that code for viral receptors and, when possible, to identify the chromosomes that carry genes involved in some steps in viral maturation (7).

In some hybrid cells, derived from permissive and nonpermissive parental lines, viral proliferation is inhibited by genes that originated from the nonpermissive line (1). In most cases however, the permissive state is dominant: for instance, hybrids from human and rodent cells may be infected by polyviruses, provided the human gene coding for poliovirus receptor is present (2, 14). As soon as the viral nucleic acids have penetrated the hybrid cell, the multiplication of virus does not need the presence of human genes (14, 9). Such a situation has made it possible to associate a human gene coding for this viral receptor with a metacentric chromosome (12), and then to localize it in the 19th chromosome (13). As the present time, viruses causing diseases of domestic animals are being studied, so as to identify the genic control of their receptors. This study suggests that it is possible to decompose global quantitative characters of resistance into their genetic components. The synthesis of interferon is another component.

In the future, studying interactions between the viral genes and those of the infected cell, will allow a better understanding of viral DNA synthesis steps, and then of the forthcoming maturation process (1). Furthermore, it will give an insight into the evolutionary mechanisms that confer upon viruses their infectious ability, directed either against one or many species.

Some intracellular parasites have been maintained in fused cells. These include Plasmodium gallinaceum, P. berghei (8), Theileria parva (10), Babesia rodhaina (11). In all cases except B. rodhaina, there was evidence of continued development within the heterokaryon.

CONCLUSION

The somatic cell genetics of domestic animals is now operational because all the tools required for its development are available.

Some of the established cell lines are as follows: PK 15, RP, F in pigs, MDBK in cattle, in eggs, RL in rabbits. The banding karyotypes of these are well-known. These cell lines have been the subject of intensive technical and methodological research in the following ways: (1) culture conditions, (2) deep cold live storage conditions, (3) growth and incorporation kinetics, (4) development of minimum culture medium, (5) definition of mutagenesis procedures, (6) perfection of mutant selection procedures, (7) finding of effective fusing agents for domestic animal cell hybridization.

The somatic cell genetics of domestic animals does not have as its only object the final settlement of scientific disputes, but to accumulate results with a view to fulfilling the four following purposes:

1. To incorporate molecular genetics into embryological, physiological, pathological and nutritional studies of domestic animals and to renovate them.

2. To establish gene mapping of domestic animals and to describe parts of these genes and the chromosomal structure.

3. This truly biochemical and genetic knowledge of the genomes of domestic animals will allow us to discover biochemical and genetic building patterns for quantitative characters, since the number of genes involved, their reciprocal relations and their regulations will be known.

4. To create, to manage, and to develop reference collections of mutant cells, some of which are carriers of deficient genes at loci which play a key part in quantitative characters (e.g., genes which code for hormonal receptors in the physiology of growth or reproduction, genes which code for viral specific receptors or interferon synthesis in pathology, etc.).

By hybridization with primary cells taken from an animal kept for breeding purposes, these reference strains of cells will give objective information on their genotypes.

The procedures and methods of quantitative inheritance applied to animal selection will be supplemented in the future by hybridization, whereas mathematical models which describe quantitative features will be modified.

ACKNOWLEDGMENT

The author is particularly grateful to J. Lewis and E. Pollak for their willingness to correct an early English version.

BIBLIOGRAPHY

(1) Basilico, C., Matsuya, Y., Green, M. (1970). Virology 41, 295.

(2) Belehradek, J., Jr., Barki, G. (1969). C. R. Acad. Sci. Paris, 269, 672.

(3) Caboche, M. (1976). J. Cell. Phys. 87 (3), 321.

(4) Echard, G. (1973). Ann. Gen. Sel. Anim., 5, 1.

(5) Echard, G. (1974). Chromosoma, 45, 133.

(6) Gillois, M., Echard, G. (1976). Somatic cell genetics (submitted).

(7) Green, H., Wang, R., Basilico, C., Pollack, R., Kusano, T., Salas, J. (1971). Fed. Proceedings Symp. 30, 930.

(8) Gun, F., Greenblatt, C. M. (1973). G. Trotozool, 20, 532.

(9) Holland, J. J., MacLaren, L. C., Syverton, J. T. (1959). J. Exp. Med., 110, 65.

(10) Irvin, A. D. et al. (1974). Int. J. Parasit. 4, 519.

(11) Irvin, A. D. et al. (1975). Int. J. Parasit., 5, 465.

(12) Kusano, T., Wang, R., Pollack, R., Green, H. (1970). J. Virol., 5, 682.

(13) Miller, D. A., Miller, O. J., Vaithillingam, G., Hashmi, S., Tantravahi, R., Medrano, L., Green, M. (1974). Cell, 1, 167.

(14) Wang, R., Pollack, R., Kusano, T., Green, H. (1970). J. Virol. 5, 677.

(15) In comptes Rendus activités 1974-1975. Laboratoire de Génétique Cellulaire. Centre de Recherches I.N.R.A. de Toulouse. B.P. 12. 31320 Castanet-Tolosan, France.

Transmission of genes and transmission of characters

Albert M. Jacquard
INSTITUT NATIONAL D'ÉTUDES DEMOGRAPHIQUES
PARIS, FRANCE

INTRODUCTION

In their initial invitation, Professors Kempthorne and Pollak wrote: "There is deep uncertainty about the quality and utility of the available theory of quantitative genetics."

I agree with that statement. I think that the cause of this situation is the inadequacy of the concepts that we manipulate. We are victims of our concepts. I shall just give two examples.

MENDEL'S LAWS AND "MENDELIAN CONCEPTS"

Quantitative genetics has been developed from concepts of genes, alleles, loci, and so on, which we call "Mendelian Concepts." But Mendel did not know of the existence of chromosomes. What exactly did he propose? We can summarize Mendel's ideas as follows:

1. An individual I is made from two gametes: g_F given by his father and g_M given by his mother. Each characteristic C_I of the phenotype of I is a function of the characteristics γ_F and γ_M of these gametes and of the environment E. We can write

$$C_I = f(\gamma_F, \gamma_M, E) \qquad (1)$$

This was, of course, known before Mendel; but

2. When an individual I makes a gamete g_I, the pre-Mendelian opinion was that the characteristic γ_I of this gamete is a function of the characteristic of I

$$\gamma_I = \phi(C_I) .$$

The first original contribution of Mendel was to refute this relation and to affirm that γ_I is a function of the characteristics of the previous gametes g_F and g_M and that the phenotype of I is irrelevant; that is

$$\gamma_I = \phi(\gamma_F, \gamma_M) . \qquad (2)$$

In other words there are no acquired, transmissible characters as Lamark or Darwin believed.

3. The other original contribution of Mendel was to affirm that the function ϕ is not a deterministic one, but a function with parameters which are random variables.

Equation (2) and the random character of the function ϕ are the essence of Mendelism, and we need no things such as alleles, genes or loci to express this theory.

But what precisely are the functions f and ϕ? The simplest case is what we now call the "Mendelian character," for example, the color of peas. In this case we have the following table for f.

$\gamma_M \backslash \gamma_F$	1	0
1	1	1
0	1	0

$$\underbrace{}_{C_I}$$

where 1 codes for yellow and 0 codes for green.

The relation for ϕ is

$$\phi: \gamma_I = p\, \gamma_F + (1-p)\, \gamma_M$$

where p is a random parameter such that

p = 1 with probability .5
p = 0 with probability .5 .

These simple functions fit the observations perfectly well. This is also the case for some characters of human blood, for example the ABO system. In this case the function f is given by the following table.

$$
\begin{array}{c|ccc}
Y_M \backslash Y_F & 2 & 1 & 0 \\
\hline
2 & 2 & 3 & 2 \\
1 & 3 & 1 & 1 \\
0 & 2 & 1 & 0 \\
\end{array}
$$

$$\underbrace{}_{c_I}$$

where 3 codes for (AB)
      2 codes for (A)
      1 codes for (B)
      0 codes for (O)

and the function ϕ is the same as for the color of peas.

However, such simple functions do not work when we deal with quantitative characters.

The way which has been chosen to develop quantitative genetics has been to keep the simplest function ϕ, but to take into account more than one locus. Thus, several concepts such as epistasy, pleiotropy and interaction have been introduced which lead to the introduction of new parameters in the model, allowing it to fit the data.

But in this way the reference to the genes leads immediately to a very large number of parameters. The covariance between two individuals for characters depending on two loci, for example, is expressed by so many parameters that we have no hope we can really obtain estimates of them; they are purely formal and cannot explain anything of what happens in the real world. It is then necessary to reduce this number by the use of techniques such as principal component analysis or correspondence analysis. Is it really reasonable to introduce so many parameters, if we have to get rid of them as soon as they appear?

Another way, quite different, could be explored. It consists of keeping the formal Mendelian model, in the form we have presented, without any reference to biological concepts, and to introduce directly in the central functions f and φ as many parameters as are necessary, but no more. The search for a model is then the search for functions f and φ fitting the data, not the search for an eventual genetic determination of the character we study. For example we can admit more than two values for the random parameter p, and determine the best probabilities for these values. We can also give a more complex form to the deterministic function f and choose the values of the various parameters by means of maximum likelihood techniques.

The central point is that, in this way, we leave aside, as Mendel did, the existence of chromosomes and we can get rid of difficult concepts such as identity, isoaction, recombination, phenocopy, linkage disequilibrium, and so on, which, for this type of problem, are more troublesome than helpful.

NATURAL SELECTION AND HERITABILITY

When we consider a heritable character that is subjected to selective pressure, we often do not pay sufficient attention to a relation which is trivial but which is often forgotten, in spite of its triviality.

In this case we manipulate three concepts:

1. "quantitative character," that is a character such that to each individual I there corresponds a measure C_I, with mean $\bar{C}$ and variance V_C ($\neq 0$) in a given population in a given generation,

2. "heritability," which means that the resemblance between mothers and daughters is such that

$$E(C_d) = \bar{C} + \frac{h^2}{2} (C_M - \bar{C})$$

where E is the "expected value" operator and h^2 is the classical parameter "heritability,"

3. "selective value," which is a measure w_I, for each individual I, of his capacity to transmit his genes to the next generation; we let $\bar{w}$ and V_w be the mean and variance of w_I.

As a consequence of the definitions of these three concepts, without any hypothesis, it is easy to show that between two successive generations the mean $\bar{C}$ of the character is modified by selective pressure according to the relation

$$\bar{C}_{g+1} - \bar{C}_g = \frac{h^2}{2\bar{w}} r \sqrt{V_C V_w}$$

where r is the coefficient of correlation between the parameters C_i and w_i.

If the character under study is stable, i.e., if $\bar{C}_{g+1} = \bar{C}_g$, we must have either $h^2 = 0$, or $r = 0$. In other words, a stable character must be either nonheritable (which does not mean it is not subject to a genetical determination) or selectively neutral.

Examples of useless work due to having forgotten this evidence are several studies devoted to the estimation of the heritability of the fecundity of women. Obviously the fecundity is correlated with the selective value ($r \neq 0$) and is stable (otherwise the population would "explode"), therefore h^2 must be zero. Such are the results of most surveys, but their authors express their surprise, as they were expecting quite another value.

Analyzing and precisely defining the concepts is not a "philosophical" exercise, it is really the basis of scientific activity; maybe geneticists have not entirely done their duty in this field.

What's different about chickens?

Dewey L. Harris
USDA-ARS
PURDUE UNIVERSITY, WEST LAFAYETTE, IN 47907

Earlier this week Professor Kempthorne asked me a probing question that I will attempt to answer. That question was - "What's different about chickens?" This question was conditioned by Professor Kempthorne's astute observation that there are more signs of stress among poultry breeders than among the breeders of other animal species. For example, there is more skepticism on the part of poultry breeders about the validity of our quantitative genetic and selection theory. This skepticism and concern was expressed by Nordskog (1976) in his paper earlier this week. Of course, there are two types of chickens and thus two approaches to poultry breeding. One is egg production chickens and the other is meat type chickens. Here, I will primarily be discussing the difficulties of breeding egg production poultry although the breeding of meat-type poultry encounters many of the same difficulties.

From a scientific standpoint the primary difference from large animal species has been that experiments to evaluate whether response to selection is in agreement with the predictions of that response have been more feasible in chickens and, thus, more experiments have been executed and reported. It has not been feasible to conduct such experiments in the large animal species. The data analysis work on breeding of dairy cattle for milk production, as reported in the earlier paper by Van Vleck (1976) has, in general, been very reassuring that the response to selection is consistent with that expected in that species for that trait. However, much of the experimentation in poultry has been done in unreplicated experiments, which means that it is impossible to separate the real departures from prediction from the random departures expected due to the large amount of random genetic drift which can occur in populations of finite size. Much confusion and skepticism has been generated in the attempts to interpret these unreplicated experiments.

Some of the factors that make the breeding of poultry different from the breeding of other species are of a genetic nature and some are definitely of a non-genetic nature even though they influence the manner of implementing genetic procedures. First among the non-genetic factors is that the breeding of egg production poultry is highly competitive with a few participating companies competing with each other for a reasonable share of a declining market. Because consumers are eating less eggs and because hatcherymen are more efficiently using their parent breeding stock for producing commercial chicks, the market for parent breeding stock is considerably smaller than it was in years past and is continuing to decline. This intensifies the competitive nature of the industry. Secondly, breeding is a long slow process with considerable lag between implementing an improved breeding procedure and the time that resultant improved performance will occur in commercial flocks. Only then does this improvement influence income due to increased sales or increased selling prices.

A primary characteristic of poultry relative to other animal species is a considerably higher reproductive rate. Even though this high reproductive rate can lead to much greater intensities of selection relative to other animals, it can be disadvantageous in that it makes possible the use of relatively small breeding populations for the production of the commercial individuals. As was discussed elsewhere in this meeting small breeding populations can lead to restrictions in the long-term potential from selection and, of course, there have been many strains and lines of egg production poultry that have been under continued selection for many generations. This is in addition to the fact that the primary trait of egg production poultry is rate of lay which has considerable natural selection in its history. This means that considerably more of the additive genetic variation has been exhausted, and thus, a larger percent of the remaining genetic variation is likely to be non-additive. And, of course, rate of lay for egg production is approaching an upper physiological limit of 100%. Although levels of production are not closely pressing this limit, there is undoubtedly considerable influence of this limit upon the further potential. Coupled with this, some populations of poultry may have been maintained in population sizes small enough that plateaus are being approached.

A big manifestation of the lack of complete maturity of the science of quantitative genetics as applied to the breeding of poultry is the great diversity of breeding schemes used by the major breeders of egg production poultry ranging from reciprocal recurrent selection to pureline selection only with in some cases intentional use of inbreeding and, in other cases, avoidance of inbreeding involved in the variety of programs. Most schemes seemingly are effective as evidenced by their continued use but the decisions as to the relative effectiveness of the different schemes have not been completed.

Another aspect of poultry breeding that seemingly makes it different from other animal species is the occurrence of considerable "relaxed selection effect" as evidenced in the research of Dickerson (1965). Whether this is due to antagonistic natural selection as in the work of Enfield (1976) or is due to the recombination loss of selected epistatic effects as discussed by Griffing (1960a and 1960b) or is due to some other aspects, such as the maternal transmission of pathogens which influence performance, has not been clarified. Nevertheless, it is an important part of the implementation of breeding programs that has to be seriously considered.

In recent years, there has been three occurrences that have severely disrupted and antagonized the efforts to implement logical scientific breeding programs. The first of these was the occurrence of Marek's disease as a very serious problem of the industry that required panic efforts of breeders to attempt to develop disease resistance. This was prior to the development of a vaccine for the disease that made the results of such genetic efforts drastically less important. Of course, if Marek's disease occurred within a breeders flock the increased mortality and the reduced egg production seriously reduced the possible intensities of selection and reduced the accuracy of selection for other traits due to variation associated with the morbidity effects of the disease.

The problems of either avoiding Marek's disease or breeding for resistance was closely followed by the demand of the industry for flocks which were free of the Mycoplasma organisms. The procedures involved in eradicating the Mycoplasma organisms from the breeding flocks disrupted to a large degree the breeding programs in many breeding companies. The last traumatic experience for the industry and for some breeders was the attempts to introduce an easier process for the sex identification of baby chicks by introducing the sex-linked slow feathering gene into commercial crosses. The industry has experienced some disasterous associated declines in performance. Hopefully, the poultry breeding industry will soon return to stable consistent application of our known principles of quantitative genetics and will be able to implement the additional developments in this area in its breeding programs.

BIBLIOGRAPHY

Dickerson, G. E. (1965). Experimental evaluation of selection theory in poultry. GENETICS TODAY (Proc. XI Int'l. Cong. Genetics, The Hague, 1963.) 3: 747-761.

Enfield, F. D. (1976). Selection experiments in Tribolium designed to look at gene action issues. Proc. Int'l. Conf. on Quant. Gen.

Griffing, B. (1960a). Theoretical consequences of truncation selection based on the individual phenotype. Aust. J. Biol. Sci. 13: 307-343.

Griffing, B. (1960b). Accommodation of linkage in mass selection theory. Aust. J. Biol. Sci. 13: 501-526.

Nordskog, A. W. (1976). Success and failure of quantitative genetic theory in poultry. Proc. Int'l. Conf. on Quant. Gen.

Van Vleck, L. D. (1976). Theoretical and actual genetic progress in dairy cattle. Proc. Int'l. Conf. on Quant. Gen.

Comments on the conference

R. S. Gowe
ANIMAL RESEARCH INSTITUTE, RESEARCH BRANCH
AGRICULTURE CANADA, OTTAWA, ONTARIO, CANADA

After five days of interesting and stimulating papers and discussions, several points appear to stand out clearly, from the viewpoint of one interested principally in the relationship of quantitative genetics theory to applied animal and poultry breeding.

1. The major problem of applied animal and poultry breeders, that remains mostly unresolved by the work reported here this week, is the question of the optimal selection procedures for improving populations for several traits simultaneously, where (a) the populations are moderately large with more than 100 effective parents and, (b) inbreeding is not a major problem, and (c) there are several (4 to 10 or more) important economic traits to be improved at once, and (d) some of these traits are genetically negatively correlated, and (e) some of the traits are positively correlated, and (f) some of the traits are intermediate optimum traits. Simplified quantitative genetic assumptions (additive effects only, no epistasis or dominance effects) and Hazel-Smith indices do not seem to lead to optimal response in the complex of traits required,

and workable alternate schemes did not arise from the meeting nor did there appear to be much interest in developing and testing such alternates.

2. Much more research is required (utilizing laboratory species and the more flexible domestic species such as poultry and swine) to test out alternate breeding schemes for multi-trait selection programs. Such breeding studies need to take into consideration the problems raised in point one above.

3. The characteristics of the base population and their relevance to selection or breeding schemes have not been thoroughly researched, and only one speaker (Comstock) really emphasized their importance. There seems to be clearly two types of situations requiring different models or procedures. The base population is either one that has been under selection for many generations for a complex of economic traits, some of which are fitness traits and some negatively correlated with others, or the base population is a newly synthesized population with very broad gene base, the latter type of population is one that has not been intensively selected for many generations. Research is needed on optimal selection schemes for the two different extreme situations described.

4. The resources required for selection studies involving large populations and multiple traits and different base populations are very large even if laboratory species are used. More effective coordination of the resources available is required to ensure that the limited resources available are used effectively.

5. New information is needed on optimal schemes for introducing new useful qualitative genes (such as qualitative colour genes) into populations, as well as methods for introducing new genes

responsible for small positive additive effects on the quantitative traits of concern.
6. Despite the large resources deployed by commercial breeding concerns (animal and plant), there is little of real significance being contributed by these organizations to our general knowledge of applied breeding and to quantitative genetics theory.

XII
Contributed Papers

Genetic analysis of quantitative data from the families of identical twins

Walter E. Nance, M.D., Ph.D. and *Linda A. Corey, Ph.D.*
MEDICAL COLLEGE OF VIRGINIA, RICHMOND, VA 23298

Since human beings cannot be subjected to experimentally designed mating programs or selection protocols, inferences about the genetic determination of quantitative traits in man must be drawn from a comparison of the similarity of relatives of various degrees. Previous approaches to this problem have involved the comparison of heterogeneous bodies of data on rare or abnormal human relationships such as identical twins reared apart, adopted children, or conventional and illegitimate half-siblings. We wish to describe a model in which all of the relationships that are required to clearly separate genetic and environmental effects are contained within each family unit. The model involves analysis of data from the families of identical twins. The families contain individuals who share one quarter, one half, all, and none of their genes, respectively. These and other unique relationships within the families not only permit resolution of genetic and environmental effects, but also allow a partitioning of genetic variance into additive, dominance, epistatic and maternal components.

Although they are genetically related to each other in the same way as half-sibs, the offspring of identical twins are usually raised in separate homes by their biologic parents in a manner that would appear to differ in no important way from the families of single born individuals. In contrast to conventional half-sibs resulting from the death of a parent, divorce, or illegitimacy, all of the parents of monozygotic twin half-sibships are usually available for analysis. There are no systematic differences among the half-sibs in age or parity and the expected sizes of the two sibships within each half-sibship are equal since they are ascertained through parents of the same age and sex.

Data from the offspring may be partitioned by a nested analysis of variance into Among half-sibship, Between sibships-within half-sibship, and Within sibship mean squares. In order to solve for the Among, Between-within, and Within components, the appropriate coefficients must be calculated using a weighting procedure to adjust for unequal family size. The genetic interpretation of the variance components from the analysis of variance is identical to that of a conventional half-sib analysis with two important exceptions. First, we estimate three environmental variance components corresponding to the three orthogonal partitions of the data. Second, the availability of data from the offspring of both male and female monozygotic twins permits estimation of the variance component attributable to maternal effects. If present, genetically determined maternal effects will augment the covariance of the offspring of female identical twins, who are born to genetically identical mothers and will also inflate the Between sibship-within half-sibship component for the offspring of male twins whose mothers are genetically unrelated. Since the genetic expectations for the Within sibship mean square are the same for male and female twins, the estimates may be pooled yielding a total of five equations from the offspring data.

Two additional equations of estimation may be obtained from an analysis of variance of data from the twin parents as well as from the five distinct parent offspring relationships. Several of these relationships are quite novel. For example, the relationship between an offspring and his twin parent differs from that between the offspring and the twin aunt or uncle only because of the increased environmental similarity in the former comparison.

The effects of assortative mating and common environment are two possible sources of bias that can profoundly influence analysis of quantitative traits in man, particularly psychological traits. However, we know of no previous attempts to separate these potential causes for correlation between husband and wife. In the present model, this resolution can be achieved. Since offspring of an identical twin are related to the spouse of the co-twin only because they share the same half-sibship environment, the covariance between them provides an estimate of the Among half-sibship environmental component (V_{EH}). The correlation between husband and wife reflects similarity of the home environment (V_{ES}) and the and the half-sibship environment (V_{EH}) as well as effects of assortative mating. Finally, the correlation between spouses of identical twins will reflect the half-sibship environment and a component equal to the square of the correlation attributable to assortative mating. Note that the effect of nonrandom mating on the correlation between spouses will be positive regardless of whether matings are assortative or disassortative. By combining these three equations, effects of common environment and assortative mating can be separated.

The model can be readily extended to estimate the effects of X-linked genes by partitioning data into half-fraternities and half-sororities. If this is done, a total of 10 equations may be derived from the offspring data alone, providing estimates of additive and dominance components of X-linked and autosomal genes, respectively. Similarly, five equations may be derived from parental data, as well as a total of nine distinct equations from the parent offspring relationships. It is interesting to speculate about possible differences in the expression of X-linked gene effects in males and females. In males, effects of X-linked genes are necessarily additive because of haploidy. In females, on the other hand, most if not all effects of X-linked genes would be expected to appear in the dominance component because of random X inactivation, a genetic mechanism which creates nondirectional dominance at the cellular level. The present model may provide a better fit for data on traits that are influenced by X-linked genes because it allows for both additive and dominance effects.

Although genetic-environmental interactions and genetic-environmental covariances are not explicitly estimated, we believe their effects can be detected by repeating the analysis following adjustment of data for specific environmental variables. If the adjustment alters both the estimated genetic variance as well as the environmental variance, interactions should be suspected. Adjustment of sibship or half-sibship means should disclose genetic-environmental covariances, while adjustment of individual scores would be expected to reveal genetic-environmental interactions.

To obtain error estimates of relevant genetic and environmental variance components, mean squares are used instead of components and an iterative weighted least squares solution is obtained. An important feature of the model is that the variance-covariance matrix of mean squares used to obtain the iterated estimates contains both weights and constraints; the latter to adjust for non-independence of the highly interrelated mean squares used in the analysis.

In preliminary studies of several marker traits, we have confirmed the largely additive determination of total ridge count and have found a major maternal effect on birth weight. The model would appear to hold great promise for the analysis of psychological traits where, for example, demonstration of a significant maternal effect on certain subtest scores could make a constructive contribution to the current controversy over the genetic and environmental determinants of psychological test scores which could serve to redirect our efforts to improve the educational performance of children.

Genetic analysis of human serum cholesterol

Charles F. Sing and John D. Orr
DEPARTMENT OF HUMAN GENETICS
UNIVERSITY OF MICHIGAN MEDICAL SCHOOL,
ANN ARBOR, MI 48104

ABSTRACT

Approximately sixty percent of the variability of ln cholesterol adjusted for age, replicate, and socioeconomic status in Tecumseh, Michigan, is genetic.

INTRODUCTION

Serum cholesterol ranks with blood pressure, glucose tolerance, and smoking habits as a predictor of coronary heart disease (Truett et al., 1967). Coronary heart disease, hypercholesterolemia, hypertension, and diabetes each tend to aggregate in families (Epstein, 1975). Whether there is a common genetic and/or environmental etiology has not been established. Research is proceeding on the assumption that understanding the genetics of risk factors will improve attempts at early detection of disease-prone individuals (prediction) and facilitate a better understanding of the biochemical defects which may underlie coronary heart disease (causation). We are conducting studies on a population sampled without regard to age or health status to identify the relative contribution of genetic and environmental factors to prediction of the serum cholesterol levels of normal individuals.

Studies of continuously varying traits in man to obtain a partition of genetic and environmental influences on deviations from the population mean present statistical problems not usually encountered in experimental plant and animal studies. At least two general statistical strategies exist for handling quantitative human data. They are the analysis of covariances among relatives and the use of maximum likelihood procedures and probability models which have recently been developed (Elston and Stewart, 1971, and Lange et al., 1976). We are carrying out both strategies using appropriate subsets of the data available from the Tecumseh community. When one attempts to insure the validity of the correspondence between the biological and the statistical parameterizations of the phenotypic variance, use of the covariance approach can result in an inefficient utilization of data. Sampling large pedigrees to take advantage of likelihood procedures can overcome the problems of correlated observations and unbalanced sampling of various genetic and environmental effects but invariably results in redefinition of the population of inference. A summary of our analyses to date is reported here.

Sample and Analysis Procedure

From 1959 to 1970 data were collected in the town of Tecumseh, Michigan, to define predictors of health and disease in a total community. From this study, we obtained 14,310 serum cholesterol measurements from 4,619 males and 4,730 females. These data were collected in two rounds, the first from 1959 to 1960 and the second from 1963 to 1966. A total of 2,457 males and 2,504 females appeared in both rounds, the remaining individuals only in one. Non-fasting cholesterol as mg per 100 ml of serum was determined by the method of Abell et al. (1952). Different frequency distributions and regressions on age justify separate analyses of males and females. Each cholesterol observation was first transformed by natural logarithm (ln) to remove skew and then adjusted by a third-order polynomial on age within round. If an individual was measured in both rounds, the two age-adjusted ln cholesterol observations were averaged. These adjusted ln cholesterol values were used in analysis of the role of education, occupation and income as sources of cholesterol variation (Orr et al., 1975). The adjusted ln cholesterol was analyzed on each of the three socioeconomic factors, separately and in combination. The data then were adjusted for education, occupation and income in preparation for analyses of genetic markers and estimation of polygene effects. Only individuals typed for all twelve available marker systems were included in the sample for genetic analyses (3,112 males and 3,254 females). Single locus phenotype frequencies and adjusted ln cholesterol distributions for these subsamples did not differ significantly from the original samples. Genetically independent sibships residing in separate households have been defined for use in an intraclass correlation analysis.

Contribution of Concomitants

Adjustment for age reduced the variance of a ln cholesterol measurement by 29% for both sexes. The difference in ln cholesterol between rounds accounted for less than 1% of the observational variance but was statistically significant. Averaging replicates reduced the variance of an observation in males and females by 7% and 9%, respectively. The socioeconomic variables contributed significantly to prediction of age-round-replicate adjusted ln cholesterol in males, but not in females. High cholesterol in males was associated with more education and higher income; sales, managerial and office, and professional people had higher than average cholesterol, while unskilled, unemployed, and clerical workers had lower than average cholesterol. Education, occupation and income adjustments reduced the variance of an observation approximately 1% in males. Altogether, the concomitants studied accounted for a reduction in variance of approximately 37% in males and 39% in females.

Contribution of Genetic Markers

A nested analysis of variance conducted for each of twelve polymorphic genetic markers identified four--Gm, haptoglobin, secretor, and ABO--which gave statistically significant variation among phenotypes for one or both sexes (Sing and Orr, 1976). The reduction in the variance of an individual's serum cholesterol due to fitting these four marker phenotypes was 1.1% for males and 0.4% for females. The average difference between the high and low phenotypes for individual marker loci in Tecumseh was 4.30 mg/100 ml for males and 6.28 mg/100 ml for females. The average difference between high and low phenotypes for all two locus combinations was 8.78 and 8.90 mg/100 ml for males and females, respectively. The additivity of marker effects defined by single locus analyses is summarized in Table I (from Sing and Orr, 1976).

TABLE I

No. of Positive Single Locus Effects in Four Locus Phenotype	Males		Females	
	No.	Mean	No.	Mean
0	341	193.2	383	198.3
1	1,105	197.4	1,111	199.2
2	1,169	202.5	1,198	203.7
3	437	202.8	480	205.5
4	60	210.7	82	206.8
Linear Regression		3.72		2.79

The conclusion that each locus has a small additive effect is supported by the consistent rank order of means in both sexes. Correlated gene frequencies and/or epistasis could account for deviations from linearity. The concordance of each of the four single locus effects between sexes and the consistency of effects over a number of studies for ABO, secretor, and haptoglobin suggest that at least four polymorphic loci randomly distributed throughout the genome are directly involved in cholesterol metabolism. The linkage disequilibrium required to keep a linked cholesterol gene from randomizing its alleles with respect to the marker alleles across populations would need to be of the order found for MNSs and Rh.

Contribution of Polygenes

The marker results are consistent with the polygene hypothesis of causation. The total contribution, ignoring markers, of additive genetic variance to adjusted ln cholesterol estimated from sibs living apart was 54% and 76% for males and females, respectively (Sing and Orr, in manuscript). Compared to the contribution of the locus determining hypercholesterolemia (Sing et al., 1975), this is a major source of genetic variability in normal individuals.

BIBLIOGRAPHY

[1] Abell, L. L., Levy, B. B., Brodie, B. B. & Kendall, F. E. (1952). A simplified method for the estimation of total cholesterol in serum and demonstration of its specificity. J. Biol. Chem. 195, 357-66.

[2] Elston, R. C. & Stewart, J. (1971). A general model for the genetic analysis of pedigree data. Hum. Hered. 21, 523-42.

[3] Epstein, F. H. (1975). Familial aggregation studies of coronary heart disease risk factors. In Report by the Task Force on Genetic Factors in Atherosclerotic Disease. Washington: Government Printing Office, 33-40.

[4] Lange, K., Westlake, J. & Spence, M. A. (1976). Extensions to pedigree analysis. III. Variance components by the scoring method. Ann. Hum. Genet. 39, 485-91.

[5] Orr, J. D., Sing, C. F. & Block, W. D. (1975). Analysis of genetic and environmental sources of variation in serum cholesterol in Tecumseh, Michigan: II. The role of education, occupation, and income. Soc. Biol. 22, 34-43.

[6] Sing, C. F., Chamberlain, M. A., Block, W. D. & Feiler, S. (1975). Analysis of genetic and environmental sources of variation in serum cholesterol in Tecumseh, Michigan. I. Analysis of the frequency distribution for evidence of a genetic polymorphism. Amer. J. Hum. Genet. 27, 333-47.

[7] Sing, C. F. & Orr, J. D. (1976). Analysis of genetic and environmental sources of variation in serum cholesterol in Tecumseh, Michigan. III. Identification of genetic effects using twelve polymorphic genetic blood marker systems. Amer. J. Hum. Genet. In press.

[8] Truett, J., Cornfield, J. & Kannel, W. B. (1967). A multivariate analysis of risk of coronary heart disease in Framingham. J. Chron. Dis. 20, 511-24.

Likelihoods on complex pedigrees for quantitative traits

E. A. Thompson
STATISTICAL LABORATORY
UNIVERSITY OF CAMBRIDGE, ENGLAND

M. H. Skolnick
DEPARTMENT OF MEDICAL BIOPHYSICS AND COMPUTING
UNIVERSITY OF UTAH, LOGAN, UT 84321

INTRODUCTION

In segregation analysis [Morton (1962)] the inheritance of characteristics is studied through data on nuclear families. Pedigree analysis allows data on large pedigrees to be utilized. Similarly the study of quantitative characteristics has often been restricted to studies of correlations between members of nuclear families; again, pedigree analysis allows us to extend the scope of such a study. In particular we propose to study the inheritance of quantitative characteristics in large Utah Mormon genealogies. Genealogical records are being computerized as part of an epidemiological program.

Likelihood Methods on Pedigrees

Elston and Stewart (1971) discuss the derivation of the probability of a set of phenotypes observed on a large pedigree under a given mode of inheritance, and hence the computation of a likelihood for that mode of inheritance. Amongst other models they discuss a (hypothetical) quantitative trait whose phenotypic value is normally distributed about some underlying genotypic value, which is in turn determined by an infinite number of equally additive loci. However, implementations of methods of pedigree analysis have with a few exceptions been restricted to qualitative traits on pedigrees of restricted structure.

Cannings et al. (1976, 1977) have generalized the methods of Elston and Stewart to compute probability functions and likelihoods on pedigrees of arbitrary size and complexity. (The latter paper may also be consulted for references to previous methods of computing such probability functions). The basic idea of the method of Cannings et al. is the recursive condensation of phenotype information on an increasingly large section of the pedigree into functions on subsets of the remaining individuals. A feature of the implementation of the method is that the likelihood for a mode of inheritance may be derived as a function of the parameters of the underlying model, enabling the maximum likelihood estimates to be found without repeated computation, and the form of the likelihood surface to be examined.

For the purposes of the above method a model is specified by a <u>transition array</u> giving the probability distribution of offspring genotype for given parental genotypes, and <u>penetrance functions</u> giving the probability of an observed phenotype conditional on underlying genotype. With these specifications models involving a variable number of genotypes (and hence loci) and an arbitrary number of phenotypes may be proposed and their likelihoods computed. The primary comparison of models may be either through the transition arrays or through the penetrance functions. The latter method involves a comparison of alternative schemes of genotype-phenotype correspondence for a fixed underlying model of genotype determination, and appears to be more suited to a study of quantitative traits.

A Model for Longevity

The above methods may be exemplified by an analysis of the inheritance of longevity in the Mormon population of Utah. We assume a fixed finite number, s, of equally additive loci, and compare models ranging from one where all genotypes give the same phenotype distribution and all observed variance is "environmental" to ones where the majority of observed phenotypic variance is accounted for by differences between genotypes and within-genotype phenotypic variances are small. This provides a direct analogy with previous studies of heritability [Skolnick (1975) and references therein], but allows this concept to be extended to the analysis of data on a large pedigree. We shall consider primarily $s = 5$; the eleven genotypes, present in population frequencies $\binom{2s}{r}$ for $r = 0, 1, \ldots, 10$, provide an approximation to the infinitely additive model. The effect of variation in s will also be considered. For this study we are interested primarily in post-reproductive mortality, and to avoid problems of differential selection shall assume that all genotypes give the same probability distribution for age-at-death below 40.

Our null hypothesis H_o is that of no inheritance of longevity, or that all eleven genotypes have the same penetrance function, this being simply the observed sex-specific distributions of age at death. In order to parameterize the problem we propose also a secondary null hypothesis H_o^* in which the eleven genotypes have as penetrance function at ages over 40 the best-fitting sex-specific distribution for a random variable $Y = a-Z$, where Z has a log-normal distribution. We may then specify series of alternative hypotheses where the eleven genotypes provide log-normal distributions of differing means and variances subject to the restriction that the resulting total distribution, mixed according to the population genotype frequencies, has the same mean and variance as the distribution for the hypothesis H_o^*. There are several possible single-parameter alternatives for these series of distributions, which will be detailed more fully elsewhere. By varying the ratio of between-genotype and within-genotype contributions to the variance, but maintaining the total, we compare hypotheses of varying heritability.

A preliminary analysis has been made on a small subset of the Mormon genealogy, chosen to ensure no bias through the absence of death ages for still-living family members. The heritability of longevity is low and problems arise since in the absence of inheritance effects the likelihood will be maximal for that hypothesis which best fits the population data. It is for this reason that we choose H_0 as our basic hypothesis for then, this having perfect population fit, we cannot mistakenly infer inheritance. However, unless mixed hypotheses also fitting the population values can be found we may make the reverse error. Further analysis is required before firm conclusions can be drawn. We do not of course propose that we have an exact description of the determination of longevity, but if longevity is to some extent inherited, pedigree data must provide more information than parent-offspring correlations alone. This information can be extracted provided some model is assumed, and when the model is a sufficiently accurate description of observed data the information obtained will be of practical use.

ACKNOWLEDGEMENTS

This research was supported by National Institutes of Health grants HD-10267 and CA-16573.

BIBLIOGRAPHY

(1) Cannings, C., Thompson, E.A. & Skolnick, M.H. (1976). The recursive derivation of likelihoods on pedigrees of arbitrary complexity. *Adv. Appl. Prob.*, in press.

(2) Cannings, C., Thompson, E.A. & Skolnick, M. (1977). Probability functions on complex pedigrees; in preparation.

(3) Elston, R.C. & Stewart, J. (1971). A general model for the genetic analysis of pedigree data. *Hum. Hered.* 21, 523-542.

(4) Morton, N.E. (1962). Segregation and linkage. *Methodology in Human Genetics*. San Francisco: Holden Day.

(5) Skolnick, M.H. (1975). *The Construction and Analysis of Genealogies from Parish Registers with a Case Study of Parma Valley, Italy*. Ph.D. Thesis, Stanford University.

Genetic integration of morphometric traits in randombred house mice

Larry Leamy
DEPARTMENT OF BIOLOGY
CALIFORNIA STATE UNIVERSITY, LONG BEACH, CA 90840

Multivariate analyses of quantitative characters such as mammalian dental and/or skeletal dimensions typically start with matrices of phenotypic correlations (r_p). The present study involves an alternative approach--that of a multivariate analysis of separate genetic and environmental correlations among a battery of morphometric traits in randombred house mice.

The original total population of randombred mice consisted of 200 families, each with 2 parents of age 5 months, and 6 offspring divided into sublitters of 2 mice at each of three ages: 1-month, 3-months, and 5-months (Leamy, 1974). Fifteen osteometric and 3 external metric characters used were as follows: Sk_L, skull length; P_L, palate length; ZF_L, zygomatic fenestral length; M_L, mandible length; Sk_W, skull width; Z_W, zygomatic width; IO_W, interorbital width; In_L, innominate length; Il_L, ilium length; OF_L, obturator foramen length; Sc_L, scapula length; F_L, femur length; Ti_L, tibia length; H_L, humerus length; RU_L, radioulna length; Ta_L, tail length; B_L, body length; W, body weight. Heritabilities of the 18 traits previously were calculated from the regressions of offspring of each age group on the parents, but only the 5-month offspring were used in the present study.

Genetic correlations (r_A) were calculated first for the separate sexes as follows: $r_A = 1/2(\text{covxy'} + \text{covx'y})/\sqrt{(\text{covxx'})(\text{covyy'})}$, where x and y are the values of a pair of traits in parents, and x' and y' are the values for the same traits in the offspring. The genetic correlations exhibited sufficient agreement between the sexes to warrant pooling. Environmental correlations (r_E) were also first calculated for each pair of traits, x and y, for the separate sexes (and then later pooled) by solving the following: $r_P = h_x h_y r_A + e_x e_y r_E$, where h_x and h_y are the square roots of the heritabilities of the two traits, and $e^2 = 1 - h^2$ (Falconer, 1960).

Because the matrices of genetic and environmental correlations so generated contained some values exceeding 1, and in addition were not positive, semi-definite, clustering techniques (Sneath and Sokal, 1973), rather than component or factor analysis, were used. Initial clustering proved inadequate (generating very low values of cophenetic correlations), however, and the correlations were "scaled" by multidimensional scaling (MDS) techniques into final (Euclidean) distances more suitable for clustering. MDS is a general ordination technique (Shepard, et al., 1972) which can depict patterns in a matrix of data by a geometrical picture of points, the relative distances between the points being an indication of their relative magnitude of association.

Two-dimensional solutions of MDS of the genetic and environmental correlations, respectively, are shown in Figures 1A and 2A. In the genetic (Fig. 1A) solution, differentiation can be seen between skull lengths and other dimensions, widths and lengths, and "axial" versus "appendicular" elements. Specific character groups indicated by the dashed lines (Fig. 1A) are those derived from the clustering of the final (6-dimensional) solution, described below. Minimal spanning trees for the original correlations (Fig. 1B) and final distances (Fig. 1C) depict the strength of the association among the characters (the thickest lines denoting the highest magnitude of association) and also show the effect of the final MDS

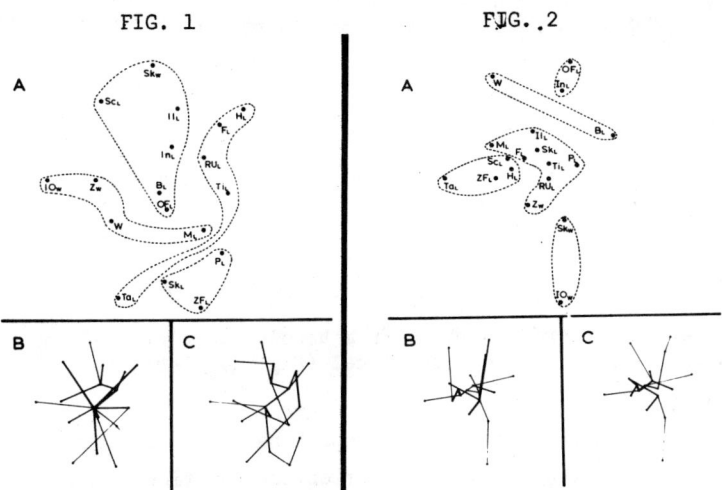

FIG. 1 and 2. Two-dimensional MDS solutions for the original genetic (Fig. 1A) and environmental (Fig. 2A) correlations, and minimal spanning trees for the original correlations (B) and final distances (C) in each case.

solution in scaling these associations. Results of MDS for the environmental correlations (Fig. 2) indicate a contrast of the 3 widths and 3 other dimensions (W, In_L, OF_L) from the remaining bulk, Ta_L and IO_W being obvious outliers.

Figure 3 shows the dendrogram produced from Ward's hierarchical clustering of the final genetic distances. Limb length, girdle, skull width and skull length clusters are apparent, and differentiation between axial and appendicular clusters is striking. Ward's clustering of the final (7-dimensional) environmental distances (Fig. 4) shows basically a different pattern of five clusters.

The four genetic clusters form remarkably acceptable basically functional groups very much reminiscent of the morphologically integrated groups of Olson and Miller (1958). The limb length and axial clusters presumably serve locomotor functions, whereas M_L, Z_W, IO_W, and W are most probably linked together by the masseter muscle complex into a functional chewing complex. These

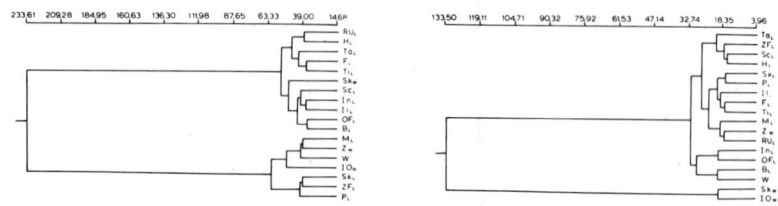

Fig. 3 and 4. Dendrograms derived from Ward's clustering of the final genetic (Fig. 3) and environmental (Fig. 4) distances.

clusters suggest that there are genes which have been selected for developmental pathways which affect entire functional complexes. They also provide a pattern of expectation of the correlated response (in other characters) given either artificial or natural selection for any one of the characters. The environmental clustering did not produce apparent anatomical or functional groups, but did emphasize the difference between genetic and environmental sources of covariation.

BIBLIOGRAPHY

(1) Falconer, D. S. (1960). Introduction to Quantitative Genetics. Edinburgh: Oliver and Boyd.

(2) Leamy, L. (1974). Heritability of osteometric traits in a randombred population of house mice. J. of Hered. 65, 109-120.

(3) Olson, E. C. & Miller, R. L. (1958) Morphological Integration. Chicago: Univ. of Chicago Press.

(4) Shepard, R. H., Romney, A. K. & Nerlove, S. B. (1972). Multidimensional Scaling, vol. I. New York: Seminar Press.

(5) Sneath, S. A. & Sokal, R. R. (1973). Numerical Taxonomy. San Francisco: Freeman and Co.

Coefficients of constraint as a generalization of coefficients of identity

M. Gillois and *Cl. Chevalet*
LABORATOIRE DE GÉNÉTIQUE CELLULAIRE
CENTRE DE RECHERCHES DE TOULOUSE I.N.R.A.
B.P. 12, 31320 CASTANET-TOLOSAN, FRANCE

ABSTRACT

When genes act on the mating (homogamy) or on the probability of reproduction (selection) of individuals, the identity situations do not give sufficient information to define the conditional laws for the genic situation. The constraint situation must be introduced. The constraint coefficients for small homogamous populations are equivalent to the inbreeding coefficients for small panmictic populations.

IDENTITY AND DEPENDENCE BETWEEN GENES

Two genes are identical if and only if they are descended in the Mendelian way, without any mutation, from one unique ancestral gene. This leads to a binary relation of equivalence, which induces on every set of genes a breakdown into disjoint subsets referred to as "identity-classes." Genes belonging to the same class are identical. The knowledge of the nature of one gene allows the determination of the nature of all the others, because a material relation exists between them; they represent the unmodified copies of one unique model, the ancestral gene (Malécot, 1948; Gillois, 1964, 1965).

Two genes belonging to two distinct identity classes are nonidentical. It is necessary to have more information so as to decide if, for example, these two genes are independent or not. In general, we may say that two nonidentical genes are independent either because it is the most simple and effective hypothesis in the absence of further information, or there exists supplementary information that the population is panmictic.

Two genes are independent if the knowledge of the nature of one does not give any information that would specify the nature of the other. Uncertainty is measured by the following expression:

$$H = \Sigma_i \, p_i \, \text{colog} \, p_i$$

(Brillouin, 1956) where the p_i's are probabilities of events. Either this uncertainty is defined a priori (H) or it is conditioned, that is to say, it depends on the knowledge of the nature of a gene and on information establishing a link between this gene and the one which is of unknown nature (Gillois, 1966a). If two genes are rigorously dependent, the quantity of conditional uncertainty leads to $H_c = 0$, if the two genes are subordinate, it leads to $H_c \neq H$, and if the two genes are independent, it leads to $H_c = H$.

THE RELATIONS OF DEPENDENCE AND CONSTRAINT

2.1 Definitions

The preceding definition of uncertainty gives rise to the following definition (Gillois, 1966a): "Two genes are rigorously dependent (or constrained) if the knowledge of the nature of one gives all the necessary information to determine the nature of the other."

This binary relation is an equivalence relation which imposes on a set of genes (homologous ones in particular) a breakdown into disjoint subsets referred to as "constraint classes." Two genes belonging to two separate "constraint classes" are independent.

A class of rigorously dependent genes (constrained genes) is not necessarily a set of isoactive genes. Two heteroactive genes can be rigorously dependent. Thus, it is necessary to introduce the relation of "isoactive dependence" (isoconstraint), a relation of equivalence that realizes on the sets of constrained genes a breakdown into disjoint subsets referred to as "isoconstraint classes." Two genes of the same isoconstraint class are such that the knowledge of the nature of one is sufficient to define the equivalent nature of the other. Two constrained genes belonging to two isoconstraint classes are "heteroconstrained."

It should be noted that the sets of identical genes are all included in the sets of isoconstrained genes.

In the same way as inbreeding and kinship coefficients are defined with respect to the identity relation, three "constraint coefficients" may be defined for any pair of genes in a population. For two homologous genes of one zygote, F_1, F_2 and F_3 (with $F_1 + F_2 + F_3 = 1$) will stand for the probabilities that they are isoconstrained, heteroconstrained, and nonconstrained, respectively.

2.2 Joint Laws of The Genetic Random Variables (Gillois, 1966a, 1967)

The partitioning of the set of genes by the constraint relation, and then by the isoconstraint relation, allows the a priori definition of the joint laws of the genetic random variables.

This can be illustrated in the case of one locus with two alleles A and a. Let X and X^+ be two random variables associated with the two genes G and G^+ of one random zygote.

With probability F_1, G and G^+ are isoconstrained. The conditional joint law of X and X^+ is thus defined by the conditional marginal probability p_1 that one gene is A: $P_1(A,A) = p_1$, $p_1(a,a) = 1 - p_1$, $P_1(A,a) = P_1(a,A) = 0$.

With probability F_2, G and G^+ are heteroconstrained. Therefore the pair (G,G^+) is either in state (A,a) or in state (a,A). With biallelism both events have the same probability 1/2, thus $P_2(A,A) = P_2(a,a) = 0$, $P_2(a,A) = P_2(A,a) = 1/2$.

With probability F_3, G and G^+ are nonconstrained. In this situation of independence, the joint law is deduced from the conditional marginal probability, p_3, that one gene is A, according to the Hardy-Weinberg law: $P_3(A,A) = (p_3)^2$, $P_3(A,a) = P_3(a,A) = p_3(1 - p_3)$, $P_3(a,a) = (1 - p_3)^2$.

It must be noticed that the conditional probabilities p_1 and p_3 generally differ from the overall probability $p = F_1 p_1 + \frac{1}{2} F_2 + F_3 p_3$ that one gene is A, except in such cases as panmixia where F_2 is zero (Gillois, Bouffette and Bouffette, 1969a).

RESULTS AND COMMENTS

3.1 Genotypic Frequencies

Probabilities of genotypes, usually written as:

$$P(AA) = f\,p + (1 - f)p^2, \quad P(Aa) = (1 - f)\,2\,p\,(1 - p), \ldots$$

according to the definition of Malécot's (1948) inbreeding coefficient f, are now written in the following way:

$$P(AA) = F_1\,p_1 \qquad + F_3\,(p_3)^2$$
$$P(Aa) = F_2 \qquad\qquad + F_3\,2\,p_3(1-p_3)$$
$$P(aa) = F_1(1 - p_1) + F_3(1 - p_3)^2$$

Such a formulation allows the description of any change in the genotypic frequencies, and not only a decrease in the frequency of heterozygotes. This makes it possible to deal with partial assortative mating in small populations, in which case knowledge of the mating having led to one zygote is information like identity, which allows the description of the joint law describing two homologous genes in a zygote (Gillois, 1966b; Gillois, Bouffette and Bouffette, 1969a).

3.2 Genotypic and Gametic Correlations

Working with the joint laws associated with constraint coefficients, Gillois, Bouffette and Bouffette (1969b) have obtained a general expression of genotypic covariances between relatives in a small population subjected to partial assortative mating, and could express the gametic correlation r as a function of F_1 and F_2. Such a result can be seen as a generalization of both Wright's (1922) and Malécot's (1948, 1969) approaches.

3.3 Evolution of Constraint Coefficients in a Finite Population With Partial Assortative Mating

The coefficients of constraint F_1, F_2, F_3 and all those that can be defined between other sets of genes will change from generation to generation. The knowledge of the mating system, of numbers of males and females and of the mutation rate allows the calculation of the values of these constraint coefficients. Thus, the evolution of these small populations can be described by constraint coefficients analogously to the way in which panmictic populations are described by identity coefficients.

In all possible cases of partial gametic or zygotic genotypic or phenotypic homogamy, the recurrence equations and their general solutions have been given by Gillois, Bouffette and Bouffette (1969b). The most noteworthy fact in these earlier works seems to be that the nonconditioned probability that one gene is A varies from generation to generation in constrast to what is observed in panmictic populations of finite size.

ACKNOWLEDGMENT

The authors are particularly grateful to J. R. Sedcole, J. Lewis and E. Pollak for their willingness to correct an early English version.

BIBLIOGRAPHY

(1) Brillouin, L. (1956). Science and Information Theory. Academic Press, New York.

(2) Croize-Pourcelet, J. (1970). Etude de l'évolution de petites populations homogames par la théorie de la constrainte; expériences analogiques de contrôle. Ann. Génét. Sél. anim., 2, 37-52.

(3) Gillois, M. (1964). La relation d'identité en Génétique. Thèse doctorat d'Etat. Fac. Sci. Paris. 294 p.

(4) Gillois, M. (1965). Relation d'identité en génétique. Ann. Inst. Henri Poincaré, B, 2 (1), 1-94.

(5) Gillois, M. (1966)a. La relation de dépendance en génétique. Ann. Inst. Henri Poincaré, B, 2 (3), 226-278.

(6) Gillois, M. (1966)b. L'homogamie dans une population d'effectif limité. Ann. Inst. Henri Poincaré, B, 2 (4), 299-347.

(7) Gillois, M. (1967). Les lois conjointes des variables aléatoires génétiques. Ann. Génét., 10, (4), 203-206.

(8) Gillois, M., Bouffette, J., Bouffette, A. (1969)a. Etude des populations d'effectif limité homogames, phénotypiques et panmictiques. Ann. Inst. Henri Poincaré, B, 5 (1), 59-86.

(9) Gillois, M., Bouffette, J., Bouffette, A. (1969)b. Covariance génotypique a priori dans les populations homogames. Ann. Inst. Henri Poincaré, B, 5 (1), 87-99.

(10) Malécot, G. (1948). Les mathématiques de l'hérédité. Masson et Cie, Paris.

(11) Malécot, G. (1969). Consanguinité panmictique et consanguinité systématique (coefficients de Wright et de Malécot). Ann. Génét. Sél. anim., 1, 237-242.

(12) Wright, S. (1922). Coefficients of inbreeding and relationship. Amer. Nat., 56, 330-338.

A general approach to dependence relationships among genes with some applications

A. Gallais
STATION D'AMÉLIORATION DES PLANTES FOURRAGÈRES
INSTITUT NATIONAL DE LA RECHERCHE AGRONOMIQUE
86600 LUSIGNAN, FRANCE

INTRODUCTION

Definition of some Dependence Relationships among Genes

The concept of constraint relationships introduced and developed by Gillois (1966, 1967, 1969) allows the study of populations with departure from random mating where, besides inbreeding, there can exist other constraints on the union of genes, due to the system of mating or to the structure of the initial population. This concept can be extended to non-homologous genes. To be more general we use the expression "dependence relationships" to designate all types of dependence between homologous or non-homologous genes. The identity by descent relationship is a particular dependence relationship between homologous genes.

A dependence relationship can be defined in the same way among non-identical homologous genes and among non-homologous genes. "Two non-identical homologous genes or two non-homologous genes drawn from one or two zygotes will be said to be dependent if they derive from a non random association". This definition is more general than those which we gave for the study of a diploid population in linkage disequilibrium and of an autotetraploid population with panmictic disequilibrium (Gallais, 1974 a,b).

In these studies the dependence derived by descent from the initial population. The new definition allows the inclusion of the case of genes which can become dependent by constraints in the system of mating. The partition of the set of homologous genes into three subsets, identical genes, dependent non-identical genes, independent genes, leads to some interesting properties of the expectations of the genetic parameters and allows a separation of what is due to inbreeding and what is due to other constraints. To be more general it would be necessary to distinguish four sets of genes: identical, isoconstraint non-identical, heteroconstraint, and independent.

The above definition of the dependence relationships among two genes can be extended to more than two genes drawn from one or two zygotes. The identity by descent relationship, the dependence relationship among homologous non-identical genes, and the dependence relationship among non-homologous genes induce a partition of the sets of zygotes and of pairs of zygotes into states of dependence of rank 1 and rank 2 respectively. States of dependence between more than 2 zygotes could also be considered. With each state, we can associate a probability or coefficient of dependence. Coefficients of rank 1 are useful for expressions of population structure and of means and those of rank 2 are useful for expressions of frequencies of the different pairs of zygotes and of second order statistics. Coefficients of rank 3 are useful for third degree statistics. We have shown (Gallais, 1976) that coefficients of rank 3 and 4 are also useful in studying populations of interacting genotypes.

FORMULATION OF A GENERAL PROCEDURE TO OBTAIN MEANS AND COVARIANCES

To simplify we assume that the structure in probability of each set of states of dependence of rank 1 is constant. This is the case for inbred populations derived from an initial population in panmictic and linkage disequilibrium. We summarize here only the general procedure which leads to the expressions for the mean and covariances between two zygotes for any situation. There are four steps.

1) We have to list for a given rank (1,2) all states of dependence between 2, 3,..., n genes, with n = $2\nu r l$ ν = degree of ploidy, r = rank of the state, l = number of loci considered. For a general approach the parental origin of the genes must be considered.

2) According to the number of genes drawn in each zygote involved, we can associate with each state a parameter of the factorial decomposition of the genotypic value (Kempthorne, 1957). For example, for one gene drawn we associate the parameter α (additivity), for two genes drawn in a zygote, β (diallelic interaction)...and so on according to the genetic effects under consideration. The parameters are appropriately subscripted to identify the genes by their classes of dependence. For states of rank 2 they form products in pairs. Note that the genetic parameters are defined in a reference population in panmictic and linkage equilibrium.

3) Consider the expectation of the parameter or product of 2 parameters associated with each state. According to the work of Bouffette (1967) and Gallais (1970, 1974 a,b) it can be shown that an expectation will be zero if among the genes involved at least one is independent of the others (for all dependence relationships considered).

4) Each nonzero conditional expectation is multiplied by the appropriate coefficient of dependence and by the number of ways of drawing the number of genes involved. The sum of all such products gives the first i.e. the mean and second order noncentral moment from which is calculated the expression for the covariance. The expression of the variance can be obtained as a covariance of an individual with itself. The extension of the results for one locus for a subset of loci to the entire set

of loci involved in the character considered must be made cautiously because panmictic disequilibrium can induce correlations between loci or subsets of loci.

We can thus generalize our work on panmictic and linkage disequilibrium in diploids or in autotetraploids. Our early requirement of one generation of random mating to form the initial population can now be lifted. However we examine here only the results for panmictic disequilibrium for diploids, without epistasis. States of dependence of rank 2 useful for the covariance expression are given in Table I. We note that there appears a covariance between additive effects; there are three types of covariances between additive effects and dominance effects, and for the dominance there are 10 different expectations. This leads to 15 terms instead of 5 with panmictic equilibrium in the initial population.

TABLE I

Case of One Locus. States of Dependence of Rank 2, Order 2, 3 and 4 which lead to a non-zero expectation

On either side of the vertical line is shown the distribution of the genes taken in the zygotes. A letter represents a class of identity. Dependent genes i, j or k, l are written $\overline{ij}$ or $\overline{kl}$ if in the same zygote and $i^*|j^*$ or $k^\circ|l^\circ$ if one in each of two zygotes. The parental origin of the genes is not considered.

Order 2	Order 3	Order 4			
$i\|i$	$i\|ii$	$ii\|ii$	$ij^*\|il^*$	$i^*i^*\|j^*j^*$	$ii\|\overline{jk}$
$i^*\|j^*$	$i\|\overline{ij}$	$ii\|jj$	$i^*j^\circ\|k^*l^\circ$	$\overline{ij}\|\overline{kl}$	
	$i^*\|j^*j^*$	$ij\|ij$	$\overline{ij}\|\overline{ij}$	$ii\|\overline{ij}$	

SOME APPLICATIONS

We concentrate here our comments on the mean and variance of crosses (single or double) among non-inbred plants or among completely inbred lines (See Gallais, 1973, 1974 a).

Expected Values of Crosses

Without epistasis, and whatever the panmictic disequilibrium, all types of hybrid varieties (single, three-way, double crosses) derived from inbred lines or not, have the same expected value. Under random mating with linkage disequilibrium and epistasis, we have already shown (Gallais, 1973, 1974 a) that the expected value of a double cross (in the classical definition) will differ from that of a single cross. The same result also holds for crosses between inbred lines. Concerning results for three-way crosses, their values can also be greater or less than that of a single or a double cross. Schnell (1973), with a more specific calculation, also gives the result that three-way crosses can in some situations be better on average than single crosses, and he reviews (Schnell, 1975) some data which seem to support this conclusion.

Variance among Crosses

Under random mating with linkage disequilibrium, but even without epistasis we have already shown that the variance among crosses can increase or decrease with advanced generations of random mating. In the same situation, using inbreeding, the variance among single crosses could be greater (as classical) or less than the variance among three-way or double crosses. It can be mentioned that complete inbreeding suppresses the parameters due to panmictic disequilibrium at one locus.

Combining the studies on means and variances it appears that with linkage and panmictic disequilibrium the best hybrid varieties will not necessarily be single crosses.

Bias in Combining Ability Variances

Due to panmictic or linkage disequilibrium the G. C. A. and S. C. A. variances can be over or under estimated in comparison with the estimates in a population in equilibrium. In the case of linkage disequilibrium, even without epistasis we have shown that from one to several generations of random mating the two variances could increase or decrease. However, in the case of biallelism it appears that S. C. A. variance can only decrease. With our approach the effect of several generations of random mating prior to inbreeding (for example self-fertilization) can also be easily studied in some simplified situations.

CONCLUSION

The generalization of the dependence relationships allows the study of different situations of interest for plant or animal breeders. Indeed in a real population we must always assume some linkage and panmictic disequilibrium. The statistics can also be easily studied whatever the ploidy, and we have shown (Gallais, 1976 a) that this approach allows the study of populations of interacting individuals.

The cost of this generalization is an increase in the number of parameters, thus making difficult and expensive experiments capable of estimating all the variance components. Experimental mating designs leading to estimates of functions of these components must be found. A merit of the general approach is that it facilitates the study of simplified situations. Our application to some problems of plant breeding shows that a reduced model can lead to some original conclusions which seem to correspond to observed unexplained results. Furthermore, for a given experiment, i.e., for a given system of mating, there is an interdependence among some coefficients of dependence and a few of them alone could usefully explain and predict the observed results on means and covariances. In an autotetraploid species, where even with the assumption of panmictic equilibrium and no epistasis there are 61 parameters for the covariances between arbitrary inbred relatives, we have obtained a good statistical explanation of the results with only a few coefficients (Gallais, 1976 b). It is also possible for some systems of mating to try to find principal components as in multivariate analysis.

It must be noted that the increase in the number of parameters is partially due to the models starting from one locus effects to study a multi-locus character. Theoretical development allowing a more direct and global approach to such a character must be stimulated.

BIBLIOGRAPHY

Bouffette, J. (1966). Expression de la covariance génotypique chez les tétraploïdes. Thèse 3ème cycle, Fac. Sci., Lyon.

Gallais, A. (1970). Covariances entre apparentés quelconques avec linkage et épistasie. I. Expression générale. *Ann. Génét. Sél. Anim.*, 2, 417-427.

Gallais, A. (1973). Some factors affecting the strategy of breeding for yield in cross fertilized plants. *Proc. 1st Meeting Sect. Biometrics in Plant Breeding of Eucarpia*, Inst. Angew. Genet., Hannover : Rundfeldt H. & Wricke G. 9-19.

Gallais, A. (1974 a). Covariances between arbitrary relatives with linkage and epistasis in the case of linkage disequilibrium. *Biometrics*, 30, 429-446.

Gallais, A. (1974 b). Covariances between arbitrary relatives in autotetraploïds with panmictic disequilibrium. *Genetics*, 76, 587-600.

Gallais, A. (1976 a). Effect of competition on means, variances and covariances in quantitative genetics with an application to general combining ability selection. *Theoretical & Applied Genetics*, 47, 189-195.

Gallais, A. (1976 b). An experimental check of quantitative genetics on an autotetraploid plant, *Medicago sativa*, L., with special reference to the identity by descent relationship. *International Conference on Quantitative Genetics*, Iowa State University, August 16-21.

Gillois, M. (1966). La relation de dépendance entre gènes non-identiques. *Ann. Biol. anim. Bioch. Biophys.*, 6, 117-120.

Gillois, M. (1967). La notion de génotype. *Ann. Génétique*, 10, 4, 201-202.

Gillois, M. (1967). Les lois conjointes des variables aléatoires génétiques. *Ann. Génétique*, 10, 4, 203-205.

Gillois, M., Bouffette, J. & Bouffette, Ar. (1969). Etude des populations d'effectif limité homogames phénotypiques et panmictiques. *Ann. Inst. Henri Poincaré, Nouv. ser.*, Sect. B, 5, 1, 69-86.

Gillois, M., Bouffette, J. & Bouffette, Ar. (1969). Covariance génotypique a priori dans les populations homogames. *Ann. Inst. Henri Poincaré, Nouv. ser.*, Sect. B, 5, 1, 87-99.

Kempthorne, O. (1957). *An Introduction to Genetic Statistics*. New York. J. Wiley.

Schnell, F.W. & Geiger, H.H. (1973). Possible genetic causes of differences between the mean yields of various types of hybrids. *Proc. 1st Meeting Sect. Biometrics in Plant Breeding of Eucarpia*, Inst. Angew. Genet., Hannover : Rundfeldt H. & Wricke G., 5-8.

Schnell, F.W. (1975). Type of variety and average performance in hybrid maize. *Z. Pflanzenzüchtung*, 74, 177-188.

Diallel crosses in relation to breeding systems

S. O. Fejer
RESEARCH STATION, AGRICULTURE CANADA,
OTTAWA, ONT. K1A 0C6

INTRODUCTION

Differences in genetic variation can be expected in various species with different breeding systems as a consequence of differences in their evolutionary history. Wright (1956) summarized his views on this, based on animal work, and in plants Stebbins (1950) and Gustafsson (1954) discussed differences between cross- and self-fertilizers and clonally propagated species. Lerner (1954) speculated on the relation of such differences to heterosis and homeostasis in plants and animals. A convenient way of collecting information on genetic variance components is the diallel cross, even though perhaps not the most suitable one in the present context, because of the limited number of parents and consequent restriction of the sample. First statistical treatment of diallels was provided by Mather (1949), further developed by Kempthorne (1952) and later by many others. Being only a general practitioner in this field, my only pretext for this talk is my first-hand experience with diallels of economic plants with various breeding systems, again not the most suitable organisms because of their directed evolution.

FORAGE PLANTS

Of the economic plants, forages are perhaps the most suitable ones because of their limited history of selection. The usual method of polycross breeding is essentially a diallel cross brought down to later generations to ensure ample seed supply. The loss of heterosis, or inbreeding depression in these cross-fertilized plants was well demonstrated in New Zealand perennial ryegrass (Lolium perenne) by Corkill (1956), exactly following Sewall Wright's formula. Corkill was able to fix this heterosis in hybrid Lolium, exploiting the unique climate of New Zealand by combining the summer- and winter-growing habits of two species. To my mind this was a prime example of epistasis or non-additive gene interaction, and I found similar differentiation within perennial ryegrass, explained by interaction between morphological yield components but also dependent on environment (Fejer 1955). We devised selection indexes for optimum combinations (Glenday and Fejer 1956), including competitive conditions, but found mainly additive genetic variance in a diallel of some of the extreme parents of Corkill's synthetic (Fejer 1958). Torrie (1957) showed in partly the same material that with parents preselected for general combining ability (GCA), dominance and epistatic effects as measured by specific combining ability (SCA) were greater similar to the classical work of Sprague and Tatum (1942) in maize. Later, I found similar differences in New Zealand Dactylis for autumn and spring vigor, with higher SCA if grown under pasture than under hay management (Fejer 1966). In legumes, GCA was much higher than SCA in Trifolium medium (Fejer 1967), and in Medicago sativa in Canada (Fejer 1971). In summary, additive genetic effects prevailed in forages except when exhausted by selection.

FRUIT CROPS

Most of the work is confined to Ottawa, where Watkins and Spangelo first compared diallels of vegetatively propagated strawberries with high epistasis and seed-propagated apples (1970) where plant survival showed no epistasis but preselected fruit quality showed some, explained by the theory of Wright (1956). We found epistasis also in the highly successful Sparkle × Valentine strawberry cross (Fejer et al. 1975) and in clonally propagated raspberry (Fejer and Spangelo 1972) also dominated by one cross, Lloyd George × Newman. However, SCA was also high in seed-propagated gooseberry (Spangelo et al.1970), indicating that the adaptive peaks of Wright could occur in artificial selection of clones. In apples, the new Ottawa seed-propagated hybrid rootstocks showed mainly GCA (Fejer and Spangelo 1974), however new information on scab resistant scion and vegetatively propagated rootstocks material (Fejer 1976) showed SCA also in survival traits, and history of selection was overruled by strong effects of individual parents. This also shows the bias introduced by using small samples in diallels.

CEREALS

Any recent issue of Crop Science will show an increasing number of papers on the subject without resolving the issue which is particularly important in self-fertilizing cereals because of the controversy about advantages of hybrid cultivars. I restrict myself to Ottawa results, where in oats Sampson (1976) found midparent values reliable predictors in non-stress environments, and that GCA effects of selection index values were useful in identifying best parents. In barley, we found surprising amounts of heterosis (Fejer and Fedak 1976) in diallels of 2-row and 6-row spring barley, similar to earlier results with spring × winter crosses (Fedak and Fejer 1975). This could be exploited by hybrid methods but based on the work of Aastveit (1964), who found that heterosis in 2 × 6-row barley was caused by non-allelic interactions and could retain it over 8 generations, and encouraged by the reported success of Redden and Jensen (1974) in recurrent selection, we started a selection experiment over 6 generations. Retaining heterosis by conventional methods would be preferable, especially as they can be accelerated by single seed descent and in barley by haploid methods. The outcome is not clear yet although we found some epistasis over the years (Fejer and Jui 1977) but recurrent selection seemed to have little additional advantage in this case. Techniques for better recognition of potential parents in the field may be more urgent than theory (cf. Dragavtsev 1972, Fasoulas 1973). However the fact that heterosis is prevalent in self-fertilized cereals is showing the difficulties of generalizations about breeding systems in cultivated plants, and is in contrast to previous views (Lerner 1954), undoubtedly based on the lack of inbreeding depression in self-fertilizers.

ACKNOWLEDGEMENTS

Thanks are due for encouragement to the late Dr. A. Jánossy, Tápiószele, Hungary, Past-President, and Dr. G. Wricke, Hannover, Germany, President, 2nd Meeting of Biometrics Section, EUCARPIA, Gödöllö, Hungary, 1975, where an earlier version of this paper was read.

BIBLIOGRAPHY

(1) Aastveit, K. (1964). Heterosis and selection in barley. Genetics, 49, 259-64.

(2) Corkill, L. (1956). The basis of synthetic strains of cross-pollinated grasses. Proc. 7th Internat. Grassl. Congr., 427-38.

(3) Dragavtsev, V.A. (1972). Experimental comparison of three principles in the estimation of genotypic variability of quantitative characters in plant populations. Genetika (Moscow) 8, 28-34.

Complete bibliography available on request. Most of the papers associated with the author appeared in recent volumes of Can. J. Plant Sci. and Can. J. Genet. Cytol.

(4) Fasoulas, A. (1973). A new approach to breeding superior yielding varieties. Dept. Genet. Plant Breedg. Univ. Thessaloniki, Pub. No. 3., 1-42.

(5) Fedak, G. & Fejer, S.O. (1975). Yield advantage in F_1 hybrids between spring and winter barley. Can. J. Plant Sci. 55, 547-53.

(6) Fejer, S.O. (1955). Genotype-environment interactions in Lolium perenne. Nature 175, 944.

(7) Fejer, S.O. (1958). Genetic and environmental components of the productivity of perennial ryegrass (Lolium perenne. L.). New Zealand J. Agr. Res. 1, 86-103.

(8) Fejer, S.O. (1966). Selection methods for breeding hay-pasture-varieties of forage plants. Proc. 10th Internat. Grassl. Congr., 618-25.

(9) Fejer, S.O. (1967). Diallel crosses in Trifolium medium. Can. J. Genet. Cytol. 9, 799-804.

(10) Fejer, S.O. (1971). Diallel crosses of 'Vernal' alfalfa plants selected for vigor in association with timothy grass. Can. J. Genet. Cytol. 13, 729-35.

(11) Fejer, S.O. (1976). Combining ability and correlations of winter survival, electrical impedance and morphology in juvenile apple trees. Can. J. Plant Sci. 56, 303-9.

(12) Fejer, S.O. & Fedak, G. (1976). Heterosis and combining ability in a diallel cross of six-rowed barley selections. Proc. 3rd. Internat. Barley Genet. Symp.

(13) Fejer, S.O. & Jui, P. (1977). Genetic variance components in a barley diallel cross over 6 generations of single-seed-descent propagation. (In preparation).

(14) Fejer, S.O. & Spangelo, L.P.S. (1972). Components of yield and their inheritance in raspberry. Proc. 18th Internat. Hort. Congr., 21-2.

(15) Fejer, S.O. & Spangelo, L.P.S. (1975). Growth of Ottawa hybrid seedling rootstock topworked with McIntosh and Quinte apples and its relation to early yield. Can. J. Plant Sci. 54, 101-4.

(16) Fejer, S.O., Spangelo, L.P.S. & Modderman, L.L. (1975). Strawberry yield improvement in recurrent selection of 'Sparkle' × 'Valentine' crosses. Z. Pflanzenzüchtg., 74, 55-61.

(17) Glenday, A.C. & Fejer, S.O. (1956). The use of discriminant functions in the selection of pasture plants, with particular reference to the Lolium species. Proc. 7th Internat. Grassl. Congr., 461-70.

(18) Gustafsson, A. (1954). Mutations, viability and population structure. Acta Agr. Scand. 4, 601-32.

(19) Kempthorne, O. (1952). The design and analysis of experiments. Wiley, New York.

(20) Lerner, I. M. (1954). Genetic homeostasis. Oliver & Boyd, Edinburgh.

(21) Mather, K. (1949). Biometrical genetics. Methuen & Co. Ltd., London.

(22) Redden, R.J. & Jensen, N.F. (1974). Mass selection and mating systems in cereals. Crop Sci. 14:345-50.

(23) Sampson, D.R. (1976). Choosing the best parents for a breeding program from among eight oat cultivars crossed in a diallel. Can. J. Plant Sci. 56, 263-74.

(24) Spangelo, L.P.S., Hsu, C.S., Fejer, S.O. & Watkins, R. (1970). Combining ability and interrelationships between thorniness and yield traits in gooseberry. Can. J. Plant Sci. 50, 439-44.

(25) Sprague, C.F. & Tatum, L.A. (1942). General vs. specific combining ability in single crosses of corn. J. Amer. Soc. Agron. 34, 923-32.

(26) Stebbins, G.L. Jr., (1950). Variation and evolution in plants. Columbia University Press, New York.

(27) Torrie, J.H. (1957). Evaluation of general and specific combining ability in perennial ryegrass (Lolium perenne. L.). New Zealand J. Sci. Tech. 38A, 1025-35.

(28) Watkins, R. & Spangelo, L.P.S. (1970). Components of genetic variance for plant survival and vigor of apple trees. Theor. Appl. Genet. 40, 195-203.

(29) Wright, S. (1956). Modes of selection. Amer. Naturalist 90, 5-24.

Model building in quantitative genetics

W. Y. Tan and M. P. Mi
MEMPHIS STATE UNIVERSITY, MEMPHIS, TN 38152
and UNIVERSITY OF HAWAII, HONOLULU, HI 96844

INTRODUCTION

This paper illustrates how to build up quantitative genetic models by using Comstock and Robinson's Design I for animal experiments (see Comstock and Robinson (1952)). Some possible applications are indicated. For deriving the models, the following assumptions are made, some of which will be relaxed in our future work:

(1) The population is diploid and each locus has only two alleles, B_i and b_i for the ith locus, $i=1,2,\ldots,r$,

(2) There are no genetic-environmental interaction,

(3) There are no linkage or linkage is at equilibrium,

(4) There are no epistatic effects among loci,

(5) The population is at panmixia,

(6) There are no selection, no mutation, no immigration.

Further, the following notations will be adopted, unless otherwise stated: $X \sim B(1,p)$ denotes the point binomial distribution and $(X_1, X_2) \sim \text{Mult}(1; p_1, p_2)$ multinomial distribution (see Tan and Chang (1972)).

A Statistical Model For Comstock And Robinson's Design I

Let $y_{i(s)}, i=1,2,\ldots,\ell$, be the phenotypic values of a random sample of sires from a sufficiently large sire population and $(y_{ij(f)}, j=1,2,\ldots,m)$, $i=1,2,\ldots,\ell$, the phenotypic values of ℓ independent random samples of dams from a sufficiently large dam population. Denote by z_{ijk} the phenotypic value of the kth offspring from the mating of the ith sire and the jth dam of the ith sample. Suppose that the frequency of B_μ is $p_\mu, \mu=1,2\ldots r$ in both sire and dam populations. Then the genotypic values together with its frequencies of the three genotypes $B_\mu B_\mu$, $B_\mu b_\mu$ and $b_\mu b_\mu$ are given by:

Genotype	$B_\mu B_\mu$	$B_\mu b_\mu$	$b_\mu b_\mu$
Genotypic value	a_μ	d_μ	$-a_\mu$
Frequency	p_μ^2	$2p_\mu q_\mu$	q_μ^2, with $q_\mu = 1-p_\mu, \mu=1,2,\ldots r$

Let $g_{\mu(i)}$ be the genotypic value at the μth locus for $y_{i(s)}$, $g_{\mu(ij)}$ the genotypic value at the μth locus for $y_{ij(f)}$, $(X_{1\mu(i)}, X_{2\mu(i)}) \sim \text{Mult}(1; p_\mu^2, 2p_\mu q_\mu)$ and $(X_{1\mu(ij)}, X_{2\mu(ij)}) \sim \text{Mult}(1; p_\mu^2, 2p_\mu q_\mu)$ independently for $\mu=1,2,\ldots,r$, $i=1,2,\ldots\ell, j=1,2,\ldots m$. Then, following Tan and Chang (1972),

$$g_{\mu(i)} = 2a_\mu X_{1\mu(i)} + (a_\mu + d_\mu) X_{2\mu(i)} - a_\mu$$

$$g_{\mu(ij)} = 2a_\mu X_{1\mu(ij)} + (a_\mu + d_\mu) X_{2\mu(ij)} - a_\mu, \mu=1,2\ldots r,$$ so that

$$y_{i(s)} = \sum_{\mu=1}^{r} g_{\mu(i)} + e_{1i} \quad \text{and} \quad y_{ij(f)} = \sum_{\mu=1}^{r} g_{\mu(ij)} + e_{2ij},$$

MODEL BUILDING IN QUANTITATIVE GENETICS 845

where e_{1i} and e_{1ij} are the continuous random variables for environmental disturbances. It is clear that $(X_{1\mu(i)}=1, X_{2\mu(i)}=0)$, $(X_{1\mu(i)}=0, X_{2\mu(i)}=1)$ and $(X_{1\mu(i)}=0, X_{2\mu(i)}=0)$ if and only if $g_{\mu(i)}=a_\mu$, $g_{\mu(i)}=d_\mu$ and $g_{\mu(i)}=-a_\mu$. Similar relations hold for $(X_{1\mu(ij)}, X_{2\mu(ij)})$ and $g_{\mu(ij)}$. This implies that $P_r\{g_{\mu(i)}=a_\mu\}=P_r\{X_{1\mu(i)}=1, X_{2\mu(i)}=0\}=p_\mu^2=P_r\{g_{\mu(ij)}=a_\mu\}=P_r\{X_{1\mu(ij)}=1, X_{2\mu(ij)}=0\}$, $P_r\{g_{\mu(i)}=d_\mu\}=P_r\{X_{1\mu(i)}=0, X_{2\mu(i)}=1\}=2p_\mu q_\mu=P_r\{g_{\mu(ij)}=d_\mu\}=P_r\{X_{1\mu(ij)}=0, X_{2\mu(ij)}=1\}$, and $P_r\{g_{\mu(i)}=-a_\mu\}=P_r\{X_{1\mu(i)}=0, X_{2\mu(i)}=0\}=q_\mu^2=P_r\{g_{\mu(ij)}=-a_\mu\}=P_r\{X_{1\mu(ij)}=0, X_{2\mu(ij)}=0\}$. For representing the genotypic value $g_{\mu(ijk)}$ of z_{ijk} at the μth locus, we let $B_{\mu(ijk)} \sim B(1;\tfrac{1}{2})$, $(X_{1\mu(ijk)}, X_{2\mu(ijk)}) \sim \text{Mult}(1;\tfrac{1}{4},\tfrac{1}{2})$, $\mu=1,2,\ldots r, i=1,2\ldots \ell$, $j=1,2\ldots m, k=1,2\ldots n_{ij}$, independently, and independent of the $(X_{1\mu(i)}, X_{2\mu(i)})$'s and the $(X_{1\mu(ij)}, X_{2\mu(ij)})$'s.

Then, it is readily seen that $z_{ijk} = \sum_{\mu=1}^{r} g_{\mu(ijk)} + e_{3ijk}$, where e_{3ijk} is the random variable for the environmental effect and $g_{\mu(ijk)} = X_{1\mu(i)}\{(a_\mu-d_\mu)X_{1\mu(ij)} + (a_\mu-d_\mu)B_{\mu(ijk)} \times X_{2\mu(ij)} + d_\mu\} + X_{2\mu(i)}\{[(a_\mu-d_\mu)B_{\mu(ijk)}+d_\mu]X_{1\mu(ij)} + [2a_\mu X_{1\mu(ijk)} + (a_\mu+d_\mu)X_{2\mu(ijk)} - a_\mu]X_{2\mu(ij)} + [d_\mu - (a_\mu+d_\mu)B_{\mu(ijk)}](1-X_{1\mu(ij)} - X_{2\mu(ij)})\} + (1-X_{1\mu(i)}-X_{2\mu(i)}) \times \{(a_\mu+d_\mu)X_{1\mu(ij)} + (a_\mu+d_\mu) \times (1-B_{\mu(ijk)})X_{2\mu(ij)} - a_\mu\}$,

The above formula for $g_{\mu(ijk)}$ provides a basic relationship between parents and offspring and is in fact the well-known principle of independent segregation of genes. For example, if both the sire genotype and the dam genotype are $B_\mu b_\mu$, then $(X_{1\mu(i)}=0, X_{2\mu(i)}=1)$ and $(X_{1\mu(ij)}=0, X_{2\mu(ij)}=1)$ so that $g_{\mu(ijk)} = 2a_\mu X_{1\mu(ijk)} + (a_\mu+d_\mu) X_{2\mu(ijk)} - a_\mu$. Or equivalently,

$P_r\{g_{\mu(ijk)}=a_\mu\} = P_r\{X_{1\mu(ijk)}=1, X_{2\mu(ijk)}=0\} = \frac{1}{4}, P_r\{g_{\mu(ijk)}=d_\mu\}$
$= P_r\{X_{1\mu(ijk)}=0, X_{2\mu(ijk)}=1\} = \frac{1}{2}, P_r\{g_{\mu(ijk)}= -a_\mu\} = P_r\{X_{1\mu(ijk)}=0, X_{2\mu(ijk)}=0\} = \frac{1}{4}$. Other cases can be illustrated similarly.

Some Possible Applications of the Models

The above models not only allow for the introduction of the maximum likelihood method in quantitative genetics but also make it possible for the study of effects of departure from normality and other assumptions on the classical statistical inference procedures in quantitative genetics. Using the above models, we shall be able to approximate the joint distribution of linear combinations of mean squares by finite series of Laguerre polynomials (see Tan and Wong (1976)). This allows us to examine the effect of nonnormality on the classical inference procedures in quantitative genetics. These results will be given in two forthcoming papers. Because of the limitation of time and space, we shall not pursue any further here.

References

Comstock, R.E. and Robinson, H. (1952), "Experiments for estimation of the average dominance of genes affecting quantitative characters" in Heterosis, pp. 494-516, Iowa State College Press.

Tan, W.Y. and Chang, W.C. (1972), "Convolution approach to the analysis of quantitative characters in self-fertilized populations", Biome trics 28, 1073-1090.

Tan, W.Y. and Wong, S.P. (1976), "On the Roy-Tiku approximation to the distribution of sample variances from nonnormal universes", to appear in Journal of American Statistical Association (under revision).

Response and variance of response to selection

Y. *Park* and R. *Nassar*
DEPARTMENT OF STATISTICS
KANSAS STATE UNIVERSITY, MANHATTAN, KS 66502

INTRODUCTION

The effect of selection in finite populations on the ultimate probability of fixation of a favorable allele has been studied extensively by Kimura (1957) and others. Little is known, however, about the effect of drift and selection on the change in gene frequency over time. In this study we present prediction equations for response and variance of response to selection in finite populations.

Theory.

For a locus with two alleles the change per generation in gene frequency of a favorable allele at time t for no dominance and random mating finite population can be expressed as

$$E(\Delta P)_t \simeq \frac{s}{2}[E(P_t) - [E(P_t)]^2 - \sigma_{P_t}^2] \qquad (1)$$

If we let $\sigma_{P_t}^2 \simeq P(1-P)[1-(1-1/2N_e)^t]$, we get the approximate solution

$$\mu_{P_t} = e^x/(1+e^x) \qquad (2)$$

where $x = [s/2\ln(1-1/2N_e)][(1-1/2N_e)^t - 1] + \ln(\mu_{P_o}/1-\mu_{P_o}) \qquad (3)$

and μ_{P_o} is the initial gene frequency. Equations (2) and (3) apply directly to within full sib family selection and random mating if s is halved and N_e changed accordingly (N_e' in (5) is twice the number of parents). In this case the solution is given by (2) with

$$x = [s/4\ln(1-1/2N_e)][(1-1/2N_e)^t - 1] + \ln(\mu_{P_o}/1-\mu_{P_o}) \quad (4)$$

Equation (2) has been tested extensively for the two cases in (3) and (4) and found to agree well with results from simulation and transition probability matrix (representative results are given in Table 1). In the prediction equations, N_e was computed from the general expression

$$N_e = N_e'/\{1+s(h+(1-2h)\mu_{P_o})(1-2\mu_{P_o})\} \quad (5)$$

where h is the dominance parameter (h=1/2 for additive gene effect) and N_e' the effective population size for no selection. The selection coefficient was computed from the first generation as $2ui/\sigma_p$ where 2u is the difference in effect between the homozygous genotypes at a locus and i the selection differential in phenotypic standard deviations (σ_p). It was interesting to discover that the prediction equations agreed also well with results for dominance (h≠1/2) when (5) was used to compute N_e.

TABLE I.

Comparisons of Formulae (3) and (6) for Mass Selection and (4) for Within Full Sib Family Selection with Results from Simulation and Transition Probability Matrix

	Mass Selection $N_e'=20$, s=.25, $\mu_{P_o}=1/2$				Within Full Sib Selection $N_e'=8$, s=.216, $\mu_{P_o}=.25$	
	Gene Freq.		Var. Gene Freq.		Gene Freq.	
Gen.	Sim.	Formula	Sim.	Formula	Matrix	Formula
1	.52	.52	.009	.006	.26	.26
2	.53	.53	.013	.012	.27	.27
3	.54	.55	.018	.018	.28	.28
4	.55	.56	.021	.024	.29	.29
5	.57	.57	.028	.029	.30	.30
10	.62	.63	.055	.052	.33	.33
15	.66	.69	.071	.068	.36	.36
20	.71	.73	.079	.079	.37	.38
25	.75	.76	.081	.085	.38	.39
30	.77	.78	.090	.088	.39	.40

From equations (2) and (3) and the fact that $E(\Delta P)_t \simeq \frac{s}{2} E(P_t(1-P_t))$ one obtains that the variance of gene frequency at time t for mass

selection is
$$V_{P_t} \simeq \mu_{P_t}(1-\mu_{P_t}) - e^x[1-1/2N_e]^t/(1+e^x)^2 \quad (6)$$
Expression (6) compares well with results from simulation (Table 1).

Selection for a Quantitative Trait.

The application of the theory to selection for a quantitative trait seems to hold best for an F_2 population derived from two inbred lines $[\mu_{P_o} = 1/2$ at all loci]. Under the assumption of equal gene effects, no epistasis, no dominance and linkage equilibrium the expected genotypic response to selection can be expressed as

$$E(\Delta G)_t = 2nu\ [\mu_{P_t} - 1/2] \simeq R(\mu_{P_t} - 1/2) \quad (7)$$

where n is the number of loci and R the difference between two extreme inbred lines as discussed by Wright (1968). For variance of genotypic response it can be shown that

$$V\Delta G_t \simeq 8\sigma_A^2[V_{P_t} - V_{P_o}] = 8\sigma_A^2[V_{P_t} - 1/8N_e], \text{ for } \mu_{P_o} = 1/2 \quad (8)$$

in (7) and (8) $s \simeq 8ih^2\sigma_p/R$ and σ_A^2 = additive genetic variance.

In (7), R was assumed to represent the difference between the extremes of inbred lines (R=2nu). With a large number of loci the probability is practically zero of obtaining two extreme inbred lines, that is one fixed + and the other - at all loci concerned. This being the case, it is very likely that R underestimates 2nu. It is of interest to note, however, that an R biased downward is likely to have little effect on predicted response, since if R decreases, s increases and, therefore, μ_{P_t} in (7) increases. A simple computation shows that, for s=.1, $N_e = 20$ and R=80% of the true range (2nu), $E(\Delta G)_t$ in (7) is decreased by .25% at generation 5, 2.6% at generation 20 and 7% at generation 50. The effect on the variance (8), however, is that $V\Delta G_t$ is decreased by 1%, 8.4% and 22.1% respectively. For predicting variance of response at later generations, the estimation of s as in (8) might not be satisfactory if R is underestimated by much. Other possibilities of estimating s are being considered.

More work is needed on studying the magnitude of bias in (7) and (8) if the assumptions of equal gene effects, no dominance and no linkage are relaxed. In some instances one can predict the direction and magnitude of bias. If gene effects are unequal over loci, as is likely to be the case, $8ih^2\sigma_p/R$ would give an average s over all loci and can be shown to overestimate the true $\bar{s}$ by 100 $\sigma_u^2/\bar{u}^2$ percent. It is seen that the downward bias in R and the bias upward in $\bar{s}$ (due to variable u's) work against each other in affecting changes in (7). Hence, in practice (7) might not be such an unreasonable prediction equation. The variance formula, however, is expected to underestimate the true variance as a result of $\bar{s}$ being overestimated due to the downward bias in R and to variability in u's. Linkage (assuming a near balance between coupling and repulsion in the F_2) is likely to have little effect on response and variance of response in the early generations. In later generations, however, linkage would be expected to decrease response and increase the variance of response. The magnitude of the change depends of course on the intensity of linkage. The presence of unidirectional dominance will cause the true response to be less than predicted from (7) since

$$E(\Delta G)_t \simeq R[\mu_{P_t} - \frac{1}{2}] - 2\Sigma a_i u_i [\frac{1}{4} - \mu_{P_t}(1-\mu_{P_t}) + V_{P_t} - V_{P_o}] \quad (9)$$

This can be evaluated if an estimate of $\Sigma a_i u_i$ is available. A general expression for variance of response with dominance is not as easy to obtain. The effect of dominance, however, would be to increase the variance of response in the very early generations of selection and to reduce it in later generations.

BIBLIOGRAPHY

[1] Kimura, M. (1957). Some problems of stochastic process in genetics. Ann. Math. Stat. 28, 882-901.

[2] Wright, S. (1968). Evolution and the Genetics of Populations. Vol. 1. Genetic and Biometric Foundations. Chicago: University of Chicago Press.

Variance-covariance structure of group means with overlapping generations

D. L. Johnson
INSTITUTE OF ANIMAL GENETICS
WEST MAINS ROAD, EDINBURGH, SCOTLAND*

INTRODUCTION

In order to obtain the accuracy of estimates of realised genetic parameters from a finite population undergoing selection, it is necessary to have information on the variance-covariance structure of the response in successive years of the selection programme. This problem has been discussed in detail by Hill [1] for the case of a finite population with non-overlapping generations.

The main contribution to the variance of response comes from genetic drift. For a population in which the generations overlap, the effective population number and hence the rate of genetic drift has been determined, for specific models, under the assumption that the relationships of individuals in the population have stabilised [2]. However the pattern of genetic drift in the early generations, when the asymptotic formula is not applicable, is not clear.

A formal method for predicting the response to selection in populations with overlapping generations has been presented by Hill [4]. Because of genetic drift the variance of response generally

* Now at Ruakura Agricultural Research Centre, Hamilton, New Zealand

increases with time and the responses in different years become correlated. The aim of this paper is to determine the variance-covariance structure of this response in successive years. Matrix methods are used, the basic matrix being the same as that used by Hill [4] specifying the passage of genes between the different age groups and sexes.

We consider a selection experiment for some quantitative trait with additive gene action. It is assumed that genetic variances and covariances do not change within the population during selection. Thus we assume that there are many genes influencing the trait, each with a small effect. We ignore the complications of close linkage and high inbreeding.

THE MODEL

We consider a population in which males are kept for r years (or breeding seasons) and females are kept for s years. Let M_i be the number of males of age i, $1 \leq i \leq r$, and F_j the number of females of age j, $1 \leq j \leq s$. We assume random mating within the entire population, that deaths occur at random and that there is a random distribution of family size from surviving parents.

Selection is practised on male and female progeny so that only superior individuals are kept to age 1 as potential breeders and there is no further selection at later ages. Progeny born at time 0 are the first selected individuals. The quantitative trait undergoing selection has phenotypic variance σ^2 and heritability h^2. The M(t) males and F(t) females born at time t have mean performances $x_m(t)$ and $x_f(t)$ respectively. A proportion $q_m(t)$ of the males born at time t are selected for breeding and $s_m(t)$ denotes the selection differential. For females the parameters are $q_f(t)$ and $s_f(t)$.

Let p_{1i} ($1 \leq i \leq r$) and $p_{1,r+j}$ ($1 \leq j \leq s$) be the expected proportion of genes in male progeny which are derived from sires of age i and dams of age j respectively. Similarly $p_{r+1,i}$ and $p_{r+1,r+j}$ are the expected proportions of genes in female progeny derived from the various parental age groups. Following [4] we define a matrix P of order r+s with elements $p_{k\ell}$, $1 \leq k, \ell \leq r+s$, where

$$p_{k,k-1} = 1 \quad k \neq 1, r+1$$
$$p_{k\ell} = 0 \quad k \neq 1, r+1 \;,\; \ell \neq k-1$$

Let $z_i(t)$ be the mean breeding value of males of age i at time t if $1 \leq i \leq r$ and females of age i-r if $r+1 \leq i \leq r+s$. Then $z_i(0) = \mu$, $1 \leq i \leq r+s$, where μ is the mean of the base population from which the selection line is chosen. Let $\underline{s}(t)$ be the vector with $s_1(t) = s_m(t)$, $s_{r+1}(t) = s_f(t)$ and zeros elsewhere. Then we can express the vector of breeding values $\underline{z}(t+1)$ at time t+1 in terms of $\underline{z}(t)$ as

$$\underline{z}(t+1) = P\,\underline{z}(t) + h^2 \underline{s}(t) + \underline{e}(t) \qquad t \geq 0 \qquad (1)$$

where $E\underline{e}(t) = \underline{0}$. The errors $e_1(t)$ and $e_{r+1}(t)$ represent the deviations from the regressions of breeding value on phenotype, thus

$$\text{var } e_1(t) = h^2(1-h^2)\sigma^2/M_1$$
$$\text{var } e_{r+1}(t) = h^2(1-h^2)\sigma^2/F_1 \qquad t \geq 0 \;.$$

The $e_i(t)$, $i \neq 1, r+1$, arise from the fact that some individuals die before the end of their reproductive life. We have for $1 < i \leq r$

$$\text{var } e_i(t) = \text{var}\left[z_i(t+1) - z_{i-1}(t)\right]$$

$$= \begin{cases} h^2\sigma^2[1-\alpha_m(t-i+1)](\frac{1}{M_i} - \frac{1}{M_{i-1}}) & t > 0 \\ h^2\sigma^2/M_i & t = 0 \end{cases}$$

where $\alpha_m(t)$ represents the proportional reduction in the genetic variance, due to selection, among males born at time t if $t \geq 0$ and is zero for $t < 0$, and similarly for females.

Define Z_t and W_t to be the variance-covariance matrices of the vectors $\underline{z}(t)$ and $h^2\underline{s}(t) + \underline{e}(t)$ respectively, conditional upon the observed selection differentials. Then W_t is diagonal with

$$(W_t)_{11} = \frac{h^4\sigma^2}{M(t)} + \text{var } e_1(t) = \frac{h^2\sigma^2}{M_1}\left[1 - h^2(1-q_m(t))\right]$$

$$(W_t)_{r+1,r+1} = \frac{h^2\sigma^2}{F_1}\left[1 - h^2(1-q_f(t))\right]$$

and

$$(W_t)_{ii} = \text{var } e_i(t) \qquad i \neq 1, r+1 \;.$$

From (1) it follows that

$$Z_{t+1} = P\,Z_t\,P' + W_t \qquad t \geq 0 \qquad (2)$$

with $Z_0 = 0$ and that

$$Z_{t+1} = \sum_{i=0}^{t} P^i W_{t-i} P'^i \tag{3}$$

(' denotes matrix transpose).

Let $\underline{p}'_m$ and $\underline{p}'_f$ denote rows 1 and r+1 respectively of the matrix P. Then $\underline{p}'_m \underline{z}(t)$ is estimated by the mean $x_m(t)$ of a sample of M(t) progeny. The variance of this estimate depends on the distribution of family sizes but for most purposes may be taken equal to $\sigma^2/M(t)$. Thus

$$\operatorname{var} x_m(t) = \underline{p}'_m Z_t \underline{p}_m + \sigma^2/M(t) \tag{4}$$

and similarly

$$\operatorname{var} x_f(t) = \underline{p}'_f Z_t \underline{p}_f + \sigma^2/F(t) \quad .$$

The first terms of these variances are due to genetic drift. The covariance between male and female offspring means is given by

$$\operatorname{cov}(x_m(t), x_f(t)) = \underline{p}'_m Z_t \underline{p}_f \quad .$$

Suppose that $1 \leq t < \tau$ then from (1) we obtain

$$\underline{z}(\tau) = P^{\tau-t} \underline{z}(t) + \sum_{i=0}^{\tau-t-1} P^i \left[h^2 \underline{s}(\tau-i-1) + \underline{e}(\tau-i-1) \right] \quad .$$

This leads to the covariances between offspring means at different time periods,

$$\operatorname{cov}(x_m(t), x_m(\tau)) = \underline{p}'_m P^{\tau-t} Z_t \underline{p}_m + (P^{\tau-t})_{11} h^2 \sigma^2/M(t)$$

$$\operatorname{cov}(x_f(t), x_f(\tau)) = \underline{p}'_f P^{\tau-t} Z_t \underline{p}_f + (P^{\tau-t})_{r+1, r+1} h^2 \sigma^2/F(t)$$

$$\operatorname{cov}(x_m(t), x_f(\tau)) = \underline{p}'_f P^{\tau-t} Z_t \underline{p}_m + (P^{\tau-t})_{r+1, 1} h^2 \sigma^2/M(t)$$

$$\operatorname{cov}(x_f(t), x_m(\tau)) = \underline{p}'_m P^{\tau-t} Z_t \underline{p}_f + (P^{\tau-t})_{1, r+1} h^2 \sigma^2/F(t)$$

the last terms in each of the above covariances resulting from the covariance between the selection differentials and means at time t.

ASYMPTOTIC RESULTS

The matrix P is stochastic, it has a single eigenvalue of unity and all others are of smaller absolute value. The vector $\underline{v}$ defined by

$$v_i = \begin{cases} \sum_{j=i}^{r} (p_{1j} + p_{r+1,j}) & 1 \leq i \leq r \\ \sum_{j=i}^{r+s} (p_{1j} + p_{r+1,j}) & r+1 \leq i \leq r+s \end{cases}$$

is a left eigenvector for P. The matrix $A = \lim_{t \to \infty} P^t$ is given by

$$A = \frac{1}{2L} \underline{1} \underline{v}'$$

where

$$L = \frac{1}{2} \sum_{i=1}^{r+s} v_i$$

is the generation interval (average age of parents of newborn progeny) and $\underline{1}$ is the vector with each element equal to unity.

Suppose that the same number of male progeny are recorded each year so that $q_m(t) = q_m$ and $\alpha_m(t) = \alpha_m$ for $t \geq 0$, and similarly for females. (The term α_m is given by $h^2 i_m (i_m - a_m)$ where i_m and a_m are the standardised selection differential and truncation point respectively, corresponding to the proportion q_m selected). Then the matrix W_t is constant for $t \geq \max(r-1, s-1)$, equal to W say. From (3) it follows that

$$\lim_{t \to \infty} (Z_{t+1} - Z_t) = A W A' = \frac{1}{4L^2} (\underline{v}' W \underline{v}) \underline{1} \underline{1}'.$$

The asymptotic drift per generation is then $h^2 \sigma^2 / N_e$ where N_e, the effective population size, is given by

$$\frac{1}{N_e} = \underline{v}' W \underline{v} / 4h^2 \sigma^2 L$$

$$= \frac{1}{4L} \left[\frac{1 - h^2(1 - q_m)}{M_1} + (1 - \alpha_m) \sum_{i=2}^{r} v_i^2 \left(\frac{1}{M_i} - \frac{1}{M_{i-1}} \right) \right.$$

$$\left. + \frac{1 - h^2(1 - q_f)}{F_1} + (1 - \alpha_f) \sum_{j=2}^{s} v_{r+j}^2 \left(\frac{1}{F_j} - \frac{1}{F_{j-1}} \right) \right].$$

If replacements are selected at random, as in a control population, then $q_m = q_f = 1$ and $\alpha_m = \alpha_f = 0$, thus

$$\frac{1}{N_e} = \frac{1}{4L} \left[\frac{1}{M_1} + \sum_{i=2}^{r} v_i^2 \left(\frac{1}{M_i} - \frac{1}{M_{i-1}} \right) + \frac{1}{F_1} + \sum_{j=2}^{s} v_{r+j}^2 \left(\frac{1}{F_j} - \frac{1}{F_{j-1}} \right) \right]$$

EXAMPLE

The important term in equation (4) is $\underline{p}'_m Z_t \underline{p}_m$ representing the genetic drift contributing towards the variance of the male offspring mean at time t. It is interesting to compare this drift with a uniform drift of $th^2\sigma^2/LN_e$ over the same period of time. We illustrate with a population of sheep in which males have progeny at 2 years of age only and females are kept for 4 breeding seasons having progeny first at 2 years of age. We take $M_1 = 5$, $M_2 = 5$. The numbers of females and lambing percentages are given in table I.

TABLE I

age of ewe (j)	F_j	lambing %	$P_{1,2+j} = P_{3,2+j}$
1	65	0	0.00
2	60	80	0.11
3	55	100	0.13
4	50	120	0.14
5	45	120	0.12

Of the male and female progeny born each year 5% and 60% respectively are selected for breeding on the basis of a quantitative trait with heritability $h^2 = 0.3$. The results are presented in table II.

TABLE II

t	$\underline{p}'_m Z_t \underline{p}_m / h^2\sigma^2$	t/LN_e
1	0.0512	0.0052
2	0.0369	0.0104
3	0.0566	0.0156
4	0.0523	0.0207
5	0.0659	0.0259
6	0.0664	0.0311
7	0.0745	0.0363
8	0.0779	0.0415
9	0.0840	0.0467
10	0.0888	0.0519

The generation interval is $L = 2.77$ and the effective population size is $N_e = 69$. The real drift differs considerably from the drift that would be predicted using the asymptotic formula from the outset. The difference between the two estimates is initially variable but it can be shown that it asymptotes to a constant value which in this example is equal to $0.037h^2\sigma^2$.

DISCUSSION

Most of the sampling variance of group means is contributed by genetic drift. For a population with overlapping generations the difference between the real drift and the approximation using a uniform rate of drift is initially variable but eventually settles down to a constant difference. This is similar to the pattern of selection response as discussed in [4].

So far we have ignored environmental effects common to all individuals. If a control population is maintained the response is the difference between the mean of the selected population and the mean of the control population. Thus the drift from the control will also contribute to the total variance of response [3].

The parental age distribution, and hence P, is assumed to remain constant. Also age-specific mortality rates are constant. One restrictive assumption in this theory is that the expectation of progeny is the same for each parent in the population of a given age. However the effects of selection on genetic variances and covariances in later generations may be more serious than the problem of differential fertility.

ACKNOWLEDGMENTS

Thanks are due to Dr W.G. Hill at the Institute of Animal Genetics, Edinburgh, for his comments and suggestions and for the use of facilities at the Institute. I am grateful to the Ministry of Agriculture and Fisheries, New Zealand, for financial support while on study leave in Edinburgh.

BIBLIOGRAPHY

[1] Hill, W.G. (1971). Design and efficiency of selection experiments for estimating genetic parameters. **Biometrics 27**, 293-311.

[2] Hill, W.G. (1972). Effective size of populations with overlapping generations. **Theor. Pop. Biol. 3**, 278-289.

[3] Hill, W.G. (1972). Estimation of genetic change. I. General theory and design of control populations. Anim. Breed. Abst. 40, 1-15.

[4] Hill, W.G. (1974). Prediction and evaluation of response to selection with overlapping generations. Anim. Prod. 18, 117-139.

Errors in predicting genetic gain from mass selection

Fan H. Kung
DEPARTMENT OF FORESTRY
SOUTHERN ILLINOIS UNIVERSITY, CARBONDALE, IL 62901

INTRODUCTION

Mass selection is the selection of individual plants or animals based on their phenotypic expression. The breeder either culls a certain fraction of the population or saves individuals above a cutoff point. The effectiveness of mass selection logically depends on two factors. The first one is selection differential, or to what extent those individuals selected exceed the population mean. The second one is heritability, or how much the outward appearance is controlled by genetic factors. Genetic gain equals the product of selection differential and heritability.

This paper presents a simulation study on the empirical error of predicting genetic gain in mass selection.

SIMULATION MODEL

The simulation model is $X_i = G_i + E_i$

where X_i = Phenotypic value of the i-th individual
 $i = 1$ thru 100
 G_i = Genetic value of the i-th individual, obtained from a pseudo random number generator, with normal distribution

$\mu = 0, \quad \sigma^2 = Kj$

$Kj = 5$ thru 95 in steps of 5 or 10

Ei = Environmental contribution to the i-th individual, obtained from a pseudo random number generator, with normal distribution $\mu = 0, \quad \sigma^2 = 100 - Kj$

The Xi were sorted, carrying along Gi and Ei. A fraction of the data points were truncated. The mean of X in the selected group was used as the selection differential. Heritability is defined as Kj/100. Then multiplying selection differential by heritability, we obtained a predicted genetic gain. The true genetic value of the selection is the mean value for G. In short, the error E is expressed as $E = \bar{x}(Kj/100) - \bar{G}$. The number of individuals saved in the simulation ranged from 5 to 95, out of a total of 100. The procedure was repeated 5 times.

RESULT AND DISCUSSION

The numbers of positive errors in 5 simulations for various selection proportions and heritability are presented in Table I. The null hypothesis is that the probability of a positive error is 1/2. It can be seen from the subtotals that on the average the hypothesis is acceptable for various levels of selection proportion but not for various levels of heritability. The correlation coefficient between the average percent of positive errors and heritability was -0.93. A closer look at the table revealed that the main body of the table had a saddle point near the central region. The upper right hand region and the lower left hand region had excessive positive errors. On the contrary, the upper left hand region and the lower right hand region had fewer positive errors. Genetic gain would be overestimated in either one of the following conditions: (a) low selection proportion with high heritability, (b) high selection proportion and low heritability. Genetic gain would be underestimated when both selection proportion and heritability are high or low. The formula seemed to give an unbiased estimate of genetic gain when selection proportion was medium (S = .3∿.7) and heritability was medium high (h^2 = .4∿.8).

The standard deviations of prediction errors among five simulations at various levels of selection proportion and heritability are shown in Table II. The standard deviation decreased as the proportion of selection increased. At low and medium selection proportions (S < .6) the standard deviation seemed to reach a maximum when the genetic variance is about twice the environment variance.

The empirical results from simulation suggest that the commonly accepted formula, $\Delta Y = Ih^2$, needs some correction, especially when the selection goes to the extremes of selection proportion and heritability. A better prediction formula from the statistician would be appreciated by most plant and animal breeders.

TABLE I

Error Rate of Predicting Genetic Gain in Mass Selection

Variance		\multicolumn{19}{c}{No. of Individuals Selected from a Total of 100}	Sub-	
Vg	Ve	5	10	15	20	25	30	35	40	45	50	55	60	65	70	75	80	85	90	95	Total	%
							No. of Positive Errors in 5 Simulations															
5	95	1	1	2	3	2	3	4	3	4	4	2	4	4	4	4	4	5	5	5	64	67
10	90	3	2	2	2	2	3	3	3	2	2	3	3	4	3	5	5	5	5	5	62	65
20	80	2	2	3	2	1	2	2	2	2	2	2	5	3	4	4	5	5	5	5	57	63
30	70	2	2	2	2	1	2	2	1	1	2	2	3	4	4	5	5	5	5	5	57	63
40	60	3	2	2	2	2	2	2	2	3	3	4	4	4	5	5	5	5	5	5	62	69
50	50	2	2	4	2	2	4	4	4	4	2	4	3	3	4	3	5	5	5	5	59	65
60	40	3	3	4	4	3	3	3	2	4	4	1	2	2	1	1	0	0	1	1	42	47
70	30	3	4	3	3	4	3	2	3	3	2	3	1	1	1	0	0	0	0	0	41	46
80	20	3	4	4	2	2	2	3	2	2	1	2	2	1	0	1	0	0	0	0	36	40
90	10	3	4	3	3	1	3	1	3	1	2	1	2	1	1	0	0	0	0	0	32	36
95	5	4	4	3	3	1	3	1	1	1	2	1	2	1	1	2	1	0	0	0	31	34
Subtotal		29	30	31	27	24	29	28	27	28	27	26	31	28	28	29	30	30	30	31	543	
Percent		53	55	56	49	44	53	51	49	51	49	47	56	51	51	53	55	55	55	56		52

TABLE
II

Standard Deviation of Prediction Error at Various
Proportion of Selection and Variance Ratios

Proportion of Selection &	Genetic Variance									
	5	10	20	30	40	50	60	70	80	90	95
	Environment Variance									
	95	90	80	70	60	50	40	30	20	10	5
	Standard Deviation of (Predicted Gain − Genetic Mean)									
5	.81	1.18	1.25	2.55	2.05	2.81	2.99	3.06	1.82	1.14	1.30
10	.63	.79	1.46	1.83	2.19	1.66	2.70	1.82	1.97	1.23	.98
15	.68	1.09	1.57	1.67	1.53	1.70	1.81	2.19	1.81	1.27	.82
20	.62	.97	1.32	1.54	1.30	1.12	1.53	1.69	1.47	1.24	.78
25	.66	.78	.98	1.41	1.00	.91	1.55	1.32	1.20	1.03	.69
30	.45	.75	.92	.82	.92	.57	1.42	1.29	1.22	.82	.52
35	.46	.66	.66	.52	.65	.50	1.13	1.08	.96	.77	.46
40	.42	.60	.52	.57	.30	.45	.89	1.00	.85	.61	.41
45	.40	.46	.49	.60	.39	.41	.77	.86	.83	.46	.32
50	.34	.32	.41	.44	.53	.54	.68	.82	.59	.35	.31
55	.32	.42	.30	.41	.49	.62	.65	.62	.53	.41	.38
60	.29	.41	.26	.37	.33	.53	.62	.44	.53	.46	.33
65	.28	.32	.35	.39	.43	.47	.46	.38	.48	.42	.33
70	.28	.32	.37	.38	.32	.52	.43	.31	.46	.42	.33
75	.22	.30	.34	.29	.38	.45	.31	.31	.31	.44	.34
80	.26	.35	.28	.35	.35	.35	.31	.30	.38	.44	.32
85	.25	.25	.36	.36	.40	.39	.36	.34	.37	.27	.27
90	.23	.23	.39	.42	.36	.40	.42	.37	.33	.29	.19
95	.23	.23	.39	.43	.47	.43	.36	.39	.34	.23	.20

Analysis of binomial data in quantitative genetics experiments

Walter A. Becker
WASHINGTON STATE UNIVERSITY, PULLMAN, WA 99163

INTRODUCTION

In many of the experiments on binomial data one of the principal assumptions involves the simple binomial sampling variance. This sampling variance is valid only under certain conditions and it is the intention of this paper to examine the effects on the variance when three of the conditions are not met.

CONDITIONS UNDERLYING BINOMIAL SAMPLING VARIANCE

Simple sampling means random sampling where the probability, p, of success is the same for each event in the sample, the success of each event is completely independent of the success of all other events in the sample, and all samples are drawn from the same population (Yule & Kendall, 1950).

Under these conditions the mean of the proportion of successes is $\mu = p$, and the variance $\sigma^2 = pq/n$ where $q = 1 - p$, and n = number of events or observations within a sample.

Samples from Different Populations

If we remove the condition that samples must come from the same population while maintaining the other two conditions, the mean is $\mu = (p_1 + p_2 + \ldots p_k)/N = p_o$ where p_k = proportion of successes in kth sample, N = number of samples, and p_o is mean value of the varying samples, p, and $\sigma^2 = p_o q_o/n + [(n-1)/n] \sigma_p^2$.

$p_o q_o/n$ is the variance due to chance fluctuations and σ_p^2 is due to real differences amongst the samples. For derivations of these and succeeding formulas see Yule and Kendall (1950).

For example, the proportion of blood spots in chicken eggs was analyzed by Becker and Bearse (1973) with a nested analysis of variance with three levels for sires, dams/sires and progeny/dams. The expectation of MS_W (mean square for progeny/dams) is $pq/n + [(n-1)/n] \sigma_W^2$ where σ_W^2 estimates 1/2 additive genetic variance, 3/4 dominance variance, < 3/4 epistatic variance and all the environmental variance within families.

Heterogenous Events Within A Sample

Relaxing the condition that the probability of success is the same for each event in the sample results in (the other two conditions remain): $\mu = p_o$, $\sigma^2 = p_o q_o/n - \sigma_p^2/n$ where p_o is the mean chance of a sample and σ_p^2 is the variance due to the probability of each event within a sample being different. For example, if we mate a sire to a series of dams, each producing n fertile poultry eggs, the probability of each egg within a dam hatching differs because of genetic and environmental variability. With a number of sires the hatchability analysis involves a nested design analysis of variance with sires and dams within sires. The progeny response is measured as the proportion of hatch of fertile eggs for each dam. Each dam will have an effect both genotypic and maternal on the hatchability of her progeny. In the one-way layout analysis of variance, the expectation of MS_D contains the binomial sampling variance, pq/n, the reduction of the variance because of σ_p^2/n ($\sigma_p^2 = \sigma_W^2$ where σ_W^2 is the variance between full sibs), and the increase in the variance due to the samples (or dams) being different.

Correlation Between Events in the Sample

When the condition of independence between events in a sample is removed with the other two conditions held intact, the variance is $\sigma^2 = \frac{pq}{n} [1 + r(n-1)]$, where r is the correlation between the success of events in the sample.

Referring to the chicken hatchability example, each embryo is correlated genetically within the sample because they are full sibs and possibly correlated environmentally because the eggs of a dam are usually set in close proximity to one another within the incubator. Similar problems arise when investigating the genetics of infectious diseases.

DISCUSSION

In most quantitative genetic analyses, the three conditions of simple sampling are not met. The interpretation of the genetic model for quantitative genetic experiments using binomial data must take into account the effects on the simple binomial variance by the removal of the three conditions, a circumstance not considered completely by Becker and Marsden (1972) in a genetic study of blister rust resistance of Western White Pine.

The arc sine square root transformation is based upon the simple binomial sampling variance. When the variance of the proportion is different than the simple binomial sampling variance, the use of this transformation is probably not warranted.

Further difficulties arise when using the analysis of variance because of unequal numbers (see Gabriel, 1963) and the estimation of the underlying normal distribution on the binomial scale (Dempster and Lerner, 1950; Van Vleck, 1972).

BIBLIOGRAPHY

[1] Becker, W. A. & Bearse, G. E. (1973). Selection for high and low percentages of chicken eggs with blood spots. Brit. Poul. Sci. 14, 31-47.

[2] Becker, W. A. & Marsden, M. A. (1972). Estimation of heritability and selection gain for blister rust resistance in Western White Pine. Biology of Rust Resistance in Forest Trees, USDA Misc. Publ. No. 1221, 397-409.

[3] Dempster, E. R. & Lerner, I. M. (1950). Heritability of threshold characters. Genetics 35, 212-236.

[4] Gabriel, K. K. (1963). Analysis of variance of proportions with unequal frequencies. J. Amer. Stat. Assoc. 58, 1133-1157.

[5] Van Vleck, L. D. (1972). Estimation of heritability of threshold characters. J. Dairy Sci. 55, 218-225.

[6] Yule, G. U. & Kendall, M. G. (1950). An Introduction to the Theory of Statistics. 14th ed. London: Griffin, 386-409.

A new interpretation of recombination frequencies in genetics

John H. Ursell
MATHEMATICS DEPARTMENT
QUEEN'S UNIVERSITY, KINGSTON, ONT., CANADA

INTRODUCTION

Until now the recombination proportions (called "frequencies" usually) for two traits have been taken to imply that the genetic material for those two traits are on different chromosomes or that they are on the same chromosome separated by a certain "distance". This "distance", the map distance between the two traits on the genetic map of that chromosome, is measured in units each equal to 1% recombination. The "distance" is assumed to show the physical distance between the genetic material for the two traits fairly accurately. The "distances" do not add like ordinary numbers nor by any fixed rule. Indeed, one cannot predict the "sum" of any two "distances", since the value will vary depending on the traits used to specify the components to be added. Such a situation, where no numerical result is surprising, is one in which very little is known or the representation of the known observed facts loses most of the information.

A NEW HYPOTHESIS

The present author suggests a new biological assumption, that the genetic material for many pairs of traits may overlap somewhat along the length of a chromosome. The present author calls this the hypothesis of the overlapping traits.

The new assumption means that if some interval of one parental chromosome is replaced by an interval from the other parental chromosome, there will be several traits of the offspring affected and the traits will be affected to differing extents.

For simplicity, consider only a few intervals on the chromosome and only a few traits and use an array of numbers to show the fraction of a trait's genetic material in each interval along the chromosome. We need one array for each parent and we assume that those two arrays are the same.

TABLE I

PARENT A

INTERVALS		Traits		
		1	2	3
	1	0.2	0.6	0.2
	2	0.4	0.4	0.2
	3	0.4	0.0	0.6

PARENT B

INTERVALS		Traits		
		1	2	3
	1	0.2	0.6	0.2
	2	0.4	0.4	0.2
	3	0.4	0.0	0.6

Assume that for interval 1, any offspring is equally likely to follow parent A or parent B, that is, to have genetic material for that interval copied from A or B. Similarly for interval 2 and interval 3. What happens for one interval is assumed not to affect or be affected by what happens for another interval. What happens for each of the 3 intervals are 3 independent events. Thus there are 2x2x2 = 8 equally likely possibilities, namely,

RECOMBINATION FREQUENCIES IN GENETICS 871

INTERVAL 1	A	B	A	B	A	B	A	B
INTERVAL 2	A	A	B	B	A	A	B	B
INTERVAL 3	A	A	A	A	B	B	B	B

To find out the traits manifested by a particular offspring, add up the array entries for one trait from each parent and take the parent corresponding to the larger sum; then do the same for another trait; etc. The majority rules, so-to-speak.

Thus, for the case BAA, for trait 1 parent A contributes 0.8, parent B contributes only 0.2, so this offspring follows parent A for trait 1. For trait 2, the amounts are 0.4 from A, 0.6 from B. For trait 3, the amounts are 0.8 from A, 0.2 from B. Offspring falling in case BAA will follow A for trait 1, B for trait 2, A for trait 3.

We find that of the 8 equally likely types of offspring, 3 are such that both traits 1 and 2 follow A and 3 are such that both traits 1 and 2 follow B.

(Recombination Proportion of traits 1 and 2) =
$$1 - \frac{3}{8} - \frac{3}{8} = \frac{1}{4}$$

Similarly,

(Recombination Proportion of traits 2 and 3) =
$$1 - \frac{2}{8} - \frac{2}{8} = \frac{1}{2}$$

(Recombination Proportion of traits 1 and 3) =
$$1 - \frac{3}{8} - \frac{3}{8} = \frac{1}{4}$$

The resulting map distances and map:

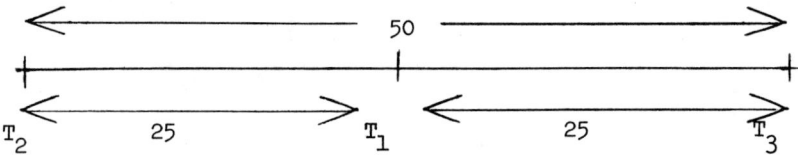

25 + 25 = 50, so here the map distances add arithmetically.

Other relationships between map distances can be illustrated, for example, a form of addition which takes the expected number of double crossovers into account (A. R. G. Owen, 1950).

ACKNOWLEDGEMENTS

The present author wishes to acknowledge his indebtedness to his late mother, Sybil E. Ursell, for displaying the laws of genetics to him in large-scale rabbit breeding. She bred hundreds, even thousands, of rabbits in the Second World War for meat and fur. The present author as a small child took a personal interest in every rabbit and he helped name each one.

This research was done in March to June, 1972, and written up in October, 1972.

BIBLIOGRAPHY

(1) Owen, A. R. G. (1950). The theory of genetical recombination. Advances in Genetics, Vol. 3. New York: Academic Press, 117-157.